Advanced Human–Robot Collaboration in Manufacturing

Lihui Wang · Xi Vincent Wang · József Váncza ·
Zsolt Kemény
Editors

Advanced Human–Robot Collaboration in Manufacturing

Springer

Editors
Lihui Wang
Department of Production Engineering
KTH Royal Institute of Technology
Stockholm, Sweden

Xi Vincent Wang
Department of Production Engineering
KTH Royal Institute of Technology
Stockholm, Sweden

József Váncza
Research Laboratory on Engineering
and Management Intelligence
SZTAKI Institute for Computer Science
and Control
Eötvös Loránd Research Network
Budapest, Hungary

Zsolt Kemény
Research Laboratory on Engineering
and Management Intelligence
SZTAKI Institute for Computer Science
and Control
Eötvös Loránd Research Network
Budapest, Hungary

ISBN 978-3-030-69177-6 ISBN 978-3-030-69178-3 (eBook)
https://doi.org/10.1007/978-3-030-69178-3

This Springer imprint is published by the registered company Springer Nature Switzerland AG
The registered company address is: Gewerbestrasse 11, 6330 Cham, Switzerland

Preface

Human–robot collaboration (HRC) in the manufacturing context aims to realise an environment, where humans can work side by side with robots in close proximity. In such a collaborative setup, humans and robots share the same workspace, the same resources, and in some cases, the same tasks. The main objective of collaboration is to integrate the best of two worlds: strength, endurance, repeatability and accuracy of robots with the intuition, flexibility and versatile problem solving and sensory skills of humans. Using HRC, higher overall productivity and better product quality can be achieved.

In the past decade, research efforts on HRC have been numerous. Varying approaches to facilitating multimodal communication, dynamic assembly planning and task assignment, adaptive robot control and in situ support to operators have been reported in the literature. Nevertheless, HRC is still undergoing intense evolution, and even some of the "common grounds" of understanding of the area have yet to be consolidated. Several parallel interpretations exist, for example, regarding the relationships between robots and humans, such as coexistence, interaction, cooperation, and collaboration. The roles of humans when working with robots still call for more clarity. The lack of standards and safety solutions results in low acceptance of most human–robot combinations in practical application. A systematic review and analysis on this subject is, therefore, needed.

This book presents a set of innovative solutions to manufacturing problems based on the symbiosis between humans and robots, in the form of integrated human–robot collaborative systems. The objective is to better understand the dynamic behaviour of HRC systems and manufacturing processes, and provide on-demand adaptive control to industrial robots and in-situ support to human workers. The timeliness and relevance of the book to the research community are underscored by the fact of increasing attention from researchers worldwide in recent years. Nevertheless, there has been a lack of a dedicated book to this very topic, presenting novel ideas and decision algorithms in developing practical solutions. With the increasing complexity and dynamism in today's manufacturing practice, more precise, robust and practical approaches and systems are needed to support real-time HRC, also as an integrated part of larger scale production systems. Recent development and innovations in this area provide a pool of focused research efforts, and motivate the introduction of this book to a broad readership, from academic researchers to practicing engineers. The book can be viewed as consisting of two major groupings, with the first presenting a broad-based review of the key areas of research in HRC, whereas the second is focused on presenting an in-depth treatment of relevant methodologies and algorithms, leading to a better understanding of HRC in real situations.

The first three chapters on literature survey form Part-I of the book. Chapter 1 aims to outline structural properties and mechanisms of collaborative systems from an agent-oriented point of view, and to provide a reference of terms and concepts which make many different views of the domain comparable. Further chapters of this book, as well as numerous application examples, known industrial solutions and standards, are positioned within this framework to connect theoretical waypoints and practical findings. The HRC literature is further extended in Chapter 2 to the latest developments of gesture recognition for HRC. Covering some of the most important technologies and algorithms of gesture recognition, this chapter is intended to provide an overview of the gesture recognition research and explore the possibility of applying gesture recognition in HRC. In this chapter, an overall model of gesture recognition for HRC is highlighted. After reviewing technical components, a case study of a practical example is presented. Chapter 3 then introduces a safety strategy and a framework for HRC. First, a taxonomy of different HRC relations is identified, thereby also giving a specific example to the abstract considerations of Chap. 1. Leaning on this taxonomy, the safety framework and strategy are developed towards a personalised solution in an HRC cell. A dynamic HRC layout approach is also introduced, based on the actual speed of human and robot, and the distance between them. The feasibility of the approaches is evaluated via the implementation in an HRC-based assembly cell in a quantifiable manner. The operator's biometric data are also included in the HRC control loop. It is proven achievable to conduct a personalised HRC safety strategy based on the human stakeholder's role, physical conditions, speed and other characteristics.

Part-II of the book focuses on human safety in HRC and is constituted from three chapters. Targeting safety in HRC, Chap. 4 reports an approach for real-time collision detection and collision avoidance in an augmented virtual environment, where a virtual model of a robot and real images of a human captured by depth cameras are superimposed for monitoring and collision detection. Four safety strategies are described: the system can alert an operator, stop a robot, move the robot away, or modify the robot's trajectory away from an approaching operator. The method developed is tested in realistic applications of collaboration between robots and humans in an assembly cell, with its performance analysed. Using the gradient of a collision avoidance measure to formulate hard constraints of a quadratic programming (QP) problem, Chap. 5 assigns strict priority to avoiding collisions and specifies other QP objectives with soft task priorities. Through experiments performed on a dual-arm robot, it shows that the proposed solution can generate safe robot motion while fulfilling other task specifications. In addition to collision detection and avoidance, Chap. 6 introduces a dynamic safety system for safe collaboration with industrial robots, thus enabling fenceless, safe, flexible, efficient and collaborative robotic workcells in industrial applications. It is especially useful for meeting the need of small batches in industry with rapid changes to part handling.

Issues in dynamic planning and monitoring are covered in Chaps. 7 through 10, and organised into Part-III of the book. Chapter 7 deals with investigating data acquisition from shop-floor operating resources in production environments towards effectively monitoring their availability, based on the generation of contextual information facilitating seamless HRC. The concept of a context-aware monitoring system is presented along with a prototype software implementation. Context drives the use of smart services both for monitoring and supervision purposes, as well as for enabling smart control and production planning. A use case is also presented to demonstrate the proposed approach along with an implemented prototype. Context-aware monitoring is extended in Chap. 8 to include dynamic assembly planning and task assignment. In this context, a set of evaluation criteria is used to evaluate alternative assembly sequences generated from a product CAD model, which are then mapped onto tasks. The task sequence can then be assigned to production site manufacturing resources for execution. A multi-criteria decision-making model is used while a hierarchical workload model supports task assignment to operators, robots or both. IEC 61499 function blocks are used as a common data model. Scheduled activities are wrapped into function blocks and dispatched for execution. A planning and control cockpit is introduced in Chap. 9 as a portal for HRC assembly. Key functionalities supporting the collaboration between human operators and industrial robots include subsystems in charge of human safety, robot execution control and high-level planning. The portal serves as the main symbiotic orchestrator capable of adaptive assembly planning and execution control of a hybrid assembly station. Chapter 10 introduces theoretical backgrounds and current technological solutions of workcell calibration, and demonstrates the principles in a case study. Although nominal models of individual workcell components are mostly available, the details of secondary importance are often ignored. Their exact placement within a unified workcell model, as well as their adjustment to the real production scene, remain to be solved. This matching of nominal component models requires calibration methods which rely on measured data and guarantee the required fidelity of a calibrated virtual representation, which is detailed in this chapter.

In Part-IV of the book, the aspects of adaptive robot control are shared by Chaps. 11 to 14. Human motion tracking, recognition and prediction for robot control are the focal point of Chap. 11. Specifically, to account for the variability and heterogeneity of human workers in HRC, a context-aware deep convolutional neural network is introduced to identify the task-associated context for inferencing human actions. To improve the accuracy and reliability of human motion trajectory prediction, a functional unit-incorporated recurrent neural network is adopted to parse the worker's motion patterns and forecast the worker's future motion trajectory. Collectively, these techniques enable online robot action planning and execution for collaborative assembly operations. Furthermore, the focus of Chap. 12 deals with recent developments of programming interfaces to efficiently instal collaborative robotic systems in manufacturing applications. The aim of this chapter is to introduce different programming interfaces and paradigms that can be exploited for easy programming of robots by non-experts. Novel interfaces that build upon and extend the current state of the art are presented. The chapter also gives a detailed qualitative matching against different criteria required to be considered while developing programming interfaces. As a new modality for robot control, Chap. 13 presents an approach to haptically controlling an industrial robot without using external sensors. This "sensorless" haptic control approach is enabled by the dynamic models of the robot for cases when only joint angles and joint torques are measurable. To have better interaction performance, accurate modelling of the dynamic model of the robot both in the pre-sliding and sliding regimes is used to estimate the joint torques where the friction model is thoroughly explored. The estimated external force applied to the robot by an operator is transferred into reference position and speed of the robot by an admittance controller, and adaptive admittance parameters are adopted to make human–robot interaction more natural and easier with accurate positioning and control with smooth movement. Brain robotics is then introduced in Chap. 14. This chapter reports a framework that can facilitate the interactions between a human and a robot, achieved by capturing and analysing the brainwave signals of the human. The system developed provides the ability for robot operators to control the robot using their own brain signals. The system can cooperate with other channels of communication (gesture, voice, etc.) to strengthen the collaboration between the human and the robot during shared assembly operations.

Part-V of this book is dedicated to multimodal in situ decision support for humans working with robots. Providing efficient worker assistance is of key importance in successful HRC as it can optimise the cognitive load of the workers. In this context, Chap. 15 introduces a novel worker assistance suite, which supports bi-directional information exchange and multimodal communication. The structure of the system and its interfaces are discussed and the implementation is detailed. Using multiple channels of communication allows to fit this context to the worker's preferences, which in turn helps increase efficiency and quality of production. For example, Augmented Reality (AR) provides an interactive approach to combining the physical manufacturing environment with computer-generated virtuality. However, it is still challenging to establish an efficient AR system due to the complexity of the actual manufacturing environment and the difficulties in HRC. In Chap. 16, a feasible AR system is developed with a novel closed-loop structure. A feasible AR-based human–robot interaction method is explained based on a HoloLens device, together with an advanced compensation mechanism towards accurate robot control. The work is validated through case studies and quantifiable performance assessment on system response time, compensation accuracy and other characteristics. Chapter 17 integrates the aforementioned methods into a unique HRC system. In particular, the chapter focuses on the role of demonstrators and presents three perspectives related to the use of demonstrators in bridging the gap between current knowledge and the work practice on the shop floor. Three industrial HRC demonstrators are reported to provide the envisioned benefits for the addressed industrial requirements gathered from three companies. Finally, Chap. 18 formulates conclusions, and presents identified research challenges and future work. It also addresses the societal impact of using collaborative robots in industry, and their contribution to society.

Altogether, the 18 chapters provide a systematic overview of some recent R&D achievements on human–robot collaboration in manufacturing. We believe that this research field will remain active and in the focus of interest for years to come, and hope that readers find this book informative and useful.

Stockholm, Sweden — Lihui Wang
Stockholm, Sweden — Xi Vincent Wang
Budapest, Hungary — József Váncza
Budapest, Hungary — Zsolt Kemény
December 2020

Acknowledgements

The co-editors would like to take this opportunity to express their deep appreciation to all the authors for their significant contributions to this book. Their commitment, enthusiasm and technical expertise are what made this book possible. We are also grateful to Springer for supporting this book project, and would especially like to thank Anthony Doyle, Senior Editor for Engineering, and Amudha Vijayarangan, Project Coordinator, for their constructive assistance and the earnest cooperation, both with the publishing venture in general and the editorial details.

Without the financial supports of the following funding agencies and research projects, this book would not have reached its current state and richness in content. The co-editors and chapter authors are thankful to all the founders listed below.

- Bridge Human–Robot Interaction "TeachBots" (Industrial)
- CMMI-1830295, US National Science Foundation
- HoliSafeMRK (FFG 864872)
- Lern4MRK (AIT)
- SYMBIO-TIC (Project #637107), EU Horizon 2020 Programme
- GINOP-2.3.2-15-2016-00002 grant, Hungary
- Research on prime exploitation of the potential provided by the industrial digitalisation (ED_18-22018-0006), Hungary
- National Lab for Autonomous Systems, Ministry for Innovation and Technology and the National Research, Development and Innovation Office, Hungary

The co-editors and the chapter authors would like to thank the industrial partners who contributed to the design, development and implementation of the aforementioned research projects, particularly the three demonstrators of SYMBIO-TIC as outlined in Chap. 17. The authors would also like to express their sincere gratitude towards the following individuals who have helped in the compilation of this book in one way or another: Mingtian Lei, Carolin Schaffert, Argyri Seira and Yongqing Zhao.

Acknowledgements

Contents

Editors and Contributors

About Editors

Professor Lihui Wang is a Chair Professor of Department of Production Engineering, KTH Royal Institute of Technology. He gained his Bachelor degree from Tsinghua University (China) in 1982, followed by his Master and PhD from Kobe University (Japan) in 1990 and 1993, respectively. He is the Editor-in-Chief of Journal of Manufacturing Systems, International Journal of Manufacturing Research, and Robotics and Computer-Integrated Manufacturing. He has published over 500 journal and conference papers, 31 book chapters and 10 books.

Dr. Xi (Vincent) Wang is an Associate Professor in the IIP Department of Production Engineering, KTH Sweden. He is working with the division of Sustainable Manufacturing Systems (SPS). Vincent received his PhD and Bachelor degrees in Mechanical Engineering from the University of Auckland (New Zealand) and Tianjin University (China), in 2013 and 2008, respectively. Vincent's main research focus includes Cloud-based manufacturing, sustainable manufacturing, robotics, computer-aided design, and manufacturing systems. He also serves as the managing editor of the International Journal of Manufacturing Research (IJMR), Associate Editor of SME Journal of Manufacturing Systems (JMS), and Array—Open Access Journal by Elsevier.

Dr. József Váncza is Head of the Research Laboratory on Engineering and Management Intelligence in the Institute for Computer Science and Control (SZTAKI) of the Eötvös Loránd Research Network, Budapest, Hungary, as well as Associate Professor at the Budapest University of Technology and Economics (BME). He received his MSc (1984) and PhD (1994) degrees from BME and the Hungarian Academy of Sciences, respectively. He is Fellow of the International Academy for Production Engineering (CIRP) where he served as Chair of STC O. He is author of 200+ scientific publications, including 25 book chapters, and editor of 4 books.

Dr. Zsolt Kemény is a senior researcher at the Laboratory on Engineering and Management Intelligence in the Institute for Computer Science and Control (SZTAKI) of the Eötvös Loránd Research Network, Budapest, Hungary. He received his computer science MSc and PhD from Budapest University of Technology and Economics, in 1998 and 2004, respectively. Aside from research focusing on robotics, automation and manufacturing, he is also teaching mechatronics, cyber-physical production systems, and robotics. He is author or co-author of 77 scientific publications.

Contributors

Sharath Chandra Akkaladevi Department of Robotics and Automation Systems, Profactor GmbH, Steyr-Gleink, Austria;
Institute of Networked and Embedded Systems, Alpen-Adria Universität Klagenfurt, Klagenfurt, Austria

Kosmas Alexopoulos Laboratory for Manufacturing Systems and Automation, University of Patras, Patras, Greece

Naresh Chowdary Chitturi Department of Robotics and Automation Systems, Profactor GmbH, Steyr-Gleink, Austria

Gábor Erdős Research Laboratory on Engineering and Management Intelligence, SZTAKI Institute for Computer Science and Control, Eötvös Loránd Research Network, Budapest, Hungary

Robert X. Gao Department of Mechanical and Aerospace Engineering, Case Western Reserve University, Cleveland, OH, USA

Leopold Hiesmair Department of Robotics and Automation Systems, Profactor GmbH, Steyr-Gleink, Austria

Michael Hofmann Department of Robotics and Automation Systems, Profactor GmbH, Steyr-Gleink, Austria

Magnus Holm School of Engineering Science, University of Skövde, Skövde, Sweden

Gergely Horváth Research Laboratory on Engineering and Management Intelligence, SZTAKI Institute for Computer Science and Control, Eötvös Loránd Research Network, Budapest, Hungary

Markus Ikeda Department of Robotics and Automation Systems, Profactor GmbH, Steyr-Gleink, Austria

Csaba Kardos Research Laboratory on Engineering and Management Intelligence, SZTAKI Institute for Computer Science and Control, Eötvös Loránd Research Network, Budapest, Hungary

Zsolt Kemény Research Laboratory on Engineering and Management Intelligence, SZTAKI Institute for Computer Science and Control, Eötvös Loránd Research Network, Budapest, Hungary

András Kovács Research Laboratory on Engineering and Management Intelligence, SZTAKI Institute for Computer Science and Control, Eötvös Loránd Research Network, Budapest, Hungary

Jessica Lindblom School of Informatics, University of Skövde, Skövde, Sweden

Hongyi Liu Department of Production Engineering, KTH Royal Institute of Technology, Stockholm, Sweden

Sichao Liu Department of Production Engineering, KTH Royal Institute of Technology, Stockholm, Sweden

Sotiris Makris Laboratory for Manufacturing Systems and Automation, University of Patras, Patras, Greece

Timo Malm VTT Technical Research Centre of Finland Ltd, VTT, Finland

Ilari Marstio VTT Technical Research Centre of Finland Ltd, VTT, Finland

Abdullah Mohammed Department of Production Engineering, KTH Royal Institute of Technology, Stockholm, Sweden

Jari Montonen VTT Technical Research Centre of Finland Ltd, VTT, Finland

Nikolaos Nikolakis Laboratory for Manufacturing Systems and Automation, University of Patras, Patras, Greece

Balázs Pataki Department of Distributed Systems, SZTAKI Institute for Computer Science and Control, Eötvös Loránd Research Network, Budapest, Hungary

Andreas Pichler Department of Robotics and Automation Systems, Profactor GmbH, Steyr-Gleink, Austria

Matthias Propst Department of Robotics and Automation Systems, Profactor GmbH, Steyr-Gleink, Austria

Bernard Schmidt School of Engineering Science, University of Skövde, Skövde, Sweden

Richard Senington School of Engineering Science, University of Skövde, Skövde, Sweden

Konstantinos Sipsas Laboratory for Manufacturing Systems and Automation, University of Patras, Patras, Greece

Sándor Tóth SANXO-Systems Ltd., Budapest, Hungary

József Váncza Laboratory on Engineering and Management Intelligence, SZTAKI Institute for Computer Science and Control, Eötvös Loránd Research Network, Budapest, Hungary;

Department of Manufacturing Science and Technology, Budapest University of Technology and Economics, Budapest, Hungary

Lihui Wang Department of Production Engineering, KTH Royal Institute of Technology, Stockholm, Sweden

Peng Wang Department of Electrical and Computer Engineering, University of Kentucky, Lexington, KY, USA

Wei Wang School of Engineering Science, University of Skövde, Skövde, Sweden

Xi Vincent Wang Department of Production Engineering, KTH Royal Institute of Technology, Stockholm, Sweden

Yuquan Wang Department of Production Engineering, KTH Royal Institute of Technology, Stockholm, Sweden;
The Chinese University of Hong Kong, Shenzhen, China

Jianjing Zhang Department of Mechanical and Aerospace Engineering, Case Western Reserve University, Cleveland, OH, USA

Part I
Literature Survey

Chapter 1
Human–Robot Collaboration in Manufacturing: A Multi-agent View

Zsolt Kemény, József Váncza, Lihui Wang, and Xi Vincent Wang

Abstract Recent years have witnessed a growing interest in reintroducing the advantages of the human workforce into industrial production processes while keeping the benefits of machines already proven in production automation. Both industry and academia exhibit intense interest in the topic of combining human and robotic resources in collaborative production environments. Nevertheless, the domain of human–robot collaboration is still undergoing intense evolution—nothing proves it more than the diversity of fundamental approaches, world-views, and gaps in standardisation that all hint at the fact that even the overall understanding of the domain is yet to be consolidated. It is, thus, not realistic to expect that a single comprehensive morphological work would reconcile today's multitude of views on human–robot collaboration, and this is not the goal of this introductory chapter either. Instead, the chapter gives an overview of the domain relying on a single selected paradigm, namely, multi-agent systems. This choice is based on the assumption that this branch of distributed artificial intelligence, having matured over several decades of research and application, provides feasible perspectives and terminological waypoints for collaborative settings under the structured circumstances of industrial production. The chapter aims to outline structural properties and mechanisms of collaborative systems from an agent-oriented point of view, and aims to provide a reference of

Zs. Kemény (✉) · J. Váncza
Research Laboratory on Engineering and Management Intelligence, SZTAKI Institute for Computer Science and Control, Eötvös Loránd Research Network, Budapest, Hungary
e-mail: zsolt.kemeny@sztaki.hu

J. Váncza
e-mail: jozsef.vancza@sztaki.hu

J. Váncza
Department of Manufacturing Science and Technology, Budapest University of Technology and Economics, Budapest, Hungary

L. Wang · X. V. Wang
Department of Production Engineering, KTH Royal Institute of Technology, Stockholm, Sweden
e-mail: lihuiw@kth.se

X. V. Wang
e-mail: wangxi@kth.se

L. Wang et al. (eds.), *Advanced Human–Robot Collaboration in Manufacturing*,
https://doi.org/10.1007/978-3-030-69178-3_1

terms and concepts which make many different views of the domain comparable. Further chapters of this book, as well as numerous application examples, known industrial solutions and standards, are positioned within this framework to connect theoretical waypoints and practical findings.

1.1 Introduction

Automation has, beyond doubt, profoundly changed the character of industrial production as it brought about efficiency, predictability and consistent quality on a scale never seen before. However, these benefits came at a cost. Mastering uncertainty in automation is, namely, expensive [1]—if possible at all in a given case—and both engineering common sense and production economy favour the removal of uncertainty from production processes right away. This has its impact on the human workforce in automated production: human behaviour does not fit well into conventional automation practice, and is often seen as a mere disturbance by automation engineers [2]. Thus, humans are either isolated from processes, or forced to act as just one of the automated components of a production system. Moreover, automation has entered ranges of speed and other process conditions that can expose humans to lasting overstrain or serious hazards—health and safety requirements, too, keep robots and humans apart in today's mainstream of automated production [3]. As a result, the most important strengths of humans (i.e., fast assessment of complex situations, in-depth understanding of contexts, and finding solutions to unforeseen problems) remain largely unused, barring automated production from a major potential of flexibility and resilience in the face of disturbances.

For decades, the technological and market expectations have made automated solutions worthwhile. However, recent trends are gradually shifting the conditions for an optimal bottom line, and more often, combining the strengths of robots and humans in the same production setting is becoming either a necessity (e.g., due to the complexity or diversity of tasks being beyond machine-tractable), or an anticipated advantage (e.g., better resource efficiency by widening process tolerances). At the same time, the inclusion of humans in automated production is becoming reality as technological and scientific advances bring the compensation of disadvantageous human traits (higher error rate, limited rationality, lower predictability, physical and mental workload constraints) within reach [1], and makes their physical involvement in robotised processes sufficiently safe [3].

Although many scientific and technological achievements are becoming mature enough for industrial requirements, the evolution of a meaningful and productive synthesis of robotic and human resources in production is still expected to be a gradual, prolonged process, spanning several decades. This is, on the one hand, due to the complexity of the topic and the multitude of contributing fields to be consolidated, and, on the other hand, due to the research and development goals themselves being reformulated as preceding results and changing industrial demands unfold the horizon of relevant challenges.

It can be expected that, as *human–robot collaboration* (HRC) gains a foothold, its feasible forms and supporting technologies will greatly diversify. Already today, it is worthwhile to think of a continuous *spectrum* of human/robot involvement in production, ranging from the human-only to the machine-only end. Also, a variety of co-action and organisation forms will appear with various degrees of autonomy of the humans and robots participating in the process, be it through discrete coupling points of human-only and robot-only processes, or with simultaneous, continuously coupled robot and human involvement. Evolving organisational complexity will enable coordination of multiple humans and robots, in accordance with higher process goals and plans. Many of the theoretically possible forms of human–robot collaboration have not yet been exemplified in research, not to speak of industrial practice. This, however, should not keep one from considering a comprehensive picture of the domain that includes untried and less researched possibilities—an expanding and evolving domain does, after all, require a frequent look ahead. In order to serve as a point of reference for future research, too, this chapter extends the recapitulation of today's state of the art and perceived trends by a conceptual framework for collaborative production settings, leaving the door open for interpreting and solving upcoming challenges.

Economic and engineering rationality makes it likely that the typical industrial production environment retains much of its structured nature in the era of human–robot symbiosis as well. Under such simplified conditions, some paradigms, especially of *artificial intelligence* (AI), and distributed, *multi-agent AI* in particular, can very well prove successful even if they have not met far-flung goals in more general application contexts [4–7]. Some other areas, along with a more accurate judgement of their actual relevance, are still rather sparsely charted territory, such as "counternarratives" [8], or gambling behaviour closer to the shop-floor level. It can be but noted here that alternative ways of discussing HRC in an overall context, too, exist. For instance [9], espouses a resource-oriented view and handles all crucial issues of HRC as various facets of a generic *operational management* problem.

Collaborative and converging research in manufacturing science and technology, computer science including AI, as well as information and communication technologies resulted in the emergence of *cyber-physical production systems* (CPPSs) [10]. CPPSs consist of autonomous and cooperative elements and subsystems that are connected within and across all levels of production, from processes through machines up to production and logistics networks. CPPS is the main technological driver of the Fourth Industrial Revolution, frequently noted as Industry 4.0. The main traits of CPPSs are (1) intelligence and smartness of elements, (2) connectedness that enables harnessing the data, knowledge and services of others, as well as (3) responsiveness, a continuously ongoing interplay and mapping between the status of physical system components and their virtual counterparts. Clearly, any HRC setting in a manufacturing environment is a genuine instance of CPPS, too.

In what follows, Sect. 1.2 proposes a *generic multi-agent framework for HRC* in manufacturing. Next, the three main abstraction and organisation levels of this framework are presented: agent capabilities (Sect. 1.3), individual human and robotic agents (Sect. 1.4), as well as organisational aspects (Sect. 1.5), respectively.

Section 1.6 discusses key issues of planning and execution control, whereas Sect. 1.7 focuses on relevant topics of human safety and ergonomics. Section 1.8 concludes this introductory chapter which populates the proposed framework with examples taken from all the other chapters of this book.

1.2 A Multi-agent Framework for Human–Robot Collaboration—Basic Concepts

Production systems and environments with HRC can be regarded as being comprised of subsystems or components that affect changes in each other, or, in turn, are changed by another component. In a structured environment as production systems, the aforementioned changes occur in accordance with some form of control, i.e., most of them were designed to be brought about in specific ways. Such compound systems can be described using various formalisms—however, due to the structured and designed character of production systems, a specific branch of distributed artificial intelligence, namely, *multi-agent systems* [4, 11], appear to fit the purpose particularly well, especially in view of its possible levels of granularity and its machine-tractable logical apparatus for describing components, their relations, and operations performed within the system [12]. Therefore, the terms used throughout this chapter will be defined and interpreted in the framework of multi-agent systems.

In a general case, it is assumed that any number of robots and/or humans can be involved in the same production process, and their levels of organisation (measurable, e.g., by the amount and relation of coherent actions exerted at a time or in a directed sequence) can vary. It is, therefore, necessary to define terms that describe inter- and intra-agent characteristics at various levels of granularity. Within the context of production systems, three levels of organisation and granularity can be distinguished, along which the definition of further terms is organised here (see also Fig. 1.1).

In a bottom-up approach, first to be observed are the *capabilities* within an agent that make it act in a defined or desired way. This term refers to any ability of the agent to react to, or reflect on inputs from the environment, and exert actions, i.e., produce output towards its environment. Capabilities are, again, organised into several layers according to their degree of abstraction and the computational or cognitive comprehensiveness they require. Although capabilities are manifested in agents, they can still be extracted, transformed or injected into agents, i.e., they can be in many regards independent of the agent in which they find deployment. Therefore, Sect. 1.3 presents capabilities, as far as possible, detached from specific agent perspectives.

Above the capabilities, the structural hierarchy continues with the *individual agent*. Agents are regarded as the basic building blocks which have a definite *identity* in relation to their surroundings due to their own persistent characteristics, and by their relations to their environment. This is an important property in a multi-agent structure: while it is possible to move capabilities around fairly free of restrictions, the identity of an agent is to retain recognisable persistency throughout time,

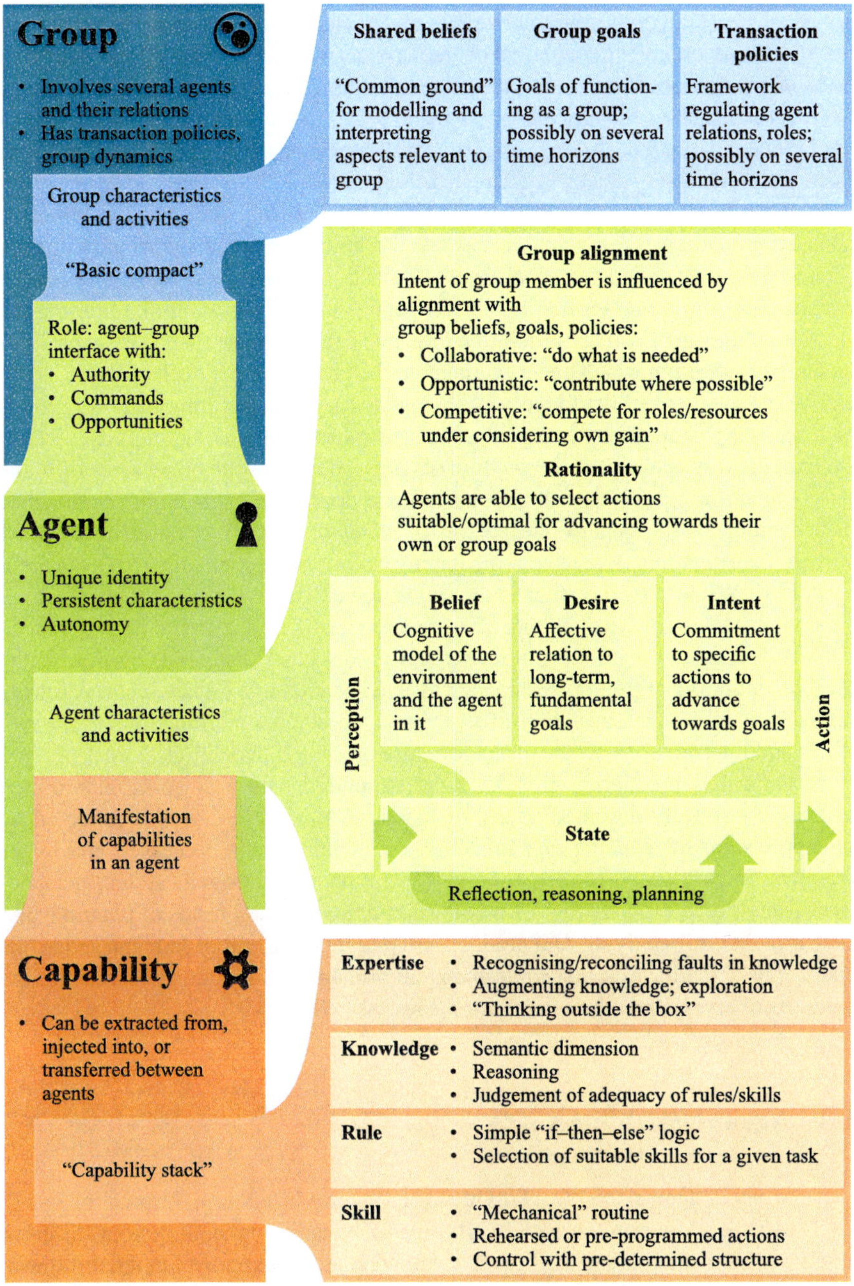

Fig. 1.1 General overview of terms on various levels of multi-agent systems

across various situations, regardless of other—possibly time-varying—agent properties such as autonomy or contextual characteristics within its operating environment. Section 1.4 describes individual agents, the fundaments of their functioning, where Sect. 1.4.3 discusses *sensing and perception*, Sect. 1.4.4 presents the terminology of *modelling* and *awareness*. Sect. 1.4.2 elaborates the concept of agent *autonomy*, and Sect. 1.4.6 discusses issues of *communication*.

As soon as several agents are part of a production setting, one more level must be considered, namely, the *group*. On this level, relations between agents gain impact. These can either define temporal, spatial or action-related *dependencies* between agents, or assign a *role* to the individual agent as part of the group. Examining the relation of the agent and the group to a role, one can recognise that a role represents an *interface* through which agents connect to the group context. Roles, just as skills, can be independent of the identity of the agent currently fulfilling it. Also, groups have their own characteristics regarding structure and dynamics. Certain cases justify a hierarchical structure, i.e., the nestling of (sub)groups that can be set apart from the rest by function, or such specific dependencies as degrees of interaction, or coherence in behaviour. In this regard, the pursuit of *common goals*, by consensus or other ways of coordination, is key in distinguishing a random set of agents from a coherent *team* [13]. While a team can act as a single agent within a larger group, its inner subdivision into individual agents often remains justified due to the heterogeneous composition of the team (e.g., humans and robots being involved), or due to the level of autonomy of the constituting agents having a notable impact on the resulting probability of a coherent action of the team ("rogue" or malfunctioning agents). Roles and interrelations on the group level are presented in detail in Sect. 1.5.

Specific to a production context is the structured nature of goals, determined by the purposefully designed nature of industrial production. This is reflected by the approaches to *planning and execution* of manufacturing processes, and by its definitive impact on all levels of the aforementioned multi-agent system perspective. Therefore, a separate Sect. 1.6 collects the fundamental terms of planning and execution control applying to a production system comprised of humans and artificial agents. Finally, with a specific focus on human agents in an HRC setting, the concepts of *safety* and *ergonomics* are discussed in Sect. 1.7.

1.3 Agent Capabilities

Capabilities in agents can be subdivided into several levels of a "capability stack" [2, 14, 15]. Building on the SRK (skill–rule–knowledge) taxonomy, four different levels can be distinguished: skills, rules, knowledge and expertise. In what follows, after defining each kind of capability, examples of their realisation in HRC are given, focussing on the assembly domain.

1.3.1 Skills

Skills are capabilities for input–output (typically sensory–motor) actions that are often referred to as "highly automatic", indicating that these are on the level of simple control tasks, implemented as such in robots. In humans, skills are acquired and "well-rehearsed" capabilities that are executed without thinking, associated with the least mental workload. Assessment of the capabilities of the human workforce often makes use of the term *skill level*. In the agent capability stack, this term can be translated into the relative amount of skills among all capabilities which the given worker deploys for the given task. If a worker is more "skilled" for the task, more actions are executed via rehearsed and automatic skills, and both task assignment (see Chap. 9 [16]) and delivery of work instructions (see Chap. 15 [17]) can be adapted accordingly.

The notion of skills can be extended to the domain of robotics, too. Specifically [18] proposes a generic concept for modelling robotic skills. Here, a relatively small set of skills is derived from current factory worker instructions, and is transferred to industrial mobile manipulators. The pre- and post-conditions (i.e., predictions) of executing a particular skill are a priori given, and procedures may continuously check some conditions during execution as well. Execution is based on the input parameters and the actual state of the world, whereas the skill affects a change in the world state (see also Fig. 1.2).

In this book, Chap. 12 [19] provides an extensive review of robotic skills (using the terms *recipe* and *skill* analogously), and conceives them as the key to programming-free HRC in assembly. In an action-centric representation, direct relations between human and robot skills are established via their post-conditions (i.e., consequences). The abstraction of the knowledge of an assembly process (see also the next subsections) facilitates learning by interactive demonstration.

The notion of robotic skills greatly facilitates not only high-level robot programming, but also the reconfiguration and repurposing of automated workcells [18, 20]. Recently, the ability of transferring human skills to robots has become a research thrust for autonomous robots and human–robot collaboration. In particular [21],

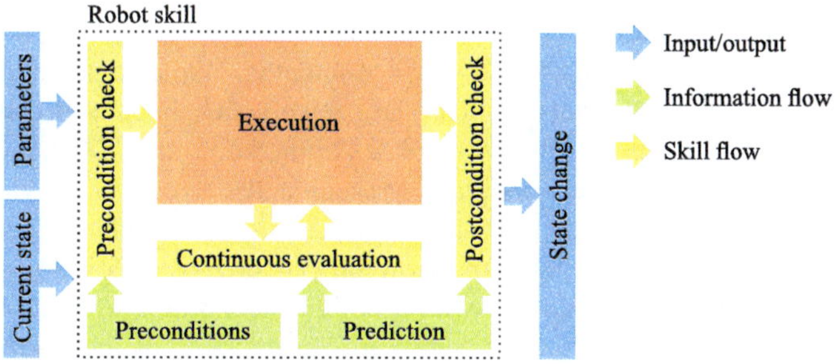

Fig. 1.2 A generic concept for robotic skills [18]

provides a survey of *skill transfer learning* (STL) methods across all domains of HRC.

1.3.2 Rules

Rules represent a higher level of capabilities, as they require the evaluation and judgement of inputs, and the selection of an applicable action (which, in turn, is mostly carried out via applying skills). Rules follow if–then–else patterns, and are usually easy to formalise and automate, making them suitable for robots and other automation components. Rules also have an important place in the human capability stack, especially if the structured nature of a situation or process makes a rigid if–then–else formalisation sensible. Rules can also bridge known weaknesses of the human mind (e.g., in preventing intuitive but unsuitable motor skills from going off in emergency situations), but, on the other hand, may also be misused to refuse responsibility. The latter can also occur in situations where false automatic interpretation and application of the wrong control leads to failure which could have been prevented if a responsible human operator resorted to higher levels of the capability stack.

1.3.3 Knowledge

Knowledge is the level where rules and judgements become comparable through a *semantic dimension.* Knowledge also enables *inductive thinking.* Both abilities are crucial when a situation—possibly even an unprecedented or ambiguous one—requires an agent to reason upon rules and judgements before blindly following them (if there are any applying to the apparent situation). An important aspect of both knowledge and reasoning processes is that they can be made transferable across agent boundaries if shared semantic conventions provide a common ground of interpretation, i.e., knowledge and reasoning processes "make sense" for other agents in the system as well, such as by a shared *ontology*, i.e., system of concepts, their definitions and their interrelations [22, 23]. Much of modelling and reasoning on the level of knowledge can be formalised, and partly automated. While some authors argue how much true knowledge-level capability has been practicably put to use in any artificial system so far [2], the structured nature and limited diversity of a manufacturing environment do set favourable conditions for such endeavours. Where knowledge-level capabilities of an artificial agent are, however, likely to underperform, or fail entirely, are situations insufficiently or incorrectly covered by existing knowledge.

In this book, Chap. 7 [24] discusses in detail the role of ontologies and presents how a domain ontology can set the common context for dynamic decisions of a robot, a human and a safety controller. This ontology captures the primary temporal and spatial relationships of the events, objects and states in an assembly cell where several human agents collaborate with a robot. In Chap. 8, the world of assembly is represented in an ontology which provided the basis for dynamic planning and task assignment decisions [25].

1.3.4 Expertise

Expertise goes one step further: in addition to merely having knowledge and being able to reason upon knowledge, it adds the abilities of (1) recognising and—if possible—reconciling faults in existing knowledge, (2) augmenting knowledge by explorative actions and (3) adding difficult-to-formalise qualities to dealing with knowledge, such as "thinking outside the box", or applying intuition in the face of uncertain, unknown, or unrecognisable situations. Developing stand-alone expertise is, clearly, out of reach of present-day artificial systems, as is the precise judgement of whether unexpected human behaviour is the manifestation of intuition or a lapse of cognitive faculties. Artificial agents can, however, augment human mental processes in interactive settings by appropriately guiding human attention in situations where the human mind is known to underperform (e.g., due to overload of low-level information) or fail frequently (e.g., due to cognitive biases or emotionally imposed limits).

In this book, Chap. 11 presents a deep-learning-based method to collect tacit expertise which facilitates the recognition and prediction of various human actions in an assembly cell [26]. The method, which parses the human worker's motion pattern and forecasts future motion trajectories, can account for the variability and heterogeneity of human workers during the assembly process. Hence, the robot agent can find answers to questions like "which tool or part should be delivered to which location next?", and perform online action planning.

1.4 Human and Robotic Agents

1.4.1 Basic Terms Characterising Agents

Agents are regarded as the atomic building blocks of multi-agent systems. An agent is an entity which can be distinguished from its surroundings by its own persistent characteristics, and by its relations to its surroundings. Agents can be involved in *interactions*, i.e., receive *inputs* (often referred to as *sensing* or *perception* when physical surroundings are involved), and produce *outputs* (*action* is often used as a synonym in physical cases). Agents normally have *states*, i.e., persistent characteristics that are changed either by the agent itself, or by the *inputs* received from the environment. Agents can produce *output* which, in turn, can affect state changes in other agents. The relation of an agent's inputs, state changes and outputs is referred to as the *behaviour* of the agent. Concrete instances of state changes or emitting output are often referred to as *events* of the multi-agent system. An event view is of considerable practical importance when the coherent action of several agents is required to elicit collective behaviour or execution of production steps in a technologically required sequence, governed by *precedence* relations.

The literature distinguishes *passive agents* and *active agents* with respect to the directed (goal-oriented) nature of their behaviour.

Passive agents (sometimes referred to as *objects*) are governed by simple laws applying to their inputs, states and outputs. Even if they may exhibit more complex or stochastic characteristics, no further abstract goal can be found to govern their behaviour. In a production setting, workpieces, simple machines or artefacts inherent to the environment can be modelled as passive agents.

The behaviour of *active agents*, on the other hand, can be interpreted as a more or less coherent pattern of taking and analysing input, modifying the agent's own state and producing output *in order to advance towards a goal* which is typically more abstract or more comprehensive than the individual relations of inputs, state changes and outputs. In a production setting, both humans and robots are considered active agents, regardless of their other characteristics such as autonomy, awareness, or integration into coordinated planning and execution control through a role (these terms will be explained in subsections further below).

In general, it is assumed that active agents apply some form of *practical rationality* while shaping their behaviour in favour of a more abstract goal. A possible formal representation of such practical rationality is the *belief–desire–intention* (BDI) model [1, 12, 27, 28]. All three of these terms can be regarded as part of an agent's state, corresponding to specific aspects of the agent's relation to the rest of the system. *Belief* is a *cognitive* quality of the agent's state, a hypothetical model of relevant parts of the multi-agent system—this can be regarded as the agent's model of reality. *Desire* is a higher level goal of the agent, often classified as an *affective* aspect—this may literally hold for human agents, while in artificial agents, it may be a representation of a higher level preference, to be optimised by the agent's behaviour in the long run. *Intention* is a conative aspect of an active agent, i.e., a persisting decision or *commitment* in favour of a selected plan. While the mechanisms behind decisions are different in humans and machines, the existence of decisions and plans, as well as their manifestation in agent behaviour is likewise assumed for both.

The BDI model is especially valuable for multi-agent systems with artificial agents and humans, as it forms the basis for formal *transactional logic* which is (1) computable, being first-order, and is (2) reasonably expressive in comprehending aspects that are characteristic of the more complex human agents. Relying on BDI in practical reasoning, formal assessment and planning of mixed-group commitment become possible, which is essential when larger human–robot groups work together in a production setting.

1.4.2 Terms for Agent Autonomy

The previous subsection already defined basic terms of agent behaviour—for the given agent, it is, however, likewise important how certain ways of behaviour come about and are executed in a given setting. Important in this regard are decisions or continuous controls that contribute to the behaviour, and the degree to which the agent relies on its own state and capabilities to make this work. The agent's independence on external incentive for shaping its behaviour is referred to as *autonomy* [23, 29]. While definitions regarding the degree of autonomy vary, they often relate assessment of autonomy to the dependence of planning and execution levels on commands or guidance from external sources (mostly: other agents). Barber et al. [30] use, for example, the concise definition "an agent's active use of its capabilities to pursue some goal without intervention, oversight, or control by any other agent", arguing that agents with any degree of autonomy must be viewed as goal-directed (referred to as *desire* in the BDI model).

The impact on an agent's decisions can vary from instance to instance, and numerous approaches are known for defining degrees of autonomy. In this regard, Mostafa et al. provide a comprehensive review [31]. In a discrete approach, one can distinguish *commands* that are binding, and *opportunities*, the consideration of which is subject to the agent's discretion (similarly to the mathematical programming terms *constraints* and *criteria*, respectively). Extending this, one can define priorities which can either be negotiated as commeasurable values (and possibly span a continuous spectrum), or form a hard hierarchy. Also, the degree of autonomy may vary on a goal-by-goal basis, or with respect to selected properties [31]. Autonomy can also be adjustable by the agent's own assessment of a situation (internal adjustment), or by preliminary agreements, subordinate mechanisms, or negotiation (external adjustment) [31]. Adjustable autonomy plays an important role in the forming of coherent team behaviour in pursuit of common goals [32, 33]. The latter, however, is considered part of group dynamics, i.e., a form of collective behaviour of several agents, and will be dealt with in Sect. 1.5 observing the group/team level of organisation.

In this book, Chap. 12 presents a high-level robot programming framework which is designed to set up industrial human–robot environments realising various levels of shared autonomy among human and robotic agents [19]. Chapter 3 discusses autonomy with special respect to human safety which—if the need arises—should override autonomy [34].

1.4.3 Sensing and Perception

A typical physical agent has no direct access to the actual relations and states of its environment, but obtains information from its surroundings by *sensing*, and relies on this in reconstructing an assumed depiction of the world by *reasoning* [35]. In this regard, suggestive parallels to an agent's functioning can be found in control engineering, i.e., in the way a controller with a state observer works [36]: the controller (corresponding to the agent) has no direct access to the state of the controlled plant (the agent's environment), but uses a state observer with an assumed plant model that estimates the state of the plant based on inputs which actually are functions of the state of the real plant. The controller's output (comparable to agent action) is then generated by some control law, taking into account the estimated plant state. One of the most widespread approaches to state estimation is the Kalman filter [37], which, by default, estimates the state only for a linear plant model with a given (assumed) structure and parameter set. Control engineering, however, also has parallels in building a model of reality—this is referred to as identification. While control engineering rarely deals with complexity and discrete logical structures to a degree they prevail in robotic multi-agent systems, an overall attitude can still be gleaned from the domain.

Sensors deployed in the HRC environment can be categorised into two families: contact-based and contactless. *Contact-based* sensing is primarily used for human gesture recognition as well as physical parameter measurement at a human–robot interface or robot–object points-of-contact. In comparison, *contactless* sensing finds its major applications in measuring geometric information suitable for localisation, mapping and tracking, such as distance and angle, as well as providing a basis for context-aware interpretation of the environment. It has also been applied to gesture recognition and movement tracking. A comprehensive review of both basic approaches is presented in [38]. Note that another categorisation advocates *image-based* and *non-image-based* sensing technologies, as discussed in Chap. 2 of this book [39].

Of special and growing importance is the *fusion* of various sensory signals. One actively explored potential application of data fusion in HRC assembly is the multimodal fusion of human commands, for example, the integration of both hand gestures and voice, to further improve the human command recognition rate and contribute to programming-free robot control. Despite the increasing use of gestures and voice commands in HRC for robot control, they are less natural or practical in busy and noisy work environments. Instead of defined gestures and voice commands, recognition and prediction of human motions through deep learning provide better context awareness and less interruption to normal performance induced by signalling gestures [40]. Details of this approach are also discussed in Chap. 11 [26]. Chapter 12 presents—as an element of a programming-free planning and control framework—a so-called perception reasoning system, which interprets incoming sensory information and recognises the actual situation (including actions performed by humans) in an HRC assembly environment.

Finally, even though the application of external sensing technologies in HRC is almost ubiquitous, there are rare exceptions which rely on signals implicitly available on the servo level of robot control. Chapter 13 presents a *sensorless* haptic control approach which relies on the dynamic model of a robot whose joint angles and torques are measurable [41, 42].

1.4.4 Awareness

The functioning of active agents with some degree of autonomy is characterised by making decisions, and taking according actions in order to advance towards some goal—this has already been explained through the BDI model in a previous subsection. Most decisions are not random choices taken at will, but are directed, based on a depiction of the agent's environment—possibly including itself—which is assumed to depict reality in a way that choices appearing in it to favour a goal will also be of good utility in the agent's real context (this fundamental assumption being a pivotal point in an agent's belief system). To this end, the agent may maintain a hypothetical *model* of its environment representing characteristic relations in it, and keep track of the assumed *state* of the environment (note that in many cases, model and state are not cleanly separable). An agent's possession of any representation of reality is constituting its *awareness* [35], with more specific terms as *intra-personal awareness* (the agent's representation of itself), *inter-personal awareness* (picture of other agents and relations with them), *situation awareness* (depiction of a surrounding situation) or *context awareness* (interpreted place of the agent and other entities in the given situation) denoting specific subsets of an agent's picture of reality.

Both humans and artificial agents have to establish awareness, maintain belief systems and make decisions in the presence of finite sensory and cognitive resources, observing a general preference of resource efficiency. Also, one must keep in mind that agent behaviour with the highest success rate—or highest utility in a multi-agent setting—does not necessarily coincide with acting out the best-founded decisions possible with the most accurate representation of reality attainable. Hoffman et al. [43, 44] elaborate this in an evolutionary context, using the term *fitness* for the measure of successful action outcomes. Parallels between natural selection pressure and "engineering common sense," especially in production systems subject to external competition, suggest similar conclusions for artificial agents with resource limitations and efficiency concerns.

Awareness is a central topic of a number of contributions in this book. In particular, Chap. 7 discusses context awareness in manufacturing and presents a context-aware monitoring system supporting human–robot collaboration [24]. Context-aware delivery of work instructions is discussed in Chap. 15 [17]. The right interpretation of sensory information gathered from human movements and the prediction of further trajectories is only possible if the system is aware of the context defined by the actual operations (see Chap. 11) [26]. As emphasised in Chap. 10, proper workcell control is only possible if the controller is aware of the state of all relevant items in the workcell. This provides an important motivation for (re-)calibration [45].

1.4.5 Sharing Views Through Pivotal Models

Humans and artificial agents, thus, pursue comparable approaches in observation and construction of beliefs about their environment, and the parallels have an impact on the ways agents *share* their world-views. Without reliance on the direct injection of a model, the sharing of a world-view has to go the tedious way of perception and inner reconstruction of a picture of the world. Its use is, then, subject to the assumption that other agents acting apparently in possession of the same world-view do share a corresponding inner model as well. There are fundamental limits to sharing views that preclude an explicit guarantee of a perfect match across individual agents (Sect. 1.4.6 gives further explanation), however, the reliance on some pivotal, and mutually accepted, representation of reality can give substantial practical support in sharing world-views.

In a manufacturing context, a shared picture of the world (both regarding underlying relations and states) more and more often relies on a *digital twin* (DT). The DT is a *virtual representation* that is able to closely follow the characteristics and changes of a real environment [46].

The importance of digital twins within CPPSs—and thus also in HRC—is hard to underestimate. Digital twins provide the passages between the real and virtual worlds of manufacturing. According to a generally accepted definition, "a digital twin is the digital representation of a unique asset (product, machine, service, product–service system or other intangible asset) that comprises its properties, condition and behaviour by means of models, information and data" [47]. Some authors distinguish between digital twins and *digital shadows*. The latter incorporate data and information gained during the usage or operation phases of an individual product or production system.

Designing, planning, operating, supervising and controlling manufacturing environments where humans and robots collaborate requires accurate digital models that are updated nearly in real time. However, here, more is needed than a kind of replica of the devices, equipment, processes or human operators. The DT should combine and align all relevant aspects of modelling the function, structure and behaviour of the robotic cell including the workers, together with capturing the symbiotic interplay of the human and robotic agents. This includes representing the multimodal and bidirectional channels of communication and control as well. In the HRC context, the basis of the digital twin can be a *linkage mechanism*, which is capable of representing the geometric and kinematic relations of all static and moving objects in the workcell [48] (see Fig. 1.3). As planning and execution proceed, the initial model is gradually enriched with new elements obtained with perception or deduction. Time and again, the DT should be efficiently *calibrated* to the real environment by using the actual measured data, so that both robot programming and operator instructions can automatically adapt to the actual situation (see also Chap. 10 [45]).

The application of digital twins and shadows, together with *predictive engineering,* can lead to anticipatory rather than reactive planning and control in a HRC environment. High-fidelity models of phenomena of interest can be constructed, by

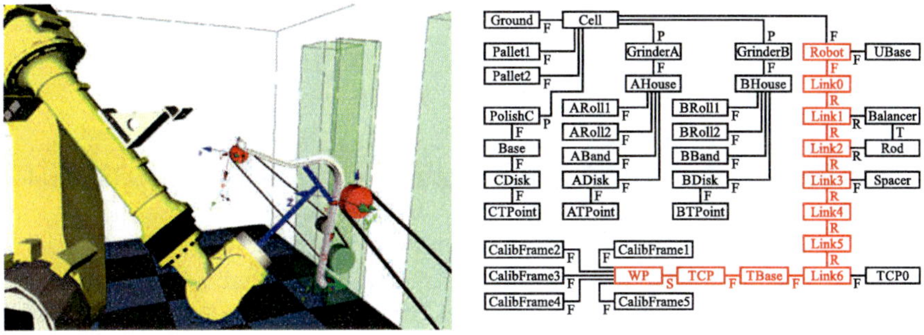

Fig. 1.3 Digital model (left) and kinematic graph (right) of a robotic workcell [48]

which future spaces can be explored and more appropriate decisions can be made. In particular, the growing demand for adaptive, *off-line programmed* (OLP) robotic cells can be met, thus eliminating the laborious and time-consuming online programming work. By improving *twin fidelity* (i.e., *closeness* of the digital twin to reality) through multi-level calibration methods, an accurate digital twin of the as-built robotic cell can be established, which ensures the sufficient accuracy of robotic operations planned off-line.

The model of the human worker should be an integral part of the digital twin [49]. HRC environments pose a need to elaborate and populate models of the human workforce and develop task execution control and communication approaches that can establish individual worker preference profiles, pick up transient changes in the state of the individual worker agents, and tune both communication and acquired models accordingly [38].

1.4.6 Sharing Information with Communication

Artificial agents can rely on a relevant view, or *presentation* of a coherent common source for making decisions that determine their behaviour. Moreover, insight gained from local perception can contribute updates or enhancements to a shared representation. These possibilities are of a major advantage when artificial agents have to establish *group awareness* [35] with shared beliefs—direct sharing can, namely, supply explicit guarantees for alignment of beliefs of the agents involved.

With humans, the situation is different, inasmuch as the content of a human mind cannot be accessed directly in any direction. In fact, many, if not most, artificial agents, too, are subject to comparable limitations, either due to their architecture not supporting direct access, or due to lack of guaranteed interoperability of inner representation. In such cases, the sharing of a world-view consists in obtaining information about proposed common views via perception, reconstructing a model internally via reasoning, and using it in the assumption that others apparently sharing the same views share a fitting inner model as well. This situation sets fundamental limits to shared beliefs (such as the *mutual knowledge paradox* [35]), and necessitates a number of agent capabilities for making group integration practicable:

(1) Agents must be able to *communicate*, i.e., be able to pass relevant information between each other relying on specific forms of perception and action which are intended to convey information and mutually recognised as such by participating agents (by prior agreement or design). Communication between agents uses changes of specific properties of a medium or structure in the environment, referred to as *communication channel*. Like other agent capabilities, communication capacity is subject to limitations [50], and it is justified to prefer utility over accuracy, as is the case with perception in general [44]. Agent activities conveying information are not necessarily dedicated to communication only. Information can also be transmitted as a by-product of other actions in the multi-agent setting, e.g., having completed a certain assembly step does not necessarily need to be explicitly communicated to other teammates if this is perceptible otherwise by relevant participants. In other cases, possible faults of the communication channel have to be considered, including corruption of transmitted information or time delays. Also, the generic terms *interaction* or *interference* are used [50]—these, however, more often refer to actual occurrences of influence or exchange between agents, rather than a capability of a single agent.

(2) Agents need to be able to model the characteristics and state of other agents in the multi-agent setting, and must be able to integrate relevant perceptions into an assumed picture of other agents [1, 33]. Such abilities, especially modelling and reasoning, are referred to as *theory of mind*. While it is still far from realistic for today's artificial agents to approach human abilities in this regard [51], it must be noted that the structured and purpose-designed nature of production environments does lower the challenges considerably. Such relaxations, however, do not eliminate the requirement of a collaborative agent being *predictable* [33]. In a structured setting, this has much in common with being *rational*, which refers to the ability of making optimal decisions with respect to some goal, measured by indicators. Rationality, however, can be *bounded*, either by resource limitations, or by other intrinsic characteristics of the agent's capabilities.

(3) Socially competent agents must be able to *model their own position* in the multi-agent setting, assess their own goals within the group context, and be able to *negotiate* goals, which involves specific forms of communication in accordance to a mutually accepted protocol.

The capabilities above enable an agent to connect with other agents and perform consistent transactions even if direct, objective reliance on a shared model is not possible. However, as mentioned earlier, the effectiveness of agent-to-agent communication depends on the individual agents' assumptions being correct, otherwise the transaction may be misinterpreted and related goals may be missed. Also, it is important to stress that agent-to-agent communication assumes the *unanimous intent of all parties involved to communicate*. Actions falsely interpreted by an agent as communication, therefore, do not count as communication due to a lack of appropriate intent in the other agent carrying out the misinterpreted action.

This subsection examined communication from the perspective of an *individual agent's capabilities* which enable it to share information deriving from its own state, and receive information stemming from the state of other agents. Deploying these capabilities results in acts of communication which already involve *multiple agents*. Communication, therefore, forms an important aspect of interrelations between agents as well—the relevance of communication will be once more highlighted in this latter context further below, in Sect. 1.5.

This book explores a number of important aspects of information sharing and communication in human–robot collaboration. In particular, Chap. 2 presents the latest developments in gesture recognition, identification and tracking [39]. Chapter 9 presents a cockpit integrated with a human–robot interface that facilitates communication via different channels [16]. Tracking of human motion and the interpretation of human intent are reported in Chap. 11 in the context of a work which is ultimately aimed at anticipating the needs and providing robotic service to human workers in assembly [26]. Chapters 12 and 15 present multimodal architectures where audio, gesture, haptic and touch channels of communication as well as their fusion help to transcend traditional ways of interacting with and programming of robots [17, 19]. Haptic control with sensors is discussed in detail in Chap. 13 [41], while the emerging technology of using brainwaves is exposed in Chap. 14 [52]. Chapter 16 [53] suggests a novel way to put *augmented reality* (AR) into the service of communicating with robots whereas Chap. 17 reports on deployed pilot applications [54].

1.4.7 Human-Specific Features

While artificial robotic agents can be designed and built to meet the requirements of a production setting, humans have some characteristics and limitations which cannot be fundamentally altered by training or long-term experience. Aside from physical limitations, humans have evolved to handle *mental workload* within a certain range. Standards under ISO 10075 [55] refer to *mental stress* (in the neutral sense of stimuli, originating in task requirements, physical conditions, as well as social, organisational and societal factors), and resulting *mental strain reactions* of the human (activation and warming-up as facilitating effects, and mental fatigue, monotony, reduced vigilance and satiation as impairing effects). Awareness and mental capabilities differ from human to human, and also vary for a given human over time, and their representation in artificial models is often of *probabilistic* nature [56–58], i.e., humans have limited predictability, and their integration into a multi-agent setting needs special measures, with more elaborate means of determining their instantaneous attitude and capacities.

In addition, humans are not entirely rational in multiple ways. Human reasoning more often produces faults, especially when working under pressure (e.g., time pressure in *real-time* situations requiring guaranteed response in limited time, or under high mental stress due to stimulatory overload), and the judgement of a situation, or even reasoning itself, can be compromised by *cognitive biases* which act as fixed heuristics with evolutionary backgrounds [59].

Humans also have a considerable *affective* or *emotional* dimension that responds to their general well-being and situation within their environment. This has an impact on the overall *engagement*, i.e., the attitude towards and commitment to common goals, and the degree of participation in processes towards the goals [60, 61]. The affective factors contributing to engagement can be related to the degree of freedom and control over the working situation (comparable with the concepts of *autonomy* and *authority*), and the perceived embedding of the individual into a group context (team alignment, acceptance of goals, rewards, fairness, meaningfulness, self-availability, social safety). Since affective factors have a profound influence on human behaviour (both experienced and anticipated), the exchange of *affective cues* is a vital part of human communication, and should be taken into account when artificial agents and humans are integrated into the same setting. Terms and definitions related to affect and engagement show much variation, but comparative surveys do provide a systematic insight into the area [3, 60, 61].

As for the interaction of human workers with their environment, recent studies indicate that the increasing complexity of manufacturing environments and processes expose operators to heavy cognitive load that may impact their performance and emotions as well [62]. In such cases, traditional paper-based work instructions may poorly facilitate operations as they both take time, and more importantly, prior experience, to apply well. Therefore, the application of multi-modal and digital work instruction systems becomes more common in advanced industrial environments, resulting in better quality, higher success rate and performance via advanced ways of information distribution [63]. Typical multi-modal instruction systems apply video, 3D graphics, voice and even augmented reality (AR) solutions to support operators in their work. These points are elaborated in Chap. 15 in this book [17].

Advanced human–robot interfaces, especially those conveying work instructions, often provide adaptability to the given worker in discrete steps, namely, by skill-level categories. Adaptation to the worker's current (and changing) fitness for the current task is, however, not part of industrial practice. The devices keeping track of the worker's awareness primarily serve safety purposes only. Chapter 11 shows a new direction of research where tracking the worker facilitates the prediction of the human's next movements and operations [26]—the system essentially "reads" the intention of the human agent this way.

As another emerging research direction, *brainwave-driven robot control*, can enhance symbiotic HRC in assembly [64]. In *brain robotics*, electroencephalography (EEG) is used to analyse and understand the behaviour of the human brain, the results of which are used to communicate with robots and control the robots directly. "Think ahead" combined with human motion prediction can contribute to timely cognition and assistance by the robots. Furthermore, mental stress and even psychological discomfort leading to any potential accident can be monitored and diagnosed via the brainwaves of human workers, which can be collected by sensors embedded in a safety helmet. Although not being without ethical concern, this technology could contribute to safety assurance in workplaces. Brain robotics in the service of human–robot collaborative assembly is presented in detail by Chap. 14 of this book [52].

1.5 Organisation: Roles, Multiplicity, Team Composition

The top level of the multi-agent organisation perspective observes several agents within the same setting, including their interrelations and overall organising mechanisms. Given the focus on human–robot collaboration, it can be assumed that at least two agents are involved, one of them likely being a human, and the other being an artificial agent. In fact, most of today's research in robotics and manufacturing science rarely goes beyond this, rather minimalistic, scenario. Therefore, binary relationships have received ample coverage in recent research, and there is much literature a framework of concepts can be built on.

While much of what applies to a mixed human–robot setting retains validity in a general multi-agent context, this subsection will keep the focus on physical agents, specifically on humans and robots involved in the same group setting.

1.5.1 Interrelation Tightness Characteristics Between Physical Agents

Human–robot interrelation is often addressed in literature by defining levels of collaboration, typically establishing a one-dimensional ordering of characteristic classes of cases. While the terms *coexistence*, *cooperation* and *collaboration* are frequently used throughout literature, their actual interpretation varies greatly, and the scientific community has yet to establish consistent common ground in this regard [65]. Moreover, some of the numerous definitions for levels of collaboration integrate these terms with aspects beyond a pure description of the given class of interrelations, and make, e.g., the motivation for choosing the given level of interrelation part of the definition itself. Such is the case with the presence or absence of common goals—their inclusion into the definition of the interrelation level essentially brings the questions of "why" and "how" under one roof [66].

Certainly, this does not make it easy to find one's way among multiple, sometimes contradicting, interpretations. Therefore, it may be best to establish a *framework of aspects* of human–robot interrelation first, which makes different definitions of coexistence, cooperation and collaboration easier to compare and position within a wider picture of the domain. The aspects listed below serve the *description* of any human–robot interrelation as a phenomenon only, drawing a boundary to matters of an agent's BDI, or group context. The latter may very well have their due impact in certain definitions, however, they must be recognised as factors external to the mere form of human–robot interrelation.

Temporal interrelation expresses how major time intervals of activity of the agents involved relate to each other. Depending on the scenario and its conceptual framework, the term "activity" may include ongoing physical action, observation, or any form of "on-line" state in general. Likewise, choices of what is regarded as a major time interval can depend on the given scenario, but it can be generally understood as the time interval relevant to the examination of the given instance of interrelation. With these preliminaries, temporal interrelation can be *non-concurrent* (with the major activity intervals of the agents being disjoint in time), or *concurrent* (with the intervals at least partly overlapping). Examining specific actions of the participating agents, one can further subdivide the concurrent case into *asynchronous* (agents have to complete certain tasks by a given point of the major activity interval, but can freely choose their particular timing within these boundaries), and *synchronous* (the timing of the actions of participating agents has to be aligned in a certain way). Synchronous temporal interrelation can, then, take on the form of strictly *sequential* operation (i.e., a given action of an agent has to follow the completion of another specific operation of the other agent), constraints can apply on the relative timing of *beginning or ending* the actions, or the operation can be *simultaneous* (one action interval being some form of subset of the other). Various forms of *temporal logic* are well-suited for describing the temporal relation of agent activities in synchronous interrelations [67], and finds frequent use in distributed artificial intelligence, including multi-agent systems with physical agents [68, 69].

Spatial interrelation refers to the relation of the workspaces of the agents. The spatial interrelation is *remote* [70] if the workspaces of the agents are entirely disjoint, and separated by considerable distance to a degree that can fundamentally shape the form of interrelation between agents (e.g., by time lags or transmission disturbances). In the case of *separation*, the aforementioned distance-induced characteristics no longer are an issue, however, the workspaces of the agents are entirely disjoint in a static way. In *co-location*, workspaces can have *dynamically changing boundaries* while still maintaining separation in any instant, or they can *overlap*, with or without *physical contact* of the agents allowed. Chapters 3 and 11 of this book illustrate spatial interrelation possibilities with practical examples.

If several physical agents are connected with some form of interrelation, and act in the same system, it is inevitable for them to influence each other. Having an effect on each other is generally referred to as *interaction*. Two main aspects of interaction play a key role in characterising agent interrelation: the *interaction intent*, and the *interaction medium*.

Interaction intent can take on two main forms: (1) agents can have an effect on each other already as a mere *consequence of their actions*, having an effect itself not being among their committed goals (i.e., the interaction itself not being intended), or (2) having an effect on other agent(s), i.e., changing their state, is a necessary part of the agent's actions taken to advance towards a committed goal (i.e., *deliberate interaction*). In the distinctly purpose-centred domain of collaborative robotics in industrial settings, the term "interaction" often refers to deliberate effects only, and unintended or not goal-determined influence is regarded as interference or disturbance of the multi-agent system that often has to be mitigated. This emphasis on deliberate interaction is duly reflected throughout this book, with details of this interpretation being put forth in Chap. 3 [34]. Deliberate interaction, again, can take two different forms: (1) the agent can either affect the operating surroundings of other agents to elicit changes of their state or behaviour (*deliberate influence* in general), or (2) the agent can transmit information to other agents and thereby effect state changes in them. The latter is referred to as a *communication intent* (note that actual communication does require a corresponding intent on the side of the receiver as well). Communication forms such a substantial part of all interaction in human–robot interrelations that some scenarios are reasonably close to permitting an interchangeable use of the terms "interaction" and "communication" for pragmatic simplicity (see, e.g., Chaps. 13 and 15). At this point, the reader is also referred to Sect. 1.4.6 where the *capability* of an individual agent to communicate is dealt with in detail. In the scope of interrelations, one can see this capability put to work, in the sense of the *act* of communication becoming an aspect of agent-to-agent interrelation.

Interaction medium describes the property or part of the multi-agent system, along which the influence is exerted during interaction. As is the case in physical multi-agent systems in general, human–robot interrelations draw a most distinctive boundary between all things tangible and non-tangible regarding interaction medium. Thus, the most basic two categories are *contactless* (also referred to as *non-physical*) requiring no physical contact to take place for interaction, and *contact-based* (or *physical*) involving physical contact. Especially in human–robot collaborative settings where the health and safety of humans override all other priorities, the non-physical or physical nature of interaction medium has fundamentally different safety implications as well, as elaborated in Chap. 3 [34].

1.5.2 Commonly Identified Classes of Interrelation Tightness in HRC

Literature distinguishes classes of interrelations along certain patterns of their characteristics, and assigns them a specific name [65]. As mentioned in the previous subsection, sources vary greatly regarding the definitions given for these "named" interrelation classes, therefore, no comprehensive comparison is aimed for at this point. Instead, some of the more commonly used interpretations are presented with respect to the place they occupy in the space of interrelation tightness characteristics. Included are also cases that are clearly outside the domain of human–robot collaborative settings—a closer examination of these cases is still useful for drawing a border between what is and what is not regarded as HRC (see Fig. 1.4).

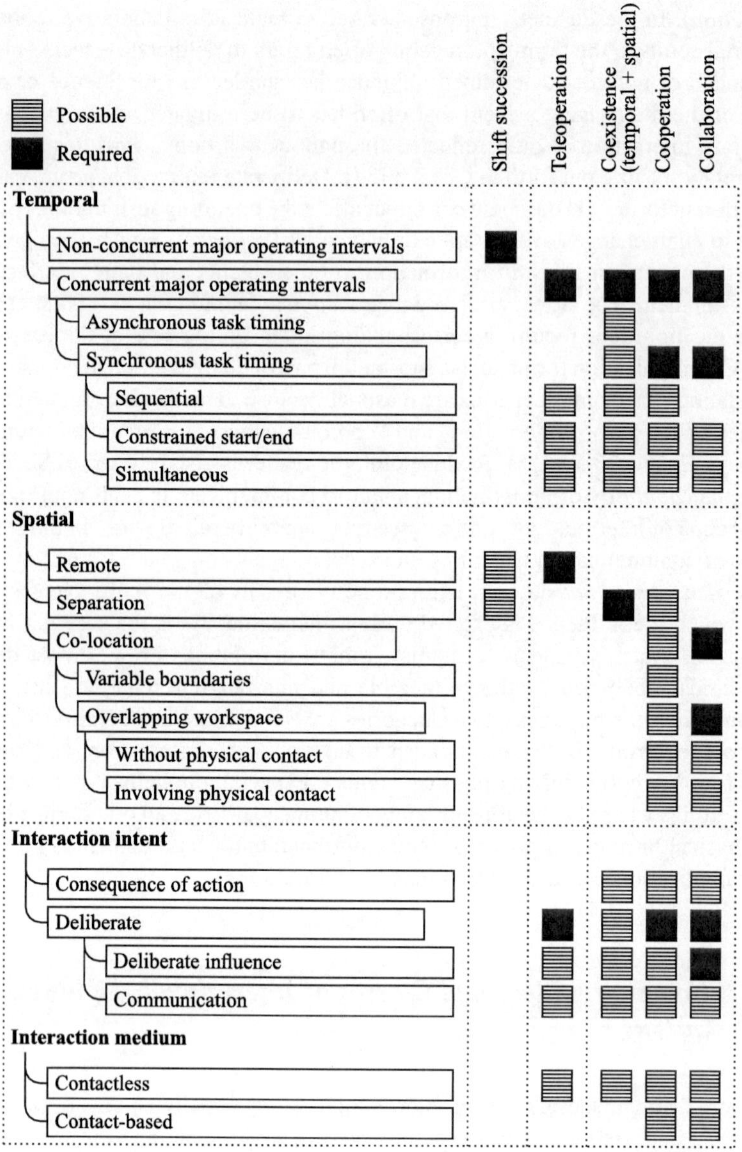

Fig. 1.4 Morphology of interrelation tightness characteristics with some of the key interrelation classes

Due to its very nature, HRC assumes that agents involved in the collaborative setting have concurrent activities, i.e., their major activity intervals overlap. Without this characteristic, only a kind of *shift succession* is possible where each agent works on its own, and terminates its activities completely before leaving the results to the next agent. In this case, the only way agents can have an effect on each other is by shaping the scene, the object of the work, and related next tasks they leave behind—these are usually handled by the next agent as task descriptions, technological circumstances, etc., where the identity and state of the preceding agent are of marginal, if any, relevance. This is typical for the construction industry where work often requires different specialists to move to the site of the task, set up and perform their work, and leave the scene before the next specialist continues the construction. Similarly, different production resources along a production line have little to no temporal (or spatial) coupling directly depending on their identities and their specific states. All these scenarios clearly lie outside the HRC domain.

Even if major activity intervals of the agents overlap in time, they may be at a spatial distance from each other that burdens their interrelation with time lag, noise or other disturbances that have a major impact on any form of coupling between their behaviours. This case is marked with "remote locations" in the spatial dimension. An example for this case is *teleoperation*, where, typically, a human takes the role of a master, and a robot acts as a slave which, if needed, has enough autonomy to locally compensate for the flaws of the master commands received. This, too, is beyond HRC, even though some approaches deployed in teleoperation can successfully be adopted in HRC, too.

Where the landmark is usually set in literature for the "lower" end of the HRC interrelation spectrum is the form labelled as *coexistence*. In some cases, a distinction is made between temporal coexistence (essentially, concurrent major operating intervals) and spatial coexistence (anything closer than "remote"), but many sources actually mean temporal *and* spatial coexistence by this term. If the tightness of interrelation does not go beyond the minimum required for this class, coexistence refers to concurrent operation with fixed spatial boundaries between agents. Chapter 3 [34] explains one possible definition of coexistence in more detail.

In *cooperation*, the actions of the agents augment each other (in such a purpose-oriented domain as industrial production, the agents' actions are required to fit as elements into a more comprehensive plan). However, synergies between actions mostly exist via connections between outcomes and preconditions only, leaving much freedom to the agents themselves in their particular timing. In other words, their actions do follow a synchronous temporal pattern (constraints apply to start and end times of certain actions), but are not subject to continuously coordinated execution. Agents, thus, still "work on their own" while they share a part of their (kinematically feasible) workspace. The workspaces actually used can be kept apart by fixed or time-varying separation, some cases can even include an overlap. Nevertheless, the main objective of workspace allocation in cooperation is to keep the agents, as far as possible, "out of each other's way" in accordance with the loose coupling of their processes. Avoiding interference, as well as maintaining the necessary coupling of agent actions, requires some form of deliberate interaction. While the "keeping clear of each other" directive implies the prevalence of contactless (non-physical) interaction, the possibility of physical contact is not excluded due to the relaxation of spatial separation. Chapter 3 [34] gives an example of a more specific interpretation of the term *cooperation* in the context of safety implications.

In *collaboration*, agents act together, mostly in simultaneous action, in a coordinated way (coordination being ensured at least to some degree implicitly by the interrelation of the agents directly involved in the actions). This has two implications for other characteristics of interrelation tightness: (1) due to the tangible nature of the working environment and tasks defined therein, a declared overlap of actively used workspaces is necessary, and (2) deliberate interaction, mostly in some form of communication, is essential in ensuring the tight, coordinated coupling of the agents' ongoing actions. Collaboration is recognised as the tightest class of interrelations possible between two physical agents, as explained further in Chap. 3 [34].

1.5.3 Autonomy and Authority in Agent Interrelations

The *degree of authority* delegated to an agent is another important characteristic in a relationship of two agents. Authority refers to the amount of decisions an agent is allowed (and required) to make by itself [32], and thus designates how much of its own *autonomy* an agent is to exploit in a given relation. The authority dimension is where several of the first occurrences of the *role* concept can be found, with a possible interpretation as the agent's contextual interface to a setting involving more than one agent. The authority level in a given collaborative relation can, in the most general case, be changed during execution [32, 71], however, current manufacturing practice usually retains the constraint of authority allocation prior to execution of a discrete task. In the quantisation of authority levels in relations, a simple leader–follower partitioning prevails [30], however, finer levels of obligation can also be identified. Based on a review by Mostafa et al. [31], five categories are possible by the distinction of *executing commands* (without authority regarding the command) from *supporting the other agent* (seizing facilitating opportunities):

(1) *Master* (issues binding commands to other agents),
(2) *Guide* (leads task execution but leaves mode of contribution to the other agent's choice),
(3) *Associate* (negotiates task execution with the other agent, regarding it as an equal partner),
(4) *Assistant* (supports other agent's task execution with own decisions, using opportunities at hand),
(5) *Servant* (obliged to execute commands issued by other agents).

In some cases, the classification of collaboration modes makes it worthwhile to combine authority with the level of collaboration, such as the distinction of the agent role classes *active*, *supportive* and *passive* [72]. The aforementioned collaboration and authority levels are sufficient for cases where roles are assigned beforehand, and

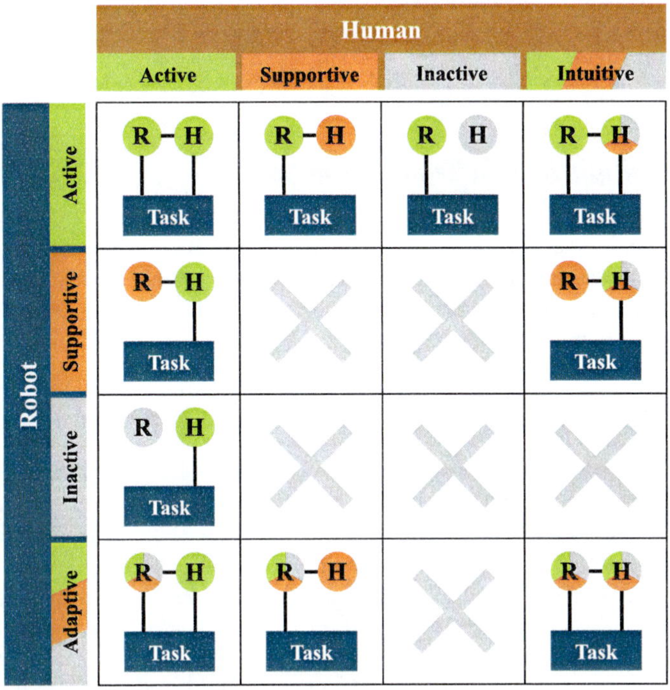

Fig. 1.5 A possible classification of roles in human–robot collaboration, based on [72]

remain unchanged during task execution. A flexible transfer of authority, as often witnessed in human-only teams, would require a separate set of cases, denoted in Fig. 1.5 as "adaptive" for robots and "intuitive" for humans, reflecting differences in the way role changes come about in different types of agents.

A similar, but finer grained classification of collaborative cases by autonomy and activity is given by Mostafa et al. [31]—here, it is presented in a concise form generalised for any (artificial or human) agent:

(1) Agent performs autonomously,
(2) Agent decides on seeking input from other agent(s),
(3) Agent always seeks input from other agent(s),
(4) Agent decides on seeking commands from other agent(s),
(5) Agent performs commands of other agent(s),
(6) Agent leaves performing to other agent(s).

Note that in the last case, the agent does not deny a command but leaves the entire execution to the other agent by design, i.e., it remains passive.

Relations between agents can be characterised by further modalities, such as the physical and cognitive dimensions affected by their interaction [65]. These are, however, so diverse and so much determined by the given application context that a full morphology covering most of the industrial production would be beyond the scope of this review.

1.5.4 Multiplicity and Team Formation

While most industrial collaborative cases still presume a one-to-one human–machine relation, it is anticipated that future manufacturing settings will involve several agents, humans and robots alike, in a variety of collaborative patterns. Therefore, it is necessary to widen the horizon of possible multiplicity cases. Yanco et al. [13] rely on binary relations while extending multiplicity to *one*, *many* or *team*, both for humans and artificial agents, resulting in 9 different patterns under the assumption that (1) humans and robots are distinguished as such, and multiples or teams are composed of agents of the same kind (i.e., robot-only or human-only). This classification scheme (see also Fig. 1.6) is still subject to restrictions, and is, in this regard, in line with such multi-agent approaches as the FIPA (Foundation for Intelligent Physical Agents) specifications [73] which draw a clear border between artificial agents contained in their own subsystem, and human users connecting to it.

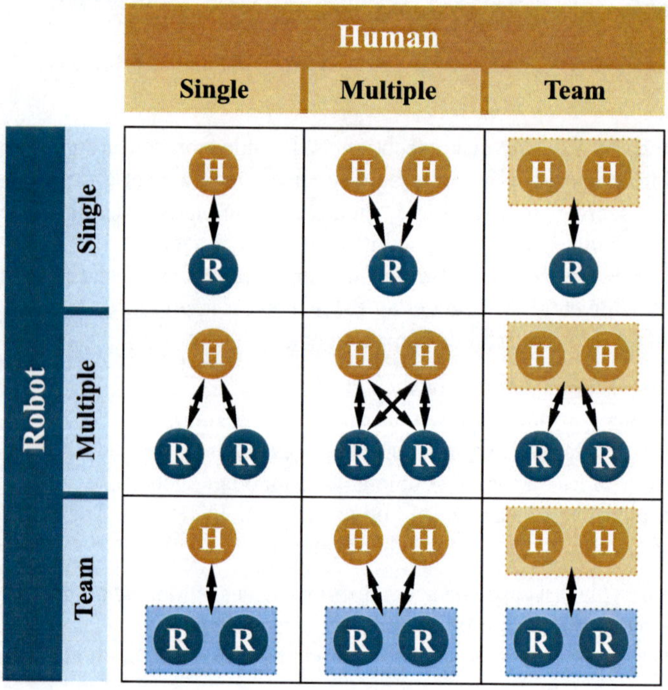

Fig. 1.6 Classification of multiplicity cases in human–robot collaboration based on [13]

Nonetheless, the morphology of Yanco et al. includes one more important concept associated with multiplicity: the *team*. This formation is set aside from a random set of agents by the presence of some organising agreement among team members. Klein et al. [33] refer to this as a *basic compact* to which all team members subscribe, i.e., team coherence is not built on individual agent-to-agent relations alone, but rather on a "common ground" agreement spanning the entire team (Ferber et al., for example, use the distinction *organisation-centred multi-agent system*, or OCMAS, as opposed to *agent-centred multi-agent system*, or ACMAS [74]). The team agreement sets appropriate constraints on coordination patterns, agent relations that contribute to coordinated operation (see a distinction of *direct supervision*, *standardisation of work* and *mutual agreement* patterns by Monostori et al. with binary agent relations marked [4]), investment and reward policies that make it worthwhile for rational agents to keep investing in common goals [33], or boundaries for *competitive* transactions (e.g., bidding for tasks by cost and capacity) which are either tolerated or preferred in meeting group goals [4, 75]. The team agreement also defines agent roles with their corresponding authority level and other properties.

In a coordinated team, the *role* of an agent becomes more distinctly an interface of the agent towards the rest of the team, which determines how the agent can and is required to contribute to the advances towards a shared goal [50]. The organisation-centred *agent–group–role* (AGR) model [74] builds on this perspective and is capable of efficiently capturing, e.g., simultaneous presence in several teams or timing of agent interaction and common activities. Similar approaches placing agent participation and individual relations into a formalised team context are likely to retain the interest of future research in human–machine collaborative manufacturing applications, due to their potential of addressing performance and fairness guarantees, and giving a comprehensive organisation-level view even in collaborative–competitive environments.

1.6 Planning and Execution Control

A key aspect in industrial production is the subdivision of the transformation or assembly of materials to products into *processes* which are to be completed in a desired way with the help of adequate means of modelling, planning and execution control that relies on sensing and acting capabilities. In the early days of technological development, the production resources were fairly simple, and process design had the formation of the product itself in the focus. This is less and less the case as production resources are becoming more complex, are gaining a higher potential of autonomy, and have an increased amount of different inner states—today, processes performed *on* products need elaborate methods to map them onto processes performed *by* the production resources [76]. In the case of human–robot collaborative production, this becomes even more pronounced, due to (1) the complexity of the entire production system being increased by employing several—partly or fully—autonomous agents, and due to (2) the increased state space of production resources owing to the entry of humans into the picture (with the intention of building on their awareness and higher level cognitive capacities at the cost of less predictable responses). For processes performed *by* agents, two dimensions of subdivision give the most important guidance: distinction by their place in an abstraction hierarchy, and relation of concrete process evolution paths with regard to the nominal process plan.

1.6.1 Levels of Planning

In complex production systems, one can observe a hierarchy along which processes and their handling are understood. This is commonly subdivided into *operational* (lowest), *tactical* (intermediate) and *strategic* (highest) levels, the distinction typically being made along degrees of abstraction, order of aggregation or some hierarchical dependence of process goals. The distinction of these levels is more a matter of system engineering decisions rather than exact definitions, however, some characteristics can be named as a "rule of thumb".

The *operational* level deals with carrying out processes on an instance-by-instance basis, with the horizon of attention rarely including other execution passes of the same process or relations across processes. From an agent capability perspective, this can be translated into the prevalence of *skill* usage, and only deploying more abstract capabilities without necessarily refining them.

The *tactical* level stretches over several process execution instances or over several parallel processes. On the tactical level, process parameters and execution decisions can already be adjusted to make ends meet with respect to goals or preferences over several processes—this is very similar to applying *rules* in agent capabilities. However, what is mostly understood as the tactical level rarely goes beyond parameter adjustment or discrete re-planning—the underlying model structures are not largely changed or even revised.

The latter is usually delegated to the *strategic* level which accommodates an elaborate aggregation and analysis of individual process characteristics, usually across wide horizons in time and organisational structure. This is performed specifically with the purpose of surveying the validity of underlying process models, and changing the model structure itself, too, when this is deemed necessary. In an agent, such activities are performed on the top levels of the capability stack, namely, *knowledge* (via the ability of validity assessment) and *expertise* (via ability of autonomous exploration and problem-solving).

In a multi-agent production setting, it is important to point out that both a complex multi-agent system, as well as the agents involved, have their own operational, tactical and strategic horizons and goals to pursue, and those of an agent are mostly nestled into those of the multi-agent context. In order to make the entire multi-agent system work as desired, the priorities and characteristics of the nestled abstraction levels need to be properly harmonised. In such cases, the planning process itself is likewise so complex that a substantial part of the problem-solving work necessarily relies on the cognitive potential of a human expert. In recent years, *mixed-initiative* planning environments have gained importance which synthesise the complementing advantages of human and artificial resources for thorough and more efficient planning of complex assembly processes.

In some cases, the characteristics of individual agents, too, can notably raise the complexity and related computational burden of planning. Such can be the case with the kinematic redundancy of robots (i.e., having more kinematic degrees of freedom than the task space dimensions exploited). Redundancy itself offers additional opportunities for *optimisation* and *adaptation* due to a continuum of solutions being available. However, redundancy resolution (i.e., picking one specific solution from the available continuum), especially in view of a complex operational environment with interlinked technological and geometric constraints, may call for computationally demanding iterative approaches that (1) preclude real-time adaptive planning and control, and (2) potentially escalate the problem to iterating between different modelling perspectives or to planning across multiple hierarchical levels due to the constraints and feasibility implications involved [77].

In this book, the key issues of dynamic assembly planning and the assignment of tasks to human and robotic agents are discussed in detail in Chap. 8 [25]. The plans are then realised by schedules where tasks are assigned to actual agents and other resources in time (see Chap. 9 [16]). An efficient method for real-time re-planning of robot motion initiated by collision avoidance is presented in Chap. 5 [78]. Note that assembly plans provide the basis also for generating work instructions as shown in Chap. 15 [17].

1.6.2 Execution Control

Regarding adherence to or divergence from a nominal process path, a more detailed view is justified, compared to fully automated production settings. This is due to two circumstances: (1) application cases involving HRC most often exploit the added robustness of human intervention to succeed in cases which would have crossed a process failure threshold in a rigid machine-only setting, and (2) the anticipated advantages of human inclusion come at the cost of humans themselves introducing sources of deviation on occasion which has to be recognised and properly addressed by the production system. In this regard, four process evolution paths can be distinguished [72], as also shown in Fig. 1.7:

Nominal process execution is the case when the process remains within the expected bounds of events/parameters, and succeeds. This does not automatically mean that human involvement can be dispensed with—in cases like handling free-form workpieces (e.g., processing natural raw materials of indefinite shape), the active participation of a human may very well be needed to keep the process (e.g., grinding of a surface) within the nominal limits. Typically, such nominal process execution is characterised by reliance on *skills* or *rules*, even if the latter means different degrees of complexity and variability for artificial and human agents.

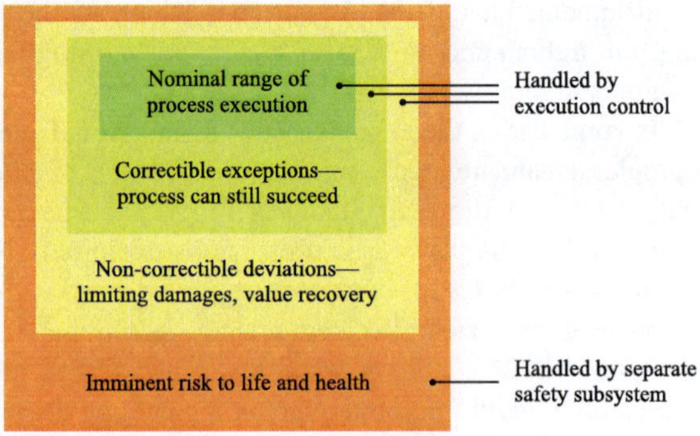

Fig. 1.7 Classification of process execution paths based on [72]

Correctible exceptions with respect to a nominal process course require measures which are outside the state/decision space of a pre-generated process execution plan, however, it still remains possible to steer the process to successful termination (e.g., removal of unforeseen artefacts by human personnel, or adjustment requested due to misaligned motion of human agent). As such cases are "out of the ordinary," tackling them requires higher-level capabilities of the agent in charge of mitigation, most typically comparable with *rules* and *knowledge* deployment.

Non-correctible deviations depart from the nominal process course to a degree that does not allow successful completion of the process (i.e., the current workpiece is destined for scrap or re-work anyway, or damage to equipment is impending that will result in considerable downtime). In this case, mitigation actions are usually aimed at limiting imminent damages, with the preference of enabling separate value-recovering measures as much as possible (note that, in this case, avoiding major downtimes can also be expressed as value). Given the possible ad-hoc nature of the necessary measures, the humans involved will certainly have to utilise their higher-level capabilities (i.e., their *knowledge* and *expertise*), while artificial agents can facilitate timely mitigation by immediate support via sensory, pre-processing or acting means that are less available to humans (e.g., fast data processing, quick retraction of obstacles from intervention area, or automatic highlighting of critical spatial features).

Imminent risk to life and health of human personnel is the most serious type of deviation from nominal process course. The binding requirement of protecting human life and health at any cost delegates the recognition and response to such cases to a separate safety subsystem that is, by regulatory requirements, disjoint from any other component of the production system, and has absolute authority in intervention with ongoing processes. Due to the required "hard proof" of suitability, safety subsystems are kept simple, and their response thresholds will necessarily represent a rigid limitation to degrees of freedom in the design and operation of collaborative assembly systems.

Advanced execution control, as discussed in Chap. 7, requires *handling of exceptions, emergency and recovery* [24]. Current industrial practice does not explicitly foresee the integration of normal production operations with the handling of exceptions and emergencies or with recovery operations. Even though some methodologies address early fault prevention and detection, this is not necessarily equal to continuous feedback regarding operation results. Not only does this often result in the costly temporal and spatial separation of training from work execution, but it also hampers orientation of the worker in ad-hoc work situations, e.g., in construction of individual products, or in the maintenance of poorly documented legacy equipment. Since training and quality check are typically separated from task execution, gradual skill development, live assistance and real-time feedback are obstructed. Industry-proof integrated processes of learning by demonstration, as shown in Chap. 12 [19], must be elaborated where performance evaluation results (including those of quality checks) are fed back to support the improvement of the overall team. In any such *learning-by-doing* scheme, the safety requirements must be continuously observed.

1.7 Safety and Ergonomics

1.7.1 Human Safety

When humans and robots coexist in the same environment, safety protection of human operators in real time is paramount. Therefore, human safety will always remain the first concern in HRC. However, even though safety should be guaranteed during execution control, the preparations require strategic thinking. Up-to-date perception, cognition and control methods for safe human–robot interaction are summarised in [79], taking a system theoretic perspective. The challenge is not only passive collision detection in real time but *active collision avoidance* at runtime by monitoring operators and controlling the robot. Principally, in a one-to-one human–robot setting, one of the safety strategies presented in Fig. 1.8 can be applied [80].

As for the enabling technologies of safety monitoring and protection, vision-based approaches using 3D models linked to depth sensors are predominant [38]. Alternatively, torque- and force-based methods can be relied on as well. These methods can limit the force potentially applied to the human body, but are not appropriate to anticipate and avoid physical collision situations, since forces and torques can only be detected once physical contact has already occurred. Hence, torque- and force-based methods are rather applied in combination with vision-based techniques.

Certainly, current distance-based collision-free HRC can ensure human safety but not overall work efficiency. Hence, [81] presents a *context awareness*-based collision-free human–robot collaboration system that can provide human safety and assembly efficiency at the same time. The system can plan robotic paths that avoid colliding with human agents while still reaching target positions in time. Human operators' poses can also be recognised with low computational expenses to further improve assembly efficiency. An efficient transfer-learning-based human pose recognition algorithm is also adapted and tested.

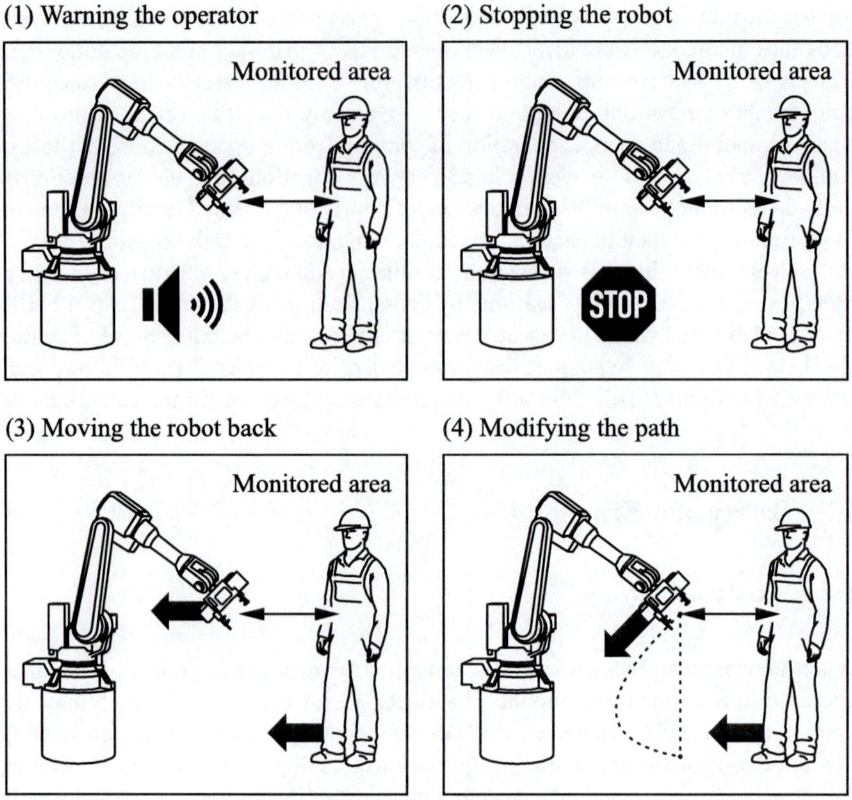

Fig. 1.8 Four safety strategies for active collision avoidance (Adopted from [80])

Although advanced methods and mechanisms have been developed in the past years, most of these methods are established in the lab environment. The hardware utilised are prototype-level devices that cannot be transferred to industry directly. More mature devices at a higher *Safety Integrity Level* (SIL) are needed to further improve and exploit the research results. Moreover, the feasibility of the HRC solutions has been evaluated well, but the safety performance needs to be assessed systematically. For example, stability, robustness, response time, redundant safety, backup solutions and emergency handling need to be evaluated in a structured and standardised way [82].

This book provides ample material about the main aspects of warranting human safety in an HRC environment. Departing from the classification of typical industrial human–robot interaction scenarios, Chap. 3 suggests an overall safety strategy which can then be refined to a personalised safety framework, complying with the characteristics of the participating human and robotic agents [34]. A systemic approach to safety is also presented in Chap. 6 where a safety system configurator is used to define adaptive and dynamic safety systems [83]. While effective safety standards are fully observed, the system can adapt its operation to the actual setup of a human–robot team. The key technology to implementing any safety strategy, collision detection and avoidance, is discussed in Chap. 4 along with an efficient method suitable for real-time operation [84].

1.7.2 Ergonomics

It is a common understanding that the combination of complementary strengths of humans and robots leads not only to increased productivity, flexibility, adaptability and robustness, but also to improved *ergonomics*, and to more attractive working conditions. This is mostly thanks to the robot which, by undertaking heavy-duty and repetitive operations, relieves humans of substantial unergonomic loads. Standards for ergonomic human–system interaction already exist [85], but for the time being, these recommendations have not instigated strategic thinking the way it happened in the case of safety. Most work focuses on mental workload, as discussed above (see Sect. 4.7).

However, with the advance of digital technology, the gap between human and machine communication is narrowing. Diversification of the workforce and machinery has become a forefront issue in advanced manufacturing industries. In reality, concerns exist about changes in the interaction between various people and machines due to differences in age, attributes and skills, and the impact of such changes on society, particularly the working environment [86]. In order to respond to this challenge, it is necessary to create a sophisticated mechanism in which humans and machines can mutually assist and evolve together for sustainable growth. This problem is addressed by a multi-mediation process called *multiverse barrier free* between humans and machines, as a future model for systematically harmonising their interactions [58].

In this book, as typical in contemporary research of HRC, ergonomics is still not in the main focus of discussions. Chapter 12 presents an ergonomic robot teaching technology using capacitive sensors [19], and Chap. 17 analyses the pilot demonstration systems from the perspective of ergonomics, too [54].

1.7.3 Regulatory Framework

The potential of collaborative work of teams of multiple human and robotic agents creates such a new scenario of industrial production which requires not only novel scientific insights and technological solutions, but also a new *regulatory framework*. Standards and directives aim to standardise the design and prevent engineering failures from the design phase on. In general, ISO 13855 [87] defined the positioning of safeguards with respect to the approaching speeds of parts of a human body. As one type of machinery, robotic cells shall have the minimum distances to a hazard zone from the detection zone or from actuating devices of safeguards. However, differences between HRC and conventional industrial manipulation require that safety and reliability standards be rethought.

ISO standards 10218-1 and -2 [88, 89] amended with ISO/TS 15066 [90] define requirements for human–robot collaboration with an emphasis on safety. Current standards for collaborative robotic environments already take an integrated approach with the compound of control, robot, end-effector, workpiece and task being simultaneously subject to assessment. This is a favourable starting point for further enhancement towards a complex multi-agent team. ISO/TS 15066 also distinguishes four modes of collaborative operation in a shared safeguarded area (safety-rated monitored stop, hand guiding, speed and separation monitoring, power and force limiting), taking the first step in removing rigid cell barriers of robotised workstations. These modes of operation will certainly determine possible ways of robots participating in collaborative teams that include humans.

Nonetheless, ISO 10218 and ISO/TS 15066 deal with physical (mechanical) aspects only—mental and emotional issues remain largely unspecified. *Mental workload* is though the subject of standards under ISO 10075 [55] which take it as pairs of (1) stimuli, originating in task requirements, physical conditions, as well as social, organisational and societal factors, and (2) effect, resulting in mental strain reactions of the human (activation and warming-up as facilitating effects, and mental fatigue, monotony, reduced vigilance and satiation as impairing effects). ISO 10075 clarifies that working environments should optimise mental stress to avoid both monotony and other forms of underload, as well as overload, fatigue and burnout. In this context, work equipment, ambient and social environment, and tasks are usually regarded as separate categories; however, it is expected that the introduction of social collaborative robots, and complex work environments that act as a team, will cross over the category limits, requiring a new integrated perspective in assessing, as well as designing such systems. While ISO 10075 does not specifically address such views, the perspectives of the standard are a consistent starting point for further synthesis. ISO 10075 largely remains on the level of the individual, and higher-level perspectives are taken by ISO 26000 giving "Guidance on social responsibility" [91]. These ethical guidelines are, however, too generic to be subjected to formal auditing or to provide firm specifications for human–robot teams working on the shop floor.

All in all, current standards do not yet address many key issues of mixed teams of human and robotic agents. Nevertheless, they outline some of the key aspects of a roadmap for such socially aware collaborative environments that set the baselines for a safe, anthropometrically optimal, mentally and socially satisfying environment for HRC. Intensive standardisation activities are ongoing and more realistic standards are anticipated.

1.8 Conclusions

In industrial practice, collaborative human–robot workstations exist already. However, they do not exhibit all characteristics of a working environment hosting multiple collaborating humans and robotic agents. As discussed above, many elements of such an environment are already available, such as advanced methods of perception, sensor fusion and situation recognition, digital twin, planning and robot control, off-line robot programming, automated collision avoidance, as well as advanced human–robot interfaces. Nevertheless, it is still an open question how to turn a group of human and robotic agents into a team, how to organise and harmonise teamwork, how to facilitate autonomous decisions, change levels of autonomy and responsibility and leave more abstract decisions to humans, once conscious judgement is needed.

In this introductory chapter, a novel multi-agent view was suggested to give a first orientation to the above issues in a systematic way. As it was emphasised, in HRC, agents need to be aware of each other's states and intentions, as well as the state of objects, to successfully work together or safely coexist within the same physical space. For robots, a proper degree of awareness depends on the adequate implementation of communication or information acquisition channels and their introduction into their modelling, planning and control. Being more autonomous but less predictable and reliable actors, humans need more support in being aware of the current situation. All these are prerequisites of shared task execution and adaptive team formation in collaborative environments. Agent participation and individual relations in a formalised team context are likely to retain the interest of future research in human–machine collaborative manufacturing applications, due to their potential of addressing performance and fairness guarantees, and giving a comprehensive organisation-level guidance for human–robot collaboration.

Acknowledgements Research for work presented in this chapter has been in part funded by the Hungarian National Research, Development and Innovation Office (NKFIH) under Grant GINOP-2.3.2-15-2016-00002 "Industry 4.0 Research and Innovation Centre of Excellence", and the ED_18-22018-0006 grant titled "Research on prime exploitation of the potential provided by the industrial digitalisation" supported by the National Research, Development and Innovation Fund of Hungary, financed under the public funding scheme according to Chapter 13. §(2) of the Scientific Research, Development and Innovation Act.

References

1. J. Bütepage, H. Kjellström, D. Kragic, Anticipating many futures: Online human motion prediction and generation for human–robot interaction, In *Proceedings of the 2018 IEEE International Conference on Robotics and Automation (ICRA)* (Brisbane, Australia, 2018), pp. 4563–4570
2. M.M. Cummings, Man versus machine or man + machine? IEEE Intell. Syst. **29**(5), 62–69 (2014)

3. G. Charalambous, S. Fletcher, P. Webb, Identifying the key organisational human factors for introducing human–robot collaboration in industry: an exploratory study. Int J Adv Manuf Technol **81**(9–12), 2143–2155 (2015)
4. L. Monostori, J. Váncza, S.R. Kumara, Agent-based systems for manufacturing. CIRP Ann. **55**(2), 697–720 (2006)
5. J. Váncza, Agent theory, in *CIRP Encyclopedia of Production Engineering*, ed. by L. Laperrière, G. Reinhart (Springer, 2014), pp. 33–40
6. L. Wang, A. Haghighi, Combined strength of holons, agents and function blocks in cyber-physical systems. J. Manuf. Syst. **40**, 25–34 (2016)
7. L. Wang, B. Schmidt, M. Givehchi, G. Adamson, Robotic assembly planning and control with enhanced adaptability through function blocks. Int J Adv Manuf Technol **77**(1–4), 705–715 (2015)
8. A. Coman, D.W. Aha, Cognitive support for rebel agents: social awareness and counternarrative intelligence. in *Proceedings of the Fifth Conference on Advances in Cognitive Systems* (Troy, NY, USA, 2017), pp. 1–6
9. S.E. Hashemi-Petroodi, S. Thevenin, S. Kovalev, A. Dolgui, Operations management issues in design and control of hybrid human–robot collaborative manufacturing systems: a survey. Annu. Rev. Control. **49**, 264–276 (2020)
10. L. Monostori, B. Kádár, T. Bauernhansl, S. Kondoh, S. Kumara, G. Reinhart, O. Sauer, G. Schuh, W. Sihn, K. Ueda, Cyber-physical systems in manufacturing. CIRP Ann. **65**(2), 621–641 (2016)
11. G. Weiss (ed.), Multiagent systems: a modern approach to distributed artificial intelligence, in *Intelligent Robotics and Autonomous Agents series* (The MIT Press, 2000). ISBN 978-0262731317
12. A.S. Rao, M.P. Georgeff, A model-theoretic approach to the verification of situated reasoning systems, in *Proceedings of the 13th International Joint Conference on Artificial Intelligence* (Chambéry, France, 1993), pp. 318–324
13. H.A. Yanco, J. Drury, Classifying human–robot interaction: an updated taxonomy, in *Proceedings of the 2004 IEEE International Conference on Systems, Man and Cybernetics (IEEE Cat. No. 04CH37583),* vol. 3 (The Hauge, Netherlands, 2004), pp. 2841–2846
14. M. Cummings, Informing autonomous system design through the lens of skill-, rule-, and knowledge-based behaviors. J Cogn Eng Decis Mak **12**(1), 58–61 (2018)
15. J. Rasmussen, Skills, rules, and knowledge: signals, signs, and symbols, and other distinctions in human performance models. IEEE Trans. Syst. Man Cybern. **13**(3), 257–266 (1983)
16. N. Nikolakis, K. Sipsas, S. Makris, Cockpit: a portal for symbiotic human–robot collaborative assembly. Ibid. Chapter 9
17. G. Horváth, Cs. Kardos, A. Kovács, B. Pataki, S. Tóth, Worker assistance suite for efficient human–robot collaboration. Ibid. Chapter 15
18. M.R. Pedersen, L. Nalpantidis, R.S. Andersen, C. Schou, S. Bøgh, V. Krüger, O. Madsen, Robot skills for manufacturing: from concept to industrial deployment, Robot. Comput.-Integr. Manuf. **37**, 282–291 (2016)
19. S.Ch. Akkaladevil, M. Plasch, M. Hofmann, L. Hiesmair, M. Ikeda, N. Chitturi, A. Pichler, Programming-free approaches for human–robot collaboration in assembly tasks. Ibid. Chapter 12
20. T. Gašpar, M. Deniša, P. Radanovič, B. Ridge, T. Rajeeth Savarimuthu, A. Kramberger, M. Priggemeyer, J. Roßmann, F. Wörgötter, T. Ivanovska, S. Parizi, Ž. Gosar, I. Kovač, A. Ude, Smart hardware integration with advanced robot programming technologies for efficient reconfiguration of robot workcells, Robot. Comput.-Integr. Manuf. **66**, 2020, 101979
21. Y. Liu, Z. Li, H. Liu, Z. Kan, Skill transfer learning for autonomous robots and human–robot cooperation: a survey. Robot Auton Syst **128**, 103515 (2020)
22. C.D. Buckingham, A. Ahmed, A. Adams, Designing multiple user perspectives and functionality for clinical decision support systems, in *Proceedings of the 2013 Federated Conference on Computer Science and Information Systems (FedCSIS 2013)* (Kraków, Poland, 2013), pp. 211–218

23. C. Schlenoff, E. Prestes, R. Madhavan, P. Goncalves, H. Li, S. Balakirsky, T. Kramer, E. Miguelanez, An IEEE standard ontology for robotics and automation, in *Proceedings of the 2012 IEEE/RSJ International Conference on Intelligent Robots and Systems* (Vilamoura, Algarve, Portugal, 2012), pp. 1337–1342
24. N. Nikolakis, K. Alexopoulos, K. Sipsas, Resource availability and capability monitoring. Ibid. Chapter 7
25. K. Sipsas, N. Nikolakis, S. Makris, Dynamic assembly planning and task assignment. Ibid. Chapter 8
26. R.X. Gao, L. Wang, P. Wang, J. Zhang, H. Liu, Human motion tracking, recognition, and prediction for robot control. Ibid. Chapter 11
27. A.S. Rao, M.P. Georgeff, Modeling rational agents within a BDI-architecture, in *Proceedings of the 2nd International Conference on Principles of Knowledge Representation and Reasoning* (Cambridge, MA, USA, 1991), pp. 473–484
28. M.E. Bratman, *Intention, Plans, and Practical Reason* (Harvard University Press, Cambridge, MA, USA, 1987). ISBN 0674458184, https://doi.org/10.2307/2185304
29. H-M. Huang, E. Messina, J. Albus, Toward a generic model for autonomy levels for unmanned systems (ALFUS). Technical report, National Institute of Standards and Technology Gaithersburg MD (2003)
30. K.S. Barber, C.E. Martin, Agent autonomy: specification, measurement, and dynamic adjustment, in *Proceedings of the Autonomy Control Software Workshop at Autonomous Agents (Agents'99)* (Seattle, WA, USA, 1999), pp. 8–15
31. S.A. Mostafa, M.S. Ahmad, A. Mustapha, Adjustable autonomy: a systematic literature review. Artif. Intell. Rev. **51**(2), 149–186 (2019)
32. M. Sierhuis, J.M. Bradshaw, A. Acquisti, R. van Hoof, R. Jeffers, A. Uszok, Human–agent teamwork and adjustable autonomy in practice, in *Proceedings of the Seventh International Symposium on Artificial Intelligence, Robotics and Automation in Space (I-SAIRAS 2003)* (Nara, Japan, 2003), paper-ID AM-26, pp. 1–9
33. G. Klein, D.D. Woods, J.M. Bradshaw, R.R. Hoffman, P.J. Feltovich, Ten challenges for making automation a 'team player' in joint human–agent activity. IEEE Intell. Syst. **19**(6), 91–95 (2004)
34. X.V. Wang, L. Wang, Safety strategy and framework for human–robot collaboration. Ibid. Chapter 3
35. B. Dunin-Kęplicz, R. Verbrugge, *Teamwork in Multi-agent Systems: A Formal Approach* (Wiley, 2011). ISBN 9781119957607
36. K. Zhou, J.C. Doyle, K. Glover, *Robust and Optimal Control* (Prentice Hall, 1996). ISBN 9780134565675
37. R.E. Kalman, A new approach to linear filtering and prediction problems. Tran. ASME–J. Basic Eng **82**, 35–45 (1960)
38. L. Wang, R.X. Gao, J. Váncza, J. Krüger, X.V. Wang, S. Makris, G. Chryssolouris, Symbiotic human–robot collaborative assembly. CIRP Ann. **68**(2), 701–726 (2019)
39. H. Liu, L. Wang, Latest developments of gesture recognition for human–robot collaboration. Ibid. Chapter 2
40. P. Wang, H. Liu, L. Wang, R.X. Gao, Deep learning-based human motion recognition for predictive context-aware human–robot collaboration. CIRP Ann. **67**(1), 17–20 (2018)
41. S. Liu, L. Wang, X. V. Wang, Sensorless haptic control for physical human–robot interaction. Ibid. Chapter 13
42. B. Yao, Z. Zhou, L. Wang, W. Xu, Q. Liu, Sensor-less external force detection for industrial manipulators to facilitate physical human–robot interaction. J. Mech. Sci. Technol. **32**(10), 4909–4923 (2018)
43. D. Hoffman, *The Case Against Reality: why Evolution Hid the Truth from Our Eyes* (W.W. Norton & Company, 2019). ISBN 9780393254709.
44. D.D. Hoffman, M. Singh, J. Mark, Does evolution favor true perceptions? in *Proceedings vol. 8651, Human Vision and Electronic Imaging XVIII, IS&T/SPIE Electronic Imaging* (Burlingame CA, USA, 2013), pp. 10–21

45. G. Horváth, G. Erdős, Zs. Kemény, Human–robot collaborative workcell calibration and control. Ibid. Chapter 10
46. B. Schleich, N. Anwer, L. Mathieu, S. Wartzack, Shaping the digital twin for design and production engineering. CIRP Ann. **66**(1), 141–144 (2017)
47. R. Stark, S. Kind, S. Neumeyer, Innovations in digital modelling for next generation manufacturing system design. CIRP Ann. **66**(1), 169–172 (2017)
48. G. Erdős, I. Paniti, B. Tipary, Transformation of robotic workcells to digital twins. CIRP Ann. **69**(1) (in print), (2020) https://doi.org/10.1016/j.cirp.2020.03.003
49. Y. Lu, C. Liu, C., K.I.-K. Wang, H. Huang, X. Xu, Digital twin-driven smart manufacturing: connotation, reference model, applications and research issues. Robot. Comput.-Integr. Manuf. **61**, 101837 (2020)
50. G. Lindemann-von Trzebiatowski, I. Münch, The role concept for agents in multiagent systems. Sozionik-Aktuell **2001**(3), 22–38 (2001)
51. S. Qi, S-C. Zhu, Intent-aware multi-agent reinforcement learning, in *Proceedings of the 2018 IEEE International Conference on Robotics and Automation (ICRA)* (Brisbane, Australia, 2018), pp. 7533–7540
52. A. Mohammed, L. Wang, Intelligent human–robot assembly enabled by brain EEG. Ibid. Chapter 14
53. X.V. Wang, L. Wang, Augmented reality enabled human–robot collaboration. Ibid. Chapter 16
54. M. Holm, R. Senington, W. Wang, J. Lindblom, Real-world industrial demonstrators on human–robot collaborative assembly. Ibid. Chapter 17
55. ISO, ISO 10075-1:2017: ergonomic principles related to mental workload—Part 1: general issues and concepts, terms and definitions. https://www.iso.org/standard/66900.html, Accessed 22 Mar 2020
56. M. Gombolay, A. Bair, C. Huang, J. Shah, Computational design of mixed-initiative human–robot teaming that considers human factors: situational awareness, workload, and workflow preferences. Int. J. Robot. Res. **36**(5–7), 597–617 (2017)
57. A. Dini, C. Murko, S. Yahyanejad, U. Augsdörfer, M. Hofbaur, L. Paletta, Measurement and prediction of situation awareness in human–robot interaction based on a framework of probabilistic attention, in *Proceedings of the 2017 IEEE/RSJ International Conference on Intelligent Robots and Systems (IROS)* (Vancouver, Canada, 2017), pp. 4354–4361
58. D. Tsutsumi, D. Gyulai, E. Takács, J. Bergmann, Y. Nonaka, K. Fujita, Personalized work instruction system for revitalizing human–machine interaction, in *53rd CIRP Conference on Manufacturing Systems* (2020, in print)
59. M.G. Haselton, D. Nettle, D.R. Murray, The evolution of cognitive bias, in *The Handbook of Evolutionary Psychology* (Wiley, 2016), pp. 968–987. ISBN 9781118755884
60. L. Sun, C. Bunchapattanasakda, Employee engagement: a literature review. Int J Hum Resour Stud **9**(1), 63–80 (2019)
61. M.B. Shuck, K.K. Wollard, A historical perspective of employee engagement: an emerging definition, in *Proceedings of the Eighth Annual College of Education & GSN Research Conference* (Miami FL, USA, 2009), pp. 133–139
62. M.S. Cole, F. Walter, A.G. Bedeian, E.H. O'Boyle, Job burnout and employee engagement: a meta-analytic examination of construct proliferation. J Manage **38**(5), 1550–1581 (2012)
63. T. Keller, C. Bayer, P. Bausch, J. Metternich, Benefit evaluation of digital assistance systems for assembly workstations. Procedia CIRP **81**, 441–446 (2019)
64. A. Mohammed, L. Wang, Brainwaves driven human–robot collaborative assembly. CIRP Ann **67**(1), 13–16 (2018)
65. I. Aaltonen, T. Salmi, I. Marstio, Refining levels of collaboration to support the design and evaluation of human–robot interaction in the manufacturing industry. Procedia CIRP **72**, 93–98 (2018)
66. J. Schmidtler, V. Knott, C. Hölzel, K. Bengler, Human centered assistance applications for the working environment of the future. Occup Ergon **12**(3), 83–95 (2015)
67. J.S. Santos, J.O. Zahn, E.A. Silvestre, V.T. Silva, W.W. Vasconcelos, Detection and resolution of normative conflicts in multi-agent systems: a literature survey. Auton. Agent. Multi-Agent Syst. **31**, 1236–1282 (2017)

68. J. Tumova, D.V. Dimarogonas, Multi-agent planning under local LTL specifications and event-based synchronization. Automatica **70**, 239–248 (2016)
69. P. Schillinger, M. Bürger, D.V. Dimarogonas, Simultaneous task allocation and planning for temporal logic goals in heterogeneous multi-robot systems. Int. J. Robot. Res. **37**(7), 818–838 (2018)
70. H. Liu, L. Wang, Remote human–robot collaboration: a cyber–physical system application for hazard manufacturing environment. J Manuf Syst **54**, 24–34 (2020)
71. M. Johnson, J.M. Bradshaw, P.J. Feltovich, C.M. Jonker, M.B. van Riemsdijk, M. Sierhuis, Coactive design: designing support for interdependence in joint activity. J Hum-Robot Interact **3**(1), 43–69 (2014)
72. X.V. Wang, Zs. Kemény, J. Váncza, L. Wang, Human–robot collaborative assembly in cyber-physical production: classification framework and implementation. CIRP Ann. **66**(1), 5–8 (2017)
73. P.D. O'Brien, R.C. Nicol, FIPA—towards a standard for software agents. BT Technol. J. **16**(3), 51–59 (1998)
74. J. Ferber, O. Gutknecht, F. Michel, Agent/group/roles: simulating with organizations, in *Proceedings of the Fourth International Workshop on Agent-Based Simulation* (Montpellier, France, 2003), pp. 12–18
75. R. Lowe, Y.I. Wu, A. Tamar, J. Harb, P. Abbeel, I. Mordatch, Multi-agent actor-critic for mixed cooperative-competitive environments, in *Proceedings of the 31st Conference on Neural Information Processing Systems (NIPS 2017)* (Long Beach, CA, USA, 2017), pp. 6379–6390
76. Cs. Kardos, A. Kovács, J. Váncza, A constraint model for assembly planning. J. Manuf. Syst. **54**, 196–203 (2020)
77. G. Erdős, A. Kovács, J. Váncza, Optimized joint motion planning for redundant industrial robots. CIRP Ann. **65**(1), 451–454 (2016)
78. Y. Wang, L. Wang, Real-time collision-free multi-objective robot motion generation. Ibid. Chapter 5
79. A. Zacharaki, I. Kostavelis, A. Gasteratos, I. Dokas, Safety bounds in human robot interaction: a survey. Saf. Sci. **127**, 104667 (2020)
80. L. Wang, B. Schmidt, A.Y.C. Nee, Vision-guided active collision avoidance for human–robot collaborations. Manuf Lett **1**(1), 5–8 (2013)
81. H. Liu, L. Wang, Collision-free human–robot collaboration based on context awareness. Robot Comput-Integr Manuf **67**, 101997 (2021)
82. H. Liu, Y. Wang, W. Ji, L. Wang, A context-aware safety system for human–robot collaboration. Procedia Manuf. **17**, 238–245 (2018)
83. I. Marstio, J. Montonen, T. Malm, Dynamic safety system for safe collaboration with industrial robots. Ibid. Chapter 6
84. B. Schmidt, Real-time collision detection and collision avoidance. Ibid. Chapter 4
85. ISO, Ergonomics of human–system interaction—Part 210: human-centred design for interactive systems, in ISO 9241-210:2019 (2019)
86. H. Kagermann, Y. Nonaka, Revitalizing human–machine interaction for the advancement of society, acatech Discussion (2019)
87. ISO ISO 13855: safety of machinery–positioning of safeguards with respect to the approach speeds of parts of the human body (2010)
88. ISO, ISO 10218-1: Robots and robotic devices—safety requirements for industrial robots—Part 1: Robots (2011)
89. ISO, ISO 10218-2: Robots and robotic devices—safety requirements for industrial robots, Part 2: Robot systems and integration (2011)
90. ISO, ISO/TS 15066 Robotics and robotic devices—collaborative robots (2016)
91. ISO, ISO 26000: guidance on social responsibility (2010)

Chapter 2
Latest Developments of Gesture Recognition for Human–Robot Collaboration

Hongyi Liu and Lihui Wang

Abstract Recently, the concept of human–robot collaboration has raised many research interests. Instead of robots replacing human operators in workplaces, human–robot collaboration is the direction that allows human operators and robots to work together. Although the communication channels between human operators and robots are still limited, gesture recognition has been effectively applied as the interface between humans and computers for a long time. Covering some of the most important technologies and algorithms of gesture recognition, this chapter is intended to provide an overview of the gesture recognition research and explore the possibility to apply gesture recognition in human–robot collaboration. In this chapter, an overall model of gesture recognition for human–robot collaboration is also proposed. There are four essential technical components in the model of gesture recognition for human–robot collaboration: sensor technologies, gesture identification, gesture tracking and gesture classification. Reviewed approaches are classified according to the four essential technical components. After the reviewed technical components, an example of gesture recognition for human–robot collaboration is provided. In the last part of the chapter, future research trends are outlined.

2.1 Introduction

2.1.1 Human–Robot Collaboration

Robotic systems have already become key components in various industrial sectors. Recently, the concept of Human–Robot Collaboration (HRC) has raised mounting research interests. The literature suggested that human operators have incomparable problem-solving skills and sensory-motor capabilities, but are restricted in force

H. Liu · L. Wang (✉)
Department of Production Engineering, KTH Royal Institute of Technology, Stockholm, Sweden
e-mail: lihuiw@kth.se

H. Liu
e-mail: hongyil@kth.se

L. Wang et al. (eds.), *Advanced Human–Robot Collaboration in Manufacturing*,
https://doi.org/10.1007/978-3-030-69178-3_2

and precision [1, 2]. Robotic systems, however, provide better resistance to fatigue, higher speed, higher repeatability and better productivity, but are restricted in flexibility. Jointly, HRC can relieve human operators from heavy tasks and establish communication channels between human operators and robots for better overall performance.

Ideally, an HRC team should work similarly to a human–human collaborative team. However, traditionally, time-separation or space-separation approaches are applied in HRC systems, which reduces productivity for both human operators and robots [1]. In order to build an efficient HRC team, human–human collaboration teams can be analysed as examples. In human teamwork and collaboration, there are two theories: joint intention theory and situated learning theory [3–6]. To apply the theories to an HRC team, there are three experiences that benefit the HRC team:

- All team members in an HRC team should have a transparent plan of execution;
- All team members in an HRC team should be aware of the context of the collaboration environment; and
- An HRC team should have structured ways of communication.

This chapter mainly focuses on the third experience, i.e., structured ways of communication.

2.1.2 *Gesture Recognition*

Gesture is one communication method. Head nodding, hand gestures and body postures are effective communication channels in human–human collaboration [2, 7]. Gestures can be categorised into three types [8]:

- Body gestures: full-body actions or motions;
- Hand and arm gestures: arm poses and hand gestures; and
- Head and facial gestures: nodding or shaking the head and winking lips.

Gesture recognition refers to the mathematical interpretation of human motions by a computing device. In order to collaborate with human operators, robots need to understand human gestures correctly and act efficiently based on the gestures. In an HRC environment, a natural way of gesture communication between robots and humans should be available.

2.1.3 *Gesture Recognition for Human–Robot Collaboration*

To recognise gestures in the HRC context, it is beneficial to investigate into a generic and simplified human information processing model. As shown in Fig. 2.1, Parasuraman et al. [9] generalised human information processing into a four-stage model. Based on the generic model in Fig. 2.1, we propose a specific model for gesture

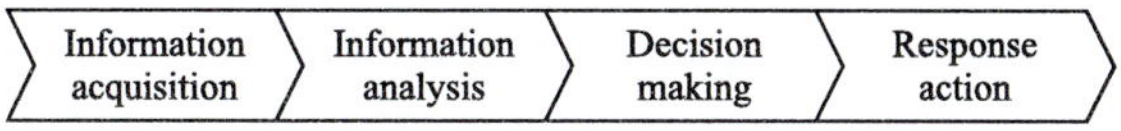

Fig. 2.1 A four-stage model of human information processing [9]

recognition in HRC. As shown in Fig. 2.2, there are five essential parts related to gesture recognition for HRC: sensor data collection, gesture identification, gesture tracking, gesture classification and gesture mapping, explained as follows.

- Sensor data collection: gesture raw data is captured by sensors.
- Gesture identification: in each frame, a gesture is located from the raw data.
- Gesture tracking: the located gesture is tracked during the gesture movement. For static gestures, gesture tracking is unnecessary.
- Gesture classification: tracked gesture movement is classified according to pre-defined gesture types. If the identified gesture is a static gesture, the gesture can be directly classified after the gesture identification step.
- Gesture mapping: gesture recognition result is translated into robot commands and sent back to operators.

The remainder of this chapter is organised as follows: Sect. 2.2 reviews enabling sensor technologies. Section 2.3 provides an overview of gesture identification

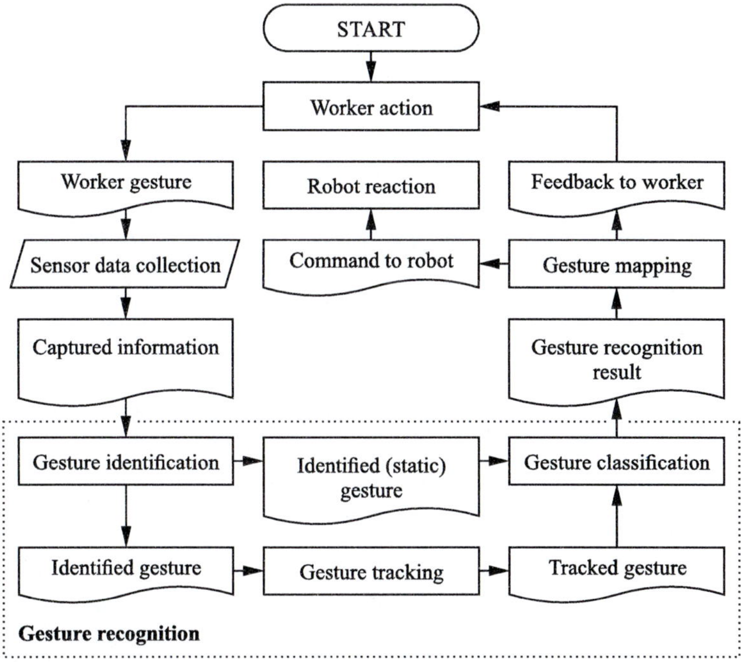

Fig. 2.2 A process model of gesture recognition for human–robot collaboration

methods. Section 2.4 discusses gesture tracking problems. Section 2.5 introduces gesture classification algorithms. Section 2.6 provides an example of gesture recognition used in HRC. Finally, Sect. 2.7 concludes the chapter with future research trends outlined.

2.2 Sensor Technologies

Before the gesture recognition process is started, gesture raw data need to be collected by sensors. In this section, different sensors in the literature are analysed based on different sensing technologies. As shown in Fig. 2.3, there are two basic categories of data acquisition: image-based and non-image-based approaches.

2.2.1 Image-Based Approaches

Technologies are often inspired by nature. As human beings, we use our eyes to recognise gestures. Therefore, for robots, it is reasonable to use cameras to "see" gestures. The image-based approaches are further divided into four categories.

- *Marker*: In marker-based approaches, the sensor is a normal optical camera. In most marker-based solutions, users need to wear obvious markers [8]. Today, we enjoy a much faster graphical processing speed compared with that 20 years ago. As a result, there are more gesture recognition sensors available.
- *Single Camera*: In the early 90s, researchers started to analyse gestures using digital cameras [10, 11]. A drawback of single-camera approaches is the restriction of view angles, which affects a system's robustness [12]. Recent research, however, applied single-camera approaches in high-speed gesture recognition

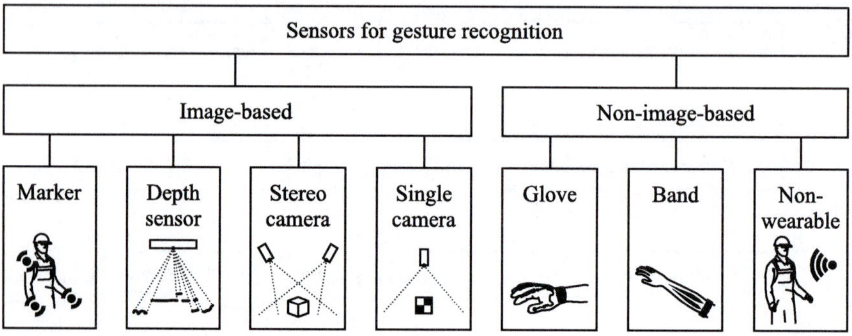

Fig. 2.3 Different types of gesture recognition sensors

[13]. The system utilises a speed image sensor and a specially designed visual computing processor to achieve high-speed gesture recognition.

- *Stereo Camera*: In order to achieve robust gesture recognition, researchers suggested stereo camera-based approaches to construct 3D vision. Here, we define stereo camera-based approaches as applications that utilise two cameras (optical stereo camera) to construct 3D depth information. Many stereo camera-based approaches follow a similar workflow [14, 15]. Although stereo camera systems improved robustness in outdoor environments, they still suffer from problems such as computational complexity and calibration difficulties [16].
- *Depth Sensor*: Recently, depth-sensing technologies have emerged rapidly. We define a depth sensor as a non-stereo depth-sensing device. Non-stereo depth sensors enjoy several advantages compared to traditional stereo cameras. For example, the problems of setup calibration and illumination conditions can be prevented [17]. Moreover, the output of a depth sensor is 3D depth information. Compared with colour information, the 3D depth information simplifies the problem of gesture identification [8]. A comparison of gesture identification accuracy by using colour and depth information can be found in [18]. There are two types of common non-stereo depth sensors: Time-of-Flight (ToF) camera and structured light sensors.
  The basic principle of ToF cameras is to identify light travel time [19]. Various literature introduced hand gesture recognition applications based on ToF cameras [18, 20]. The advantage of ToF cameras is the higher frame rate. The limitation of a ToF camera is that the camera resolution highly depends on its light power and reflection [21]. There are several structured light sensors, such as Microsoft Kinect, ASUS Xtion Pro and Apple PrimeSense. Kinect has an IR emitter, IR sensor and colour sensor. It is widely used in entertainment, education and research, in response to which a large developer community has emerged [22]. The community-wide popularity of structured light sensors has resulted in the emergence of many open-source tools and projects. Some researchers reported short distance Kinect-based hand gesture recognition systems [23, 24]. Due to resolution restrictions, currently, structured light sensors can be used for body gesture recognition and close distance hand and arm gesture recognition. For hand and arm gesture recognition over 2 m, other approaches are better used.

2.2.2 Non-image-based Approaches

Gesture recognition has been dominated by image-based sensors for a long time. Recent developments in MEMS and sensor technologies have greatly boosted non-image-based gesture recognition technologies.

- *Glove*: Glove-based gestural interfaces are commonly used for gesture recognition. Usually, glove-based approaches require wire connection, accelerometers and gyroscopes. However, a cumbersome glove with a load of cables can potentially cause problems in HRC [8]. Glove-based approaches frequently require complex calibration and setup procedures [25].

- *Band*: Another contactless technology uses band-based sensors. Band-based sensors rely on a wristband or similar wearable devices. Band-based sensors enable wireless technology and electromyogram sensors, which avoid connecting cables. The sensor only needs to have contact with the wrist; the user's hand and fingers can freely move. An example of a band-based sensor is the *Myo* gesture control armband [26]. Recently, several band-based sensor gesture control systems were published [27–29].
- *Non-wearable*: The third type of non-image-based technologies adapts non-wearable sensors. Non-wearable sensors can detect gestures without contacting the human body. Google introduced *Project Soli*, a radio frequency spectrum (radar)-based hand gesture tracking and recognition system [30]. The device is capable of recognising different hand gestures within a short distance. MIT has been leading non-wearable gesture recognition technology for years. Electric Field Sensing technology was pioneered by MIT [31]. A recent discovery by Adib et al. [32–34] from MIT introduced the *WiTrack* and *RF-Capture* systems that capture user motion by radio frequency signals reflected from the human body. The systems are able to capture human gestures even from another room through a wall with a precision of 20 cm. In summary, non-wearable technologies are promising and fast-growing sensor technologies for gesture recognition.

2.2.3 Comparison of Sensor Technologies

A comparison of different sensor technologies is provided in Table 2.1. The advantages and disadvantages of different approaches are indicated. It is clear that there is no technology that can fit all applications. Two observations can be made based on the above analyses:

- With indoor applications, depth sensor technologies are the most promising image-based technologies. Depth sensors possess advantages of easy setup calibration and easy data processing. A large application development community exists, which can provide many potential solutions.
- Non-wearable approaches are most promising among non-image-based approaches. The approaches can avoid direct contact with users. Non-wearable sensing is also a fast-growing trend.

2.3 Gesture Identification

Gesture identification is the first step in the gesture recognition process after raw data is captured from sensors. Gesture identification refers to the process of gestural

Table 2.1 Advantages and disadvantages of different sensor technologies

	Advantages	Disadvantages
Markers	Low computational workload	Markers on user body
Single camera	Easy setup	Low robustness
Stereo camera	Robust	Computational complexity, calibration difficulties
ToF camera	High frame rate	Resolution depends on light power and reflection
Structured light sensor	Fast emerging, software support for body gesture recognition	Cannot be used for hand gesture recognition over 2 m
Glove	Fast response, precise tracking	Cumbersome device with a load of cables
Band sensor	Fast response, large sensing area	Band needs to have contact with the human body
Non-wearable	Avoid contact with the human body	Low resolution, technology not mature enough

information detection and segmentation from the raw data. Popular technologies to solve the gesture identification problem are based on visual features, learning algorithms and human models.

2.3.1 *Visual Features*

Human hands and body are visually unique. In image-based gesture recognition, gestures consist of human hands or body. Therefore, it is straightforward to utilise such visual features to identify the gestures.

- *Colour*: Colour is a simple visual feature to separate a gesture from background information. However, colour-based gesture recognition systems can be easily influenced by illumination and shadows in a complex environment [35]. Due to the problems above, in recent approaches, skin colour is only considered to be one of the many cues in the gesture identification method.
- *Local Features*: In image-based gesture recognition, illumination conditions can greatly influence gesture identification quality. Therefore, many researchers avoid the problem by using a local features method. This is a detailed texture-based approach which decomposes an image into smaller regions that are not corresponding to body parts [36]. As shown in Fig. 2.4, one of the most important local features methods is Scale Invariant Feature Transform (SIFT) [37]. The SIFT method is rotational, translational, scaling and partly illumination invariant. Several similar local features approaches, for example, *SURF* and *ORB,* were proposed in later years [38, 39]. Normally, local features approaches are also only considered as one of the many cues in gesture identification.

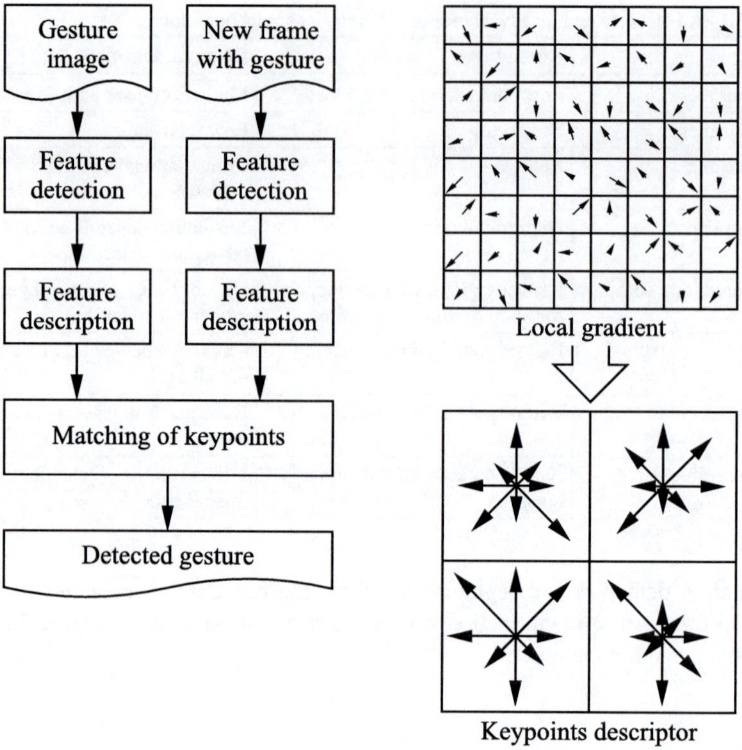

Fig. 2.4 SIFT algorithm: **a** SIFT algorithm for gesture identification; **b** SIFT feature description example [37]

- *Shape and Contour*: Another intuitive and simple way to identify gestures is to utilise the unique shape and contour of the human body. A milestone for shape detection and matching was reported by Belongie et al. [40]. They introduced a shape context descriptor method. A shape context descriptor is used for the detection of similar shapes in different images. The development of depth sensors provides opportunities to accurately measure surface shapes. The 3D models generated from the technologies enable a highly detailed representation of human body shape [41].
- *Motion*: In certain HRC environments, a human operator is the only moving object. Therefore, motion can be a useful feature to detect human gestures. Optical flow is a key technology for motion-based gesture identification. Optical flow does not need background subtraction, which is an advantage compared to shape and contour-based approaches. Several gesture recognition applications are implemented based on an optical flow method [42, 43]. Dalal and Thurau [44] introduced the well-known *Histograms of Oriented Gradients* (HOG) method. The HOG descriptors divide image frames into blocks. For each block, a histogram is computed. Among non-image-based sensors, motion-based gesture identification

is also a popular method [33, 45]. Usually, thresholding or filtering is applied to raw sensor data to identify human gestures.

2.3.2 *Learning Algorithms*

A recent trend of gesture identification is to combine learning algorithms in the identification method, especially for static gesture detection that can be represented by a single frame. The visual feature methods are based on various visual features, while learning algorithms can utilise machine learning algorithms to identify gestures from raw sensor data. Although some algorithms are based on the visual feature methods, image background removal is not always necessary for learning algorithms. Learning algorithms such as Support Vector Machine (SVM), Artificial Neural Networks (ANN) and Random Decision Forests (RDF) are widely applied in gesture recognition systems [46–48].

2.3.3 *Human Model*

Different from the aforementioned approaches, the human model approach uses an explicit human body model to recover the human body poses. Therefore, the human model approach is also referred to as a generative approach. Human-model-based gesture identification methods can largely simplify the gesture classification process. Hence, the human model approach is becoming a more and more popular solution for depth sensors [49]. Depending on different model targets, recent research in human-model-based gesture identification can be divided into two types: hand model identification and body skeleton model identification.

- *Hand Model*: There are two types of hand model approaches for hand model detection: 3D model and hand skeleton model. The 3D model approach interprets the hand detection problem as an optimisation problem that minimises the differences between a 3D hand model and the observation [50]. The hand skeleton model approach is similar to the 3D model approach, but with a more simplified hand model [51].
- *Body Skeleton Model*: To identify body gestures, a full-fledged human model can introduce too much detail into the process. Therefore, in the field of body gesture identification, the body skeleton model is a popular approach. Some of the literature regarding body skeleton processing pipeline are based on depth information gathered from depth sensors. As shown in Fig. 2.5, the body skeleton model is a simplified human body model that can preserve information about body joints and bones. Since only the most useful information is preserved from a human body, in body gesture recognition research, more and more body skeleton methods are applied [49, 53, 54].

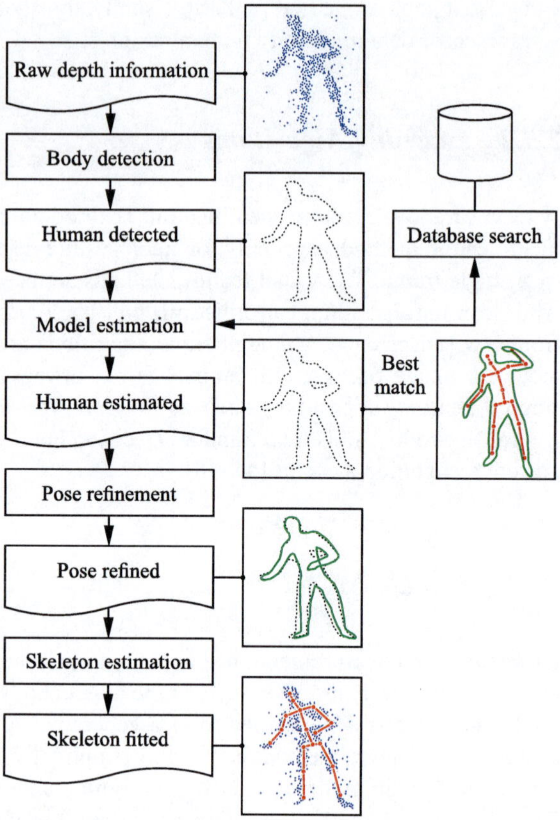

Fig. 2.5 Example of body skeleton identification [52]

2.3.4 Summary of Gesture Identification Approaches

The authors summarise the most suitable gesture identification method for each popular sensor, shown in Table 2.2. Despite understanding human body gestures, the human body model approach will also monitor human movements, which provides a secure environment for the human–robot team.

2.4 Gesture Tracking

In gesture recognition, the notion of tracking is used differently in various literature sources. We define the notion of tracking as the process of finding temporal correspondences between frames of a continuous sequence. Specifically, we focus on the gesture tracking problem that associates the gesture identified in the previous frames with the current frame. As for static gestures which can be represented by a

Table 2.2 Gesture identification methods for different sensors

Sensor	Gesture identification method
Single camera	In single-camera-based systems, visual feature method and learning algorithms method can be implemented. To achieve robust performance, learning algorithms should be applied. To achieve faster image processing, visual feature method should be applied [13]
Depth sensor	Since the human model method utilises and simplifies point cloud information, it is a better option for depth sensor-based systems [49]
Band sensor	No visual-based methods can be applied on band sensor-based systems. Usually, the collected data need basic filtering in gesture identification, and learning algorithms will be implemented in later gesture classification [27]
Non-wearable	The non-wearable sensors also receive signal data instead of images. Due to the fact that RF signals contain noises, advanced filtering and processing solutions need to be implemented in non-wearable sensor-based systems [33]

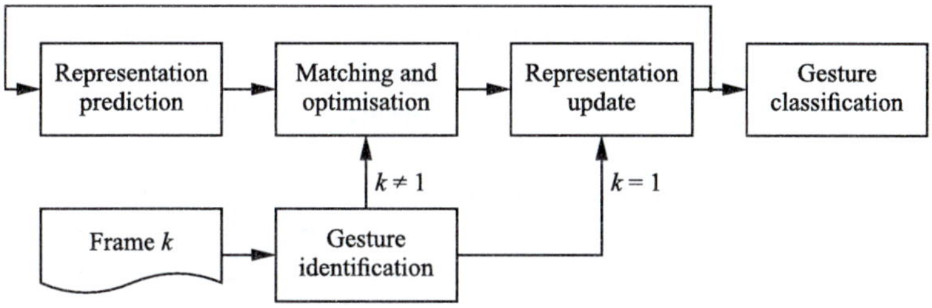

Fig. 2.6 Gesture tracking example

single frame, a gesture tracking step is unnecessary. An example of gesture tracking is shown in Fig. 2.6.

2.4.1 Single Hypothesis Tracking

Single hypothesis tracking refers to a best-fit estimation with minimum-error matching. Therefore, in single hypothesis tracking, the gesture is represented by one hypothesis only. Most of the advanced tracking algorithms are based on single hypothesis tracking technologies.

- *Mean Shift*: The mean shift tracker is a basic tracking algorithm that relies on RGB-colour histograms [55]. For each new frame, the mean shift tracker compares the Bhattacharyya distance between the target window histograms of the new frame with those of the old frame. A complete mathematical explanation can be found in [55].
- *Kalman Filter*: The *Kalman Filter* (KF) is a real-time recursive algorithm used to optimally estimate the underlying states of a series of noisy and inaccurate

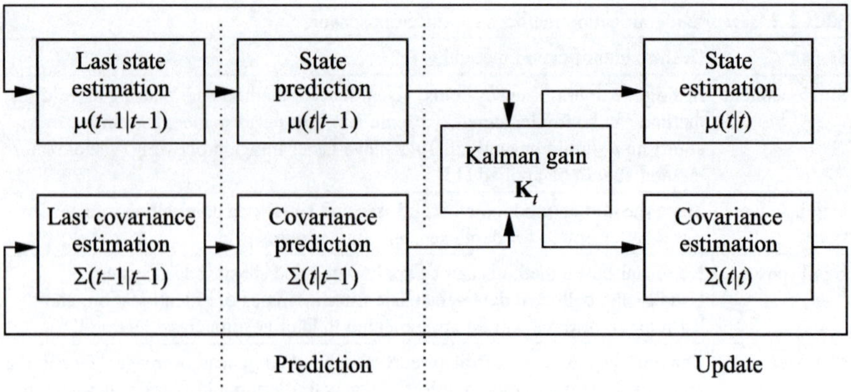

Fig. 2.7 Kalman filter process flow [59]

measurement results observed over time. The process flow of KF is shown in Fig. 2.7. A complete KF mathematical derivation can be found in [56]. Having undergone considerable evolution, KF is nowadays applied in different fields such as aerospace, robotics, and economics.

- *Kalman Filter Extensions*: The KF assumes that the changes of the state vector are governed by a linear model. The *Extended Kalman Filter* (EKF) is a functional tracking algorithm for nonlinear models of state transition and system output observation [57]. Another algorithm that solves the same problem from a different angle is the *Unscented Kalman Filter* (UKF) [58]. The UKF solves the problem by applying a deterministic weighted sampling approach. The state distribution is represented using a minimal set of chosen sample points.

2.4.2 *Multiple Hypothesis Tracking*

In many HRC scenarios, multiple human operators can work in the same workstation at the same time [60]. To track the gestures of multiple operators simultaneously, a multiple hypothesis tracking algorithm should be applied.

- *Particle Filter*: The *Particle Filter* (PF) is a popular algorithm in solving robotics problems. Different from KF, PF does not make an assumption on the posterior model. The PF representation is a nonparametric approximation which can represent a broader space of distribution. Therefore, PF can satisfy the multiple hypothesis tracking requirements [61]. An example of the PF process is shown in Fig. 2.8. Several advanced tracking algorithms also apply PF to scan a probability density function [62, 63].
- *Particle Filter Extensions*: Many researchers also attempted to combine PF with other algorithms for better performance. PF has been combined with mean shift tracker, Genetic Algorithm, PSO, Ant Colony Optimisationand other machine

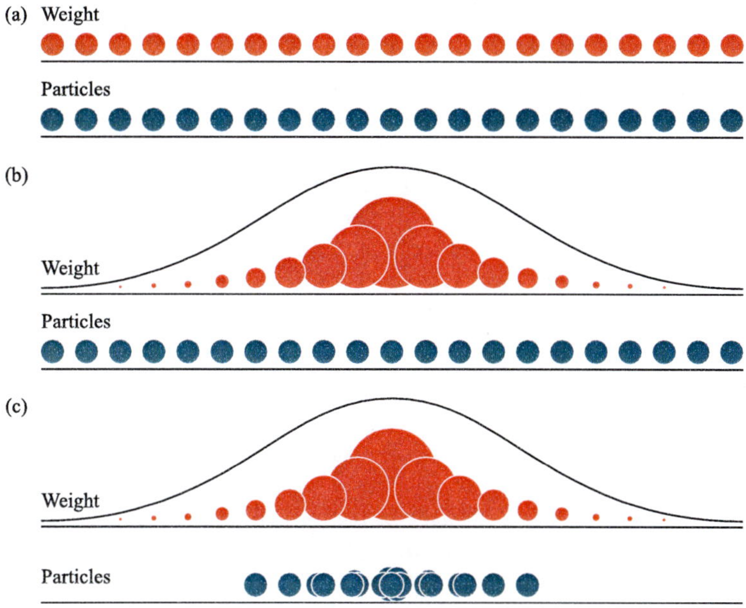

Fig. 2.8 Particles and weight factors: **a** after particle initialisation; **b** after weight factor calculation; **c** after resampling [59]

learning algorithms to solve the sample degeneracy and impoverishment problem [64]. Some other researchers also tried to improve the PF resampling strategy [65, 66].

2.4.3 Advanced Tracking Methods

Recent years saw the introduction of many advanced tracking methods. Some of these utilised part of the two kinds of tracking algorithms mentioned above. Other methods improved tracking performance by detection or learning algorithms.

- *Extended Model Tracking*: For long-term tracking problems, many tracking algorithms easily fail because the target maintains fixed models. Extended model tracking methods can save the target behaviour or appearance from the past few image frames. Therefore, more information can be secured for target estimation. The *Incremental Visual Tracker* uses an extended model to preserve more information for a tracking process [67]. Kwon et al. [63] presented *Tracking by Sampling Tracker*. In order to tackle an environment which greatly varies over time, the approach relies on several trackers running in parallel, and samples both the target states and the trackers themselves. The parallel trackers are selected and run so that adequate coverage of all potential target variations is achieved.

- *Tracking by Detection*: Another kind of tracking algorithm is built together with the gesture identification learning algorithms introduced in the earlier sections. For these tracking algorithms, a classifier or detector is provided in image frames to identify a gesture from the background information [68]. One of the representative algorithms is the *Tracking, Learning and Detection Tracker* [69]. The algorithm integrates the result of an object detector with an optical flow tracker. Another typical tracking-by-detection algorithm is the adaption of *Multiple Instance Learning* [70]. The learning algorithm can increase tracker robustness and decrease parameter tweaks.

2.4.4 Comparison of Different Gesture Tracking Approaches

Smeulders et al. [71] presented a test to evaluate the performance of different gesture tracking algorithms. The result score is the normalised F-score, which can provide us with an insight into the comparison of the tracked object bounding box and the ground truth bounding box. Hence, the higher the F-score, the better the tracking quality. As can be seen in Fig. 2.9, the test results in different video conditions are presented. *Kalman Appearance Tracker* and *Mean Shift Tracker* belong to the single hypothesis tracker type. *Tracking by Sampling Tacker* and *Incremental Visual Tracker* belong to extended model trackers. *Multiple Instance Learning Tracker* and *Tracking, Learning and Detection Tracker* belong to the tracking-by-detection methods. It is

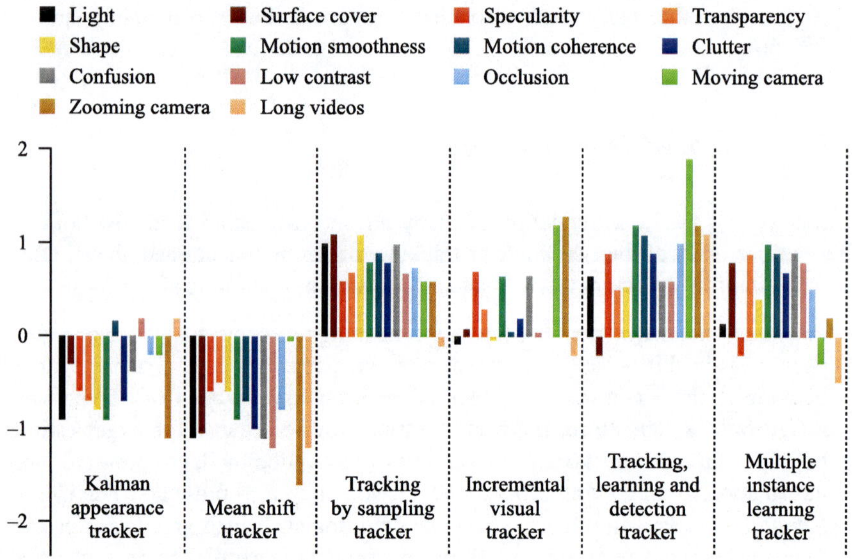

Fig. 2.9 Tracking algorithm test result in different video conditions [71]

Table 2.3 Summary of tracking approaches

Approach	Summary
Single hypothesis	Fast and simple algorithm. Suitable for one gesture tracking in a controlled environment
Multiple hypotheses	Capable of tracking multiple targets at the same time. Suitable for multiple gestures tracking in a controlled environment
Extended model tracking	Target history is saved and available for target estimation. Suitable for long-time gesture tracking task
Tracking by detection	Learning algorithm increases robustness and reduces noise. This combined approach has the preferred performance in the test. Suitable for gesture tracking in a complex environment

easy to observe that the single hypothesis trackers perform lesser than the others. A summary of different gesture tracking approaches is presented in Table 2.3.

2.5 Gesture Classification

Gesture classification is the last and the most important step in gesture recognition. Most of the human gestures are continuous gestures. One continuous gesture always consists of several frames in the sensor data. To classify continuous gestures, gesture classification has to be done after or together with gesture tracking.

2.5.1 *Hidden Markov Model*

Hidden Markov Model (HMM) is a popular gesture classification algorithm. The HMM is a combination of an unobservable Markov Chain and a stochastic process. An example of HMM is shown in Fig. 2.10. To solve the classification problem, an Expectation–Maximisation (EM) algorithm is applied [72]. Recently, many papers discussed HMM gesture recognition applications [73–76]. Some articles also combined HMM with other classification algorithms [74].

2.5.2 *Support Vector Machine*

Support Vector Machine (SVM) is a discriminative classifier defined by a separating hyperplane [78]. Classification decision boundaries can be identified by maximising a margin distance. The optimal separation hyperplane can maximise the margin of training data. The training examples closest to the optimal hyperplane are called support vectors. However, a common problem for SVM is that the number of support

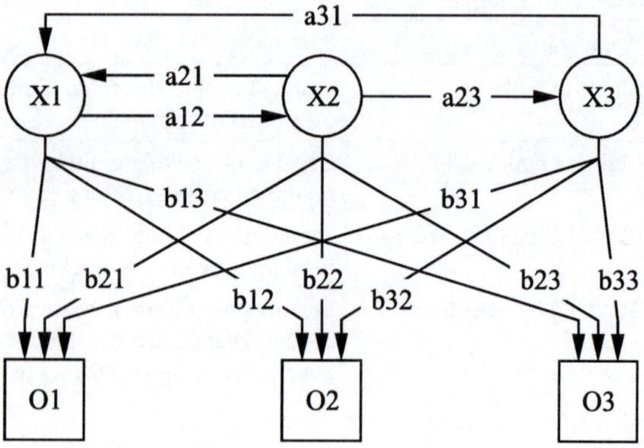

Fig. 2.10 Example of Hidden Markov Model [77]

vectors grows linearly with the growth of the training set. Some researchers proposed the *Relevance Vector Machine* (RVM) approach to solve the problem [79]. Moreover, *SVM kernel trick* was introduced by Scholkopf [80]. SVM kernel trick enables linear SVM in nonlinear classification problems. SVM kernel transforms low-dimensional training data into a high-dimensional feature space with a nonlinear method [81]. Also, many papers combine an SVM algorithm with other classification methods to improve the overall gesture classification performance [82–84].

2.5.3 *Convolutional Neural Network*

Convolutional Neural Networks (CNN) are an emerging and fast-growing branch of deep learning [85, 86]. As introduced at the beginning of this section, CNN provides convolutional layers that replace the feature extraction process. Fully connected dense layers provide classification capabilities that further classify the features extracted by convolutional layers into target categories.

An example of a CNN network is shown in Fig. 2.11. After receiving input data, a max-pooling layer is applied as the first layer to down-sample the input data by selecting the maximum value of a given data subset. The second layer takes the output of the first layer as input. The second layer is a convolutional layer, which filters the input data by convolutional kernels and outputs the extracted features. After another max-pooling layer as the third layer, two fully connected dense layers are applied as the fourth and fifth layers of the network. The fully connected dense layers receive and map the extracted features to the output categorical labels. In summary, compared with other models, CNN is a supervised learning model capable of achieving acceptable results without careful feature design and engineering [85]. Recently, many gesture recognition systems were presented that applied CNN as the gesture classification algorithm [54, 87–89].

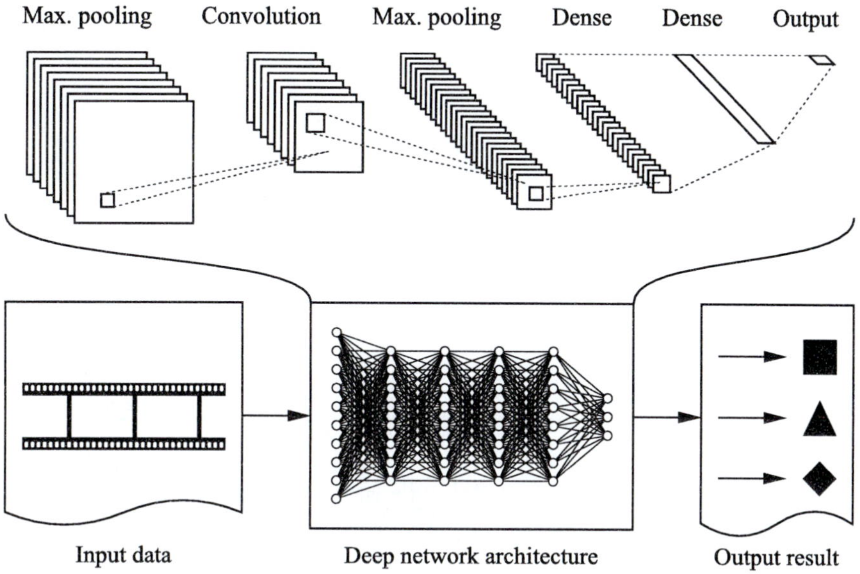

Fig. 2.11 Example of Convolutional Neural Network

2.5.4 Long Short-Term Memory

Long Short-Term Memory (LSTM) is a special unit type for RNN which is suitable for learning long sequential data. Compared with basic RNN, LSTM is capable of learning long-term dependencies from input data [90]. LSTM features cell states and gate mechanisms. With the input gate, forget gate and output gate, LSTM allows recurrent networks to selectively remember or forget information. This design increases the possibilities of linking remote causes and effects.

As shown in Fig. 2.12a, the LSTM model is formed by cells. Within a cell, several different mechanisms are designed to achieve the LSTM functionalities:

- *Forget gate*: The forget gate is used for deleting useless knowledge from the previous emitted cell state, as shown in Fig. 2.12b.
- *Input gate*: The input gate is used for adding useful knowledge to the current cell state, as shown in Fig. 2.12c.
- *Output gate*: The output gate is used for generating output selected from useful information in the current cell state, as shown in Fig. 2.12d.

Examples of utilising LSTM in HRC gesture recognition can be found in several recent papers [54, 88].

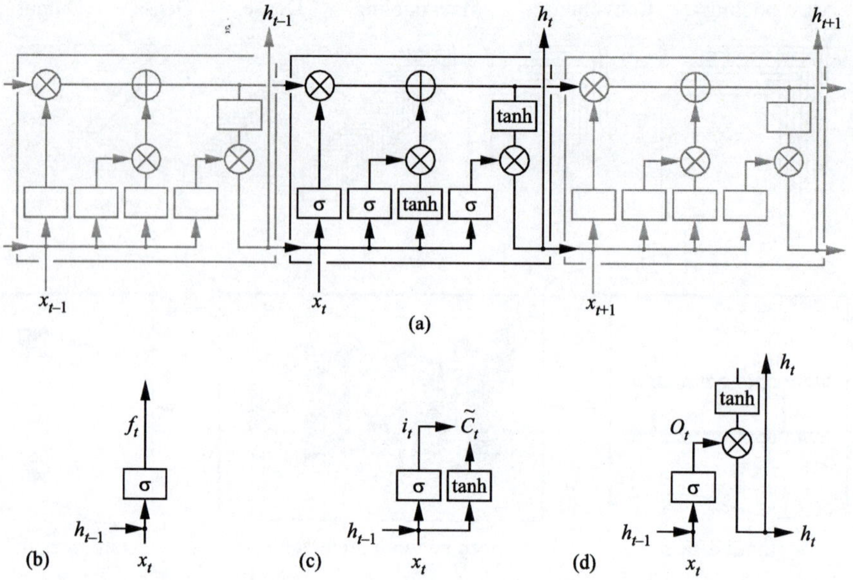

Fig. 2.12 Illustration of LSTM model: **a** LSTM model example with three cells. **b** Forget gate. **c** Input gate. **d** Output gate. Adapted from [90]

2.5.5 *Comparison of Gesture Classification Approaches*

Table 2.4. lists the advantages and disadvantages of the gesture classification approaches. One of the trends for manufacturing HRC systems is the emerging of different deep learning approaches. The primary constraint of deep learning is limited computational power. However, the exponential increase of computational power can

Table 2.4 Advantages and disadvantages of gesture classification approaches

Approach	Advantages	Disadvantages
Hidden Markov Model	Flexibility of training and verification, model transparency	Many free parameters need to be adjusted
Support vector machine	Different Kernel functions can be applied	Number of support vectors grows linearly with the size of the training set
Convolutional neural network	Can detect complex nonlinear relationships in the image-related dataset	"Black box" nature and cannot be used for small training datasets
Long short-term memory	Can detect complex nonlinear relationships in the sequence-related dataset	Computationally expensive

solve the problem in the future. The number of deep-learning-based gesture classification applications is growing rapidly. Another trend is to combine different classification algorithms for better prediction accuracy. Every classification algorithm has its own advantages and disadvantages. To utilise that, different information sources can be fused, and the classifiers can also be combined to achieve better performance in a manufacturing environment [88].

2.6 Gesture Recognition Example

The authors have tested different sensors and provide, in this section, an example as a gesture recognition case used in HRC applications. The section demonstrates the applicability to identify and classify human body gestures with the above-mentioned gesture recognition approach. The sensor used in the example is a normal 2D image sensor.

As shown in Fig. 2.13, some of the sensors can only be applied for gesture recognition in a limited area or small parts of a human body, such as the MYO sensor and the Leap Motion sensor depicted in Fig. 2.13a, b, while other sensors can be used for human pose recognition in larger areas, such as the Kinect sensor and the 2D image-based sensor shown in Fig. 2.13c, d. In this section, the authors provide an example of image-based human pose recognition, by using the 2D image-based

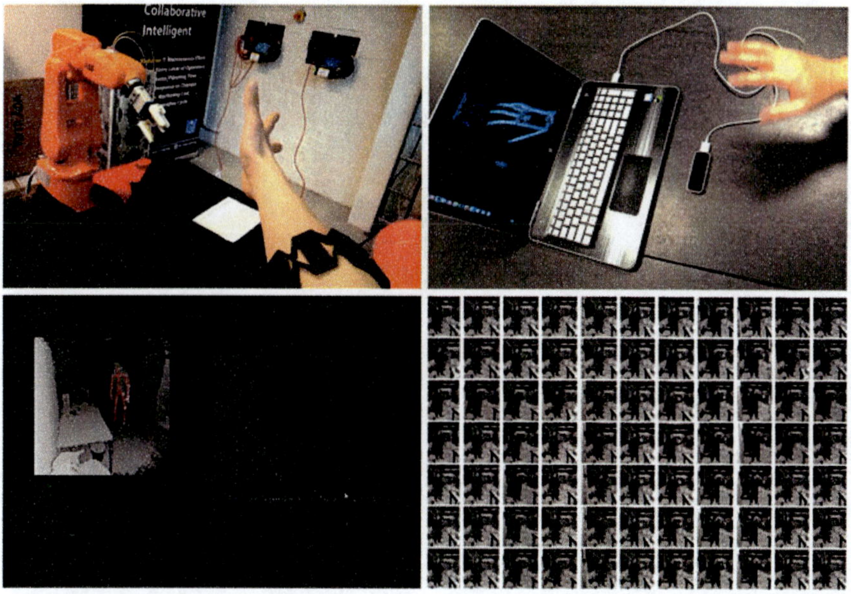

Fig. 2.13 Different sensors tested for gesture recognition: **a** MYO sensor; **b** leap motion sensor; **c** Kinect sensor; **d** 2D image-based sensor

sensor shown in Fig. 2.13d. As introduced in previous sections, human body movements can generally be classified into two types: static body posture which can be captured by only one image, and continuous body motion which needs to be captured by a video stream. In this example, the authors recognise gestures according to the static body posture. Therefore, the authors adapted a CNN-based gesture classification method. Since the focus of the example is also gesture classification, the authors mainly focus on the gesture classification part.

The first step of any machine learning problems is to collect and clean the dataset for the machine learning model training. In this example, the authors collected an HRC-specific dataset, consisting of two different postures (standing and assembly). Each of the postures includes 360 images collected from a manufacturing environment. As shown in Fig. 2.14, the authors also performed data cleaning before training, which transforms the collected colour images into 200-by-200 greyscale images. With pre-processed images prepared, the images are labelled and randomly sorted. The prepared images are further split with an 80–20% ratio into a training dataset and a test dataset to avoid overfitting. After the data cleaning process, the input for the machine learning model are 200-by-200 greyscale images, and the output of the machine learning model are two class labels of "true" or "false".

The choice of the machine learning algorithm is another crucial factor for the result accuracy. To avoid a complex feature engineering and feature selection process,

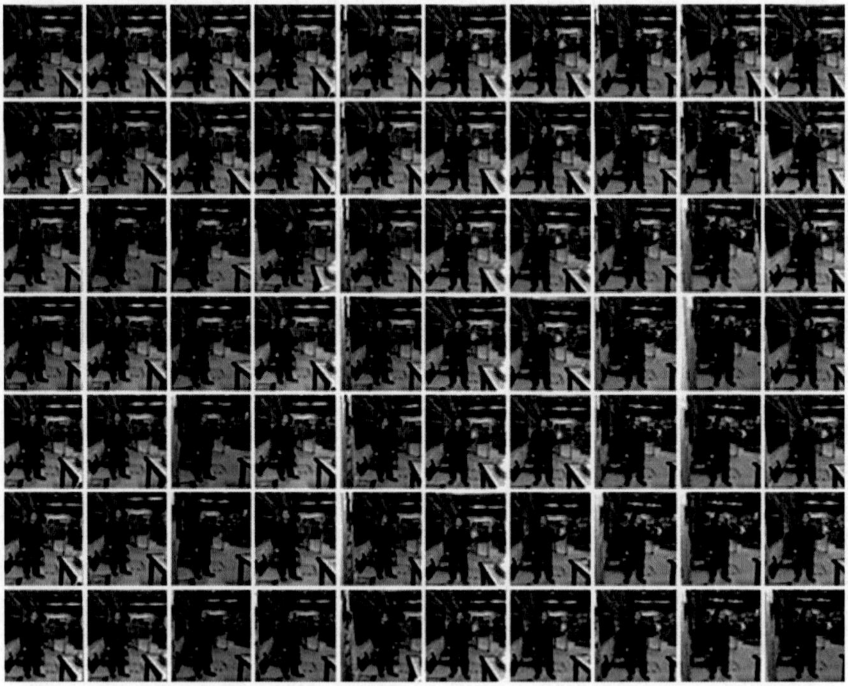

Fig. 2.14 Customised HRC dataset

Table 2.5 CNN architecture for body posture recognition

Layer (type)	Output Shape
conv2d_1 (Conv2D)	(None, 198, 198, 32)
conv2d_2 (Conv2D)	(None, 196, 196, 64)
max_pooling2d_1 (MaxPooling2)	(None, 98, 98, 64)
dropout_1 (Dropout)	(None, 98, 98, 64)
flatten_1 (Flatten)	(None, 614, 656)
dense_1 (Dense)	(None, 128)
dropout_2 (Dropout)	(None, 128)
dense_2 (Dense)	(None, 2)

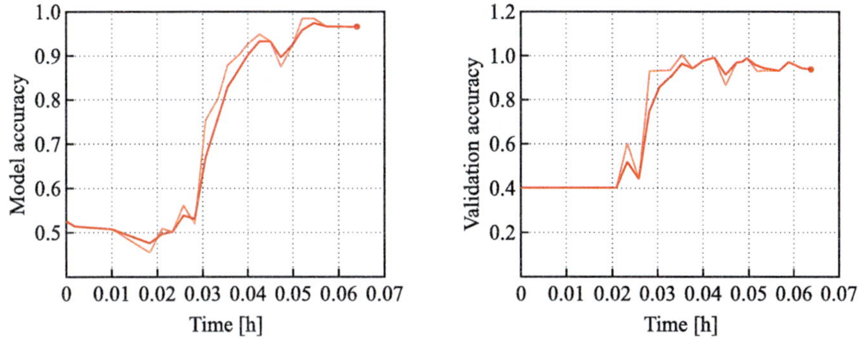

Fig. 2.15 CNN model accuracy with the training time, **a** model accuracy on the training set; **b** model accuracy on the validation set

this example applied CNN as the one-step machine learning algorithm to train and recognise the human postures. The selected CNN architecture is shown in Table 2.5. The training and test results are shown in Fig. 2.15. After 20 epochs, the selected CNN architecture achieved an accuracy of 0.96 on the training dataset. The training loss is 0.11. The test accuracy on the test dataset converged to 0.93. It is clear that the different postures are classified well, without obvious overfitting.

2.7 Conclusions and Future Trends

Although the above sections provided a general picture of gesture recognition for HRC, it is never easy to summarise such an interdisciplinary and fast-developing field in any capacity. Sensor-related technologies normally started from the hardware. Software technologies and algorithms are designed to utilise the performance of hardware. Therefore, we would list some of the predicted future trends starting with sensor technologies.

- Depth sensor and human model-based gesture recognition: due to the nature of HRC, human operators are the most important members of any HRC team. Despite understanding human body gestures, depth sensor together with the body model approach will monitor human movements, which provides a safer environment for HRC. Moreover, the body skeleton model will simplify the gesture classification process. Therefore, a simpler gesture tracking and classification method can be applied.
- Non-wearable sensor and deep-learning-based gesture recognition: although non-wearable sensor technologies have not yet attained maturity for live application in a safety-critical environment, it is still the most promising non-image-based sensor. In HRC, the human operator should be able to communicate with robots naturally. For this very purpose, nothing should be attached to the human operator's body. Non-wearable sensors still suffer from low gesture identification and classification quality. The problem can potentially be solved using deep learning methods, in addition to advances of the sensing technologies themselves.
- Hard real-time gesture recognition systems: one of the most important requirements in industry is the necessity of real-time response. Especially, in the HRC system, the safety of the human operator is always the first priority. Therefore, real-time gesture recognition systems are another future direction of importance. Currently, band and glove sensors provide the fastest response. Moreover, high-speed single-camera gesture recognition systems are also emerging recently. In gesture identification, tracking and classification, fast and effective methods can be applied.
- Multi-sensor gesture recognition systems: all the sensors have advantages and disadvantages. For instance, the band sensor has a large sensing area; Kinect has good performance in body gesture recognition. To best utilise the system performance, different gesture recognition sensors can be used in the same system.
- Algorithm combination approaches: similarly, different gesture classification algorithms also have their advantages and disadvantages. As we mentioned in the gesture classification section, the combination of algorithms can improve efficiency.
- Industrial standardisation: it is clear that an emerging technology requires industrial standards to guide practices at a factory level. In the future, the gesture recognition technology for HRC will also require more standardisation.

For a further detailed review and future trends, readers are referred to [91].

References

1. J. Krüger, T.K. Lien, A. Verl, Cooperation of human and machines in assembly lines, CIRP Ann. Manuf. Technol.**58**(2), 628–646 (2009)
2. S. Green, X. Chen, M. Billinnghurst, J.G. Chase, Human Robot collaboration: an augmented reality approach a literature review and analysis. Mechatronics **5**(1), 1–10 (2007)
3. P.R. Cohen, H.J. Levesque, Teamwork, in *Nous* (1991) pp. 487–512

4. L.S. Vygotsky, *Mind in Society: The Development of Higher Psychological Processes* (Harvard university press, 1980)
5. P.R. Cohen, H.J. Levesque, Persistence, intention, and commitment, in *Reason. About Actions Plans* (1990), pp. 297–340
6. C. Breazeal, A. Brooks, J. Gray, G. Hoffman, C. Kidd, H. Lee, J. Lieberman, A. Lockerd, D. Mulanda, Humanoid robots as cooperative partners for people. Int. J. Humanoid Robot. **1**(2), 1–34 (2004)
7. A. Bauer, D. Wollherr, M. Buss, Human–robot collaboration: a survey. Int. J. Humanoid Robot. **5**(01), 47–66 (2008)
8. S. Mitra, T. Acharya, Gesture recognition: a survey. IEEE Trans. Syst. Man, Cybern. Part C Appl. Rev. **37**(3), 311–324 (2007)
9. R. Parasuraman, T.B. Sheridan, C.D. Wickens, A model for types and levels of human interaction with automation. IEEE Trans. Syst. Man Cybern. Part A Syst. Humans**30**(3), 286–297 (2000)
10. T.E. Starner, Visual recognition of american sign language using hidden markov models (1995)
11. T. Starner, J. Weaver, A. Pentland, Real-time american sign language recognition using desk and wearable computer based video. IEEE Trans Pattern Anal Mach Intell **20**(12), 1371–1375 (1998)
12. N.R. Howe, M.E. Leventon, W.T. Freeman, Bayesian reconstruction of 3d human motion from single-camera video. NIPS **99**, 820–826 (1999)
13. Y. Katsuki, Y. Yamakawa, M. Ishikawa, High-speed human/Robot hand interaction system, in *Proceedings of the Tenth Annual ACM/IEEE International Conference on Human–Robot Interaction Extended Abstracts* (ACM, 2015), pp. 117–118
14. M. Elmezain, A. Al-Hamadi, J. Appenrodt, B. Michaelis, A hidden markov model-based continuous gesture recognition system for hand motion trajectory, in *19th International Conference on Pattern Recognition, 2008. ICPR 2008* (IEEE, 2008), pp. 1–4
15. Y. Matsumoto, A. Zelinsky, An algorithm for real-time stereo vision implementation of head pose and gaze direction measurement, in *Fourth IEEE International Conference on Automatic Face and Gesture Recognition, 2000. Proceedings* (IEEE, 2000), pp. 499–504
16. J.P. Wachs, M. Kölsch, H. Stern, Y. Edan, Vision-based hand-gesture applications. Commun. ACM **54**(2), 60–71 (2011)
17. J. Suarez, R.R. Murphy, Hand gesture recognition with depth images: a review, in *RO-MAN, 2012 IEEE* (IEEE, 2012), pp. 411–417
18. P. Doliotis, A. Stefan, C. McMurrough, D. Eckhard, V. Athitsos, Comparing gesture recognition accuracy using color and depth information, in *Proceedings of the 4th International Conference on Pervasive Technologies Related to Assistive Environments* (ACM, 2011), p. 20
19. M. Hansard, S. Lee, O. Choi, R.P. Horaud, *Time-of-Flight Cameras: Principles, Methods and Applications* (Springer Science & Business Media, 2012)
20. T. Kapuściński, M. Oszust, M. Wysocki, Hand gesture recognition using time-of-flight camera and viewpoint feature histogram, in *Intelligent Systems in Technical and Medical Diagnostics* (Springer, 2014), pp. 403–414
21. S.B. Gokturk, H. Yalcin, C. Bamji, A time-of-flight depth sensor-system description, issues and solutions, in *Conference on Computer Vision and Pattern Recognition Workshop, 2004. CVPRW'04* (IEEE, 2004), p. 35
22. J.D. Arango Paredes, B. Munoz, W. Agredo, Y. Ariza-Araujo, J.L. Orozco, A. Navarro, A reliability assessment software using Kinect to complement the clinical evaluation of Parkinson's disease, in *2015 37th Annual International Conference of the IEEE Engineering in Medicine and Biology Society (EMBC)* (IEEE, 2015), pp. 6860–6863
23. C. Wang, Z. Liu, S.-C. Chan, Superpixel-based hand gesture recognition with Kinect depth camera. IEEE Trans Multimed **17**(1), 29–39 (2015)
24. H. Liu, Y. Wang, W. Ji, L. Wang, A context-aware safety system for human–robot collaboration. Procedia Manuf. **17**, 238–245 (2018)
25. A. Erol, G. Bebis, M. Nicolescu, R.D. Boyle, X. Twombly, Vision-based hand pose estimation: a review. Comput. Vis. Image Underst. **108**(1), 52–73 (2007)

26. T. Labs, Myo (2015). https://www.myo.com/
27. Y. Zhang, C. Harrison, Tomo: wearable, low-cost electrical impedance tomography for hand gesture recognition, in *Proceedings of the 28th Annual ACM Symposium on User Interface Software & Technology* (ACM, 2015), pp. 167–173
28. N. Haroon, A.N. Malik, Multiple hand gesture recognition using surface EMG signals. J. Biomed. Eng. Med. Imaging **3**(1), 1 (2016)
29. S. Roy, S. Ghosh, A. Barat, M. Chattopadhyay, and D. Chowdhury, Real-time implementation of electromyography for hand gesture detection using micro accelerometer, in *Artificial Intelligence and Evolutionary Computations in Engineering Systems* (Springer, 2016), pp. 357–364
30. Google, Project soli (2015). https://www.google.com/atap/project-soli/
31. J. Smith, T. White, C. Dodge, J. Paradiso, N. Gershenfeld, D. Allport, Electric field sensing for graphical interfaces. IEEE Comput. Graph. Appl. **18**(3), 54–60 (1998)
32. F. Adib, C.-Y. Hsu, H. Mao, D. Katabi, F. Durand, Capturing the human figure through a wall. ACM Trans. Graph. **34**(6), 219 (2015)
33. F. Adib, D. Katabi, *See Through Walls with WIFI!*, vol. 43, no. 4 (ACM, 2013)
34. F. Adib, Z. Kabelac, D. Katabi, R.C. Miller, 3d tracking via body radio reflections, in *Usenix NSDI*, vol. 14 (2014)
35. J. Letessier, F. Bérard, Visual tracking of bare fingers for interactive surfaces, in *Proceedings of the 17th Annual ACM Symposium on User Interface Software and Technology* (ACM, 2004), pp. 119–122
36. D. Weinland, R. Ronfard, E. Boyer, A survey of vision-based methods for action representation, segmentation and recognition. Comput. Vis. Image Underst. **115**(2), 224–241 (2011)
37. D.G. Lowe, Object recognition from local scale-invariant features, in *The Proceedings of the Seventh IEEE International Conference on Computer Vision, 1999*, vol. 2 (IEEE, 1999), pp. 1150–1157
38. H. Bay, T. Tuytelaars, L. Van Gool, Surf: speeded up robust features, in *Computer Vision–ECCV 2006* (Springer, 2006), pp. 404–417
39. E. Rublee, V. Rabaud, K. Konolige, G. Bradski, ORB: an efficient alternative to SIFT or SURF, in *2011 IEEE International Conference on Computer Vision (ICCV)* (IEEE, 2011), pp. 2564–2571
40. S. Belongie, J. Malik, J. Puzicha, Shape matching and object recognition using shape contexts. IEEE Trans. Pattern Anal. Mach. Intell. **24**(4), 509–522 (2002)
41. B. Allen, B. Curless, Z. Popović, Articulated body deformation from range scan data. ACM Trans. Graph. (TOG)**21**(3), 612–619 (ACM, 2002)
42. R. Cutler, M. Turk, View-based interpretation of real-time optical flow for gesture recognition, *fg* (IEEE, 1998), p. 416
43. J.L. Barron, D.J. Fleet, S.S. Beauchemin, Performance of optical flow techniques. Int. J. Comput. Vis. **12**(1), 43–77 (1994)
44. C. Thurau, V. Hlaváč, Pose primitive based human action recognition in videos or still images, in *IEEE Conference on Computer Vision and Pattern Recognition, 2008. CVPR 2008* (IEEE, 2008), pp. 1–8
45. Q. Pu, S. Gupta, S. Gollakota, S. Patel, Whole-home gesture recognition using wireless signals, in *Proceedings of the 19th annual international conference on Mobile computing & networking* (ACM, 2013), pp. 27–38
46. R. Ronfard, C. Schmid, B. Triggs, Learning to parse pictures of people, in *Computer Vision—ECCV 2002* (Springer, 2002), pp. 700–714
47. S.-J. Lee, C.-S. Ouyang, S.-H. Du, A neuro-fuzzy approach for segmentation of human objects in image sequences. IEEE Trans. Syst. Man, Cybern. Part B Cybern**33**(3), 420–437 (2003)
48. D. Tang, H.J. Chang, A. Tejani, T.-K. Kim, Latent regression forest: structured estimation of 3D articulated hand posture, in *2014 IEEE Conference on Computer Vision and Pattern Recognition (CVPR)* (IEEE, 2014), pp. 3786–3793
49. J. Han, L. Shao, D. Xu, J. Shotton, Enhanced computer vision with microsoft kinect sensor: a review. IEEE Trans Cybern. **43**(5), 1318–1334 (2013)

50. I. Oikonomidis, N. Kyriazis, A.A. Argyros, Efficient model-based 3D tracking of hand articulations using Kinect. BMVC **1**(2), 3 (2011)
51. J. Shotton, A. Fitzgibbon, M. Cook, T. Sharp, M. Finocchio, R. Moore, A. Kipman, A. Blake, Real-time human pose recognition in parts from single depth images. Commun. ACM **56**(1), 116–124 (2013)
52. M. Ye, X. Wang, R. Yang, L. Ren, M. Pollefeys, Accurate 3D pose estimation from a single depth image, in *IEEE International Conference on Computer Vision (ICCV), 2011* (IEEE, 2011), pp. 731–738
53. Y. Li, Hand gesture recognition using Kinect, in *2012 IEEE 3rd International Conference on Software Engineering and Service Science (ICSESS)* (IEEE, 2012), pp. 196–199
54. H. Liu, T. Fang, T. Zhou, Y. Wang, L. Wang, Deep learning-based multimodal control interface for human–robot collaboration. Procedia CIRP **72**, 3–8 (2018)
55. D. Comaniciu, V. Ramesh, P. Meer, Real-time tracking of non-rigid objects using mean shift, in *IEEE Conference on Computer Vision and Pattern Recognition, 2000. Proceedings*, vol 2 (IEEE, 2000), pp. 142–149
56. R.E. Kalman, A new approach to linear filtering and prediction problems. J. Fluids Eng. **82**(1), 35–45 (1960)
57. S. Haykin, *Kalman Filtering and Neural Networks*, vol. 47 (Wiley, 2004)
58. R. Kandepu, B. Foss, L. Imsland, Applying the unscented Kalman filter for nonlinear state estimation. J. Process Control **18**(7), 753–768 (2008)
59. R. Gao, L. Wang, R. Teti, D. Dornfeld, S. Kumara, M. Mori, M. Helu, Cloud-enabled prognosis for manufacturing. CIRP Ann. Technol. **64**(2), 749–772 (2015)
60. J. Krüger, T.K. Lien, A. Verl, Cooperation of human and machines in assembly lines. CIRP Ann. Technol. **58**(2), 628–646 (2009)
61. K. Okuma, A. Taleghani, N. De Freitas, J.J. Little, D.G. Lowe, A boosted particle filter: multitarget detection and tracking, in *Computer Vision-ECCV 2004* (Springer, 2004), pp. 28–39
62. S. Oron, A. Bar-Hillel, D. Levi, S. Avidan, Locally orderless tracking, in *2012 IEEE Conference on Computer Vision and Pattern Recognition (CVPR)* (IEEE, 2012), pp. 1940–1947
63. J. Kwon, K.M. Lee, Tracking by sampling trackers, in *2011 IEEE International Conference on Computer Vision (ICCV)* (IEEE, 2011), pp. 1195–1202
64. T. Li, S. Sun, T.P. Sattar, J.M. Corchado, Fight sample degeneracy and impoverishment in particle filters: a review of intelligent approaches. Expert Syst. Appl. **41**(8), 3944–3954 (2014)
65. T. Li, T.P. Sattar, S. Sun, Deterministic resampling: unbiased sampling to avoid sample impoverishment in particle filters. Signal Process **92**(7), 1637–1645 (2012)
66. J.M. Del Rincón, D. Makris, C.O. Uruñuela, J.-C. Nebel, Tracking human position and lower body parts using Kalman and particle filters constrained by human biomechanics. IEEE Trans. Syst. Man, Cybern. Part B Cybern.**41**(1), 26–37 (2011)
67. D.A. Ross, J. Lim, R.-S. Lin, M.-H. Yang, Incremental learning for robust visual tracking. Int. J. Comput. Vis. **77**(1–3), 125–141 (2008)
68. J. Kwon, K.M. Lee, F.C. Park, Visual tracking via geometric particle filtering on the affine group with optimal importance functions, in *IEEE Conference on Computer Vision and Pattern Recognition, 2009. CVPR 2009* (IEEE, 2009), pp. 991–998
69. Z. Kalal, J. Matas, K. Mikolajczyk, P-N learning: bootstrapping binary classifiers by structural constraints, in *2010 IEEE Conference on Computer Vision and Pattern Recognition (CVPR)* (IEEE, 2010), pp. 49–56
70. B. Babenko, M.-H. Yang, S. Belongie, Visual tracking with online multiple instance learning, in *IEEE Conference on Computer Vision and Pattern Recognition, 2009. CVPR 2009* (IEEE, 2009), pp. 983–990
71. A.W.M. Smeulders, D.M. Chu, R. Cucchiara, S. Calderara, A. Dehghan, M. Shah, Visual tracking: an experimental survey. IEEE Trans. Pattern Anal. Mach. Intell. **36**(7), 1442–1468 (2014)
72. A.D. Wilson, A.F. Bobick, Parametric hidden markov models for gesture recognition. IEEE Trans. Pattern Anal. Mach. Intell. **21**(9), 884–900 (1999)

73. S. Lu, J. Picone, S. Kong, Fingerspelling alphabet recognition using a two-level hidden markov model, in *Proceedings of the International Conference on Image Processing, Computer Vision, and Pattern Recognition (IPCV). The Steering Committee of The World Congress in Computer Science, Computer Engineering and Applied Computing (WorldComp)* (2013), p. 1
74. J. McCormick, K. Vincs, S. Nahavandi, D. Creighton, S. Hutchison, Teaching a digital performing agent: artificial neural network and hidden Markov model for recognising and performing dance movement, in *Proceedings of the 2014 International Workshop on Movement and Computing* (ACM, 2014), p. 70
75. S.-Z. Yu, Hidden semi-Markov models. Artif. Intell. **174**(2), 215–243 (2010)
76. H. Liu, L. Wang, Human motion prediction for human–robot collaboration. J. Manuf. Syst. **44**, 287–294 (2017)
77. L.R. Rabiner, A tutorial on hidden Markov models and selected applications in speech recognition. IEEE Proc. **77**(2), 257–286 (1989)
78. M.A. Hearst, S.T. Dumais, E. Osman, J. Platt, B. Scholkopf, Support vector machines. IEEE Intell. Syst. Their Appl. **13**(4), 18–28 (1998)
79. M.E. Tipping, Sparse Bayesian learning and the relevance vector machine. J. Mach. Learn. Res. **1**, 211–244 (2001)
80. B. Schiilkopf, The kernel trick for distances, in *Advances in Neural Information Processing Systems 13: Proceedings of the 2000 Conference*, vol. 13 (MIT Press, 2001), p. 301
81. A. Cenedese, G.A. Susto, G. Belgioioso, G.I. Cirillo, F. Fraccaroli, Home automation oriented gesture classification from inertial measurements. IEEE Trans. Autom. Sci. Eng. **12**(4), 1200–1210 (2015)
82. K. Feng, F. Yuan, Static hand gesture recognition based on HOG characters and support vector machines, in *2013 2nd International Symposium on Instrumentation and Measurement, Sensor Network and Automation (IMSNA)* (IEEE, 2013), pp. 936–938
83. D. Ghimire, J. Lee, Geometric feature-based facial expression recognition in image sequences using multi-class adaboost and support vector machines. Sensors **13**(6), 7714–7734 (2013)
84. O. Patsadu, C. Nukoolkit, B. Watanapa, Human gesture recognition using Kinect camera, in *2012 International Joint Conference on Computer Science and Software Engineering (JCSSE)* (IEEE, 2012), pp. 28–32
85. A. Krizhevsky, I. Sutskever, G.E. Hinton, Imagenet classification with deep convolutional neural networks, in *Advances in Neural Information Processing Systems* (2012), pp. 1097–1105
86. Y. LeCun, L. Bottou, Y. Bengio, P. Haffner, Gradient-based learning applied to document recognition. IEEE Proc. **86**(11), 2278–2323 (1998)
87. H. Liu, L. Wang, Remote human–robot collaboration: a cyber–physical system application for hazard manufacturing environment. J. Manuf. Syst. **54**, 24–34 (2020)
88. H. Liu, T. Fang, T. Zhou, L. Wang, Towards robust human–robot collaborative manufacturing: multimodal fusion. IEEE Access **6**, 74762–74771 (2018)
89. P. Wang, H. Liu, L. Wang, R.X. Gao, Deep learning-based human motion recognition for predictive context-aware human–robot collaboration. CIRP Ann. (2018)
90. S. Hochreiter, J. Urgen Schmidhuber, Long short-term memory. Neural Comput. **9**(8), 1735–1780 (1997)
91. L. Wang, R.X. Gao, J. Váncza, J. Krüger, X.V. Wang, S. Makris, G. Chryssolouris, Symbiotic human–robot collaborative assembly. CIRP Ann. Manuf. Technol.**68**(2), 701–726 (2019)

Chapter 3
Safety Strategy and Framework for Human–Robot Collaboration

Xi Vincent Wang and Lihui Wang

Abstract Modern manufacturing systems call for a safe, efficient and user-friendly working environment to meet the expectation of Industry 4.0 and Smart Manufacturing, which needs new approaches with smarter functionalities and a better/safer working environment for human beings. Human–Robot Collaboration (HRC) is considered as one of the promising approaches, and it attracts major research interest in both academia and industry, as it provides a feasible solution combing the flexibility and intelligence of a human, together with the accuracy and strength of an industrial robot. However, in the past years, despite the significant development of different HRC approaches, there is still a lack of a clear safety strategy for an HRC system. Current research efforts focus more on advanced robot control methods, while the uniqueness of each human body is not considered in the planning or control loop. Thus, in this research, the taxonomy of different human–robot relations is identified first. Then, the safety framework and strategy are developed towards a personalised solution in a human–robot collaboration cell. A dynamic HRC layout approach is also introduced, based on the actual speed of humans and robots, and the distance between them. The feasibility of the proposed approaches in this research is then evaluated via the implementation in an HRC-based assembly cell in a quantifiable manner. The operator's biometric data is also included in the HRC control loop. It is proven achievable to conduct a personalised HRC safety strategy based on the human stakeholder's role, physical conditions, speed and other characteristics. A future research outlook and essential considerations are addressed at the end of the chapter.

X. V. Wang (✉) · L. Wang
Department of Production Engineering, KTH Royal Institute of Technology, Stockholm, Sweden
e-mail: wangxi@kth.se

L. Wang
e-mail: lihuiw@kth.se

L. Wang et al. (eds.), *Advanced Human–Robot Collaboration in Manufacturing*,
https://doi.org/10.1007/978-3-030-69178-3_3

3.1 Introduction

In recent years, the modern manufacturing industry requires new systems with higher intelligence, flexibility and user-friendly environment. New technologies open the door to novel interaction methods along production lines, e.g., Human–Robot Collaboration (HRC), augmented reality, and voice control. Despite the promising functionality, impact and performance introduced by the new technologies, the safety of the human operators and workers has to remain the highest priority in the next generation of manufacturing technologies and solutions, too. Among the new technologies, HRC is an important research topic as it combines the flexibility of humans and repeatability/accuracy of robots. Thus, it is expected to help achieve higher production performance by faster speed and higher quality. At the same time, a better working environment is also demanded with the help of flexible assistance for the human operator in the HRC cell, while still satisfying the binding safety requirements as mentioned before. In a conventional robot cell, humans and robots are completely separated by fences or walls in the production line, to avoid potential collision between them. In this way, the human operator is indeed protected by isolated workspaces. However, it also limits the possibility of dynamic usage and interaction with the robot. In an HRC solution, on the other hand, the interaction and coexistence of humans and robota is an essential aim. As a result, support of a shared workspace—and, optionally, shared tasks as well—is a common requirement from the design phase till the end of the HRC cell's lifecycle. Thus, the safety strategy and framework become one of the most urgent issues for HRC applications.

In the past years, multiple standards and research were proposed to establish safe systems and methods in an HRC scenario. However, most of them are general definitions without the consideration of the diversity of individual operators, especially the physiological characteristics of humans. As a matter of fact, the physiological characteristics of operators are different, and a user-friendly working environment should be adaptive to them. Thus, in this chapter, different types of relationships between humans and industrial robots are first defined, to provide a clear classification and guideline for the design and actions in an HRC robot cell. Related research work is reviewed, and a systematic approach is presented based on the personalised physiological characteristics. In the background, a mathematical model is embedded inside the proposed system to achieve control limitations on industrial robots automatically. A personalised safety strategy system is developed during implementation towards the validation of the proposed personalised safety plan for specific operators working in the HRC cell.

3.2 Classification and Definition of Human–Robot Relationship

Schmidtler et al. [1] summarised the human–robot cell in terms of working time, workspace, aim and contact. Van Oosterwyck [2] considered different workspaces as being one type of coexistence. However, such a perspective may not be fully convincing—if the human and robot are separated in both temporal and spatial scales, the relationship between robots and humans may be weak, and coexistence is far from its fully possible extent. Thus, in this research, the shared contents between the human and the industrial robot are identified as workspace, direct contact, work task, simultaneous process and sequential process. *Shared workspace* refers to whether the human and robot are working in the same working area with no physical or virtual fences or separation. *Direct contact* indicates whether there is a direct physical contact between the human and the robot. *Shared work task* represents whether the operator and the robot work on the same operation towards the same working objective. *Simultaneous process* means that the operator and the robot are working at the same time, but the task can be the same or different. In contrast, *sequential process* indicates that the operations of the human and the robot are arranged one after the other, and there is no overlap on the spatial scale. Based on these definitions, the classifications regarding four types of human–robot relationships are given below, and the sharing contents in the manufacturing environment are summarised in Table 3.1, extended based on previous work [3].

- The basic situation is *Human–Robot Coexistence* when a robot and a human are placed within the same physical space but without overlapping each other's workspace. There is no direct contact between the human and the robot. The work object might be exchanged between them, but the process is performed independently and simultaneously. Thus there is no need for the human or robot to adapt or learn from each other. The risk of direct human–robot injury is almost none because of the strict separation devices and rules. A monitored stop is applied when the human or robot enters the workplace of others.
- *Human–Robot Interaction* (HRI) occurs if a human and a robot exert some effect on each other. In a wider interpretation of the term, unintended or unplanned effects count as interaction as well. However, in cases where interaction forms a substantial part of human–robot interrelation, planned or intended forms of interaction prevail. It is, in these cases, justified to regard human–robot interaction as an identifiable type of human–robot interrelations. Here, much of the interaction typically is communication as well, i.e., intended interaction with the purpose of one participant transferring information to the other. HRI can, for example, occur when one party guides or controls the other, or any physical contact (either planned or unintended) occurs between the human and the robot. Both parties can work on the same task but do not have to be in tighter coupling than completing the task step by step in a sequential order. HRI includes two types of synergies: *non-physical* and *physical* interactions, depending on whether or not physical contact is involved. Verbal/audio communication, vision and other types of contactless

Table 3.1 Taxonomy of different human–robot relations

		Coexistence	Interaction			Cooperation		Symbiotic collaboration
			Non-physical	Physical	Hand Guiding	Serial	Parallel	
Shared	Workplace		×	×	×	×	×	×
	Task		×	×	×			×
	Direct contact			×	×			×
	Resource					×	×	×
	Serial process		×	×	×	×		
	Parallel process	×					×	×
	Communication		×	×	×	×	×	×
Adapting and learning			×	×	×			×
Risk of direct injury from robot		Almost none	Low	Medium	Medium	Low	Low	Medium

capturing are non-physical interaction, while collisions, tactile communication or force feedback count as physical interaction. Normally, the risk of injury is low for non-physical HRI as there is no physical contact between the human and the robot. In parallel, physical HRI (pHRI) is widely applied in *cobots* (collaborative robots). In this scenario, the human operator can directly touch and lead the robot to complete the process together. The risk of injury is higher than in non-physical HRI as any protective distance is eliminated. In HRI, the robot is normally restrained by speed, power and force limiting.

- As a matter of fact, *hand guiding* can be considered as a special type of pHRI. The main difference is that there are more guiding and teaching tasks for human operators compared with other pHRI solutions.
- *Human–Robot Cooperation* assumes an interactive relationship that makes it possible to harness the competence and knowledge of other system components in the hope of mutual benefit. It can be developed between a human and a robot which have their own autonomy—expressed in terms of goals, objectives, utility or profit—have the faculties for pursuing their own interest, and admit to the autonomy of others [4]. The parties can share a partially overlapping workspace but with no direct contact between them. They share the same working task but complete the task in a sequential order. As physical contact is not normally required in the cooperation mode, the injury risk is relatively low. The working process can be planned either in a serial or a parallel sequence, and the possibility of adaption and learning is minimised.
- *Human–Robot Collaboration* is the joint activity of humans and robots in a shared workspace, with the definite objective to accomplish together a set of given working tasks. It requires typically a coordinated, synchronous activity from all parties [5] where physical contact is also allowed. In any case, collaboration assumes a joint, focused goal-oriented activity from the parties who have—due to the division of labour—different capabilities and competencies. Even though the collaborative process is supervised and protected by different types of approaches, it has to be admitted that the risk of injury is also higher due to the complexity of the system and unpredictable conditions in the HRC cell.

To recapitulate, the interaction and safety risk levels of different human–robot modes are summarised in Fig. 3.1. Regarding the level of interrelation, the engagement and complexity in human–robot coexistence are low. Meanwhile, safety is easy to be guaranteed as the operator is protected by physical distance. In the human–robot cooperation mode, the interrelation level and safety risk are raised as direct contact between the operator and the robot may be involved. In human–robot interaction, the level of interrelation is not significantly increased but the safety risk is much higher, once the human and the robot start to share the same workspace with the possibility of physical contact, especially in pHRI. In a full symbiotic collaboration between a human and a robot, the work task is completed by both sides collaboratively. It is unavoidable for the human operator to be in direct contact with the robot. Thus, the safety risk level is higher than in other types of human–robot relations. Hence,

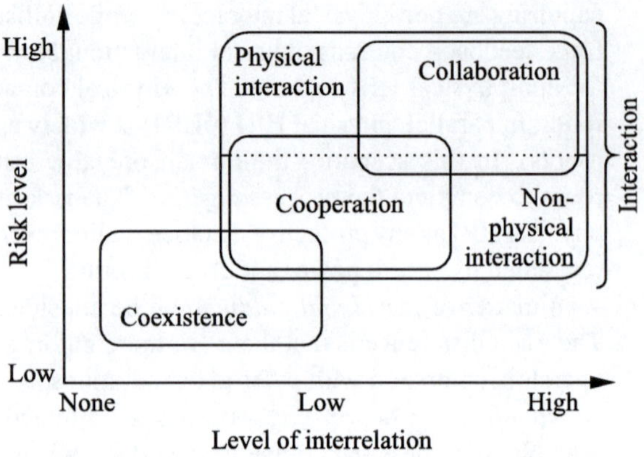

Fig. 3.1 Interrelation and risk level of different types of human–robot relationships

a reliable safety system is essential to guarantee the safety and well-being of the human operator working in an HRC cell.

3.3 Literature Review

The interaction between a human and a robot requires shared working time, workspace, aim and contact [1]. The concept of coexistence can even be considered when the human worker is located in a different workspace [2] (i.e., when *co-location* does not occur). Despite different definitions about human–robot interaction and collaboration, the safety of the human needs to be guaranteed, especially when the robot and the human are sharing the same workspace. Robla-Gómez et al. [6] presented a review of the main safety systems in industrial robotic environments that contribute to the achievement of safe collaborative human–robot work. The framework for safety in industrial robotic environments is categorised into five types, i.e., human action restriction, robot behaviour modification, injury classification, injury minimisation and collision avoidance.

Recent safety strategy approaches are summarised in Table 3.2. Ikuta et al. [7] addressed two types of safety strategies, i.e., design safety and control safety. The design strategy refers to the weight, cover, joint compliance, surface and shape of the robot. In parallel, the control strategy limits the distance, speed and posture in the HRC solution. Lasota et al. [8] further extended the safety requirements to control, planning, prediction and psychological considerations.

At the design phase, Marvel et al. [9] proposed an ontology-based task description for HRC tasks. It provides the basis and guideline for collaborative robots. Tan et al. [10] presented a design strategy for HRC. In their work, five safety designs are described considering the potential accidents during human–industrial robot collaboration. The proposed concept is demonstrated via a prototype system based on safety considerations. Design considerations are also discussed for safe

Table 3.2 Safety strategy research

Safety strategy		Ikuta et al. [7], Lasota et al. [8]
Safety design		Marvel et al. [9], Tan et al. [10], Michalos et al. [11], Maurtua et al. [12]
Safety assessment		Kulić and Croft [13–15], ISO [16] and Lacevic and Rocco [17]
Safety planning		De Santis et al. [18] and Sisbot et al. [19]
Safety control	Control strategy	Corrales et al. [20], Mohammed et al.[21] and Haddadin et al. [22]
	Vision based	Zanchettin et al. [23], Morato et al. [24] and Cherubini et al. [25]
	Torque-/force based	Cherubini et al. [25], Heinzmann and Zelinsky [26], and Koveos et al. [27]

human–robot collaborative workplaces [11]. The variability in HRC assembly is discussed including assembly processes (screwing, insertion, fixing, etc.), components (small/big/flexible parts), robotic equipment (single arm, dual arm, overhead) and station layouts. The result of the *FourByThree* project is presented based on the five safety strategies, including serial elastic actuators, friendly robot design, external monitoring system, adjustable stiffness control and real-time control architecture [12].

After the HRC design is generated, it is essential to develop a secured assessment tool considering direct and physical human–robot contact [13–15]. In fact, ISO/TS 15066 defines the biomechanical limits based on body measurements and values and the mode of operation [16]. It states the specific body areas that could be contacted, and the maximum permissible biomechanical limits associated with the contact. A mathematical model is established associated with the standard. Lacevic et al. [17] introduced the concept of *kinetostatic danger fields* to capture the complete state of the robot, i.e., its configuration and velocity. As the outcome of the mathematical model, danger fields can be a meaningful indicator for the risk in the vicinity of the robot.

After the safety assessment is completed, the next step is to conduct task planning based on the task requirements and safety/danger measurements. Metrics are proposed to standardise HRC and establish a task plan accordingly in [28]. In this work, mobile robots are utilised in five task categories considering HRC and robot autonomy at a high level. Similarly, metrics for HRC safety and dependability are summarised based on multiple categories and severity levels [18]. A zero-injury scenario is the goal of a mature HRC system to guarantee the operator's safety. Sisbot et al. [19] proposed planning tools towards robot paths that are safe, and synthesise being good, socially acceptable and legible.

During the control and manipulation phase, two types of safety strategies are deployed in the work of Carrales et al. [20], including robot stop and move-away actions. Similarly, Mohammed et al. [21] extended the safety strategy towards active collision avoidance. The action strategies include warning, robot stopping, moving away and path modification. Haddadin et al. [22] proposed the framework of detecting human–robot collision and manipulating the robot accordingly.

In the past years, many of the HRC systems have utilised a vision-based approach for safety. Zanchettin et al. [23] used a kinematic control strategy that enforces safety via an optimisation-based real-time algorithm in which safety is regarded as a hard constraint to be satisfied. Morato et al. [24] presented an exteroceptive sensing framework based on multiple *Kinect* sensors to achieve safe human–robot collaboration during assembly tasks. The approach enables the evaluation of the human–robot separation in a 3D Euclidean space, which can be used to generate safe motion goals for the robot.

Alternatively, torque/force-based approaches are also utilised to limit the force potentially applied to the human body. Heinzmann et al. [26] proposed a control scheme for robot manipulators which restricts the torque commands of a position control algorithm to values that comply to pre-set safety restrictions. These safety restrictions limit the potential impact force of the robot in the case of a collision with a human. Similarly, Koveos et al. [27] proposed mathematical methods to increase safety against unexpected, unintentional robot contacts with a human or its environment. The regulation of the interaction force profile is introduced within acceptable/safe boundaries and a prevention method is developed against undesirable after-contact collisions.

Cherubini et al. [25] combined vision- and torque/force-based approaches. The proposed safety module relies on both external wrench and image monitoring. When the safety component is greater than the corresponding one in a threshold vector, a safety stop will be triggered or waypoints are changed.

To recapitulate, in previous research, different safety strategies are proposed at different stages of the HRC system. However, there is still a lack of a structured action strategy during the control phase. Thus, in this research, a feasible safety strategy is proposed considering different scenarios in an HRC cell.

3.4 Safety Strategy for HRC

Based on the features and risk levels identified above, the action strategy for an HRC cell is presented in Table 3.3. Human actors are divided into two categories, i.e., *operators* with training and full knowledge of HRC, and *visitors* who randomly enter the HRC cell. When an operator enters the warning zone (Z1 shown in Fig. 3.2), the robot system sends out warning signals to the operator, which can be a warning sound, light, etc. The operator is notified that he/she has entered the workspace of the HRC cell, and can continue his/her process with caution.

When the operator steps closer to the robot and enters the speed reduction zone (Z2), the robot's moving speed is decreased to limit the potential energy transferred from the robot to the human body. It needs to be noted that during the execution of HRC, contacts with the face, skull, forehead, eyes, ears and larynx areas are not permissible. Therefore, even though the operator is working in the speed reduction

Table 3.3 Action strategy in HRC

	Robot action		Human action	
	If operator enters	If visitor enters	Operator	Visitor
Warning zone (Z1)	Warning signal	Warning signal	Enter with awareness	Stay with caution
Speed reduction zone (Z2)	Speed reduction	Stop	Continue individual process	Not allowed, leave safely
Speed reduction zone—the head area (Z3)	Stop	Overridden by hard stop, manual restart needed	Continue individual process	Not allowed, leave safely
Collaborative zone (Z4)	Collaborative task	Overridden by hard stop, manual restart needed	Collaborative task	Not allowed, leave safely
Stop zone (Z5)	Stop	Overridden by hard stop, manual restart needed	Continue individual process	Not allowed, leave safely

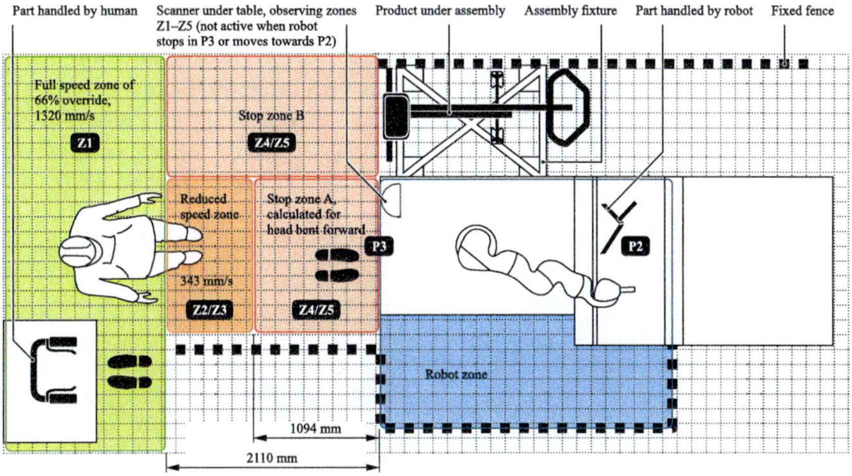

Fig. 3.2 Dynamic layout of working zones

zone, the robot needs to be stopped when the end effector is close to the head area (Z3).

In the collaborative zone (Z4), the operator works together with the robot under the monitoring and supervision of the HRC system. In most cases, a collaborative zone can be considered part of the speed reduction zone with more collaborative configurations and technical support.

In some applications, specific areas are not collaborative due to the characteristics of the product or tools. Thus, when the operator enters the stop zone (Z5), the robot movement is paused until the operator leaves the stop area.

As mentioned above, the other type of human in an HRC cell is a visitor. The role of the human is identified via the user authentication method, e.g., ID card, face recognition, password, or similar. When a visitor enters the warning zone, the system sends a signal to inform him/her that it is not allowed to move forward.

If the visitor enters the speed reduction zone, the robot stops movements until the visitor leaves the area. If the visitor entered the collaborative zone or sensitive area with the risk of collision on the head, a hard stop overrides the whole system and stops the robot completely. Compared with the normal stop zone, the robot can only be restarted by the operator after he/she guarantees that there is no visitor in the area.

Besides, the configuration of the working zone can be defined in different ways. In a typical HRC cell, the safety distance and dimension of the zone are defined based on the physical characteristics of the robot, tool, object and task. The proposed HRC system layout is shown in Fig. 3.3. The industrial robot is placed in the robot operation zone on the right, and scanners are deployed at the boundary between the human and the robot to detect the position and speed of the human operator. Stop and reduced speed zones are configured during the layout design of the HRC cell, considering the task of both the human and the robot. Conventionally, the size of different zones is a static environment without changes during operation. However, in a symbiotic HRC cell, the robot speed and task are dynamically changed and monitored by the system. Consequently, the size of the zones can be dynamically modified as well. For instance, when the moving speed of the robot decreases, the potential energy

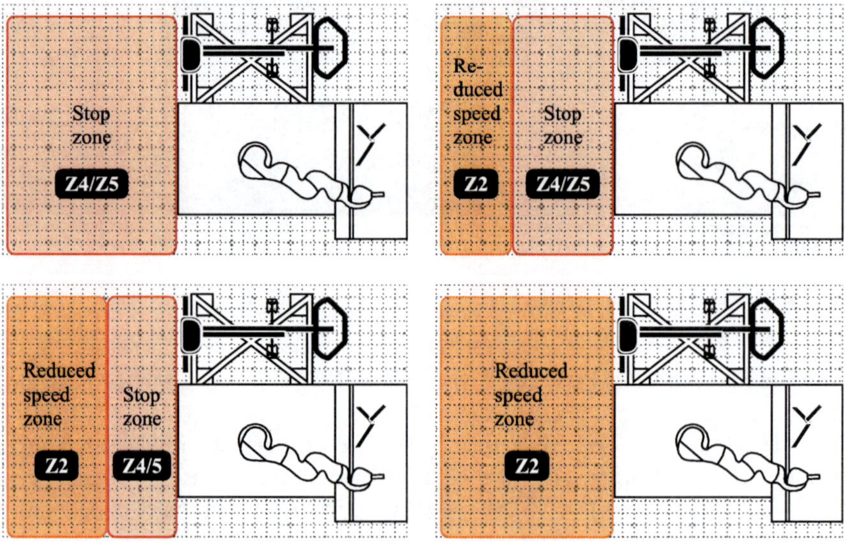

Fig. 3.3 Safety layout and dynamic work zone strategy

transferred to the human is lower. Hence, the stop zone can be smaller as well to guarantee the same level of safety while allowing the operator to work closer to the robot.

3.5 Personalised Safety Framework for HRC

As mentioned before, the HRC system must guarantee the safety of operators, especially considering the physiological characteristics. A mathematical model is useful to limit the speed and force of the robot, and eventually to avoid potential injuries to the human operator. In this research, the restrictions of ISO/TS 15066 in biomechanical limits are applied to the HRC system [16]. First, the known area (A) of contact between the robot and the body region is introduced in cm^2. The robot may be holding a reasonable small tool when the operator and the robot are working in a shared working area. The effective payload of the robot system m_L including tooling and workpiece in kg and the total mass of the moving parts of the robot M in kg are the inputs which are used for the calculation of effective robot mass m_R considering Eq. (3.1):

$$m_R = \frac{M}{2} + m_L. \tag{3.1}$$

Next, the μ factor is calculated by Eq. (3.2):

$$\mu = \left(\frac{1}{m_R} + \frac{1}{m_H}\right)^{-1}, \tag{3.2}$$

where m_H is the segment's effective mass calculated by Eq. (3.3):

$$m_H = B_0 + B_1 m_h + B_2 h, \tag{3.3}$$

where m_h and h are the human's mass (kg) and height (cm), respectively. Regarding the B_0, B_1 and B_2 values, Zatsiorsky et al. [29] presented the coefficients of multiple regression equations for estimating the inertia properties of the male body segments from the known body mass and height of the person—this is also followed in research presented in this chapter. The result is the maximum transferred energy E (in J) as the result of the formula (divided by 1000, in order to take the values in joule).

$$E = \frac{F_{\max}^2}{2k} = \frac{A^2 p_{\max}^2}{2k}. \tag{3.4}$$

The second desirable result is the maximum permissive speed limit $v_{\mathrm{rel,max}}$ (in mm/s) obtained by Eq. (3.5):

$$v_{\text{rel,max}} = \frac{p_{\max} A}{\sqrt{\mu k}}, \tag{3.5}$$

where k is the effective spring constant for a specific body region, while p is the maximum permissible pressure value. Note that those types are used in case of a transient contact. Eventually, the energy in this model is expressed as in Eq. (3.6):

$$E = \frac{1}{2}\mu v_{\text{rel}}^2, \tag{3.6}$$

where v_{rel} is the relative speed between the robot and the human body region; μ is the reduced mass of the two-body system.

After the mathematical models are established, the system architecture of the personalised safety solution is presented in Fig. 3.4. First, the operator scans the 2D QR code with a camera directly linked to the User Interface, which can be hosted on either a local PC as a software tool or mobile phone/tablet as an app. The Human–Machine Interface (HMI) then extracts the users' physiological characteristic data (e.g., weight, height and safety notes) from the database. In parallel, the safety-related operation data is fetched from the operation database (e.g., robot model, payload and possible contact area). The Safety Module then combines the specific features of the operator and the operation, and processes them based on the mathematical model outlined above. It needs to be mentioned that the communication and transfer of

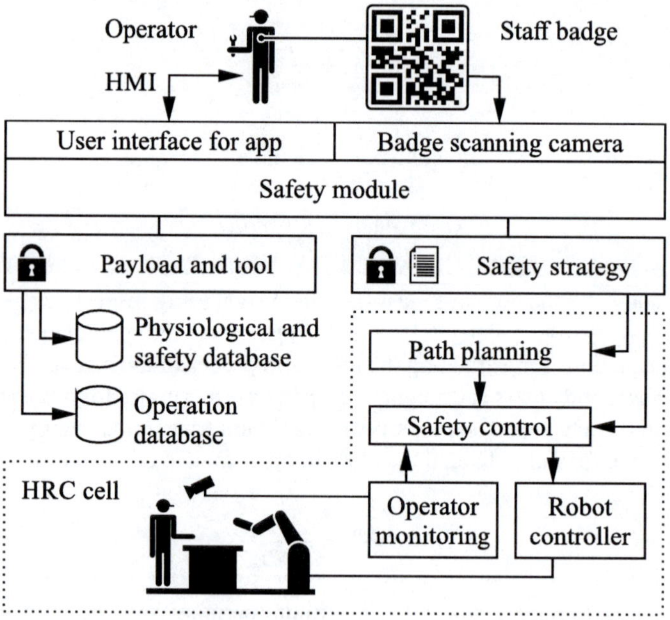

Fig. 3.4 Personalised safety framework for HRC

the physiological and safety data are protected by encrypted network communication, shielding personal and private information of the operator from exposure and unauthorised access.

Eventually, the safety strategy is generated with speed limits for different areas of contact over the human body. Then, the safety strategy is sent to the path-planning module which generates the static robot trajectory, and the safety control module which executes and dynamically modifies the robot trajectory. When the operator starts to work with the robot collaboratively, the monitoring module captures the operator's movement in the robot cell in real time, and communicates with the safety control module. Then, the result is translated to low-level control commands for execution by the robot controller. As the safe energy and speed limits have been included in the control loop already, the operator will not be injured even if he/she comes in physical contact with the robot.

3.6 Implementation

Based on the safety strategy and framework described above, an actual HRC system is established to validate the feasibility of the proposed approach. A KUKA LBR iiwa is utilised as the testing industrial robot, while laser scanners are utilised to detect the distance between the human and the robot. The layout of the HRC cell is outlined in Figs. 3.2 and 3.3. Due to the nature of the assembly process, Z2 and Z3 are merged in the same area, as are Z4 and Z5. The case study is built based on the assembly of a product on top of the figure, and the robot is mounted in the middle area. There is open space to the left side and a human can enter from this direction. In the visitor mode, a stop zone is applied beside the robot. When the visitor enters the area, a hard stop is triggered which needs system re-initialisation by an authorised operator or system administrator.

As mentioned above, the dynamic work zone mechanism is also implemented during the case study. Corresponding to the original speed of the robot, the zone configuration is shown in Fig. 3.2. When the robot slows down, the kinetic energy potentially transferred to the human operator is lower, which means the risk of human injury is lower, too. Thus, the size of the speed reduction zone is increased and that of the stop zone decreased. As a result, the human operator can work in a bigger area compared with the original layout configuration, and the stop zone is smaller. As a result, the zone configuration shown in Fig. 3.2 is changed to that in Fig. 3.3. In full collaborative mode, the whole working envelope is the speed reduction zone in which the position of the operator is monitored and diagnosed. The stop zone is unnecessary in this case, which is opposite to the visitor mode.

In this research, the personalised safety strategy is also implemented during the case study. As shown in Fig. 3.5, the human needs to scan his/her ID badge before entering the HRC cell. The QR code printed on the badge is then interpreted and linked to the user database to first identify the role of the user, and also extract basic biometric data, i.e., height and weight. It needs to be noted that the storage of personal

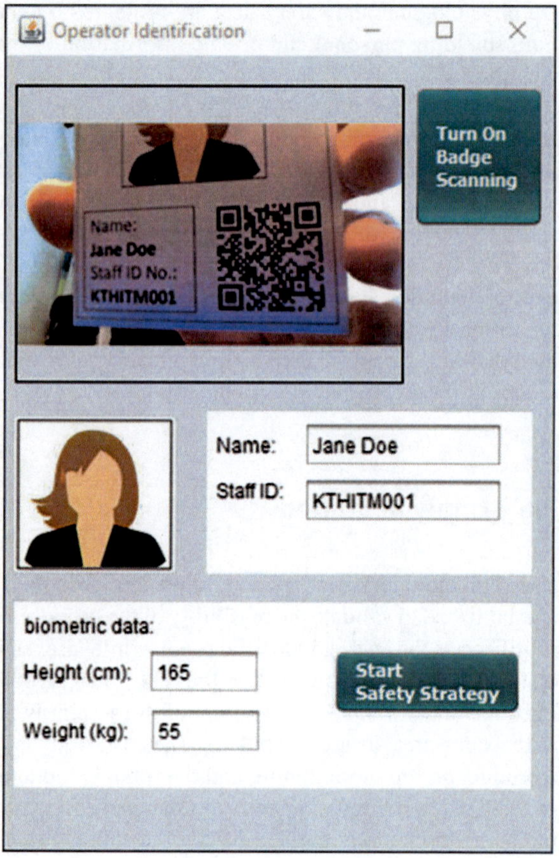

Fig. 3.5 Operator identification module

data might be against the policy of privacy of employees. Therefore, an agreement to collect and use personal information is needed before the user data is established.

After the human's role is identified, the operator's biometric data is transferred to the personalised speed strategy module. In this module, the effective mass of different areas is estimated based on the weight and height of the operator. The body area categorisation is simplified to face, chest and leg areas for the demonstration. The contact area, payload and robot mass information are stored in the associated process data. Combined with biometric data, the maximum energy that is allowed to be transferred to the human is calculated, so as to determine the maximum speed allowed in different body areas. The speed limit is then transferred to the robot controller as part of the HRC configurations, which is applied in the reduced speed zone (Z2).

The proposed safety system is implemented in the HRC Lab at KTH, Sweden. The safety system is developed based on Java, as it is a platform-independent language that can work in multiple operating systems and devices. After the operator scans the QR code printed on a standard staff badge, the system automatically identifies the unique user ID and locates the physiological characteristic data stored in the database as shown in Fig. 3.5. Here, the user can check and confirm the validity of the physical information, in this case, the operator's height and weight. If there is any change in the operator's physical condition, he/she can modify and update the data accordingly and save it in the database.

The safety strategy planning is initiated by clicking on the appropriate button. As illustrated in Fig. 3.6, the operator's data is transmitted to the system together with the operation information. In the personalised safety strategy, the most critical factor of the operation is the contact area between the human and the robot. In this research, the contact area is identified as the sharpest area on the robot to guarantee the safety of the operator, which is the tooling area held by the gripper. Thus, those parameters are then processed in the system and the results are presented on the interface. In the implementation, the body area is simplified into three main areas, i.e., face (top to the neck), chest (neck to waist) and lower body (waist to feet). It needs to be noted that it is possible to define more detailed body areas and identify related energy and speed limit. Only three main areas are shown in the case study to simplify the presentation of the proposed methods.

Comparative results from the system output are presented in Fig. 3.7. As previously discussed, to protect the safety of the human operator, the key is to limit the energy transferred from the robot to the operator. In practice, it is normal that different operators are working on the same operation. For example, operators work in the same spot on different shifts around the clock, or multiple operators work in parallel production lines at the same time. Thus, the safety strategy should be personalised as the physical nature of the operators—for instance, weight—is diverse. In Scenario 1, the operation task in the HRC is the same, and the operators are different. Naturally the weights of those three operators are diverse, i.e., 55, 77 and 90 kg, respectively. As a result, the speed limits for different body areas are calculated, shown in the system, and sent to the robot control system later. Naturally, the speed is lower when the robot is moving around the face area (6.16–6.57 mm/s) compared to the chest

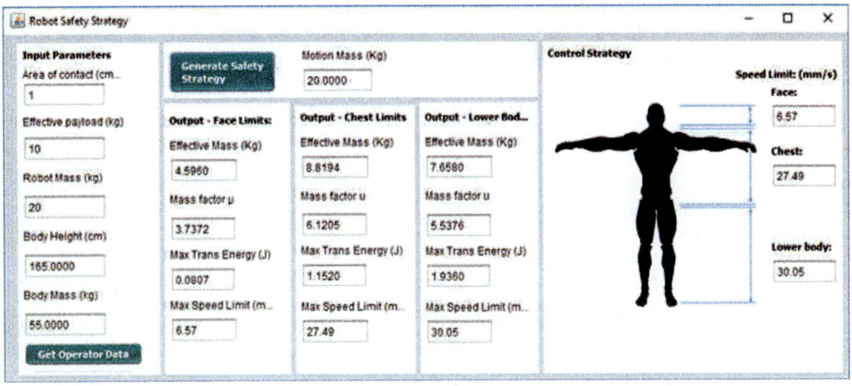

Fig. 3.6 Operator identification module (**a**), and Safety strategy module (**b**)

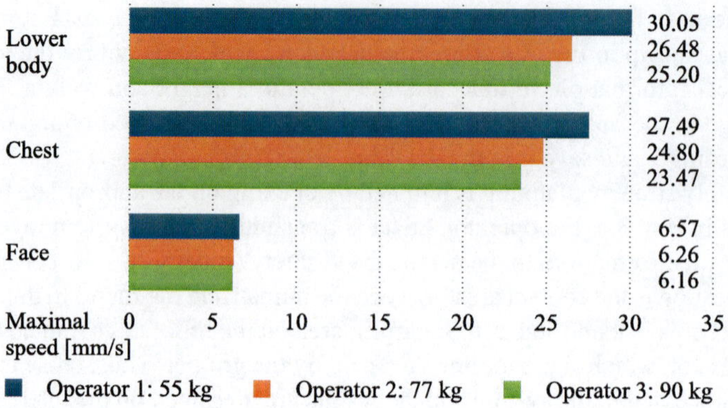

Fig. 3.7 Scenario 1—same operation and different operators

(23.47–27.49 mm/s) and lower body area (25.20–30.05 mm/s), as the face is more vulnerable than other areas. In parallel, it needs to be pointed out that the higher weight the operator has, the lower speed the robot should be limited to. It might deviate from the inherent thinking that the operator of a bigger size might be able to endure more physical impact than others. In fact, the result of safety control is the opposite because the heavier the operator is, the more energy will be transferred between the human and the robot under the same speed, which leads to more severe injury to the operator. Thus, to guarantee the safety of the operator, the speed limit in practice should be lower when an operator in the HRC cell is heavier. For instance, operator 1 (55 kg) is lighter than operator 3 (90 kg). As a result, the speed limits in the lower body area are 30.05 mm/s and 25.20 mm/s for operators 1 and 2, respectively.

Scenario 2 aims to evaluate the system result based on the same operator under different operation conditions. In this case, the most critical factors are the contact area and payload. In Fig. 3.8, the results show the impact of the contact area on transferred energy and speed. In this case, it is assumed that other parameters of operations 1 and 2 are the same, and the tool held by the robot gripper in operation 2 is sharper than that in operation 1. This means that the contact area between the human operator and robot is reduced by 50%. As a result, speed limits in all body areas decrease by 50%, and energy allowed to be transferred decreases by 75%.

3.7 Discussion and Conclusion

In the next generation of the manufacturing industry, the human still plays a leading role in the system. Despite the significant development of smart technologies and artificial intelligence, the human is still in the centre of manufacturing. Therefore, it remains essential to continuously improve the working environment and safety of humans. HRC offers the possibility of assisting human operators via industrial robots' accuracy, repeatability and strength. Thus, it is even more important than

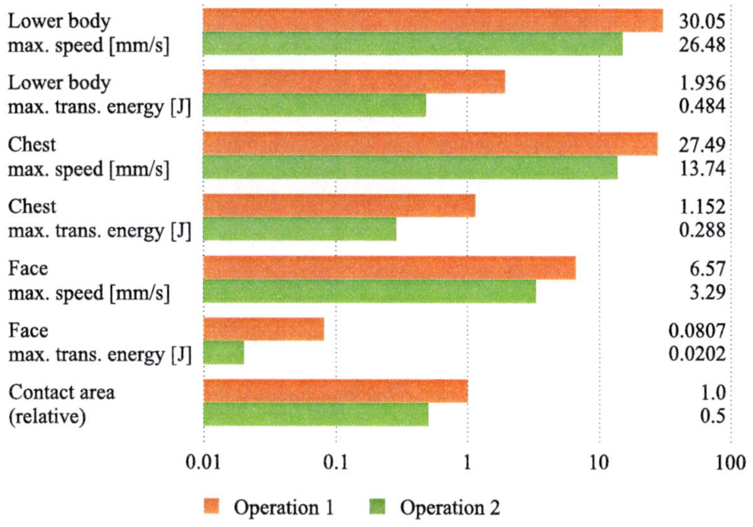

Fig. 3.8 Scenario 2—same operator and different operations

ever to properly design and implement a safe HRC system. Previous HRC safety research is established based on the assumption that all human operators are identical. Consequently, the diversity of human bodies is not reflected in the planning or control methods in the HRC cell. From the perspective of Industry 4.0, the collaborative HRC robot cell should guarantee the safety of the human operator, and meet the individual needs of the operators adaptively. Hence, in this research, the physiological difference of human bodies is included in the safety system to generate personalised safety for the operator.

The scientific contribution of the proposed work includes the clear classifications of human–robot relationships, i.e., Human–Robot Coexistence, Interaction, Cooperation and Collaboration. Moreover, a novel approach is proposed to include the operator's physiological characteristics in the safety strategy. A dynamic HRC layout approach is proposed based on the real-time speed and distance of the human and the robot. A mathematical and automatic approach is developed to accordingly limit the robot control system based on the aforementioned method. The proposed approach is then implemented and validated through a prototype system and case studies. The feasibility of the proposed work is proven and the quantifiable results among different operators and operations are discussed.

In the future, the proposed work can be directly integrated with high-level robot control methods, such as IEC 61499 function blocks. It will work as the threshold of a safety shell to ensure the safety of the HRC cell. Also, more detailed human body areas can be defined to realise more accurate safety planning for the specific operator. With the help of the monitoring methods, e.g., vision system, it is possible to identify the location and gesture of the operator, and then dynamically combine the result with the proposed safety strategy. Meanwhile, it needs to be noted that privacy and ethical considerations need to be included in future HRC research, as more biometric data are utilised in a collaborative robot cell.

Acknowledgements The authors would like to thank Carolin Schaffert and Argyri Seira for their contributions to the system implementation and programming tasks.

References

1. J. Schmidtler, V. Knott, C. Hölzel, K. Bengler, Human centered assistance applications for the working environment of the future. Occup. Ergon. **12**(3), 83–95 (2015)
2. N. Van Oosterwyck, Real Time human–robot interactions and speed control of a robotic arm for collaborative operations (2018)
3. L. Wang, R. Gao, J. Váncza, J. Krüger, X.V. Wang, S. Makris, G. Chryssolouris, Symbiotic human–robot collaborative assembly. CIRP Ann. Manuf. Technol. **68**(2), 701–726 (2019). https://doi.org/10.1016/j.cirp.2019.05.002
4. J. Váncza, L. Monostori, D. Lutters, S.R. Kumara, M. Tseng, P. Valckenaers, H. Van Brussel, Cooperative and responsive manufacturing enterprises. CIRP Ann. Manuf. Technol. **60**(2), 797–820 (2011)
5. J. Lambrecht, M. Kleinsorge, J. Krüger, Markerless gesture-based motion control and programming of industrial robots, in *2011 IEEE 16th Conference on Emerging Technologies & Factory Automation (ETFA)* (2011), pp. 1–4
6. S. Robla-Gómez, V.M. Becerra, J.R. Llata, E. Gonzalez-Sarabia, C. Torre-Ferrero, J. Perez-Oria, Working together: a review on safe human–robot collaboration in industrial environments. IEEE Access **5**, 26754–26773 (2017)
7. K. Ikuta, H. Ishii, M. Nokata, Safety evaluation method of design and control for human-care robots. Int. J. Rob. Res. **22**(5), 281–297 (2003)
8. P.A. Lasota, T. Fong, J.A. Shah, A survey of methods for safe human–robot interaction, Found. Trends® Robot. **5**(4), 261–349 (2017)
9. J.A. Marvel, J. Falco, I. Marstio, Characterizing task-based human–robot collaboration safety in manufacturing, IEEE Trans. Syst. Man Cybern. Syst. **45**(2), 260–275 (2015)
10. J.T.C. Tan, F. Duan, R. Kato, T. Arai, Safety strategy for human–robot collaboration: design and development in cellular manufacturing. Adv. Robot. **24**(5–6), 839–860 (2010)
11. G. Michalos, S. Makris, P. Tsarouchi, T. Guasch, D. Kontovrakis, G. Chryssolouris, Design considerations for safe human–robot collaborative workplaces. Procedia CIRP **37**, 248–253 (2015)
12. I. Maurtua, A. Ibarguren, J. Kildal, L. Susperregi, B. Sierra, Human–robot collaboration in industrial applications: safety, interaction and trust. Int. J. Adv. Robot. Syst. **14**(4), 1729881417716010 (2017)
13. D. Kulić, E. Croft, Pre-collision safety strategies for human–robot interaction. Auton. Robot **22**(2), 149–164 (2007)
14. D. Kulić, E.A. Croft, Real-time safety for human–robot interaction. Robot. Auton. Syst. **54**(1), 1–12 (2006)
15. D. Kulić, E.A. Croft, Safe planning for human–robot interaction. J. Robot. Syst. **22**(7), 383–396 (2005)
16. ISO, ISO/TS 15066 robotics and robotic devices—collaborative robots (2016)
17. B. Lacevic, P. Rocco, Kinetostatic danger field—a novel safety assessment for human–robot interaction, in *2010 IEEE/RSJ International Conference on Intelligent Robots and Systems (IROS)* (2010), pp. 2169–2174
18. A. De Santis, B. Siciliano, A. De Luca, A. Bicchi, An atlas of physical human–robot interaction. Mech. Mach. Theory **43**(3), 253–270 (2008). https://doi.org/10.1016/j.mechmachtheory.2007.03.003

19. E.A. Sisbot, L.F. Marin-Urias, R. Alami, T. Simeon, A human aware mobile robot motion planner. IEEE Trans. Robot. **23**(5), 874–883 (2007)
20. J.A. Corrales, G.J.G. Gomez, F. Torres, V. Perdereau, Cooperative tasks between humans and robots in industrial environments. Int. J. Adv. Robot. Syst. **9**(3), 94 (2012)
21. A. Mohammed, B. Schmidt, L. Wang, Active collision avoidance for human–robot collaboration driven by vision sensors. Int. J. Comput. Integr. Manuf. **30**(9), 970–980 (2017). https://doi.org/10.1080/0951192X.2016.1268269
22. S. Haddadin, A. Albu-Schäffer, A. De Luca, G. Hirzinger, Collision detection and reaction: a contribution to safe physical human–robot interaction, in. IEEE/RSJ Int Conf Intell Robot Syst **2008**, 3356–3363 (2008). https://doi.org/10.1109/IROS.2008.4650764
23. A.M. Zanchettin, N.M. Ceriani, P. Rocco, H. Ding, B. Matthias, Safety in human–robot collaborative manufacturing environments: metrics and control. IEEE Trans. Autom. Sci. Eng. **13**(2), 882–893 (2015)
24. C. Morato, K.N. Kaipa, B. Zhao, S.K. Gupta, Toward safe human robot collaboration by using multiple kinects based real-time human tracking. J. Comput. Inf. Sci. Eng. **14**(1), 11006 (2014)
25. A. Cherubini, R. Passama, A. Crosnier, A. Lasnier, P. Fraisse, Collaborative manufacturing with physical human–robot interaction. Robot. Comput. Integr. Manuf. **40**, 1–13 (2016). https://doi.org/10.1016/j.rcim.2015.12.007
26. J. Heinzmann, A. Zelinsky, Quantitative safety guarantees for physical human–robot interaction. Int. J. Rob. Res. **22**(7–8), 479–504 (2003). https://doi.org/10.1177/02783649030227004
27. Y. Koveos, G.A. Rovithakis, I. Agriomallos, S. Doltsinis, A minimum invasive strategy to guarantee safety under unexpected unintentional human–robot contact, in *2017 25th Mediterranean Conference on Control and Automation (MED)* (2017), pp. 334–339
28. A. Steinfeld, T. Fong, D. Kaber, M. Lewis, J. Scholtz, A. Schultz, M. Goodrich, Common metrics for human–robot interaction, in *Proceedings of the 1st ACM SIGCHI/SIGART Conference on Human–Robot Interaction* (2006), pp. 33–40
29. V. Zatsiorsky, The mass and inertia characteristics of the main segments of the human body, in *Biomechanics* (1983), pp. 1152–1159

Part II
Human Safety In HRC

Chapter 4
Real-Time Collision Detection and Collision Avoidance

Bernard Schmidt

Abstract In today's manufacturing environment, safe human–robot collaboration is of paramount importance for improving efficiency and flexibility. Targeting safety in human–robot collaboration, this chapter reports an approach for real-time collision detection and collision avoidance in an augmented virtual environment where a virtual model of a robot and real images of a human captured by depth cameras are superimposed for monitoring and collision detection. The chapter presents a distributed system which is linked to an industrial robot in a collaborative assembly cell. Four safety strategies are described: the system can alert an operator, stop a robot, move the robot away, or modify the robot's trajectory away from an approaching operator. These strategies can be activated based on configured zones and the operator's location with respect to the robot. The method developed is tested in realistic applications of collaboration between robots and humans in an assembly cell. Finally, the performance of the implemented approach is analysed.

4.1 Introduction

In a human–robot collaborative environment, when humans and robots co-exist, the safety of human operators in real-time is paramount. The challenge is not only passive collision detection in real time but also active collision avoidance along the robot trajectory by monitoring operators and controlling the robot.

Several approaches have been reported during the past years for human–robot collaboration. Bussy et al. [1], Agravante et al. [2], and Monje et al. [3] introduced a control system for a humanoid robot to carry out a joint operation with an operator. Takata and Hirano [4] proposed a planning algorithm that can adaptively allocate humans and robots in a hybrid assembly system. Pellegrinelli et al. [5] presented a framework for integrated task and motion planning capable of coordinating human

B. Schmidt (✉)
School of Engineering Science, University of Skövde, Skövde, Sweden
e-mail: bernard.schmidt@his.se

L. Wang et al. (eds.), *Advanced Human–Robot Collaboration in Manufacturing*,
https://doi.org/10.1007/978-3-030-69178-3_4

and robot tasks. Chen et al. [6] presented a simulation-based multi-objective optimisation process for allocating the assembly subtasks to both human and robot. Krüger et al. [7] summarised the advantages of human–robot collaboration in assembly lines.

A hybrid assembly system exhibits both the efficiency of robots and the flexibility of humans. Nevertheless, it may induce extra stress to human operators if the assembly lines are improperly designed. Arai et al. [8] thus assessed operator stress from the perspective of distance and difference of speed between a robot and an operator, aiming to guide the design of a productive hybrid assembly system. Hoffman [9] discusses subjective and objective metrics to evaluate human–robot collaborative fluency. Wang et al. [10] proposed a systematic approach to constructing a symbiotic human–robot collaborative environment.

Regarding safety monitoring and protection in a human–robot collaborative manufacturing environment, two approaches were popularly used, i.e., vision-based approaches such as 3D surveillance [11] via motion, colour and texture analysis, and inertial-sensor-based approaches [12] via a motion-capture suit. From a practical point of view, the latter may not be feasible for real-world applications due to its dependence on a special suit with sensors and the limitation of motion capture of the suit wearer only, with the surroundings remaining unmonitored. This may lead to a safety leak, e.g., a mobile object may hit a stationary operator.

In the area of vision-based approaches, efficient collision detection has been the focus for many years. Ebert et al. [13] proposed an emergency-stop approach to avoid a collision using an ad-hoc developed vision chip, whereas Gecks and Henrich [14] adopted a multi-camera system for detecting obstacles. Vogel et al. [15] presented a projector-camera-based approach for dynamic safety zone boundary projection and violation detection. Tan and Arai [16] applied a triple stereovision system for tracking the movement of a sitting operator wearing a suit with colour markers. However, mobile operators that might appear in a monitored area may not show consistent colour or texture. Moreover, changes in illumination may also affect the overall effectiveness of the collision detection process. Instead of markers, a range of imaging can be applied, which is more robust to changes in colour and illumination. Schiavi et al. [17] used a time-of-flight (ToF) camera in an implementation of a collision detection system. Fischer and Henrich [18] presented an approach using multiple 3D depth images for collision detection. For range imaging, laser scanners provide high resolution but with relatively low processing speed, because each point or line of the scanned scene is sampled separately. Comparatively, ToF cameras allow acquiring depth images with high speed, but with an insufficient level of pixel resolution (up to 200×200) and with relatively high cost. Lately, Rybski et al. [19] combined data from different depth sensors to construct a volumetric evidence grid for locating foreign objects and collision detection. Flacco et al. [20] presented an integrated framework for collision avoidance with the use of a single Kinect depth sensor and depth space. Morato et al. [21] presented an approach that uses multiple Kinect skeletons with application to safe human–robot collaboration. Verification has been performed in a case study with a small laboratory 5-DOF robot.

One of the common choices from available commercial safety protection systems is SafetyEYE® [22]. It calculates 3D data about a surveilled space from three cameras embedded into a sensing device and checks for intrusions into predefined safety zones. Entry into a safety zone may limit the movement speed or trigger an emergency stop of a robot in the zone. During active surveillance, the safety zones are static and cannot be changed.

More details on human–robot collaboration can be found in [23].

For the purpose of close human–robot collaboration without sacrificing its productivity, there is a need to develop a cost-efficient yet real-time safety solution suitable for a dynamic manufacturing environment where human operators co-exist with robots. Targeting the need, this chapter presents an approach for human safety protection. Its novelty includes (1) timely collision detection between depth images and 3D models in an augmented virtual environment; and (2) active collision avoidance via path modification and robot control in real time with zero robot programming.

4.2 Depth-Sensor-Based Active Collision Avoidance

For the collision detection purpose in the *Collision Avoidance System* (CAS) presented in this chapter, 3D models linked to depth sensors are used to represent a structured manufacturing environment. A cost-efficient depth camera is then chosen for surveillance of any other foreign mobile objects, including operators, which are not present in the virtual 3D models.

The concept is depicted in Fig. 4.1, in the simplified human–robot co-existing environment. The current status of the robot is obtained from the robot controller and shown via a virtual robot 3D model, whereas the appearance of the human operator is captured by the Kinect sensors via a set of depth images. The presented approach is an improvement and extension of previous research [24–26] where a new model of Kinect sensor is utilised and a distributed processing is applied.

The depth sensors selected in this system are *Microsoft Kinect v2* devices. The sensors have a spatial resolution of 512×424 pixels and a field of view of $70° \times 60°$. Kinect v2 is a multi-frequency time-of-flight ranging sensor with three amplitude-modulated infrared (IR) signals of 16, 80, and 120 MHz, respectively [27]. Using the open-source *libfreenect2* driver [28] that implements depth decoding based on [29], the sensor can operate without depth ambiguity in a range of up to 18.75 m. The depth images can be streamed and processed with a frame rate of 30 Hz. The human skeleton that can be obtained from the Kinect sensor with the use of *Microsoft Kinect SDK* is not used in this work as it limits the refresh rate to 10 Hz.

While *Microsoft Kinect for Windows Runtime* supports only a single sensor connected to the PC, libfreenect2 has been reported to work for up to 5 devices on a high-end PC using multiple separate PCI Express USB3 expansion cards [28]. However, in the presented research, each depth sensor has a dedicated host PC that performs image acquisition and processing, and constitutes a Collision Avoidance System (CAS) client.

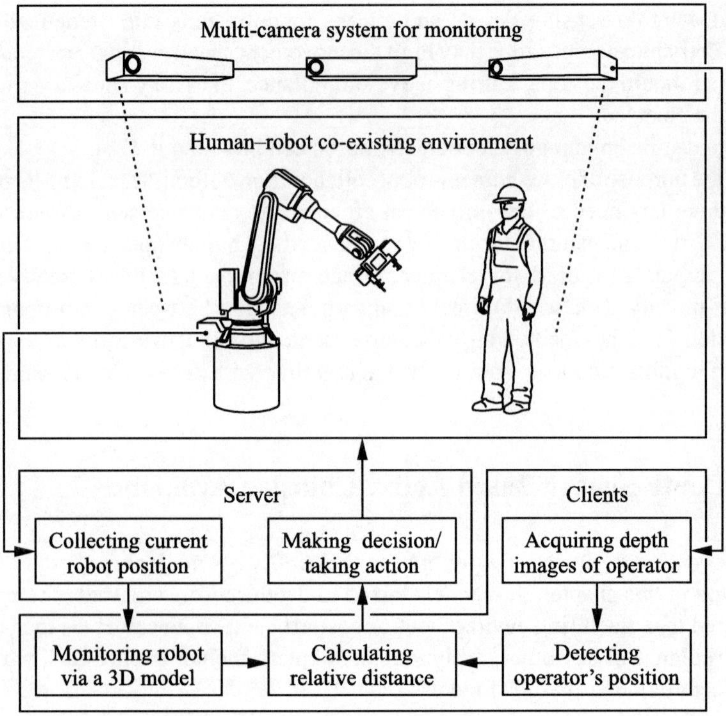

Fig. 4.1 Concept of collision detection in an augmented environment (modified from [25])

4.2.1 Calibration

The initial step in the approach developed is the calibration of all sensors used. The method applied in the process is adopted from [30] and briefly described in the following subsections.

For optics calibration of IR camera, the open-source computer vision library *OpenCV* is utilised. Several raw infrared images of known chessboard targets are chosen for obtaining intrinsic parameters and distortion coefficients, whereas the pin-hole camera model defined in Eqs. (4.1)–(4.4) is used for camera calibration and 3D reconstruction.

$$\begin{aligned} x' &= x/z \\ y' &= y/z \end{aligned} \tag{4.1}$$

$$\begin{aligned} x'' &= x' \frac{1+k_1r^2+k_2r^4+k_3r^6}{1+k_4r^2+k_5r^4+k_6r^6} + 2p_1x'y' + p_2\left(r^2 + 2x'^2\right) \\ y'' &= y \frac{1+k_1r^2+k_2r^4+k_3r^6}{1+k_4r^2+k_5r^4+k_6r^6} + p_1\left(r^2 + 2y'^2\right) + 2p_2x'y' \end{aligned} \tag{4.2}$$

$$r^2 = x'^2 + y'^2 \tag{4.3}$$

$$\begin{aligned} u &= f_x x'' + c_x \\ v &= f_y y'' + c_y, \end{aligned} \tag{4.4}$$

where (x, y, z) are coordinates in camera space, distortion coefficients $k_1 - k_6$ for radial distortion and p_1, p_2 for tangential distortion, f_x, f_y are focal lengths, (c_x, c_y) is the principal point, and (u, v) are depth image coordinates.

During the calibration process, the intrinsic parameters f_x, f_y, c_x, c_y and the distortion coefficients $k_1 - k_3$, p_1, p_2 for the IR camera can be obtained. This enables the system to convert the captured depth image to a 3D point cloud in the camera's coordinate system.

The calibration results are presented in Table 4.1. Calibration parameters are stored in YAML configuration files, for each sensor separately in a file identified with an ID of the corresponding Kinect sensor.

The pose, i.e., location and orientation, of each sensor being placed in a robotic cell needs to be obtained. Being external to a sensor, they are referred to as *extrinsic parameters*. In the calibration process based on [30], a pattern (calibration object) with a known pose with respect to the reference coordinate frame is used. The relationship between transformation matrices shown in Fig. 4.2 can be expressed in Eq. (4.5):

$$\boldsymbol{T}_O^W = \boldsymbol{T}_C^W \boldsymbol{T}_O^C, \tag{4.5}$$

where $\boldsymbol{T}_O^W$ is the transformation matrix between the calibration object coordinate frame and the world reference frame, $\boldsymbol{T}_C^W$ the transformation matrix between the camera frame and the word frame, and $\boldsymbol{T}_O^C$ the transformation matrix between the calibration object frame and the camera frame which can be resolved from a captured image. As the calibration pattern is placed in the robot working range, the work object calibration method can be used to obtain $\boldsymbol{T}_O^W$.

4.2.2 *Processing*

Overview of procedures of depth image acquisition and processing is illustrated in Fig. 4.3. Efficient processing is enabled by removing background reference obtained during the calibration process. Such a background reference image is obtained for each of the depth sensors. During the background calibration process, the robot performs some movements, and the background behind the robot can then be registered. Background calibration can also be performed with the robot positioned in an area that does not need to be monitored by the system.

After the background removal, only the foreign objects remain in the depth images, as shown in the second row of the right column in Fig. 4.3. Objects corresponding to the monitored moving robot are also removed from the depth images by subtraction

Table 4.1 IR Camera calibrated parameters

	Focal length		Image centre		Distortion coefficients				
Kinect ID	f_x	f_y	c_x	c_y	$k_1[10^{-1}]$	$k_2\ [10^{-1}]$	$k_3\ [10^{-1}]$	$p_1\ [10^{-3}]$	$p_2\ [10^{-3}]$
…764,647	359.3	361.4	254.5	216.0	0.9126	−2.998	1.325	4.976	−1.077
…741,942	362.5	362.5	257.8	205.3	1.006	−3.147	1.350	−2.372	1.642
…340,947	367.5	367.7	257.2	206.7	0.9888	−2.091	−0.6718	−1.350	−0.4781
…562,147	359.5	359.3	258.3	214.4	0.8318	−2.731	1.022	3.758	−0.3945

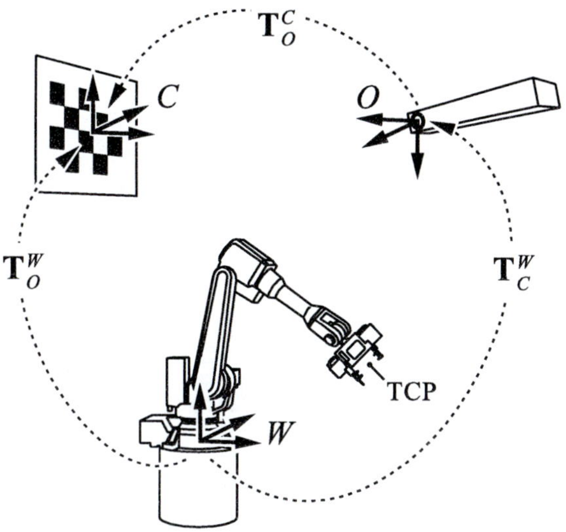

Fig. 4.2 Extrinsic parameter calibration

of a virtual depth image of the robot model. The third-row image reveals a mobile operator as the subject of interest after applying a noise removal filter and a connected-component algorithm.

To remove captured points belonging to the robot, the depth image is compared with a virtual depth image of a simplified robot model. To speed up calculations, the GPU rendering capabilities of OpenGL API are used. A simplified model (see Fig. 4.4) is built from bounding volumes (blocks and cylinders) to speed up the rendering process as well as to take into account calibration error and flexible components or components not included in available CAD models attached to the robot. To model a virtual depth camera, intrinsic and extrinsic parameters of the Kinect v2 sensor obtained during calibration are applied.

The depth image of detected mobile objects is converted to a point cloud in the robot coordinate system. Finally, the 3D point cloud of the objects is superimposed onto the 3D model of the robot in augmented virtuality, where the minimum distance between the point cloud and the 3D robot model is calculated. The bottom image in Fig. 4.3 demonstrates the result of the processing of the depth images.

4.2.3 Collision Detection

Although the size of the depth images is significantly reduced after background removal and filtering, the point cloud of mobile objects could still consist of a large amount of data. To speed up the computation of collision detection, minimum bounding boxes are applied to the 3D point cloud. In the current implementation, the

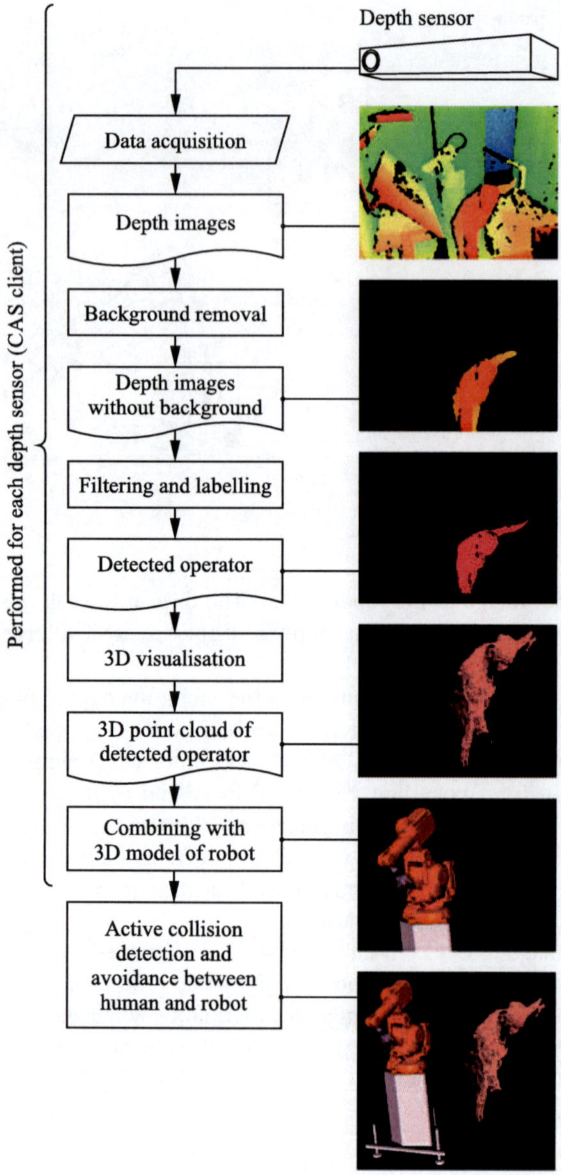

Fig. 4.3 Overview of depth image processing (Adapted from [26])

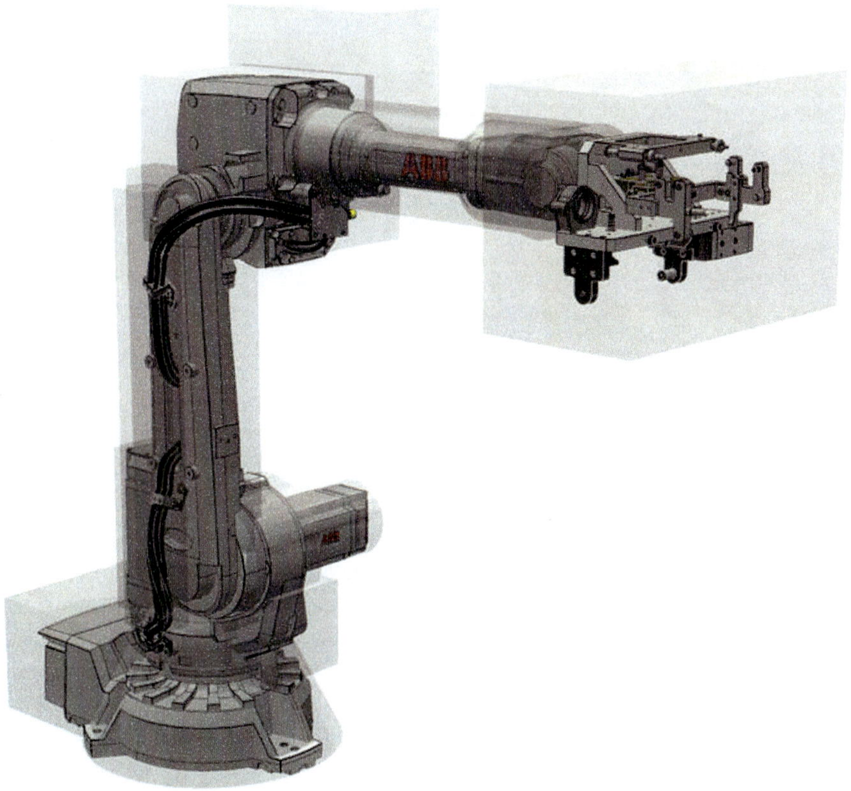

Fig. 4.4 Bounding volumes used to filter a point cloud from points that belong to the robot

axis-aligned bounding box is chosen for faster computation and easier representation by only two opposite corners. Figure 4.5 depicts the point cloud with bounding boxes at different granularity. The actual level of granularity relates to the sensitivity of the collision detection and is governed by a threshold parameter. To speed up the distance calculation, each of the subdivided boxes is treated as a minimal sphere bounding this sub-box. In this case, it only needs to calculate a distance to the centre of each sub-box.

4.2.4 Collision Avoidance

For assembly operations, we can distinguish two behaviour scenarios of collision avoidance. The first is to stop the robot movement when the close proximity of an operator has been detected, and continue the robot movement after the operator walks away. This scenario is suitable for operations with a limited degree of freedom, for

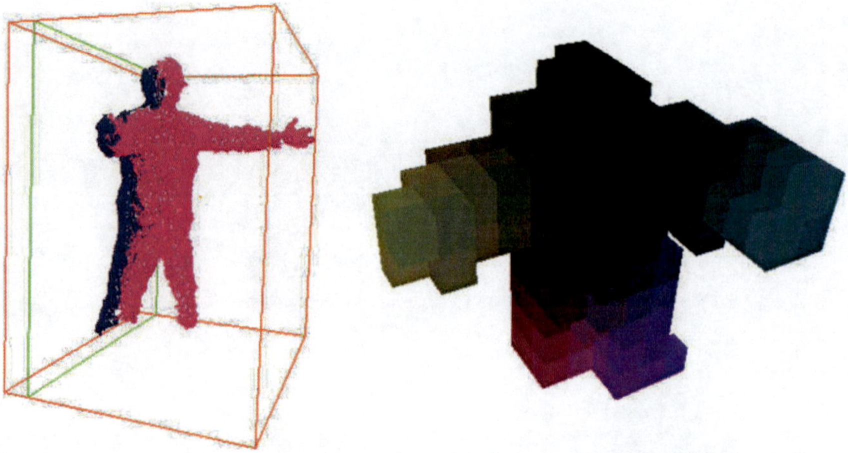

Fig. 4.5 a Point cloud in a minimum bounding box; b an array of subdivided boxes

example, inserting a pin or when moving in close proximity to the workpiece. The second scenario is applicable to such operations as transportation where the path can be dynamically modified to avoid collision with human operators and other obstacles.

As shown in Fig. 4.6, when a short distance between one of the points of interests (POI) on the robot end-effector and operator (a) is detected, the direction of movements of the end-effector is modified as p_1 to maintain a defined minimal distance. However, if operator (b) appears in proximity to the robot arm and the distance d_i

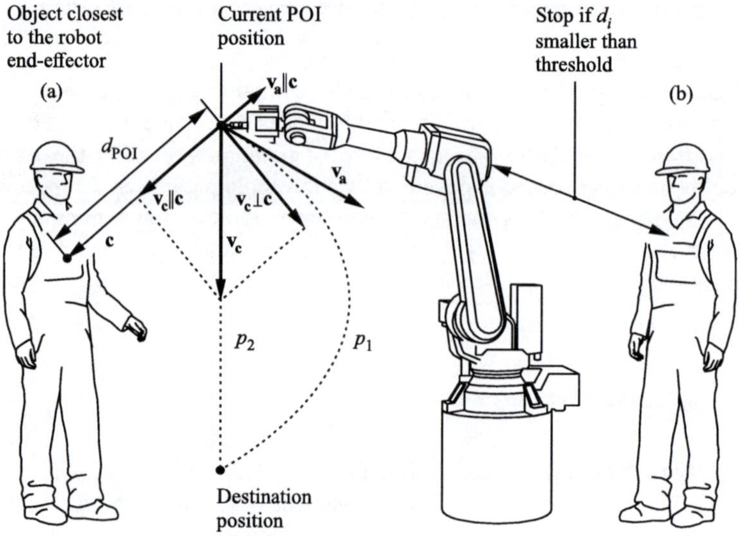

Fig. 4.6 Collision avoidance vectors (Adapted from [25])

between the operator and the robot becomes less than a safety threshold, the robot will stop moving. Once the situation is relaxed and d_{POI} is greater than the threshold, the end-effector will move back in direction p_2 to its destination position.

The representation of vectors taken into consideration for path modification is also shown in Fig. 4.6. The collision vector $\mathbf{c}$ is calculated as a vector between the POI on the end-effector and the corresponding closest point of detected obstacles. The vector representing the direction of the robot movement $\mathbf{v_c}$ is decomposed into two orthogonal components $\mathbf{v_c}\|\mathbf{c}$ and $\mathbf{v_c}\perp\mathbf{c}$, respectively, parallel and perpendicular to the collision vector $\mathbf{c}$. The parallel component is calculated in Eq. (4.6) as a dot product of the movement vector and collision vector and has the direction of the collision vector. The collision vector is a unit vector with the direction of the collision vector. Then, the perpendicular component can be calculated using Eq. (4.7). The parallel component is responsible for the movement towards the obstacle and is modified to avoid the collision. Modification is based on the distance from the obstacle and the direction of the movement, and it gives $\mathbf{v_a}\|\mathbf{c}$ as a result vector. Vector for modified movement is composed of $\mathbf{v_c}\perp\mathbf{c}$ and $\mathbf{v_a}\|\mathbf{c}$ as given in Eq. (4.8).

$$\mathbf{v_c}\|\mathbf{c} = \left(\mathbf{v_c} \cdot \hat{\mathbf{c}}\right) \cdot \hat{\mathbf{c}}, \text{ where } \hat{\mathbf{c}} = \frac{c}{\|c\|} \tag{4.6}$$

$$\mathbf{v_c}\perp\mathbf{c} = \mathbf{v_c} - \mathbf{v_c}\|\mathbf{c} \tag{4.7}$$

$$\mathbf{v_a} = \mathbf{v_a}\|\mathbf{c} + \mathbf{v_c}\perp\mathbf{c} \tag{4.8}$$

Based on a performed task and the calculated shortest distance to obstacle, one of the four safety strategies is applied to control the robot [24]. Figure 4.7 explains four cases where the four safety strategies can be employed: (1) an audio or visual warning is triggered as soon as an operator walks into the monitored area, and at the same time the speed of the robot can be reduced to prepare for a full stop; (2) a retrievable stop message is sent to the robotic system if the human steps into a defined hazard zone; nonetheless, when the human continues to move towards the robot (e.g., for inspection), and the task allows the robot to move, it will (3) move away automatically to keep a safe distance from the operator to avoid a collision; and in light of the possibility to change the path, (4) the current robot trajectory to the target is modified dynamically to prevent any collision with the operator while the robot is moving. Immediately after the operator exits the monitored area, the robot—in cases (2) and (3)—will resume the task from where it stopped. Implementing the four safety strategies ensures a minimum downtime on the robot side as it replaces the traditional static zones and emergency stops with recoverable temporary interruptions. At the same time, the operator is assured to step in and out of the robotic cell freely and securely, despite the fact that the robot is moving, which leads to enhanced productivity on the human side. The last three safety strategies are especially important for operators in human–robot cooperation and collaboration cells

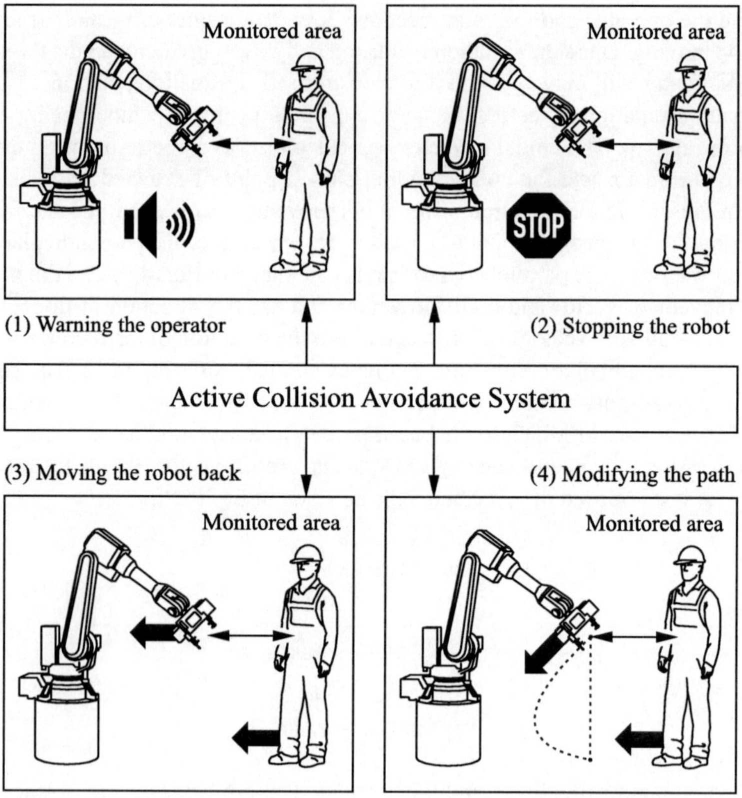

Fig. 4.7 Four safety strategies for active collision avoidance (Adapted from [24])

to perform tasks sharing the same space with robots. Switching between the three safety strategies is possible dynamically at runtime.

4.3 System Design and Implementation

To improve the performance and scalability of the CAS, it has been designed in a distributed manner. The server node is responsible for the decision-making process and manages communication, while client nodes are responsible for depth image acquisition and processing.

The CAS has been integrated within the SYMBIO-TIC [31] Demo 3 with a *4diac FORTE* implementation of an IEC 61499 function block runtime environment that integrates assembly process planning and execution. Communication with the robot is performed by a Robot Control Module function block through the robot controller's

External Guiding Module (EGM) interface. In the demonstration testbed, an ABB IRB 2600 industrial robot with RobotWare 6.08 and EGM is used.

4.3.1 CAS Server

The server node runs on a PC of Intel 2.4 GHz Corei5 CPU with 16 GB RAM and running 64-bit Windows 7 operating system. Overview of data flow and processing at the server node are visualised in Fig. 4.8. The four main processing blocks in the server are forward kinematics, differential inverse kinematics, collision avoidance algorithm, and decision-making.

The forward kinematics module uses the robot kinematic model and obtained joint values to calculate the current position of configured POI in Cartesian space, which should be used for collision detection. Each point of interest is defined in the reference frame of the robot link to which it is assigned, e.g., the robot tool centre point (TCP) is one of the POI represented in the frame corresponding to the last link, i.e., link 6. An example configuration of POI is presented in Table 4.2.

In the decision-making process, based on the received status and presence of the robot in one of the configured zones, one of the presented action scenarios is triggered. In the presented case, zones are defined in proximity to assembly stations, where it could be unsafe to control the robot to move away from the operator. When

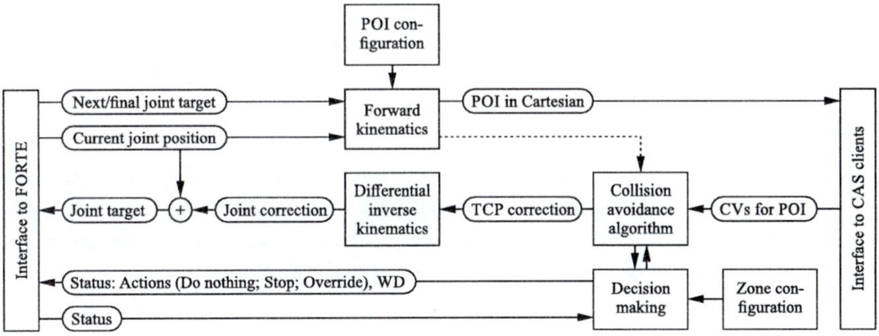

Fig. 4.8 Data flow and processing in CAS server

Table 4.2 Definition of points of interest

	Points of interest (POI)						
x [mm]	0	0	0	200	200	225	225
y [mm]	0	– 200	– 200	– 150	250	0	50
z [mm]	0	0	0	– 60	– 60	– 60	– 60
Ref. link	1	2	3	6	6	6	6

Table 4.3 Definition of zones

x [mm]	y [mm]	z [mm]	Length [mm]	Width [mm]	Height [mm]	Type
1023	−290	426	900	1400	400	2
92	−748	426	1600	−700	400	2
−835	−326	426	−600	−1400	400	2
−771	651	426	1900	500	500	2
−250	−1375	495	400	500	800	2

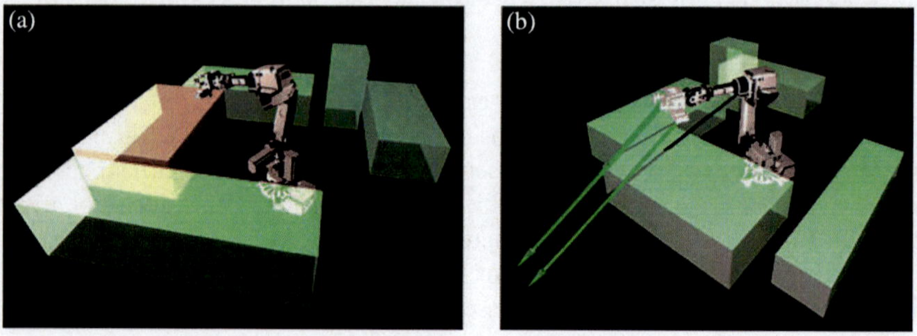

Fig. 4.9 Visualisation on the server side; **a** defined zones (green boxes), triggered zone (yellow box); **b** arrows pointing from POI to the closest points

the robot end-effector is within defined zones and a potential collision is detected, the selected scenario is to stop the robot (collision avoidance scenario 2 in Fig. 4.7). Configuration of zones used is presented in Table 4.3 and the zones themselves are visualised in Fig. 4.9.

The collision avoidance algorithm implements the procedure described earlier, preceded by a selection of the shortest collision distance found among collision vectors received from distributed client nodes. The "moving away" scenario is implemented as a special case of the "modifying path" scenario where the destination target is configured as the most recent point on the original robot path.

The differential inverse kinematics module converts corrections of the POI position obtained from the collision avoidance algorithm to corresponding corrections in joint space. Joint corrections are applied to the arm joints 1–3 only, while wrist joints remain unchanged. This simplifies calculations, as the Jacobian matrix becomes a 3 × 3 square matrix that can be analytically inverted. Moreover, this reduces the risk of singularities appearing in calculations of inverse kinematics.

4.3.2 CAS Clients

Clients are deployed on Mini PCs of Intel 3.5 GHz Corei7 CPU with 16 GB RAM and running 64-bit Ubuntu 18.4 operating system. This specification allows acquiring images from the Kinect sensor with the full frequency of 30 Hz and depth calculation in a time of 4 ms per frame. Each client node automatically detects the ID of the connected Kinect sensor and loads corresponding configuration files. Location and background calibration can be triggered using execution options or through a message from the server node.

4.3.3 Communication

The diagram of communication is presented in Fig. 4.10.

The connection between the CAS server and the RCM was realised with UDP socket communication and encoding according to ITU recommendation X.690 (ISO/ECT 8825–1) [32]. The structure of messages sent from the CAS to RCM is

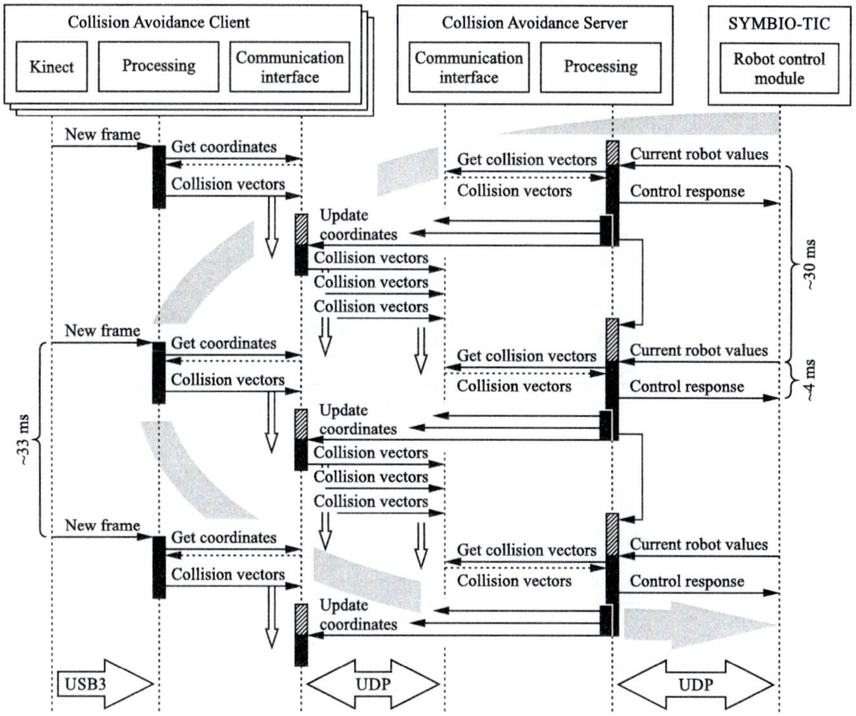

Fig. 4.10 The communication model of the implemented collision avoidance system

presented in Tables 4.4 and 4.5, while messages received from the RCM are explained in Tables 4.6 and 4.7. Joint values are encoded as 32-bit float values. Status is encoded as 16-bit signed two's complement integer. Negative values are used to transmit error messages.

The status of the RCM message consists of the following bits:

Table 4.4 Message from CAS to RCM

	Status Header	Status	Requested joints' values						
			Array header	J1	J2	J3	J4	J5	J6
Bytes:	1	2 3	4 5 6 7	8 9 10 11	12 13 14 15	16 17 18 19	20 21 22 23	24 25 26 27	28 29 30 31

Table 4.5 Status from CAS to RCM

	Status				
	Error	Miscellaneous	Stop	Heartbeat	Active
Bits:	15	14 13 12 11 10 9 8 7 6 5 4 3	2	1	0

Table 4.6 Message from RCM to CAS

	Status Header	Status	Current joints' values							Target joints' values						
			Array header	J1	J2	J3	J4	J5	J6	Array header	J1	J2	J3	J4	J5	J6
Bytes:	1	2 3	4 . . .	8. . .	12 . . .	16 . . .	20 . . .	24 . . .	28 . . .	32 . . .	36 . . .	40 . . .	44 . . .	48 . . .	52 . . .	56 . . .

Table 4.7 Status from RCM to CAS

	Status			
	Error	Miscellaneous	Gripper closed	Active
Bits:	15	14 13 12 11 10 9 8 7 6 5 4 3 2	1	0

- *Active*: bit value set indicates that the CAS is allowed to control the robot;
- *Gripper closed*: an indication that the robot is transporting a tool/workpiece, information is used to update the robot model used for collision detection;
- *Miscellaneous*: bits for future extension, e.g., different tools/workpieces to be modelled;
- *Error*: bit indicates that there occurred an internal RCM error, e.g., manual safety stop activated.

Current joint values are monitoring values corresponding to the current robot position, while target joint values correspond to the next robot target and are used in the scenario of modifying the path.

The status of the CAS message consists of the following bits:

- *Active*: bit value set indicates that the CAS detected a potential collision and is sending requested joint values of the robot, otherwise the joint values sent should be discarded;
- *Heartbeat*: bit toggled in every message to indicate that the CAS works properly,
- *Stop*: request to stop the robot in place and ignore transmitted joint values;
- *Miscellaneous*: bits for future extension, e.g., extended collaboration scenarios, error codes;
- *Error*: bit indicates that an internal CAS error, e.g., sensor failure, occurred.

4.4 System Verification

The collaborative assembly cell consists of the robot surrounded by three assembly stations and a conveyor belt. Three depth sensors are placed so that each covers all three workstations from a different angle, as presented in Fig. 4.11. To verify the placement of sensors, snapshots from IR cameras are captured while the robot is positioned in different locations. i.e., in front of each assembly station. As can be seen from snapshots presented in Fig. 4.12, all stations are covered well by the sensors.

4.4.1 Response Time

The procedure for the robot response time measurement is performed by the CAS server and covers communication delays among CAS, RCM and the robot controller in both directions, and dynamic response of the robot. During the test, logging of all messages between CAS and RCM is turned on. The server sends a periodical square-shaped signal for the target value of robot joint 1. Then, the time between a change in set value and a noticeable change in the joint motion is measured as presented in Fig. 4.13. In the current system implementation, the response time measured is around 120 ms. Merging CAS with RCM could reduce the response

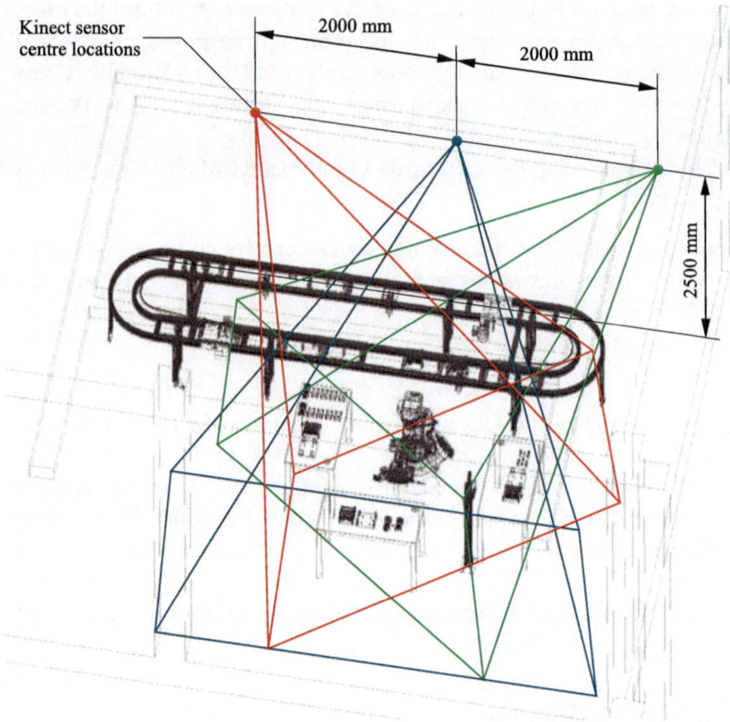

Fig. 4.11 Visualisation of sensors field of view and placement in the collaborative cell

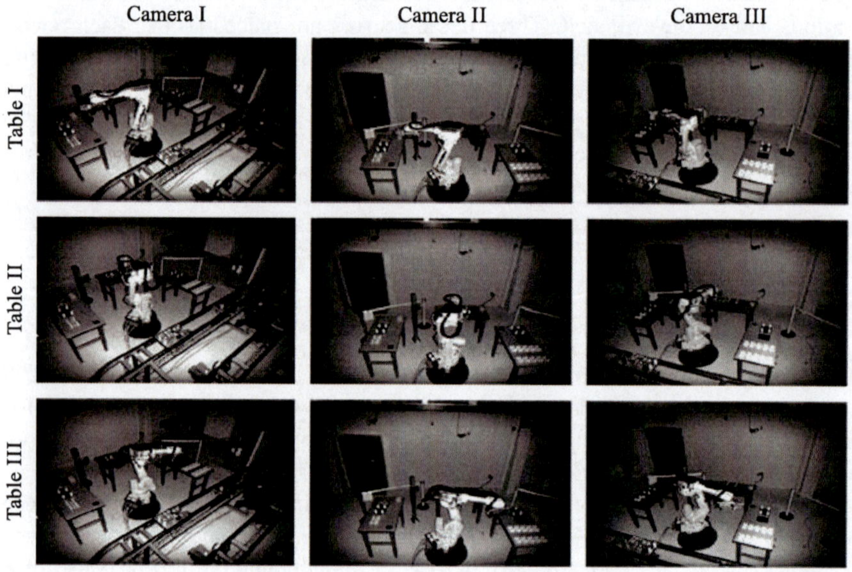

Fig. 4.12 Views from monitoring cameras with robot placed in front of three working tables

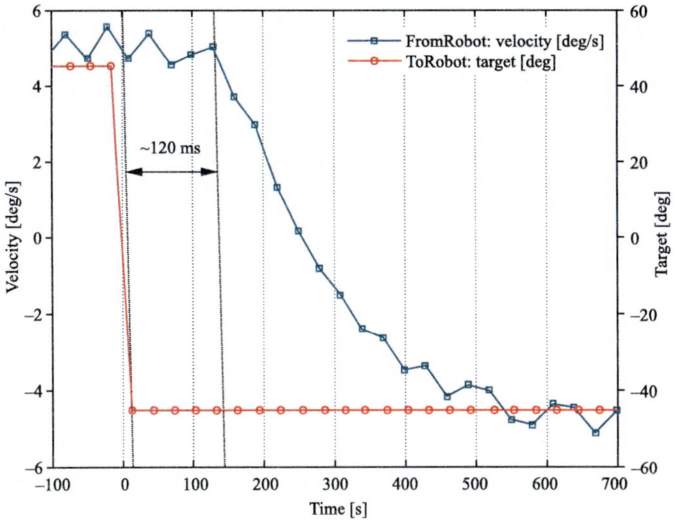

Fig. 4.13 A measure of a response time of the robot, performed from CAS

time from 120 to around 20–40 ms if the direct CAS to robot controller message exchange is implemented with a frequency of up to 250 Hz.

4.4.2 *Collision Avoidance*

During the execution of a transportation task, the operator intentionally approaches the robot to trigger the collision avoidance scenario, as visualised in Fig. 4.14. Distance thresholds are configured as follows: 0.5 m to trigger collision avoidance in the mode of slowing down, 0.4 m to activate/deactivate the moving away scenario, and 0.6 m to deactivate collision avoidance control. Results from the experiment are presented in Figs. 4.15, 4.16, and 4.17.

In the test setup, the robot's end-effector is moving with a linear velocity of approximately 400 mm/s, see Fig. 4.17, while the operator's average approach velocity measured is 1300 mm/s. At time 0.0 s, as the collision distance passed the configured threshold of 0.5 m, the CAS takes control over the robot path. Initially, it slowly stops the robot. At time 0.5 s, the robot moves with a velocity of 50 mm/s and the collision distance passes the move back threshold of 0.4 m, and the corrections in the robot position are transmitted to the robot. Robot motion corrections are applied until the collision distance is beyond the threshold of 0.6 m, which occurs at around 7.25 s. Then, the control over the robot path is released by the CAS.

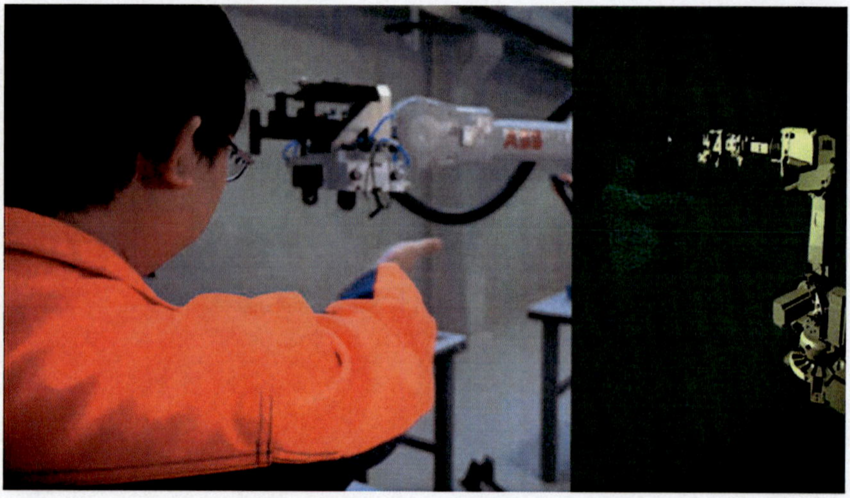

Fig. 4.14 Frame from a video recording of an experiment and corresponding visualisation from one of the CAS clients

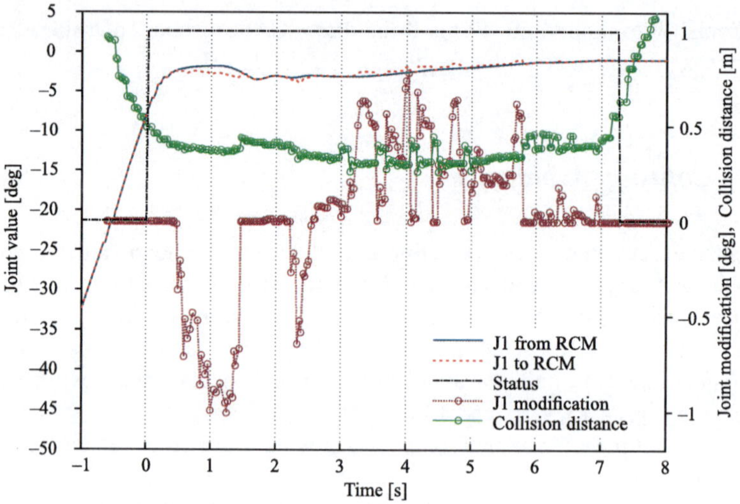

Fig. 4.15 Data from an experiment, only values of the robot's joint 1 (J1) are presented

4.5 Conclusions and Future Work

This chapter presents a distributed real-time approach for collision detection and active collision avoidance in a human–robot collaborative work cell. The aim of the solution developed is to improve the overall robotic system performance by connecting the robot's virtual 3D models with a distributed set of depth sensing and

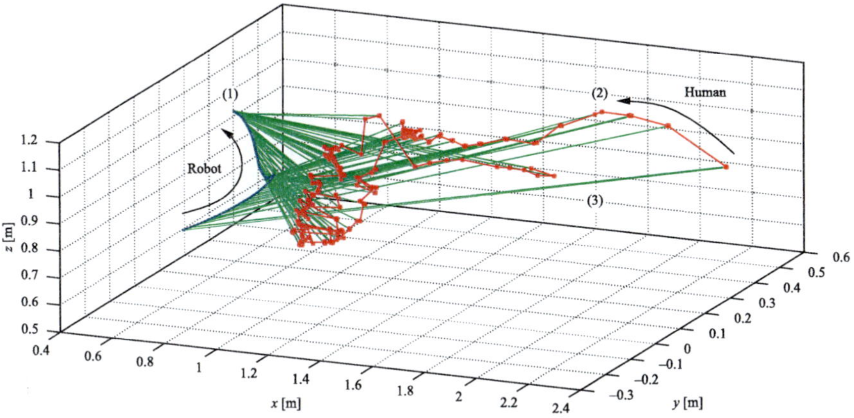

Fig. 4.16 3D visualisation of (1) robot end-effector positions (blue dots); (2) closest point positions (red asterisks); and (3) associations of corresponding points (green dotted lines)

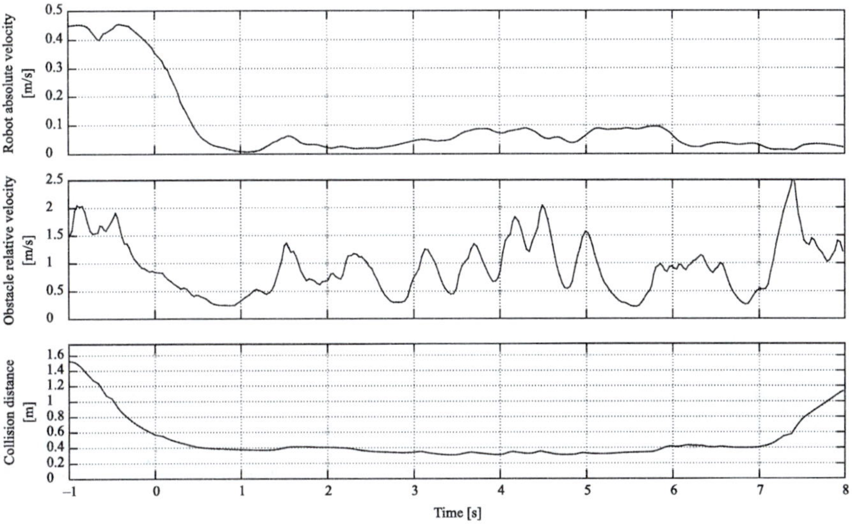

Fig. 4.17 Robot and obstacle velocities

processing units for online collision detection and avoidance in an augmented virtual environment. Instead of using the traditional emergency stops with fixed safety zones that do not allow human–robot co-existence, the reported approach detects in real time any possible collision and actively controls the robot via four safety modes: alarming a human operator and slowing down the robot, stopping the robot, moving the robot away from the approaching operator, or modifying the robot trajectory in real time. The approach facilitates flexibility and productivity. Verification has been performed in a demonstration collaborative assembly cell.

Future work will focus on improving the time performance of the collision avoidance system and extension of collision detection and avoidance capabilities to facilitate an operator intention detection and gesture control of the robot.

Acknowledgements The author expresses great appreciation to Prof. Lihui Wang and Dr Abdullah Mohammed for their valuable and constructive suggestions during the planning and development of the presented work. This work is partially supported by the EU FoF-06-2014 SYMBIO-TIC project (No. 637107).

References

1. A. Bussy, A. Kheddar, A. Crosnier, F. Keith, Human–humanoid haptic joint object transportation case study, in *2012 IEEE/RSJ International Conference on Intelligent Robots and Systems* (Vilamoura, Portugal, 2012), pp. 3633–3638
2. D.J. Agravante, A. Cherubini, A. Kheddar, Using vision and haptic sensing for human–humanoid joint actions, in *6th IEEE Conference on Robotics, Automation and Mechatronics (RAM)* (Manila, Philippines, 2013), pp. 13–18
3. C.A. Monje, P. Pierro, C. Balaguer, A new approach on human–robot collaboration with humanoid robot RH-2. Robotica **29**(6), 949–957 (2011)
4. S. Takata, T. Hirano, Human and robot allocation method for hybrid assembly systems. CIRP Ann. Manuf. Technol. **60**(1), 9–12 (2011)
5. S. Pellegrinelli, A. Orlandini, N. Pedrocchi, A. Umbrico, T. Tolio, Motion planning and scheduling for human and industrial-robot collaboration. CIRP Ann. **66**(1), 1–4 (2017)
6. C. Fei, K. Sekiyama, H. Sasaki, H. Jian, S. Baiqing, T. Fukuda, Assembly strategy modeling and selection for human and robot coordinated cell assembly, in *2011 IEEE/RSJ International Conference on Intelligent Robots and Systems* (San Francisco, CA, USA, 2011), pp. 4670–4675
7. J. Krüger, T. K. Lien, A. Verl, Cooperation of human and machines in assembly lines. CIRP Ann. Manuf. Technol. **58**(2), pp. 628–646 (2009)
8. T. Arai, R. Kato, M. Fujita, Assessment of operator stress induced by robot collaboration in assembly. CIRP Ann. Manuf. Technol. **59**(1), 5–8 (2010)
9. G. Hoffman, Evaluating fluency in human–robot collaboration. IEEE Trans. Hum. Mach. Syst. **49**(3), 209–218 (2019)
10. X.V. Wang, Zs. Kemény, J. Váncza, L. Wang, Human–robot collaborative assembly in cyber-physical production: classification framework and implementation. CIRP Ann. **66**(1), 5–8 (2017)
11. J. Krüger, B. Nickolay, P. Heyer, G. Seliger, Image based 3D surveillance for flexible man–robot cooperation. CIRP Ann. Manuf. Technol. **54**(1), 19–22 (2005)
12. J.A. Corrales, F.A. Candelas, F. Torres, Safe human–robot interaction based on dynamic sphere-swept line bounding volumes. Robot Comput Integr Manuf **27**(1), 177–185 (2011)
13. D. Ebert, T. Komuro, A. Namiki, M. Ishikawa, Safe human–robot coexistence: emergency-stop using a high-speed vision-chip, in *2005 IEEE/RSJ International Conference on Intelligent Robots and Systems* (Edmonton, Alta, Canada, 2005), pp. 1821–1826
14. T. Gecks, D. Henrich, Human–robot cooperation: safe pick-and-place operations, in *ROMAN 2005. IEEE International Workshop on Robot and Human Interactive Communication* (Nashville, TN, USA, 2005), pp. 549–554
15. C. Vogel, M. Poggendorf, C. Walter, N. Elkmann, Towards safe physical human–robot collaboration: a projection-based safety system, in *2011 IEEE/RSJ International Conference on Intelligent Robots and Systems* (San Francisco, CA, USA, 2011), pp. 3355–3360

16. J.T.C. Tan, T. Arai, Triple stereo vision system for safety monitoring of human–robot collaboration in cellular manufacturing, in *2011 IEEE/CIRP International Symposium on Assembly and Manufacturing* (Tampere, Finland, 2011), pp. 1–6
17. R. Schiavi, A. Bicchi, F. Flacco, Integration of active and passive compliance control for safe human–robot coexistence, in *2009 IEEE International Conference on Robotics and Automation* (Kobe, Japan, 2009), pp. 259–264
18. M. Fischer, D. Henrich, 3D collision detection for industrial robots and unknown obstacles using multiple depth images, in *Advances in Robotics Research*, ed. by T. Kröger, F. Wahl (Springer, Berlin, Heidelberg, 2009), pp. 111–122
19. P. Rybski, P. Anderson-Sprecher, D. Huber, C. Niessl, R. Simmons, Sensor fusion for human safety in industrial workcells, in *2011 IEEE/RSJ International Conference on Intelligent Robots and Systems* (Vilamoura, Portugal, 2011), pp. 3612–3619
20. F. Flacco, T. Kroeger, A. De Luca, O. Khatib, A depth space approach for evaluating distance to objects. J. Intell. Rob. Syst. **80**(1), 7–22 (2015)
21. C. Morato, K.N. Kaipa, B. Zhao, S.K. Gupta, Toward safe human robot collaboration by using multiple kinects based real-time human tracking. J. Comput. Inf. Sci. Eng. **14**(1), 011006 (2014)
22. Pilz GmbH & Co. KG.2016. https://www.safetyeye.com/
23. L. Wang, R.X. Gao, J. Váncza, J. Krüger, X.V. Wang, S. Makris, G. Chryssolouris, Symbiotic human–robot collaborative assembly. CIRP Ann. Manuf. Technol. **68**(2), 701–726 (2019)
24. L. Wang, B. Schmidt, A.Y.C. Nee, Vision-guided active collision avoidance for human–robot collaborations. Manuf. Lett. **1**(1), 5–8 (2013)
25. B. Schmidt, L. Wang, Depth camera based collision avoidance via active robot control. J. Manuf. Syst. **33**(4), 711–718 (2014)
26. A. Mohammed, B. Schmidt, L. Wang, Active collision avoidance for human–robot collaboration driven by vision sensors. Int. J. Comput. Integr. Manuf. **30**(9), 970–980 (2017)
27. J. Sell, P.O. Connor, The Xbox one system on a chip and Kinect sensor. IEEE Micro **34**(2), 44–53 (2014)
28. libfreenect2: open source drivers for the Kinect for Windows v2 device (2016)
29. F. Järemo Lawin, P.-E. Forssén, H. Ovrén, Efficient multi-frequency phase unwrapping using kernel density estimation, in *Computer Vision—ECCV 2016. ECCV 2016. Lecture Notes in Computer Science*, vol 9908 (Springer, Cham, 2016)
30. B. Schmidt, L. Wang, Automatic work objects calibration via a global–local camera system. Robot Comput-Integr Manuf **30**(6), 678–683 (2014)
31. SYMBIO-TIC, EU SYMBIO-TIC project—symbiotic human–robot collaborative assembly: technologies, innovations and competitiveness (2015). https://cordis.europa.eu/project/id/637107
32. ISO/IEC, ISO/IEC 8825-1: information technology—ASN.1 encoding rules: specification of basic encoding rules (BER), canonical encoding rules (CER) and distinguished encoding rules (DER)—Part 1 (2015)

Chapter 5
Real-Time Collision-Free Multi-Objective Robot Motion Generation

Yuquan Wang and Lihui Wang

Abstract Despite safe mechanical design being necessary for collaborative robots, one should not underestimate the importance of active safety due to a multi-objective control design. Active safety not only complements mechanical compliance but also enables classical industrial robots to fulfil additional task-space objectives. Through a task-space controller based on a quadratic programming (QP) formulation, the proposed approach assigns strict priority to avoid collisions and specifies other QP objectives with soft task priorities. Through experiments performed on a dual-arm robot, the proposed solution can generate safe robot motion while fulfilling other task specifications.

5.1 Introduction

From shop floors to research labs, an increasing amount of collaborative robotic applications can be found in different contexts. Different from the classical manufacturing tasks with a well-defined static environment, such as painting and welding tasks in an automotive assembly line, a future is envisioned in which robots are supposed to interact with an unpredictable and complex environment. For instance, the tailor-made humanoids from Disney research are used in face-to-face entertainment with tourists, and the collaborative robots from ABB and Universal Robots are about to work side-by-side with human workers, as in the recent development of collaborative robots in the assembly line presented by Realyvásquez-Vargas et al. [1].

Y. Wang · L. Wang (✉)
Department of Production Engineering, KTH Royal Institute of Technology, Stockholm, Sweden
e-mail: lihuiw@kth.se

Y. Wang
e-mail: yuquan@kth.se

Y. Wang
The Chinese University of Hong Kong, Shenzhen, China

L. Wang et al. (eds.), *Advanced Human–Robot Collaboration in Manufacturing*,
https://doi.org/10.1007/978-3-030-69178-3_5

Breaking the separation principle and interacting directly with humans requires a robot in the future to be able to adapt its motion to the following: (1) non-static environments due to the reconfiguration of a specific task or relocation of the robot itself; (2) humans with different intentions that are possibly changing considerably over a short period; (3) dynamic obstacles including other autonomous robots, e.g., another robot performing another task independently; and (4) fulfilling multiple task constraints, e.g., maintain a contact force with the environment while moving along the surface of a workpiece. The demands can be specifically summarised as

- Avoiding dynamic obstacles in a changing environment in real time,
- Applying collision avoidance calculation to entire bodies,
- Simultaneously fulfilling multiple task specifications that are potentially conflicting with each other.

Motion planning algorithms based on random sampling or joint space interpolation would not qualify as solutions to the above requirements. There are real-time planning and replanning methods that can find a feasible solution within 100 ms, either in simulation or experiment, see the earlier work reported by van den Berg et al. [2], and Vannoy and Xiao [3], respectively. The gap between the state-of-the-art motion planning algorithms and fulfilling the aforementioned constraints is due to the simultaneous satisfaction of requirements of real-time computation, planning with uncertain data, generating constrained motion, and avoiding collisions with dynamic obstacles.

Alternatively, different control methods can effectively produce locally optimal constrained motion based on real-time sensory feedback [4]. Since about 2010, quadratic programming (QP) approaches are widely accepted by robotics researchers to resolve multiple control objectives employing a redundant robot [5, 6]. Concerning Lyapunov stability, Ögren and Robinson [7] proposed a Linear Programming framework within which every objective is stable. In the relative space between two robot arms, Wang et al. [8] proposed time feedforward terms, different bounds as well as convergence rates for different tasks and quadratic objectives to penalise the overuse of joint commands. These redundancy resolution techniques lead to task programming approaches that enable a versatile combination of robot skills in manufacturing tasks as summarised by Pedersen et al. [9].

However, as reported by Ögren and Robinson [7], Wang et al. [8], and Kanoun et al. [10], in redundancy resolution, collision checking and avoidance are applied between simple geometric shapes [11] and the robot end-effector only. Whole-body collision checking is more often used as a black box to validate sampled configurations for planning methods [12].

This chapter proposes a task-space QP controller to continuously update the collision avoidance constraints, such that the robot manipulator keeps the closest distances between the robot and dynamic obstacles—e.g., other robots, human operators—above a user-specified threshold.

Enforcing the gradient of collision avoidance as hard constraints, experiments are performed on a dual-arm robot to show that the proposed QP controller assigns strict priorities to avoid dynamic obstacles while fulfilling multiple task specifications simultaneously.

The rest of the chapter is structured as follows. Related work is introduced in Sect. 5.2; the multi-objective problem is formulated in Sect. 5.3; the details of the QP-based solution are presented in Sect. 5.4; and experimental results are given in Sect. 5.5.

5.2 Related Work

To find the optimal control solution in an unstructured and dynamic environment, the state of the art relies on two main approaches: real-time replanning algorithms, and reactive control methods.

5.2.1 *Real-Time Replanning Methods*

Path planning constructs graphs or trees of connected configurations, which are obtained with random samples [13].

There are two popular ways to sample in high-dimensional spaces: the Probabilistic Roadmap Method (PRM) by Kavraki et al. [14], and Rapidly exploring Random Trees (RRT) by LaValle et al. [15], which are featured as a multi-query method and a single-query method, respectively. As recently reviewed by Elbanhawi and Simic [16], the key research issues addressed by numerous variants of both PRM and RRT include real-time kinodynamic planning, optimal planning, planning under uncertainty, etc.

Recent efforts have remarkably improved the computational efficiency of path planning algorithms.

TrajOpt proposed by Schulman et al. [6] offers an efficient sequential convex-optimisation-based formulation that can find locally optimal, collision-free trajectories. *RRT*^{X} introduced by Otte and Frazzoli [17] provides an asymptotically optimal RRT algorithm for dynamic environments with upper bounded complexity.

The *CHOMP* algorithm reported by Zucker et al. [18] is a representation-invariant trajectory optimisation technique that can balance between smoothness and obstacle avoidance. Based on iterative optimisation using functional gradients, it improves the quality of an initial trajectory with local smoothing. The *Tangent Space RRT* (TS-RRT) algorithm approximates the constraint manifold by constructing bounded tangent space approximations with randomised trees [19]. According to Jaillet and Porta [20], the quality of TS-RRT is affected by the possible overlap of the tree branches that approximate the tangent spaces. However, these algorithms are either not fast enough for real-time replanning, e.g., TrajOpt, or in lack of constrained motion, e.g.,

RRT^X. These limitations motivate exploiting the trade-off of what should be computed online (by a real-time controller) and what needs to be taken care of offline (by a planner).

5.2.2 Online Control Methods

Analogously to control problems in general, robot control is not exempt from the trade-off between global and local optimality.

Local optimality comes with a lower computational cost, thus offering reactiveness. As reported by different researchers over decades [13, 21, 22], constraint-based programming enables the fast prototyping and configuration of complex real-time robotic tasks. Due to modularity, it is easy to specify a wide range of tasks e.g., assembly skills reported by Polverini et al. [23]. With Convex Programming, as analysed by Schulman et al. [24] and Lengagne et al. [25], much more complex tasks can be solved at a computation time of around a few seconds—however, this is long enough to be non-reactive.

Rather than using null-space projection by Seraji and Colbaugh [26], or De Schutter et al. [22], more and more researchers implement constraint-based programming as a QP controller in favour of (1) a smooth transition between constraints set [27]; (2) straightforward integration of a large set of constraints [28], and (3) easy formulation of complex tasks specified with high degrees of freedom, e.g., the hierarchical quadratic programming for humanoid control problems reported by Escande et al. [29]. Different from modifying an existing solution, e.g., projecting the solution to a feasible direction with the invariance control approach reported by Kimmel and Hirche [30], scalarisation by minimising a weighted sum of objectives enables the simultaneous execution of multiple tasks in addition to fulfilling the safety conditions.

Despite the recently proposed priority-learning approach, as presented, e.g., by Dehio et al. [31, 32], it is common to manually assign soft priorities for different tasks in all of the aforementioned multi-objective control frameworks. Hard constraints, e.g., the proposed collision avoidance, may lead to infeasible solutions, and the controller will fail to run.

Keeping the feasibility of the constraints is a challenging problem. Taking the joint position constraint as an example, it is not possible to instantly change joint position and velocity due to limited joint acceleration. Assuming a maximum of acceleration/deceleration, Decré et al. [33] proposed viability constraints to infer the maximum allowable joint velocity, such that the joint position bounds will be fulfilled. Recently, a more rigorous treatment of the constraints has been introduced by Del Prete [34], where the exact time of constraint saturation is considered. Both [33, 34] need to check the joint position or velocity limit for a certain time horizon. However, joint limit constraints are decoupled joint by joint, in case of a generic constraint, i.e., collision avoidance, the prediction of constraint evolution is highly nonlinear. Thus, the viability constraints are not applicable for collision avoidance purposes.

On the other hand, when a robot needs to perform an intended collision, i.e., impacts by design, Wang et al. [35, 36] introduced impact-awareness to a QP controller. Simulations [36] and experimental results [35] support that the one-step-preview can generate maximal, yet feasible motion despite the impact-induced state jumps.

For the sake of computational efficiency, convex hulls [37], or convex capsules are common choices to enclose the obstacles and the robot. State-of-the-art continuous distance checking libraries, e.g., the FCL library by Pan et al. [38], allow the formulation and update of collision avoidance constraints for two groups of convex shapes in real time.

5.3 Dynamic Collision Avoidance Problem

The whole-body collision avoidance problem is formulated in terms of abstract constraints and objectives:

Problem 5.1 Let the robot joint positions and the robot joint commands be denoted by $\mathbf{q}$ and $\mathbf{u}$, respectively. The desired robot motion is defined by functionals

$$f_i : \mathbb{R}^n \rightarrow \mathbb{R}$$

belonging to the set of linear equalities I_e, the set of inequalities I_{ie}, and the set of objectives I_o. To generate $\mathbf{u}$, the following optimisation problem is solved:

$$\min_{\mathbf{u}} \sum_{i \in I_o} w_i f_i(\mathbf{q}, \mathbf{u}) \tag{5.1}$$

$$\text{s.t.} \quad f_i(\mathbf{q}, \mathbf{u}) \leq b_i, \ \forall i \in I_{ie}, \tag{5.2}$$

$$f_i(\mathbf{q}, \mathbf{u}) = b_i, \ \forall i \in I_e, \tag{5.3}$$

where b_i denotes the bounds corresponding to the constraints, and w_i denotes the weights of the objectives in I_o.

If the minimum distance between the entire robot body and the obstacles is $d(\mathbf{q}, \mathbf{u}) \in \mathbb{R}$, the allowed minimum distance between the robot and obstacles is b_d, the following inequality is added to the set I_{ie}:

$$-d(\mathbf{q}, \mathbf{u}) \leq -b_d. \tag{5.4}$$

If (5.1–5.3) is updated using the latest sensory feedback and solved in real time, the robot can reactively avoid collisions while performing the desired task. ■

The quadratic programming (QP) problem (5.1–5.3) is nonlinear. State-of-the-art redundancy resolution methods suggest to calculate the gradient of f_i for $i \in$

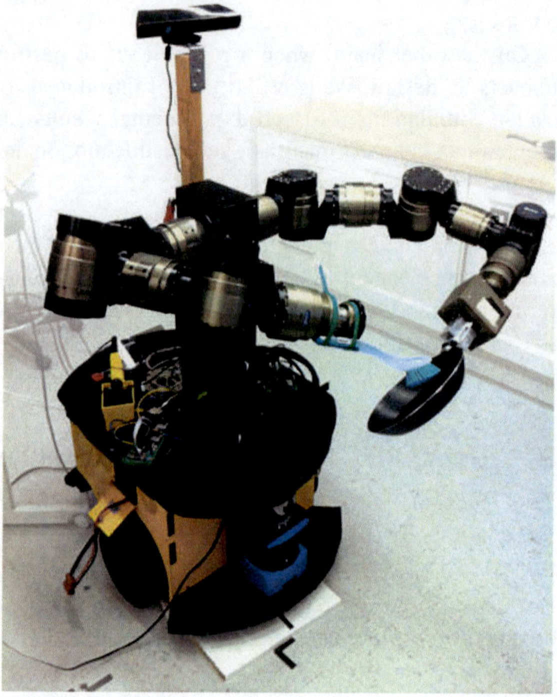

Fig. 5.1 A dual-arm robot performing a pan cleaning task

I_o, I_e, I_{ie}, and then solve for a locally optimal solution. As an example, a pan cleaning task is specified for a dual-arm robot, as shown in Fig. 5.1.

However, the *feasibility* of Problem 5.1 is not well-established due to the high *uncertainty* of the dynamic obstacles. Abrupt changes in the geometry and distance of the dynamic obstacles may make the obstacle avoidance constraint (5.4) infeasible. Sometimes, it is even difficult to find a feasible initial configuration.

When the constraints are infeasible, they may be converted to objectives. Namely, (5.1–5.3) can be transformed to the following unconstrained optimisation problem:

$$\min_{\mathbf{u}} \sum_{i \in I_o} w_i f_i(\mathbf{q}, \mathbf{u}) + \sum_{i \in I_e} w_i f_i(\mathbf{q}, \mathbf{u}) + \sum_{i \in I_{ie}} w_i f_i(\mathbf{q}, \mathbf{u}). \tag{5.5}$$

There are different ways to transform from (5.1–5.3) to (5.5), e.g., using a state machine [4], or a behaviour tree [39]. In all the cases, the guard conditions, the transition functions, and the weight w_i for each objective have to be carefully selected. To apply a hierarchy of priorities, different weights need to take different orders of magnitude, which is a hard—if not impossible—problem, and can even lead to numerical issues.

5.4 Proposed QP Controller

In the approach presented in this section, the minimum-norm QP solution to (5.1–5.3) is formulated in Sect. 5.4.1. Detailed steps to prioritise multiple tasks with soft task priorities are presented in Sect. 5.4.2, and the formulation of a generic task is given in Sect. 5.4.3. To apply strict priorities of the collision avoidance task and avoid abrupt task transition, an inequality constraint and a virtual admittance task are combined in Sect. 5.4.4. The proposed QP controller is summarised in Sect. 5.4.5, and details of commonly used tasks are listed in Sect. 5.4.6.

5.4.1 Minimum-Norm Solution by Quadratic Programming

Solving (5.1–5.3) following the Linear Programming approach by Ögren et al. [7], the robot motion is not smooth. Such an example is shown in Fig. 5.2a, where the relative motion between two robot end-effectors keeps chattering during task execution.

Hence, a quadratic term $\mathbf{u}^\top Q\mathbf{u}$ is needed to penalise the input $\mathbf{u}$:

$$\min_{\mathbf{u}} \sum_{i\in I_o} w_i \dot{f}_i(\mathbf{q},\mathbf{u}) + \mathbf{u}^\top \mathbf{Q}\mathbf{u},$$

where the gradient $\dot{f}_i(\mathbf{q},\mathbf{u})$ can be either analytical or numerical. Thus, a minimum norm locally optimal solution $\mathbf{u}$ is obtained. It is well known that the solution map of a linearly constrained quadratic problem is Lipschitz continuous with respect to the perturbations to the right-hand side of the linear constraints and the linear part of the objective function [40, 41]. The Lipschitz continuity conditions are summarised in Remark 5.1.

In the example presented in Fig. 5.2b, the conditions in Remark 5.1 are fulfilled and the robot motion is smooth compared to the results in shown Fig. 5.2a.

Remark 5.1 The locally optimal joint command $\mathbf{u}(k)$, $k = 1, 2, \ldots, n$ is Lipschitz continuous with respect to $\mathbf{q}(k)$, as long as the following hold:

1. The symmetric matrix $\mathbf{Q}$ is positive definite.
2. Eqs. (5.3–5.4) have a non-empty solution set, in other words, (5.3–5.4) are regular constraints.
3. $\dot{f}_i(\mathbf{q},\mathbf{u})$ for $i \in I_o, I_e, I_{ie}$ is Lipschitz continuous at $\mathbf{q}(k)$.
4. $\dot{d}(\mathbf{q},\mathbf{u})$ is Lipschitz continuous at $\mathbf{q}(k)$.

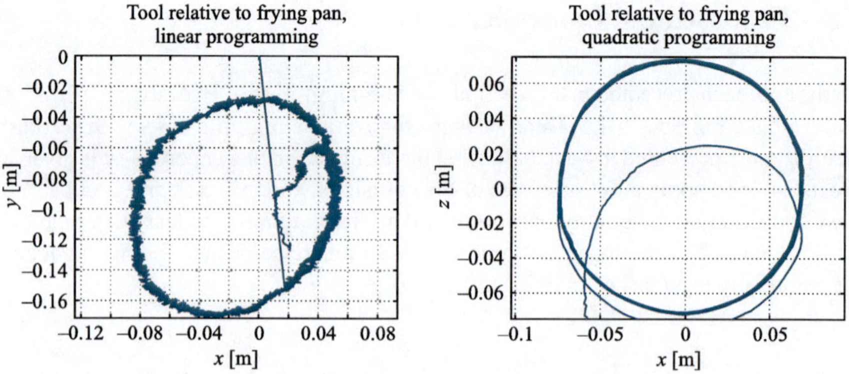

Fig. 5.2 Comparison of the robot motion generated by LP (left) and QP (right), executing a circular motion defined in the relative space between two robot arms

Fig. 5.3 Search space constrained by hyperplanes associated with the constraints $f_1 \ldots f_5$

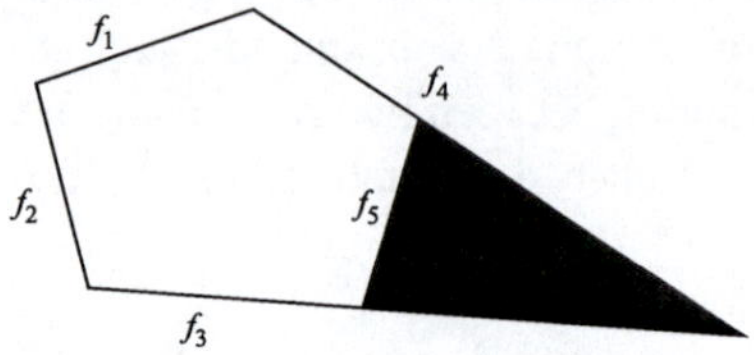

5.4.2 *Prioritising Multiple Objectives*

The difference between a constraint and a task is the priority. The constraints need to be strictly fulfilled all the time, otherwise the QP becomes infeasible, while the QP tolerates errors of the tasks that lead to a less optimal objective.

A linear constraint amounts to a hyperplane in $\mathbb{R}^n$. For multiple constraints, e.g., f_i for $i \in I_e, I_{ie}$, the intersection of these hyperplanes formulates a feasible search space. An example is outlined in Fig. 5.3: the search space is constrained by hyperplanes associated with the constraints $f_1 \ldots f_5$—introducing f_5 leads to a smaller search space by a margin marked by the shaded area.

If this search space is non-empty, then the constraints are regular. Otherwise, the underlying optimisation problem becomes infeasible. A constraint can be converted to a task and vice versa automatically by using slack variables [42]. That is, each f_i for $i \in I_e, I_{ie}$ can be converted to the set I_o. As highlighted by the framed expressions in (5.6), a slack variable μ_i is added to each of the constraints f_i, and $\|\mu_i\|^2$ is penalised in the objective; see the schematic picture in Fig. 5.4.

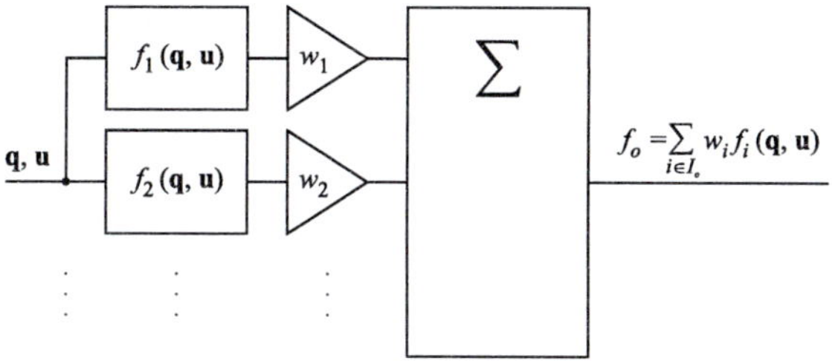

Fig. 5.4 QP controller objective as an accumulated sum of sub-objectives

$$\begin{aligned}\min_{\mathbf{u},\boldsymbol{\mu}} \quad & \mathbf{u}^\top \mathbf{Q}\mathbf{u} + \sum_{i\in I_o} w_i \dot{f}_i(\mathbf{q},\mathbf{u}) + \boxed{\sum_{i\in I_{ie}, I_e} w_i \|\mu_i\|^2} \\ \text{s.t.} \quad & -d(\mathbf{q},\mathbf{u}) + \boxed{\mu_i} \le -b_d, \\ & f_i(\mathbf{q},\mathbf{u}) + \boxed{\mu_i} \le b_i, \ \forall i \in I_{ie}, \\ & f_i(\mathbf{q},\mathbf{u}) + \boxed{\mu_i} = b_i, \ \forall i \in I_e,\end{aligned} \tag{5.6}$$

The feasibility of (5.6) is guaranteed by using the slack variables. All the constraints are breakable at the *price* specified by the weight of the slack variables. Suppose the desired error dynamics is

$$\ddot{\mathbf{e}} = -\mathbf{K}_d\dot{\mathbf{e}} - \mathbf{K}_p\mathbf{e}, \tag{5.7}$$

where $\mathbf{e} \in \mathbb{R}^n$ denotes the task-space error. The proportional and derivative gains $\mathbf{K}_p$ and $\mathbf{K}_d$, respectively, are diagonal positive definite matrices. The weights w_i and the gains $\mathbf{K}_p$, $\mathbf{K}_d$ could be chosen either manually or through learning, such as the sample-efficient learning approach by Su et al. [32]. The slack variable amounts to the residual

$$\mathbf{e}_f = \ddot{\mathbf{e}} + \mathbf{K}_d\dot{\mathbf{e}} + \mathbf{K}_p\mathbf{e}.$$

Thus, the QP controller (5.6) is equivalent to the unconstrained QP:

$$\min_{\mathbf{u},\boldsymbol{\mu}} \quad \mathbf{u}^\top \mathbf{Q}\mathbf{u} + \sum_{i\in I_o} w_i \dot{f}_i(\mathbf{q},\mathbf{u}) + \sum_{i\in I_e, I_{ie}} w_i \|\mathbf{e}_{f_i}\|^2, \tag{5.8}$$

if all the constraints are converted to tasks.

5.4.3 Generic Task Formulation

The error $\mathbf{e}$ for a generic quantity $\boldsymbol{\lambda} \in \mathbb{R}^n$ is defined as

$$\mathbf{e} = \mathbf{S}\boldsymbol{\lambda} - \mathbf{S}\boldsymbol{\lambda}^*, \tag{5.9}$$

where $\mathbf{S} \in \mathbb{R}^{n \times n}$ is a diagonal matrix with its diagonal elements being either 1 or 0, depending on whether the corresponding axis is regarded "on" or "off" in a given case. Differentiating $\mathbf{e}$ leads to

$$\dot{\mathbf{e}} = \mathbf{S}\dot{\boldsymbol{\lambda}} - \mathbf{S}\dot{\boldsymbol{\lambda}}^* = \mathbf{S}\frac{\partial \boldsymbol{\lambda}}{\partial \mathbf{q}}\dot{\mathbf{q}} - \mathbf{S}\dot{\boldsymbol{\lambda}}^* = \mathbf{S}\mathbf{J}_\lambda \dot{\mathbf{q}} - \mathbf{S}\dot{\boldsymbol{\lambda}}^* \tag{5.10}$$

$$\ddot{\mathbf{e}} = \mathbf{S}\dot{\mathbf{J}}_\lambda \dot{\mathbf{q}} + \mathbf{S}\mathbf{J}_\lambda \ddot{\mathbf{q}} - \mathbf{S}\ddot{\boldsymbol{\lambda}}^*, \tag{5.11}$$

and the residual

$$\begin{aligned} \mathbf{e}_\lambda &= \ddot{\mathbf{e}} + \mathbf{K}_d\dot{\mathbf{e}} + \mathbf{K}_p\mathbf{e} \\ &= \mathbf{S}\mathbf{J}_\lambda\ddot{\mathbf{q}} + \underbrace{\mathbf{S}(\dot{\mathbf{J}}_\lambda + \mathbf{K}_d\mathbf{J}_\lambda)\dot{\mathbf{q}} - \mathbf{K}_d\dot{\boldsymbol{\lambda}}^* + \mathbf{K}_p\mathbf{S}\mathbf{e} - \mathbf{S}\ddot{\boldsymbol{\lambda}}^*}_{\mathbf{C}_{\mathbf{e}_\lambda}}, \end{aligned}$$

whose quadratic form is given as

$$\begin{aligned} \|\mathbf{e}_\lambda\|^2 &= \mathbf{e}_\lambda^\top \mathbf{e}_\lambda \\ &= \ddot{\mathbf{q}}^\top \underbrace{\mathbf{J}_\lambda^\top \mathbf{S}^\top \mathbf{S}\mathbf{J}_\lambda}_{\mathbf{Q}_{\mathbf{e}_\lambda}} \ddot{\mathbf{q}} + \underbrace{2\mathbf{C}_{\mathbf{e}_\lambda}^\top \mathbf{S}\mathbf{J}_\lambda}_{P_{\mathbf{e}_\lambda}} \ddot{\mathbf{q}} + \mathbf{C}_{\mathbf{e}_\lambda}^\top \mathbf{C}_{\mathbf{e}_\lambda}. \end{aligned} \tag{5.12}$$

If strict priority is needed for $\boldsymbol{\lambda}$, the following gradient can be defined as an equality constraint:

$$\mathbf{S}\mathbf{J}_\lambda\ddot{\mathbf{q}} + \mathbf{C}_{\mathbf{e}_\lambda} = \mathbf{0}$$

to enforce the strict decrease of the error according to the desired error dynamics (5.7). Alternatively, if soft task priorities are preferred, additional objectives with a manually specified weight w_λ can be used:

$$w_\lambda\|\mathbf{e}_\lambda\|^2 = w_\lambda(\ddot{\mathbf{q}}^\top \mathbf{Q}_{\mathbf{e}_\lambda}\ddot{\mathbf{q}} + \mathbf{P}_{\mathbf{e}_\lambda}\ddot{\mathbf{q}}).$$

5.4.4 Whole-Body Collision Avoidance

Weights in different orders of magnitude can be chosen to enable a hierarchy of soft task priorities. However, such soft task priorities cannot strictly achieve the error dynamics, e.g., (5.7), due to the non-zero task residuals (5.12). Therefore, in case of critical situations such as collision avoidance, constraints are needed.

Suppose there are m obstacles around a robot with n links, and m pairs of a constraint and a task with different *proximity zones* allocated to each link, as shown in Fig. 5.5.

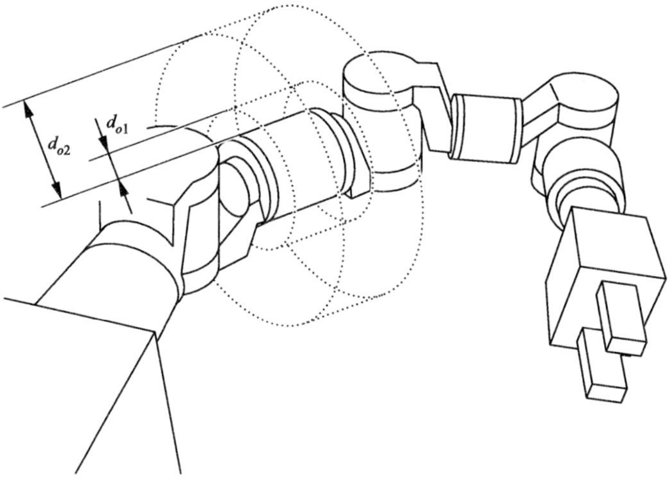

Fig. 5.5 Schematic view of a robot arm with collision avoidance zones around one of its links, marked by d_{o1} and d_{o2}

Performing continuous distance checking between the robot and an obstacle, the pair of nearest points $\mathbf{p}_r, \mathbf{p}_o \in \mathbb{R}^3$ belonging to the robot and the obstacle, respectively, is obtained. When the obstacle is closer than a threshold $d \leq d_{o1}$, the inequality constraint

$$S_i \mathbf{J}_{f_i} \ddot{\mathbf{q}} + C_{\mathbf{e}_{f_i}} \leq 0 \Longleftrightarrow \ddot{e}_j + K_{dj} \dot{e}_i + K_{pi} e_i \leq 0 \tag{5.13}$$

enforces the error dynamics

$$\ddot{\mathbf{e}} + \mathbf{K}_d \dot{\mathbf{e}} + \mathbf{K}_p \mathbf{e} = \mathbf{0},$$

where $\mathbf{e} = S_i \mathbf{n}_o d$, and $\mathbf{n}_o$ denotes the normal defined by $\mathbf{p}_r - \mathbf{p}_o$. If the distance is defined as $d(\mathbf{u}, \mathbf{q}) = \|\mathbf{p}_r - \mathbf{p}_o\|_2$, differentiating $d(\mathbf{u}, \mathbf{q})$ leads to the Jacobian $\mathbf{J}_{f_i}$:

$$\mathbf{J}_{f_i} = \frac{\partial d}{\partial \mathbf{q}} = 2(\mathbf{p}_r - \mathbf{p}_o)^\top \frac{\partial \mathbf{p}_r}{\partial \mathbf{q}}.$$

The obstacles can enter the proximity zone $d \leq d_{o1}$ with a significant velocity, which enforces an abrupt change of the gradient due to the velocity term $\mathbf{K}_d \dot{\mathbf{e}}$ of (5.13). Therefore, another proximity zone $d \leq d_{o2}$ is defined with $d_{o1} \leq d_{o2}$. When the distance to the obstacle is smaller than d_{o2}, a virtual admittance task is used to apply a virtual force that pushes the link away. Using a pre-defined virtual stiffness parameter, the virtual force is

$$\mathcal{F} = -\mathbf{K}_p \mathbf{n}_o (d - d_{o2}). \tag{5.14}$$

Assuming that the setpoint $\mathcal{F}^*$ of the virtual admittance task is zero, the robot would generate motion to mitigate the virtual force, and, therefore, move away from the approaching obstacle. The virtual admittance task is presented in detail in Sect. 5.4.6.1.

5.4.5 *QP Controller Synthesis*

Specifically, to avoid m obstacles, the proposed QP controller for a robot with n links is defined by

$$\min_{\mathbf{u}} \ \mathbf{u}^\top \mathbf{Q} \mathbf{u} + \underbrace{\sum_{i \in I_o, I_e, I_{ie}} w_i (\ddot{\mathbf{q}}^\top \mathbf{Q}_{f_i} \ddot{\mathbf{q}} + \mathbf{P}_{f_i} \ddot{\mathbf{q}})}_{\text{Objectives with } \textit{soft task priorities}}$$

$$\begin{aligned} &\text{s.t.} \\ &\text{Collision avoidance with } \textit{strict priorities}\text{:} \\ &\begin{cases} \{m \text{ collision avoidance constraints of form (5.13)} \\ \vdots \\ n \text{ sets of constraints for each link.} \end{cases}, \\ &\text{Joint position, velocity, torque constraints.} \end{aligned}$$

where each inequality constraint of the form (5.13) is associated with a virtual admittance task as mentioned before in Sect. 5.4.4.

5.4.6 *Details of Common Tasks*

The details of commonly used tasks in a QP controller are presented in this subsection: the admittance task in 5.4.6.1, the orientation task in 5.4.6.2, and the velocity task in 5.4.6.3.

5.4.6.1 Admittance Task

As force is measured locally in the end-effector frame, it is necessary to couple motion control and force regulation. Suppose a contact force $\mathcal{F} \in \mathbb{R}^3$ is generated at a position $\mathbf{p}_{\mathcal{F}}$ associated with a Jacobian $\mathbf{J}_{\mathcal{F}}$. To achieve the desired force $\mathcal{F}^* \in \mathbb{R}^3$, a contact point acceleration $\ddot{\mathbf{p}}_{\mathcal{F}} \in \mathbb{R}^n$ is generated as

$$\ddot{\mathbf{p}} = -\mathbf{K}_f(\mathcal{F} - \mathcal{F}^*) - \mathbf{K}_i \int_{t_I}^{t_I+\Delta t} (\mathcal{F} - \mathcal{F}^*)dt, \tag{5.15}$$

where the diagonal positive definite matrices $\mathbf{K}_f, \mathbf{K}_i \in \mathbb{R}^{3\times3}$ are proportional and integral gains. The integral term cancels out the steady-state error, see Sect. 12.3 of the book by Khalil [43]. The error term is defined as $\mathbf{e}_\mathcal{F} = (\mathcal{F} - \mathcal{F}^*)$. Denoting the environment stiffness as $\mathbf{K}_s$, the contact force relates to the end-effector position difference as

$$\mathcal{F} = \mathbf{K}_s(\mathbf{p}_\mathcal{F} - \hat{\mathbf{p}}_\mathcal{F}),$$

where $\hat{\mathbf{p}}_\mathcal{F}$ denotes the location of the environment constraint surface. Ideally, knowing the exact contact surface position, $\hat{\mathbf{p}}_\mathcal{F} = \mathbf{p}^*{}_\mathcal{F}$. The error dynamics associated with (5.15) becomes

$$\ddot{\mathbf{e}}_\mathbf{p} = -\mathbf{K}_f\mathbf{e}_\mathcal{F} - \mathbf{K}_i \int_{t_I}^{t_I+\Delta t} \mathbf{e}_\mathcal{F} dt \tag{5.16}$$

$$= -\mathbf{K}_f\mathbf{K}_s\mathbf{e}_\mathbf{p} - \mathbf{K}_i\mathbf{K}_s \int_{t_I}^{t_I+\Delta t} \mathbf{e}_\mathbf{p} dt. \tag{5.17}$$

The integral term guarantees that the regulation error tends to be zero. A large integral gain $\mathbf{K}_i$ can achieve (1) non-oscillatory transient force response; (2) controllable peak impulsive force without knowledge of the stiffness of the environment $\mathbf{K}_s$, see [44]. To achieve the error dynamics (5.16), the residual is defined:

$$\mathbf{e}_\mathbf{a} = \ddot{\mathbf{e}}_\mathbf{p} + \mathbf{K}_f\mathbf{e}_\mathcal{F} + \mathbf{K}_i \int_{t_I}^{t_I+\Delta t} \mathbf{e}_\mathcal{F} dt.$$

Concerning (5.11), it is easy to obtain $\ddot{\mathbf{e}}_\mathbf{p} = \mathbf{S}\dot{\mathbf{J}}_\mathcal{F}\dot{\mathbf{q}} + \mathbf{S}\mathbf{J}_\mathcal{F}\ddot{\mathbf{q}}$ and

$$\mathbf{e}_\mathbf{a} = \mathbf{S}\mathbf{J}_\mathcal{F}\ddot{\mathbf{q}} + \underbrace{\mathbf{S}\dot{\mathbf{J}}_\mathcal{F}\dot{\mathbf{q}} + \mathbf{K}_f\mathbf{S}\mathbf{e}_\mathcal{F} + \mathbf{K}_i \int_{t_I}^{t_I+\Delta t} \mathbf{S}\mathbf{e}_{\mathcal{F}dt}}_{\mathbf{C}_{\mathbf{e}_a}}.$$

Expanding its quadratic form leads to

$$\|\mathbf{e}_\mathbf{a}\|^2 = \mathbf{e}_\mathbf{a}^\top \mathbf{e}_\mathbf{a} = \ddot{\mathbf{q}}^\top \underbrace{\mathbf{J}_\mathcal{F}^\top \mathbf{S}^\top \mathbf{S}\mathbf{J}_\mathcal{F}}_{\mathbf{Q}_{\mathbf{e}_a}} \ddot{\mathbf{q}} + \underbrace{2\mathbf{C}_{\mathbf{e}_a}^\top \mathbf{S}\mathbf{J}_\mathcal{F}}_{\mathbf{P}_{\mathbf{e}_a}} \ddot{\mathbf{q}} + \mathbf{C}_{\mathbf{e}_a}^\top \mathbf{C}_{\mathbf{e}_a}, \tag{5.18}$$

where $\mathbf{Q}_{\mathbf{e}_a}$ and $\mathbf{P}_{\mathbf{e}_a}$ will be used in the QP objective.

Remark 5.2 In practice, the integral will be calculated in a discrete form. Denoting $\mathbf{I}(t) = \mathbf{K}_i \int_{t_I}^{t_I+\Delta t} \mathbf{e}_{\mathcal{F}} dt$, and hence $\frac{\partial \mathbf{I}}{\partial t} = \mathbf{K}_i \mathbf{e}_{\mathcal{F}}$, the forward difference becomes

$$\frac{\mathbf{I}(t_{k+1}) - \mathbf{I}(t_k)}{\Delta t} = \mathbf{K}_i \mathbf{e}_{\mathcal{F}}(t_k) \Rightarrow \mathbf{I}(t_{k+1}) = \mathbf{I}(t_k) + \mathbf{K}_i \mathbf{e}_{\mathcal{F}}(t_k)\Delta t,$$

which means that the integral term can be pre-calculated in each step. ∎

5.4.6.2 Orientation Task

Either Euler angles or the angle-and-axis representation, $\{\vartheta, \mathbf{r}\}$ can be used to define the desired and current orientation $\mathbf{x}^*$ and $\mathbf{x}$.

Whereas, the problem associated with Euler angles or the angle-and-axis representation is that a representation singularity occurs at 90°.

This representation problem can be addressed by the unit quaternion $\mathbb{Q} = \{\eta, \boldsymbol{\epsilon}\}$, where $\eta = \cos\frac{\vartheta}{2}$, $\boldsymbol{\epsilon} = \sin\frac{\vartheta}{2}\mathbf{r}$ and $\|\boldsymbol{\epsilon}\|^2 = 1$. The orientation error $\mathbf{R}_{se}^{*\top}\mathbf{R}_{se}$ is represented as

$$\mathbf{e}_{\mathbb{Q}} = \bar{\mathbb{Q}}_{se}^* * \mathbb{Q}_{se} = \bar{Q}_m(\mathbb{Q}_{se}^*)\mathbb{Q}_{se}, \tag{5.19}$$

where the conjugate $\bar{\mathbb{Q}}$ is defined as $\bar{\mathbb{Q}} = \{\eta, -\boldsymbol{\epsilon}\}$ and $\mathbf{Q}_m(\mathbb{Q}) : \mathbb{R}^4 \rightarrow \mathbb{R}^{4\times 4}$ denotes the *quaternion matrix*. According to [45], there exists

$$\mathbf{Q}_m = \begin{bmatrix} \eta & -\boldsymbol{\epsilon}^\top \\ \boldsymbol{\epsilon} & \eta\mathbf{I} + \hat{\boldsymbol{\epsilon}} \end{bmatrix}, \quad \bar{\mathbf{Q}}_m = \begin{bmatrix} \eta & \boldsymbol{\epsilon}^\top \\ -\boldsymbol{\epsilon} & \eta\mathbf{I} - \hat{\boldsymbol{\epsilon}} \end{bmatrix}, \tag{5.20}$$

where $\hat{\boldsymbol{\epsilon}} : \mathbb{R}^3 \rightarrow \mathbb{R}^{3\times 3}$ denotes the *cross product matrix*. Using the first- and second-order derivatives of the unit quaternion with respect to the spatial angular velocity $\hat{\omega}_{se}^s = \mathbf{R}^{-1}\dot{\mathbf{R}}$, see Remark 5.3, there are

$$\begin{aligned} \dot{\mathbf{e}}_{\mathbb{Q}} &= \bar{\mathbf{Q}}_m(\mathbb{Q}_{se}^*)\dot{\mathbb{Q}}_{se} = \frac{1}{2}\bar{\mathbf{Q}}_m(\mathbb{Q}_{se}^*)\mathbf{W}(\mathbb{Q}_{se})^\top \mathbf{w}_{se}^s \\ \ddot{\mathbf{e}}_{\mathbb{Q}} &= \bar{\mathbf{Q}}_m(\mathbb{Q}_{se}^*)\ddot{\mathbb{Q}}_{se} = \frac{1}{2}\bar{\mathbf{Q}}_m(\mathbb{Q}_{se}^*)\mathbf{W}(\mathbb{Q}_{se})^\top \dot{\mathbf{w}}_{se}^s \end{aligned}, \tag{5.21}$$

where $\mathbf{w}_{se}^s = \mathbf{J}_{se}^s\dot{\mathbf{q}}$, $\dot{\mathbf{w}}_{se}^s = (\dot{\mathbf{J}}_{se}^s\dot{\mathbf{q}} + \mathbf{J}_{se}^s\ddot{\mathbf{q}})$, and $\mathbf{J}_{se}^s \in \mathbb{R}^{3\times n}$ denotes the spatial angular velocity Jacobian. The *quaternion rates matrix* $\mathbf{W} : \mathbb{R}^4 \rightarrow \mathbb{R}^{3\times 4}$ is defined as

$$\begin{aligned} \mathbf{W}(\mathbb{Q}) &= \mathbf{D}\mathbf{Q}_m^\top = \begin{bmatrix} 0\,1\,0\,0 \\ 0\,0\,1\,0 \\ 0\,0\,0\,1 \end{bmatrix} \mathbf{Q}_m^\top \\ \mathbf{W}'(\mathbb{Q}) &= \mathbf{D}\bar{\mathbf{Q}}_m^\top = \begin{bmatrix} 0\,1\,0\,0 \\ 0\,0\,1\,0 \\ 0\,0\,0\,1 \end{bmatrix} \bar{\mathbf{Q}}_m^\top \end{aligned} . \tag{5.22}$$

If the body velocity $\hat{\omega}_{se}^e$ is needed, there exist

$$\begin{aligned} \dot{\mathbf{e}}_{\mathbb{Q}} &= \bar{\mathbf{Q}}_m(\mathbb{Q}_{se}^*)\dot{\mathbb{Q}}_{se} = \frac{1}{2}\bar{\mathbf{Q}}_m(\mathbb{Q}_{se}^*)\mathbf{W}'(\mathbb{Q}_{se})^\top \mathbf{w}_{se}^e \\ \ddot{\mathbf{e}}_{\mathbb{Q}} &= \bar{\mathbf{Q}}_m(\mathbb{Q}_{se}^*)\ddot{\mathbb{Q}}_{se} = \frac{1}{2}\bar{\mathbf{Q}}_m(\mathbb{Q}_{se}^*)\mathbf{W}'(\mathbb{Q}_{se})^\top \dot{\mathbf{w}}_{se}^e \end{aligned} . \tag{5.23}$$

When $\mathbf{R}_{se}^*$ is identical to $\mathbf{R}_{se}$, $\mathbf{e}_{\mathbb{Q}} = \{1, \mathbf{0}\}$ is obtained. Thus only the vector part $\boldsymbol{\epsilon}$ of the error $\mathbf{e}_{\mathbb{Q}}$ is needed. To achieve the desired error dynamics

$$\ddot{\mathbf{e}}_\epsilon = -\mathbf{K}_d\dot{\mathbf{e}}_\epsilon - \mathbf{K}_p\mathbf{e}_\epsilon,$$

the quadratic form of the following is minimised:

$$\mathbf{J}_{\mathbf{e}_{\mathbb{Q}}}\ddot{\mathbf{q}} + \mathbf{C}_{\mathbf{e}_{\mathbb{Q}}},$$

where $\mathbf{J}_{\mathbf{e}_{\mathbb{Q}}}$ and $\mathbf{C}_{\mathbf{e}_{\mathbb{Q}}}$ are defined as

$$\begin{aligned} \mathbf{e}_o &= \ddot{\mathbf{e}}_\epsilon + \mathbf{K}_d\dot{\mathbf{e}}_\epsilon + \mathbf{K}_p\mathbf{e}_\epsilon = \frac{1}{2}\mathbf{D}\bar{\mathbf{Q}}_m(\mathbb{Q}_{se}^*)\mathbf{W}(\mathbb{Q}_{se})^\top(\dot{\mathbf{J}}_{se}^s\dot{\mathbf{q}} + \mathbf{J}_{se}^s\ddot{\mathbf{q}}) \\ &+ \mathbf{K}_d\mathbf{D}\frac{1}{2}\bar{\mathbf{Q}}_m(\mathbb{Q}_{se}^*)\mathbf{W}(\mathbb{Q}_{se})^\top\mathbf{J}_{se}^s\dot{\mathbf{q}} + \mathbf{K}_p\mathbf{D}\bar{\mathbf{Q}}_m(\mathbb{Q}_{se}^*)\mathbb{Q}_{se} \\ &= \underbrace{\frac{1}{2}\mathbf{D}\bar{\mathbf{Q}}_m(\mathbb{Q}_{se}^*)\mathbf{W}(\mathbb{Q}_{se})^\top\mathbf{J}_{se}^s}_{\mathbf{J}_{\mathbf{e}_{\mathbb{Q}}}\in\mathbb{R}^{4\times n}}\ddot{\mathbf{q}} \\ &+ \underbrace{\frac{1}{2}\mathbf{D}\bar{\mathbf{Q}}_m(\mathbb{Q}_{se}^*)\mathbf{W}(\mathbb{Q}_{se})^\top\left(\dot{\mathbf{J}}_{se}^s + \mathbf{K}_d\mathbf{J}_{se}^s\right)\dot{\mathbf{q}} + \mathbf{K}_p\mathbf{D}\bar{\mathbf{Q}}_m(\mathbb{Q}_{se}^*)\mathbb{Q}_{se}}_{\mathbf{C}_{\mathbf{e}_{\mathbb{Q}}}\in\mathbb{R}^{4\times 1}} \\ &= \mathbf{J}_{\mathbf{e}_{\mathbb{Q}}}\ddot{\mathbf{q}} + \mathbf{C}_{\mathbf{e}_{\mathbb{Q}}}. \end{aligned}$$

Thus, the orientation task objective is formulated via $\mathbf{J}_{\mathbf{e}_{\mathbb{Q}}}$ and $\mathbf{C}_{\mathbf{e}_{\mathbb{Q}}}$:

$$\|\mathbf{e}_o\|^2 = \mathbf{e}_o^\top\mathbf{e}_o = \ddot{\mathbf{q}}^\top\underbrace{\mathbf{J}_{\mathbf{e}_{\mathbb{Q}}}^\top\mathbf{S}^\top\mathbf{S}\mathbf{J}_{\mathbf{e}_{\mathbb{Q}}}}_{\mathbf{Q}_{\mathbf{e}_{\mathbb{Q}}}}\ddot{\mathbf{q}} + \underbrace{2\mathbf{C}_{\mathbf{e}_{\mathbb{Q}}}^\top\mathbf{S}\mathbf{J}_{\mathbf{e}_{\mathbb{Q}}}}_{\mathbf{P}_{\mathbf{e}_{\mathbb{Q}}}}\ddot{\mathbf{q}} + \mathbf{C}_{\mathbf{e}_{\mathbb{Q}}}^\top\mathbf{C}_{\mathbf{e}_{\mathbb{Q}}}.$$

Remark 5.3 *(Quaternion rate and body velocities)* According to the spatial angular velocity definition (2.60) in [46] and definition (146) in [45], there exists

$$\begin{bmatrix} 0 \\ \omega^s \end{bmatrix} = 2\bar{\mathbb{Q}} * \dot{\mathbb{Q}} = 2\mathbf{Q}_m(\mathbb{Q})^\top \dot{\mathbb{Q}},$$

which in turn gives us the reciprocal relation:

$$\dot{\mathbb{Q}} = \frac{1}{2}\mathbf{W}(\mathbb{Q})^\top \omega^s.$$

According to equation (166) in [45], the second-order derivative is given by

$$\dot{\mathbb{Q}} = \frac{1}{2}\mathbf{W}(\mathbb{Q})^\top \dot{\omega}^s.$$

Based on these two relations, the first- and second-order derivatives in (5.21) are derived. ■

5.4.6.3 Velocity Task

The error definition and the first-order dynamics are given as

$$\mathbf{e} = \mathbf{S}\mathbf{v}_i - \mathbf{S}\mathbf{v}_i^* = \mathbf{J}_{\mathbf{v}_i}\mathbf{S}\dot{\mathbf{q}} - \mathbf{S}\mathbf{v}_i^* \tag{5.24}$$

$$\dot{\mathbf{e}} = \mathbf{S}\dot{\mathbf{J}}_{\mathbf{v}_i}\dot{\mathbf{q}} + \mathbf{S}\mathbf{J}_{\mathbf{v}_i}\ddot{\mathbf{q}}. \tag{5.25}$$

To achieve the error dynamics $\dot{\mathbf{e}} = -\mathbf{K}_p\mathbf{e}$, the velocity task error is expanded as

$$\mathbf{e}_{\mathbf{v}_i} = \dot{\mathbf{e}} + \mathbf{K}_p\mathbf{e} = \mathbf{S}\mathbf{J}_{\mathbf{v}_i}\ddot{\mathbf{q}} + \underbrace{\mathbf{S}(\dot{\mathbf{J}}_{\mathbf{v}_i} + \mathbf{K}_p\mathbf{J}_{\mathbf{v}_i})\dot{\mathbf{q}} - \mathbf{K}_p\mathbf{S}\mathbf{v}_i^*}_{\mathbf{C}_{\mathbf{e}\mathbf{v}_i}},$$

whose quadratic form is

$$\|\mathbf{e}_{\mathbf{v}_i}\|^2 = \mathbf{e}_{\mathbf{v}_i}^\top\mathbf{e}_{\mathbf{v}_i} = \ddot{\mathbf{q}}^\top \underbrace{\mathbf{J}_{\mathbf{v}_i}^\top\mathbf{S}^\top\mathbf{S}\mathbf{J}_{\mathbf{v}_i}}_{\mathbf{Q}_{\mathbf{e}\mathbf{v}_i}}\ddot{\mathbf{q}} + \underbrace{2\mathbf{C}_{\mathbf{e}\mathbf{v}_i}^\top\mathbf{S}\mathbf{J}_{\mathbf{v}_i}}_{\mathbf{P}_{\mathbf{e}\mathbf{v}_i}}\ddot{\mathbf{q}} + \mathbf{C}_{\mathbf{e}\mathbf{v}_i}^\top\mathbf{C}_{\mathbf{e}\mathbf{v}_i}.$$

5.5 Experimental Validation

The proposed formulation (5.8) has been validated using the dual-arm robot shown in Fig. 5.1. The robot has two *Schunk* lightweight arms, each of them having 7 degrees of freedom. As a non-trivial task, the robot is programmed to perform a task resembling pan cleaning, in which the pan and the cleaning utensil in Fig. 5.1 need to follow a circular motion relative to each other. The detailed description of the constraints is given in Sect. 5.5.1.

In addition to the experimental results, open-source simulation packages are provided to encourage the interested readers to check the details of a QP controller in Python [47] and C++ [48].

Gurobi has been chosen as the optimisation solver, and the *FCL* library to perform continuous collision checking. The feedback loop runs at 200 Hz. According to the benchmark against *QPLIB*, a summary of quadratic programming instances by Furini et al. [49], Gurobi is the best commercial solver among *CPLEX*, *XPRESS-ANT*, *SCIP*, *COU*, and *BARON* as reported recently by H. D. Mittelmann [50]. Due to the limited degrees of freedom and number of tasks, the QP controller for a manipulator typically has less than 100 optimisation variables and 100 constraints. According to the simulation and experiments conducted with the example of this chapter, Gurobi can solve such QP problems within even less than 1 millisecond. Meanwhile, convex shapes are used to capsulate the robot and the objects. According to the performance reported by Pan et al. [38], FCL requires less than a millisecond to check a PR2 robot against primitive types.

From Sect. 5.5.1, it can be seen that the proposed QP controller is able to simultaneously fulfil multiple objectives. In the given case, simulated dynamic obstacles are used to validate the collision avoidance constraints in Sect. 5.5.2. In Sect. 5.5.3, an example illustrates the conversion to an objective when a constraint becomes infeasible. In Sect. 5.5.4, a faster obstacle is used to demonstrate that the collision avoidance constraint has the highest priority over the other objectives and constraints, i.e., in order to avoid a faster obstacle, the other constraints are no longer well-satisfied as before.

5.5.1 Simultaneous Satisfaction of Objectives and Constraints

The pan cleaning task can be specified by the relative translation and orientation between the two manipulators. Three equality constraints are used to restrict the translation, and one inequality to keep the relative orientation within a cone:

$$\begin{bmatrix} f_1(\mathbf{q}) \\ f_2(\mathbf{q}) \\ f_3(\mathbf{q}) \end{bmatrix} = \mathbf{p}_1(\mathbf{q}) - \mathbf{p}_2(\mathbf{q}) - \mathbf{p}^*(t, \mathbf{x}_1, \mathbf{y}_1, \mathbf{z}_1, \mathbf{x}_2) = 0$$

$$f_4(\mathbf{q}) = \mathbf{x}_1^T \mathbf{x}_2 \leq b_4,$$

where axis $\mathbf{x}_i, \mathbf{y}_i, \mathbf{z}_i \in \mathbb{R}^3$ are columns of $\mathbf{R}_i \in SO(3)$, $\mathbf{p}_1 \in \mathbb{R}^3$ is supposed to be the contact point of the frying pan and $\mathbf{p}_2 \in \mathbb{R}^3$ is supposed to be the position of

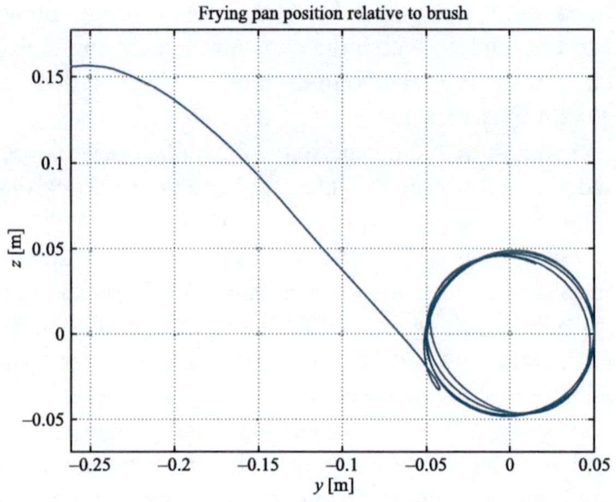

Fig. 5.6 Relative trajectory between pan and brush, following a pre-defined circular motion

the cleaning utensil tip. The offset $\mathbf{p}^*(t, \mathbf{x}_1, \mathbf{y}_1, \mathbf{z}_1, \mathbf{x}_2)$ is time-dependent, expressing how the relative motion between $\mathbf{p}_1$ and $\mathbf{p}_2$ evolves. For simplicity, in this set of experiments, the relative offset $\mathbf{p}^*(t, \mathbf{x}_1, \mathbf{y}_1, \mathbf{z}_1, \mathbf{x}_2)$ is selected to follow a circular pattern[1]. Figure 5.6 shows that the two manipulators start from different initial positions in the relative space between each other, and then maintain a circular motion in accordance with the pre-defined relative offset. The trajectory shows that the robot motion is not repetitive, due to the solution being locally optimal.

Without loss of generality, the convergence gain is chosen as $k_i = 1$ for $i = 1, \ldots, 5$. For the translation equality constraints, the bounds are set as $b_{1-3} = 0$. For the relative orientation, the bound is $b_4 = -\cos\left(\frac{\pi}{8}\right)$, which keeps the relative orientation between the two x_i axes within a cone. In the first two rows of Fig. 5.7, it can be seen that at the beginning of the experiments the norm of the translation between two manipulators converges exponentially to zero, while the relative orientation is kept below b_4. As pointed out by several textbooks, the robot needs to maintain a reasonable transmission ratio between joint space velocity and Cartesian space velocity, i.e., keep a good condition number for the sake of numerical stability. Therefore, the following singularity measure is limited with an inequality constraint:

$$f_5(\mathbf{q}) = \frac{-1}{2}\det(J_i^T J_i) \leq b_5, \quad i \in \{1, 2\}.$$

Here, only the result for the translation velocity part is listed to give it a clear physical interpretation. In the third row of Fig. 5.7, it can be seen that f_5 is kept well below the bound $b_5 = 0.00005$. Note that the absolute value of this measure varies from one manipulator to the other. Typically, an empirical value is chosen according to [13]. Here, $b_5 = 0.00005$ was chosen after several trials.

[1]The velocity depends on the joint velocity limits. In this case, it is $0.25 \cdot \pi$ rad/s.

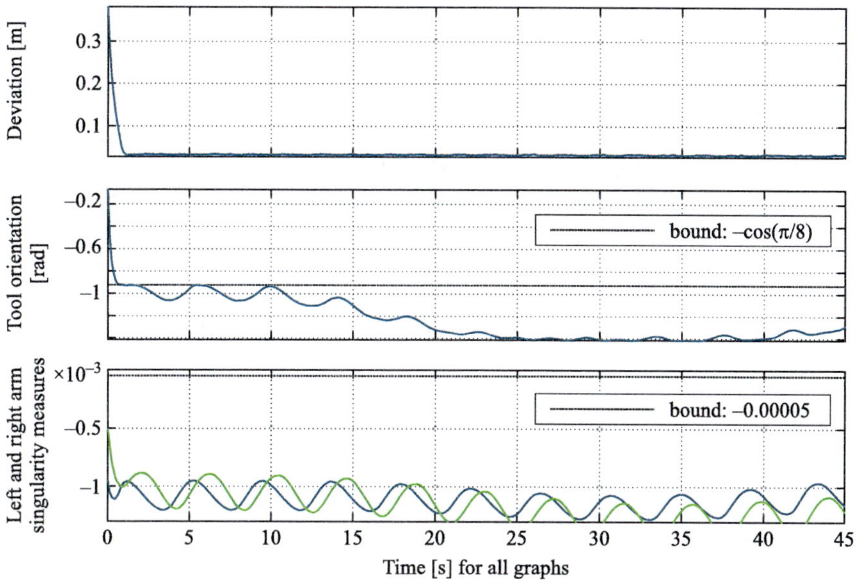

Fig. 5.7 Performance indices

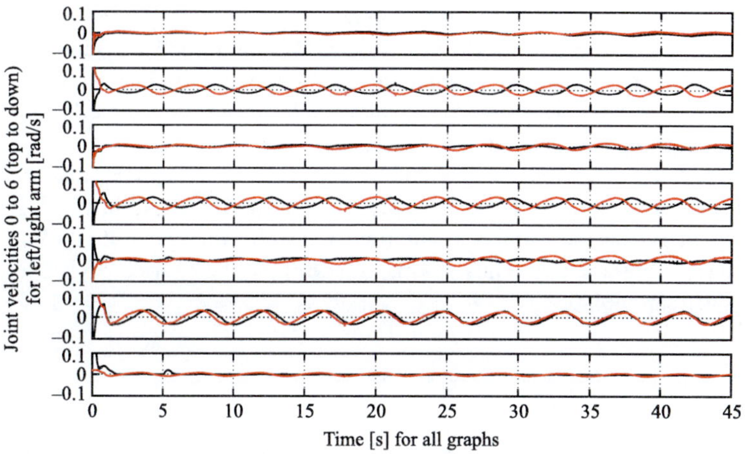

Fig. 5.8 Joint velocities of the two arms (left: red, right: black)

Given Remark 5.1, the joint velocity commands of the two manipulators are plotted in Fig. 5.8. These velocity commands are generated by solving the proposed QP controller (5.8) using sensory feedback in real time. In line with the results shown in Fig. 5.2b, smooth robot motion is achieved.

5.5.2 *Dynamic Collision Avoidance*

In this example, the obstacles are defined as two virtual planes orthogonal to the z axis. The relative distance between them is 0.3 m. The two robot end-effectors need to avoid the two moving virtual planes while performing a circular motion. In this way, "ground truth" is given to benchmark the performance of the proposed solution.

The two virtual planes are simulated to follow a sine wave with an amplitude of 0.1 m. The positions of the two virtual planes are plotted in Fig. 5.9, where the two end-effectors managed to avoid the obstacles with a clear margin. Despite that initially the obstacle constraint is not satisfied, the QP can bring the end-effector position within the bounds later on, employing the redundancy of the robot.

5.5.3 *Resolving Infeasible Constraints with Slack Variables*

As mentioned before in Sect. 5.4, a constraint can be converted to a task and vice versa if conflicting constraints occur. Using slack variables can help to easily achieve this goal and obtain smooth transitions without using complicated mechanisms, e.g., a hybrid state machine. From Figs. 5.6, 5.7, and 5.9, it can be concluded that the initial robot configurations do not fulfil the constraints $f_1 \ldots f_3$, f_4, and f_5. However, Fig. 5.8 shows that these unsatisfied constraints did not introduce discontinuities to the joint commands. The reason behind this could be explained by the results shown in Figs. 5.10 and 5.11, where the slack variables (or task residuals) associated with the collision avoidance constraints are plotted.

Checking the magnitude of the axis of Fig. 5.11, it can be seen that the slack variables shown in Fig. 5.11 are much larger compared to those in Fig. 5.10. The reason for this is that the initial configuration of the robot does not fulfil the bound for the distance to obstacle two. Thus, the magnitude of the slack variable is proportional to the constraint violation.

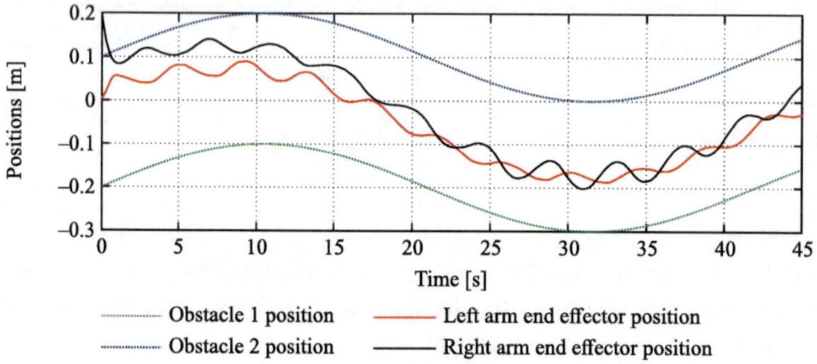

Fig. 5.9 End-effector positions along the z-axis, obtained with constraints set for avoiding obstacles following a sinusoidal motion

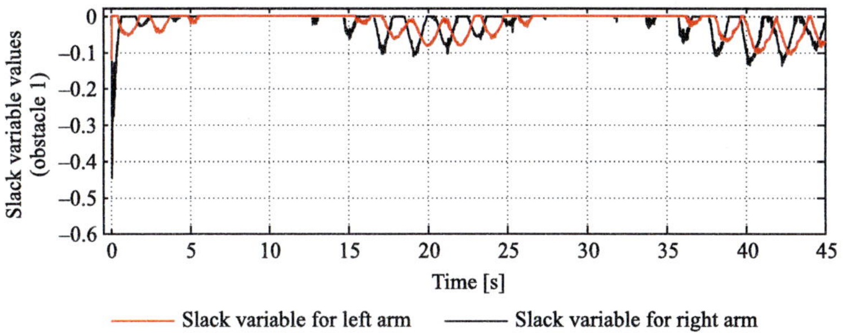

Fig. 5.10 Slack variables for avoiding obstacle one

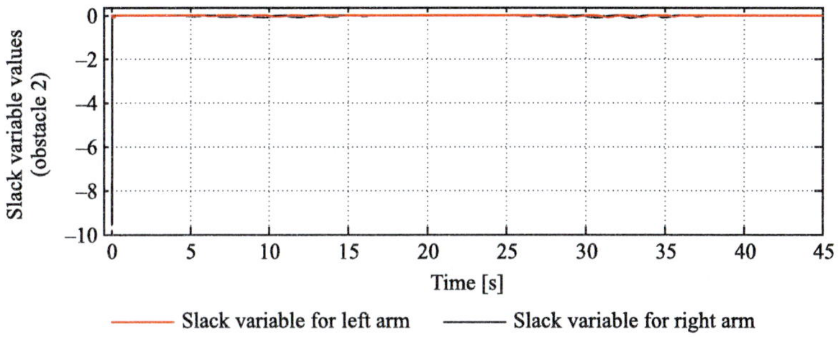

Fig. 5.11 Slack variables for avoiding obstacle two

5.5.4 *High-Priority Obstacle Avoidance Constraint*

As long as the constraint formulation (5.13) is kept, collision avoidance always dominates the other constraints and objectives. In this subsection, the dominance is validated using obstacles moving with higher velocity.

In Fig. 5.12 the obstacles one and two are moving twice as fast as in Fig. 5.9. In Fig. 5.12, the obstacle avoidance constraint is still fulfilled at the price of breaking the equality constraint that specifies the relative translation (see Fig. 5.13).

5.6 Conclusion

As robots become more and more collaborative, an increasing demand can be observed for reactively resolving robot redundancy for obstacle avoidance while accomplishing other task specifications. The QP approach proposed in this chapter easily specifies strict priorities without breaking controller feasibility. Experimental validation has demonstrated that the proposed solution can solve infeasible problems induced by conflicting constraints, and assign the highest priority to obstacle avoidance, while trying to fulfil other task objectives and constraints. Such an active-safety

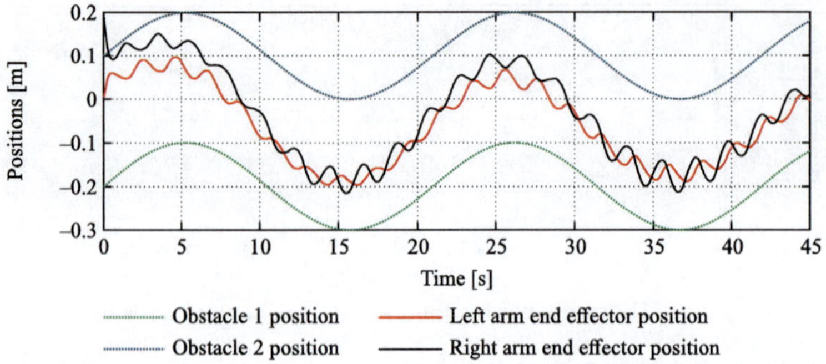

Fig. 5.12 Task example with obstacles moving faster—the frequency of the sinusoidal wave is twice of that in the case of Fig. 5.9

Fig. 5.13 Avoiding faster moving obstacles results in degradation of the relative circular motion, compared to Fig. 5.6

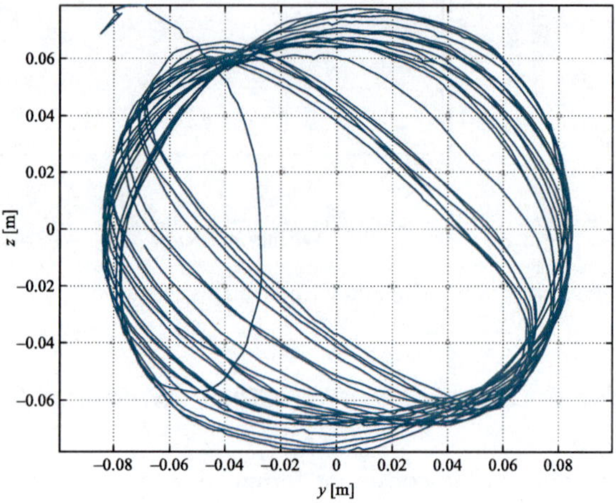

control mechanism offers the possibility of a robot to work with humans without hard separation.

Despite the spreading use of multi-objective control through a QP formulation over the last decade, open issues such as constraint feasibility, smooth task transition, and coefficients tuning are not fully addressed yet. More specifically, the field of collaborative robots is expecting: (1) collision/impact detection for under actuated robots; (2) human–robot interaction through multi-contact or impact; (3) understanding/interpreting contact properties to distinguish intentional from accidental contact; (4) a readily available QP control framework that is suitable for numerous tasks and generic robot platforms.

Acknowledgements Yuquan Wang is partially supported by the National Natural Science Foundation of China (U1613216) and Shenzhen Fundamental Research Grant (JCYJ20180508162406177) through the Chinese University of Hong Kong, Shenzhen. He gratefully acknowledges the support.

References

1. A. Realyvásquez-Vargas, K.C. Arredondo-Soto, J.L. García-Alcaraz, B.Y. Márquez-Lobato, J. Cruz-García, Introduction and configuration of a collaborative robot in an assembly task as a means to decrease occupational risks and increase efficiency in a manufacturing company. Robot. Comput. Integr. Manuf. **57**, 315–328 (2019)
2. J. Van Den Berg, D. Ferguson, J. Kuffner, Anytime path planning and replanning in dynamic environments, in *Proceedings 2006 IEEE International Conference on Robotics and Automation, 2006. ICRA 2006* (IEEE, 2006), pp. 2366–2371
3. J. Vannoy, J. Xiao, Real-time adaptive motion planning (RAMP) of mobile manipulators in dynamic environments with unforeseen changes. IEEE Trans. Robot. **24**(5), 1199–1212 (2008)
4. N. Mansard, O. Khatib, A. Kheddar, A unified approach to integrate unilateral constraints in the stack of tasks. IEEE Trans. Robot. **25**(3), 670–685 (2009)
5. O. Kanoun, F. Lamiraux, P.-B. Wieber, F. Kanehiro, E. Yoshida, J.-P. Laumond, Prioritizing linear equality and inequality systems: application to local motion planning for redundant robots, in *IEEE International Conference on Robotics and Automation, 2009. ICRA'09* (IEEE, 2009), pp. 2939–2944
6. J. Schulman, J. Ho, A.X Lee, I. Awwal, H. Bradlow, P. Abbeel, Finding locally optimal, collision-free trajectories with sequential convex optimization. Robot. Sci. Syst. **9**, 1–10 (Citeseer, 2013)
7. P. Ogren, C. Smith, Y. Karayiannidis, D. Kragic, A multi objective control approach to online dual arm manipulation, in *International IFAC Symposium on Robot Control, SyRoCo* (Dubrovnik, Croatia, 2012)
8. Y. Wang, F. Vina, Y. Karayiannidis, C. Smith, P. Ögren, Dual arm manipulation using constraint based programming, in *19th IFAC World Congress* (Cape Town, South Africa, 2014)
9. M.R. Pedersen, L. Nalpantidis, R.S. Andersen, C. Schou, S. Bøgh, V. Krüger, O. Madsen, Robot skills for manufacturing: from concept to industrial deployment. Robot. Comput.-Integr. Manuf. **37**, 282–291 (2016)
10. O. Kanoun, F. Lamiraux, P.-B. Wieber, Kinematic control of redundant manipulators: generalizing the task-priority framework to inequality task. IEEE Trans. Robot. **27**(4), 785–792 (2011)
11. R.V. Patel, F. Shadpey, F. Ranjbaran, J. Angeles, A collision-avoidance scheme for redundant manipulators: theory and experiments. J. Robot. Syst. **22**(12), 737–757 (2005)
12. S. Dalibard, A. Nakhaei, F. Lamiraux, J.-P. Laumond, Whole-body task planning for a humanoid robot: a way to integrate collision avoidance, in *9th IEEE-RAS International Conference on Humanoid Robots, 2009. Humanoids 2009* (IEEE, 2009), pp. 355–360
13. B. Siciliano, L. Sciavicco, L. Villani, G. Oriolo. *Robotics: modelling, Planning and Control* (Springer Science & Business Media, 2009)
14. L.E. Kavraki, P. Svestka, J.-C. Latombe, M.H. Overmars, Probabilistic roadmaps for path planning in high-dimensional configuration spaces. IEEE Trans. Robot. Autom. **12**(4), 566–580 (1996)
15. S.M. LaValle, J.J. Kuffner, Randomized kinodynamic planning. Int. J. Robot. Res. **20**(5), 378–400 (2001)
16. M. Elbanhawi, M. Simic, Sampling-based robot motion planning: a review. IEEE Access **2**, 56–77 (2014)
17. M. Otte, E. Frazzoli, RRT^X: real-time motion planning/replanning for environments with unpredictable obstacles, in *Algorithmic Foundations of Robotics XI* (Springer, 2015), pp 461–478
18. M. Zucker, N. Ratliff, A.D. Dragan, M. Pivtoraiko, M. Klingensmith, C.M. Dellin, J.A. Bagnell, S.S. Srinivasa, Chomp: covariant hamiltonian optimization for motion planning. Int. J. Robot. Res. **32**(9–10), 1164–1193 (2013)
19. C. Suh, T.T. Um, B. Kim, H. Noh, M. Kim, F.C. Park, Tangent space RRT: a randomized planning algorithm on constraint manifolds, in *2011 IEEE International Conference on Robotics and Automation (ICRA)* (IEEE, 2011), pp. 4968–4973

20. L. Jaillet, J.M. Porta, Efficient asymptotically-optimal path planning on manifolds. Robot. Auton. Syst. **61**(8), 797–807 (2013)
21. Y. Nakamura, *Advanced Robotics: redundancy and Optimization* (Addison-Wesley Longman Publishing Co., Inc., 1990)
22. J. De Schutter, T. De Laet, J. Rutgeerts, W. Decré, R. Smits, E. Aertbeliën, K. Claes, H. Bruyninckx, Constraint-based task specification and estimation for sensor-based robot systems in the presence of geometric uncertainty. Int. J. Robot. Res. **26**(5), 433–455 (2007)
23. M.P. Polverini, A.M. Zanchettin, P. Rocco, A constraint-based programming approach for robotic assembly skills implementation. Robot. Comput.-Integr. Manuf. **59**, 69–81 (2019)
24. J. Schulman, Y. Duan, J. Ho, A. Lee, I. Awwal, H. Bradlow, J. Pan, S. Patil, K. Goldberg, P. Abbeel, Motion planning with sequential convex optimization and convex collision checking. Int. J. Robot. Res. **33**(9), 1251–1270 (2014)
25. S. Lengagne, J. Vaillant, E. Yoshida, A. Kheddar, Generation of whole-body optimal dynamic multi-contact motions. Int. J. Robot. Res. **32**(9–10), 1104–1119 (2013)
26. H. Seraji, R. Colbaugh, Singularity-robustness and task-prioritization in configuration control of redundant robots, in *29th IEEE Conference on Decision and Control* (IEEE, 1990), pp. 3089–3095
27. H. Audren, A. Kheddar, P. Gergondet, Stability polygons reshaping and morphing for smooth multi-contact transitions and force control of humanoid robots, in *2016 IEEE-RAS 16th International Conference on Humanoid Robots (Humanoids)* (IEEE, 2016), pp. 1037–1044
28. J. Vaillant, K. Bouyarmane, A. Kheddar, Multi-character physical and behavioral interactions controller. IEEE Trans. Vis. Comput. Graphic (2016)
29. A. Escande, N. Mansard, P.-B. Wieber, Hierarchical quadratic programming: fast online humanoid-robot motion generation. Int. J. Robot. Res. **33**(7), 1006–1028 (2014)
30. M. Kimmel, S. Hirche, Active safety control for dynamic human–robot interaction, in *IEEE/RSJ Intelligent Robots and Systems (IROS)* (2015), pp. 4685–4691
31. N. Dehio, R.F. Reinhart, J.J. Steil, Multiple task optimization with a mixture of controllers for motion generation, in *2015 IEEE/RSJ International Conference on Intelligent Robots and Systems (IROS)* (IEEE, 2015), pp. 6416–6421
32. Y. Su, Y. Wang, A. Kheddar, Sample-efficient learning of soft task priorities through bayesian optimization, in *IEEE-RAS 18th International Conference on Humanoid Robots (Humanoids)* (2018), pp. 1–6
33. W. Decré, R. Smits, H. Bruyninckx, J. De Schutter, Extending itasc to support inequality constraints and non-instantaneous task specification, in *IEEE International Conference on Robotics and Automation, 2009. ICRA'09* (IEEE, 2009), pp. 964–971
34. A. Del Prete, Joint position and velocity bounds in discrete-time acceleration/torque control of robot manipulators. IEEE Robot. Autom. Lett. **3**(1), 281–288 (2018)
35. Y. Wang, D. Niels, A. Tanguy, A. Kheddar, Impact-aware task-space quadratic-programming control (2020). https://arxiv.org/pdf/2006.01987.pdf
36. Y. Wang, A. Kheddar, Impact-friendly robust control design with task-space quadratic optimization, in *Proceedings of Robotics: science and Systems*, vol. 15 (Freiburg, Germany, 2019), p. 32. 10.15607/RSS.2019.XV.032
37. O. Stasse, A. Escande, N. Mansard, S. Miossec, P. Evrard, A. Kheddar, Real-time (self)-collision avoidance task on a HRP-2 humanoid robot, in *IEEE International Conference on Robotics and Automation, 2008. ICRA 2008* (IEEE, 2008), pp. 3200–3205
38. J. Pan, S. Chitta, D. Manocha, FCL: a general purpose library for collision and proximity queries, in *2012 IEEE International Conference on Robotics and Automation (ICRA)* (IEEE, 2012), pp. 3859–3866
39. M. Colledanchise, P. Ögren, Behavior trees in robotics and AI, an introduction (2017). arXiv preprint arXiv:1709.00084
40. G.M. Lee, N.N. Tam, N.D. Yen, Continuity of the solution map in quadratic programs under linear perturbations. J. Optim. Theory Appl. **129**(3), 415–423 (2006)
41. J.F. Bonnans, A. Shapiro, *Perturbation Analysis of Optimization Problems* (Springer Science & Business Media, 2013)

42. S. Boyd, L. Vandenberghe, *Convex Optimization* (Cambridge University Press, 2004)
43. Ha.K Khalil, Nonlinear systems, in *Upper Saddle River* (2002)
44. W.T.-J. Tarn, N.X. Yunying, A. Isidori, Force regulation and contact transition control. IEEE Control Syst. Mag. **16**(1), 32–40 (1996)
45. J. Diebel, Representing attitude: Euler angles, unit quaternions, and rotation vectors. Matrix **58**(15–16), 1–35 (2006)
46. R.M. Murray, Z. Li, S.S. Sastry, S.S. Sastry, *A Mathematical Introduction to Robotic Manipulation* (CRC press, 1994)
47. Y. Wang, pyQpController (2019). https://github.com/wyqsnddd/pyQpController
48. Y. Wang, MultiObjectiveController (2019). https://github.com/wyqsnddd/multiObjectiveController
49. F. Furini, E. Traversi, P. Belotti, A. Frangioni, A. Gleixner, N. Gould, L. Liberti, A. Lodi, R. Misener, H. Mittelmann, et al., Qplib: a library of quadratic programming instances, in *Mathematical Programming Computation* (2017), pp. 1–29
50. H.D. Mittelmann, Latest benchmarks of optimization software (2017). http://plato.asu.edu/talks/informs2017.pdf

Chapter 6
Dynamic Safety System for Safe Collaboration with Industrial Robots

Jari Montonen, Ilari Marstio, and Timo Malm

Abstract In the industry, robots do tasks in bulk manufacturing that include, e.g., lifting heavy parts, repetitive movements, or working in dangerous environments. Robots are reliable, accurate, and fast to do their tasks if the task stays relatively unchanged—contrary to humans who are very flexible and can adapt to changes in the product and the production environment quite easily. The Human–Robot Collaboration (HRC) utilises the abilities of both. However, the HRC does rise new safety requirements, as humans work in close proximity of robots without fences. Speed and separation monitoring is used in fenceless environments to guarantee the safety of humans. On the other hand this may lead to capacity issues as robots either need to move at slow speed all the time or need to be stopped for unnecessary long times. The Dynamic Safety System for Safe Collaboration with Industrial Robots presents fenceless, safe, flexible, and efficient method for the HRC in industrial applications.

6.1 Introduction

The need for human operators to be able to collaborate with robots is constantly increasing. To satisfy this need, several manufacturers have introduced *collaborative robot* (cobot) models that follow the standard ISO/TS 15066:2016 [1]. Cobots need to work with limited speed and force and are lightweight in design. Traditional industrial robots are not subject to such limitations, however, they may not move whilst a human is in near proximity (ISO 10218-1:2011 and ISO 10218-2:2011 [2, 3]) unless moving with limited velocity, and/or the human is allowing movement with an active enabling device. For safety reasons, these robots are separated from

J. Montonen (✉) · I. Marstio · T. Malm
VTT Technical Research Centre of Finland Ltd, Espoo, Finland
e-mail: jari.montonen@vtt.fi

I. Marstio
e-mail: ilari.marstio@vtt.fi

T. Malm
e-mail: timo.malm@vtt.fi

L. Wang et al. (eds.), *Advanced Human–Robot Collaboration in Manufacturing*,
https://doi.org/10.1007/978-3-030-69178-3_6

operators by fencing and/or by distance, precluding flexible collaboration during normal operation.

The fenceless HRC approach with industrial robots often requires limiting the robot to unnecessarily slow speed and/or requires unnecessarily long safety distance around the robot. The *Safe Collaboration* system is addressing both of the issues enabling the robot to run safely with high speeds when there are no humans nearby, run with slower speeds only when humans are present, and eventually stop when there is a risk of human injury. This all works without human intervention, the worker's tasks can be carried out with no concerns about emergency stops or other safety-related issues.

The introduced *Safe Collaboration* system does target specifically industrial robots but the methods may be applied to collaborative robots as well. The system consists of methods, tools, and guidelines for setting up, configuring, and running fenceless HRC-systems.

6.2 Safety for Robots and Machines

6.2.1 Regulations

A robot and the associated safety system need to be built in accordance with the ISO 10218-1 [2] and ISO 10218-2 [3] standards. Some additions for collaborative robots are described in ISO/TS 15066 [1]. Other essential standards related to robot safety system are.

- ISO 13849-1:2015. Safety of machinery—Safety-related parts of control systems—Part 1: General principles for design [5].
- ISO 13855:2010. Safety of machinery. Positioning of safeguards with respect to the approach speeds of parts of the human body [6].
- ISO 12100:2010. Safety of machinery. General principles for design. Risk assessment and risk reduction [7].

The system integrator building a robotic cell needs to verify that safety-related devices in the cell and their configurations follow the standards listed above.

6.2.2 Safety Device Features Required and Utilised

Two specific functionalities are required from safety devices in the robot cell for configuring and using the described system. The robot needs to have a safety controller that has a monitoring function for allowed working areas and moving speeds of the robot. If either one of the monitoring functions detects a violation in robot movement, the safety controller raises a category 0 emergency stop described in standard ISO 13850:2015 [4]. For flexibility and functionality reasons, the controller should also have the ability to change the monitored area and speed as requested with signalling via an I/O-bus. The second major requirement is a safety-rated device monitoring the robot's surroundings for human presence. The safety device should also have interchangeable monitoring areas that can be synchronised with the robot's monitored area and speed by the system.

6.2.2.1 Robot Safety Controller

The *Safety Controller* in the robot is actively monitoring the speed and position of the tool and joints of the robot. The Safety Controller typically has dedicated functions to monitor different values and each function can be programmed to have certain allowed limits. There are also ways to program several cases to each function and those cases can be separately activated using signals from an external device. The Safety Controller has also dedicated inputs for safety signals (OSSD, output signal switching device) for connecting safety sensors and switches. The function of the OSSD is to reliably signal hazardous situations detected by sensors to the robot and to prevent potentially dangerous movements. When a safety sensor activates an OSSD, the Safety Controller stops the moving robot with an emergency stop. Some robots allow *Safe Stand Still* situations where the Safety Controller ensures the robot does not move while the OSSD is active, and the movement is allowed again once the OSSD is no longer active.

6.2.2.2 Safety Sensor

Several types of safety sensors can be applied, such as laser scanner, 3D camera, or light curtain. Laser scanners and 3D cameras usually have the option to be programmed to include only parts of their full view to safety function (often called Detection Zone, Safe Zone, or Detection Field). There may be options to have several different zone-cases that can be activated or deactivated by switching signals from an external device.

6.2.3 Dynamic Safety System

The Dynamic Safety System follows the standards in calculation of the required safety distances as described in the formula below:

$$S = K_r(T_s + T_r) + B(K_r) + C_{\mathrm{Tol}} + K_h(T_s + T_r + T_b), \qquad (6.1.)$$

where

$B(K_r)$ is the robot stop distance (in mm) from the maximal robot speed,
C_{Tol} is the sensor resolution in mm,
T_s is the sensor reaction time in s,
T_r is the robot reaction time in s,
T_b is the robot stop time (in s) from the maximal robot speed,
K_r is the maximal robot speed in mm/s, and
K_h is the constant speed of human walk in mm/s.

The calculated safety distance is added around the border of move-limits of the robot (not the distance to the robot, the tool attached, or an object held by the robot in its current configuration). The formula (6.1) returns a minimum human–robot separation distance according to ISO 10218-1 and ISO 10218-2. The distance is specified so that the robot moving at its maximum speed is stopped or brought down to safe collaboration speed[1] before a human walking towards it from the given distance can come in contact with the robot. The values to be substituted into (6.1) are specific to brands and models of the robot and safety sensors. Many of the values are constant in the setup for the devices and are given in device specification provided by the device manufacturer. Values for the maximum robot speed, robot stop distance, and robot stop time are variables depending on active configuration in the system. These values are provided by the manufacturer as a data-table and/or a chart where each value is given as a function of robot speed, e.g., Document 3HAC048645-001 [8] for ABB robots.

6.3 Dynamic Safety Control

6.3.1 System Structure

The *Dynamic Safety Control* (DSC) system contains three main software components and utilises the software provided by hardware manufacturers. The DSC system runs on a PC connected to the robotised system with a network or other signalling channels.

The *Safety System Configurator* (SSConf) software is used off-line in modelling the system and generate configurations for devices in the system. The generated configurations are transferred to respective hardware using software supplied by device manufacturers. It is also possible to use the configurator in system design to choose the best devices for the given case, before acquiring any safety devices.

The *Worker Tracking* software is used online to locate workers and other dynamic objects in and around the robotised workcell.

The *Dynamic Safety Control* software is used on-line; it follows the state of the robot and location of workers/dynamic objects reported by the Worker Tracking software. Based on the current status of the robot and location of dynamic objects, the system orders the activation of the appropriate configuration in the safety devices.

[1] If the robot can be used in collaborative mode (moving allowed alongside with human worker) the standards define force-limits that the robot needs to follow.

The order is sent to the Master hardware device controlling all the certified safety devices.

The *Master* hardware controls the change and synchronisation of safety configurations in the devices. The Master hardware functions as the interface between certified safety devices and the DSC control. The Master hardware is typically a safety-rated Programmable Logic Controller (PLC) that signals the active configuration to all safety devices.

6.3.2 *Concept*

6.3.2.1 Cell Layout

The configuration of the system begins with modelling the layout of the robot cell using the Safety System Configurator. The model shown in Fig. 6.1 is a simplified replication of the real cell built by placing the robot, safety devices, tables, and other static appliances. The layout needs to include the areas where the robot will move with the objects that will be present in the scan area of the safety sensors (e.g., table legs).

In the configuration software, the robot is represented in a simplified way, with the safety parameters required for separation distance calculation. As shown in Fig. 6.2, the software has a library of pre-configured robot models and an easy way to include more by modifying a template. In addition to robots, the software includes sensors with safety-related parameters and templates to create more. The user can either use the software to find out which sensors would be most suitable for a given layout or find the best positions for pre-purchased sensors in the layout.

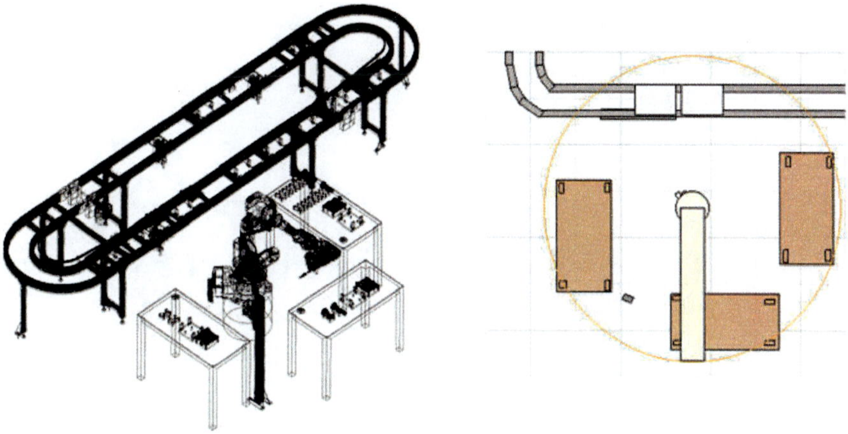

Fig. 6.1 A cell layout in CAD and modelled in the *SSConf* software

Fig. 6.2 Robot and sensor library examples as shown in the configuration software

One common problem is that static objects are blocking some parts of the scan area, thus potentially compromising safety. With the layout modelled properly, the SSConf software is able to show the required scan area and gaps caused by the occlusion.

Figure 6.3 depicts the layout with one scanner in two configurations where the scanner is in two different locations (yellow rectangle). The orange circle shows the maximum reach of the robot (1.65 m), and the turquoise circle shows the required safety distance calculated using the maximum speed of the robot (4.5 m/s). The red rasterised area displays the regions visible to the sensor within the required area. Table legs and other objects at the scan level cause the white areas in-between. While moving the scanner or any other object, the scan updates dynamically to match the new layout.

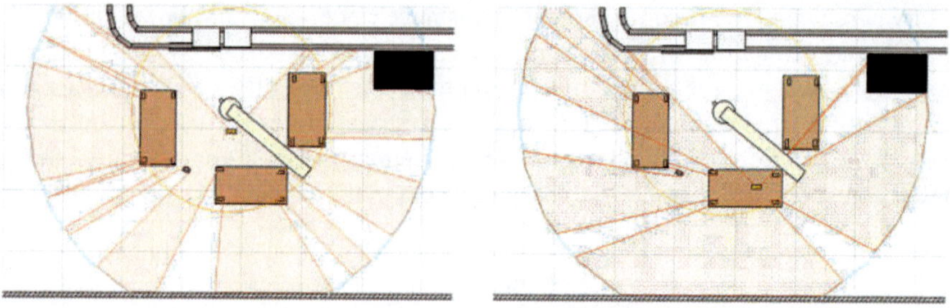

Fig. 6.3 Scan areas of a scanner with gaps

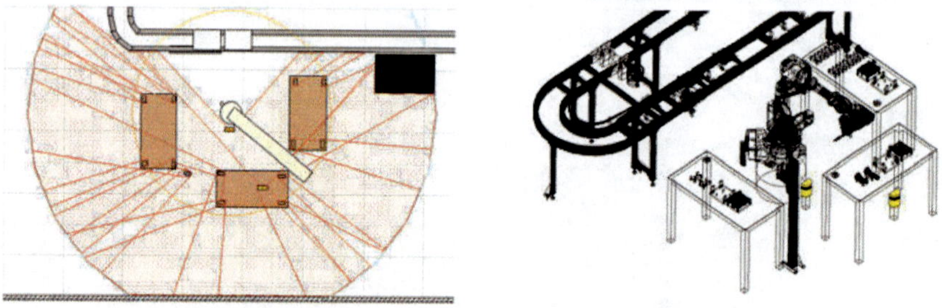

Fig. 6.4 Scan areas combined and corresponding scanner placement

Figure 6.4 shows the simultaneous use of two scanners in the same locations as shown in Fig. 6.3, i.e., one scanner near the robot and a second one placed under the table in the middle. As the figure shows, the scan areas complement each other leaving almost no gaps. The area between the robot and the conveyor on top could be fenced from human access so there would be no need to scan it.

6.3.2.2 Robot Safe Tool Zone and Safe Tool Speed

The figures in the previous subsection describe the situation when the robot is working within its full reach at its full speed. Using the safety controller integrated with the robot controller it is possible to safely restrict both the speed and the working area of the robot. Using these possibilities it is possible to allow humans to work on one table and the robot on another.

When a Safe Tool Zone (STZ) is active, the robot must be within limits. If, for some reason, the robot breaks the limits, it causes an emergency stop in a safety-rated way. This enables limiting the required scanned areas, too, as the separation distance is calculated from the STZ limit. Figure 6.5 depicts three STZs, one for each of the tables so that the robot may also transfer products between the table and the conveyor. Additionally, the robot's maximum speed can be limited in a similar way using Safe Tool Speed (STS). When the STS limit is active, the formula for separation distance returns noticeably smaller values.

The effect of activating both STZ and STS limits can be seen in Fig. 6.6, first with no limitations, then limiting the zone to #2 and then adding a maximum speed limit of 1 m/s. The last scenario demonstrates how a worker could work at tables #1 and #3 whilst the robot is doing its tasks at table #2. The last scenario also shows that the gaps caused by table legs in scanned areas become insubstantial when robot speed is limited and separation distance is smaller, resulting in that the final gaps are so small that worker can not approach the robot without being detected by the scanners. In case larger speeds would be required, a third scanner could be added to a suitable location to fill the remaining gaps.

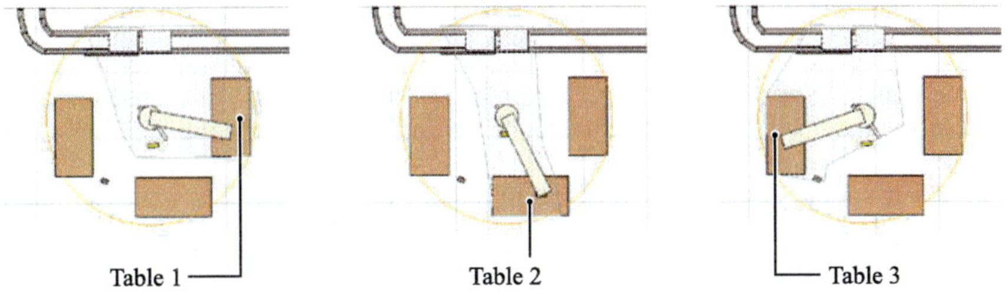

Fig. 6.5 Robot limited to tables using dedicated Safe Tool Zones

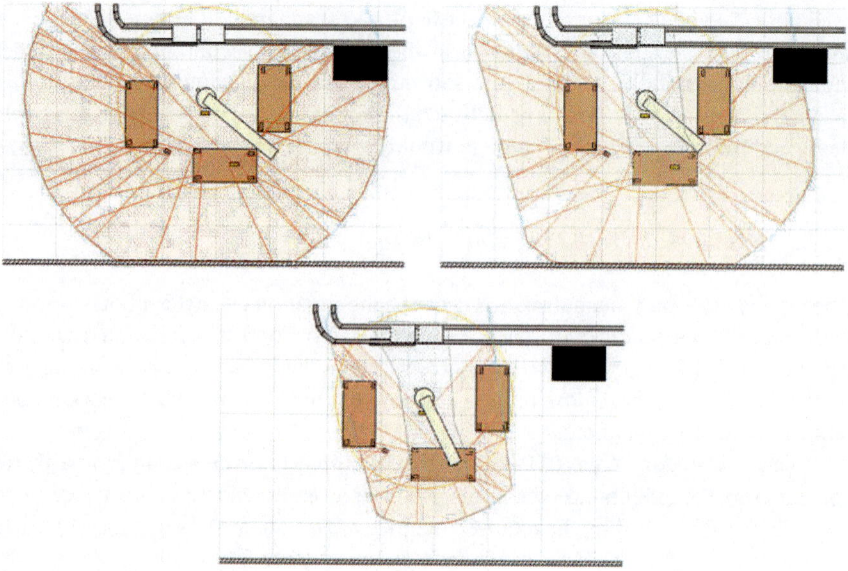

Fig. 6.6 Required scan area with no limitations, STZ active and STZ + STS active

6.3.3 System Execution

Several different combinations of Tool Zone and Tool Speed configurations are generated and activated dynamically when the layout changes. The number of configurations of a system is limited by safety device capabilities, e.g., the Safety Controller of the robot could handle a maximum of four STZs and four STSs. Safety sensors typically have 4, 8, or 16 different cases limited by the number of input signals. The example presented here applies three STZs and two STSs, resulting in six different cases, as listed in Fig. 6.7.

The robot program "knows" where it will be working during its execution, so it can control which STZ is active at which moment. When switching the STZ, both the previous and the new STZs need to be active. Safe Tool Zones do not complement each other but they cut each other. Therefore, at the moment of switching, the robot has to be in a location covered by both zones. In case there is a worker in the scan area of the new zone, the robot can stop there and wait until the worker leaves the area.

The system utilises sensors to decide which STS states should be active, and the STS is chosen based on the distance to the worker. In some cases, the laser scanners can be used for this purpose. If this is not possible, other external devices need to be added to the system. The external device does not need to be safety rated; for example, regular 3D cameras can be used, as shown in Fig. 6.8. The requirement is that the field of view covers the robot cell and its surrounding environment. Any

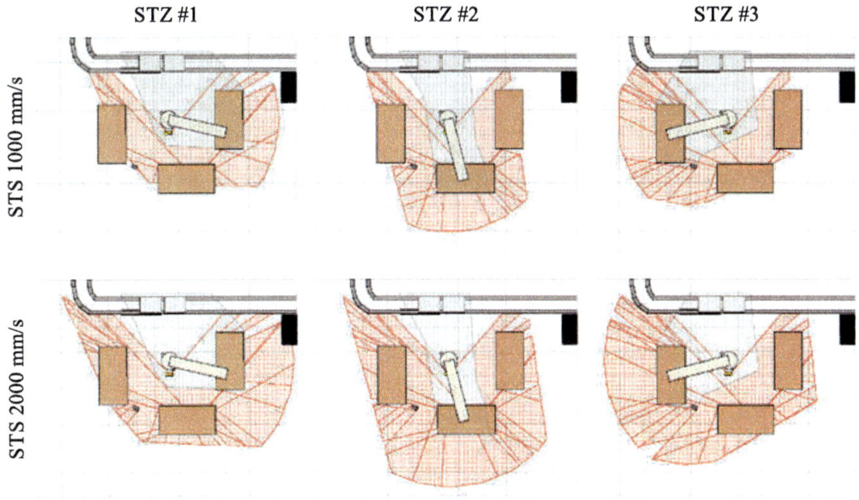

Fig. 6.7 Safety cases

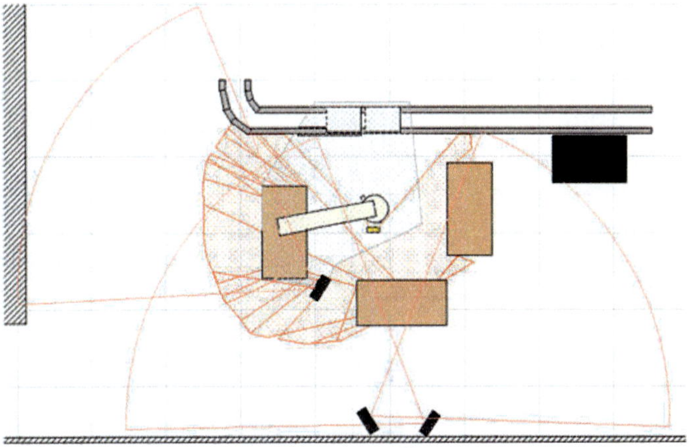

Fig. 6.8 Three external sensors covering close proximity of the cell

number of sensors can be added to fulfil the need. At runtime, the system uses the sensor data in detection of workers moving in proximity of the cell.

When the sensors detect an approaching worker, the Dynamic Safety System signals the robot to slow down dynamically in correlation to the distance of the worker to the border of the safety scan area. When the distance goes below a threshold, the safety system is prompted to observe new, lower speeds, allowing the worker to approach even closer, as shown in Fig. 6.9. When the worker moves away, higher speeds are allowed and the robot can move faster as the distance grows.

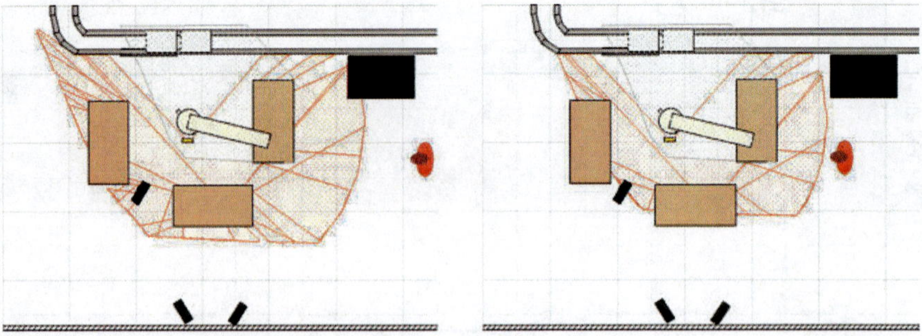

Fig. 6.9 Switching to smaller speed monitoring zones

There are two collaboration cases shown in Fig. 6.10. If the worker approaches any table other than the one where the robot is working, the robot can continue with a new limited speed, and the worker can safely work on another task (coexistence without tighter collaboration, also referred to as "semi-collaboration"). If the worker goes to the table where the robot is currently working, the robot needs to stop, or if possible, switch to collaborative mode. In the latter case, actions are signalled by the DSC in a safe way before the worker enters into the detection zone scanned by the safety sensors.

Robot stops can be monitored using the Safe Stand Still function of the Safety Controller. The collaborative mode requires extra sensors to ensure speed, force, and distance separation between the worker and the robot. A risk assessment is required in this case to ensure worker safety.

Many safety sensors have the option of a warning zone in addition to the detection zone. The warning zone in Fig. 6.11 can be used to change the STS configuration and to signal the worker of possible danger when moving near the robot. The safety sensor activates a signal when a human is detected in the warning zone, and the signal can be used to initiate other actions in the robot program and in the safety system.

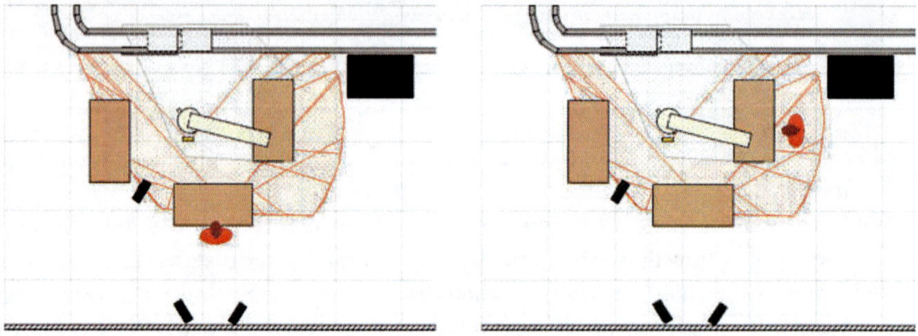

Fig. 6.10 Worker in "semi-collaboration" (coexistence) and "full collaboration" positions

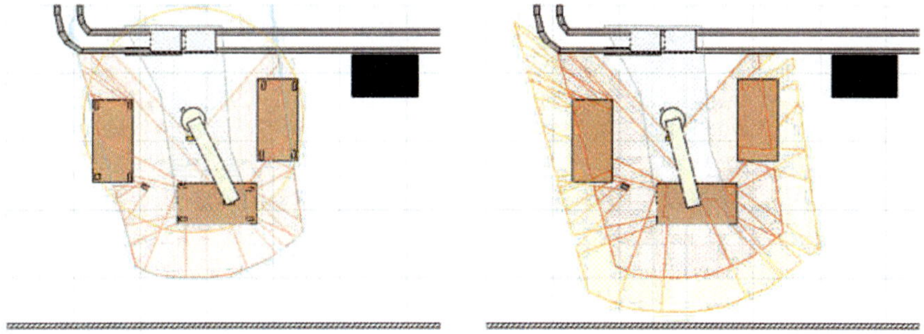

Fig. 6.11 Warning zone added around the detection zone

6.3.4 Hardware Configuration

The configurations created in the system are transferred to respective hardware using transfer files. Device manufacturers utilise their own proprietary file formats to store the configurations. The DSC system exports the configurations in human-readable formats that can be transferred to a device using a post-processor. The post-processor generates the correct format for the used device-brand. It is also possible to utilise the exported configuration data to configure a device manually in case it does not have applicable import function.

6.4 Conclusions and Future Work

Configuring a safety system for a fenceless robot cell is a cumbersome task, and integrators may be inclined to remain at a safe minimum of possibilities, and create a limited set of configurations with maximum safety distances. This results in unnecessarily large safety areas around the cell, imposing technically unjustified limitations on efficiency and collaborative functionalities. The DSC system presented in this chapter helps prevent such shortcomings by enabling an easy way to quickly generate safety areas for several different situations. The system provides the user with immediate feedback on safety-related impacts of layout changes and actions within the cell, relieving the user of repeated "manual" safety assessment. The system also offers a readily accessible way to reconfigure the areas in a few minutes in case of layout or work cycle changes.

The DSC system is instrumental in exploiting the potential of finer human–robot separation in collaborative settings. Utilising DSC, a robotic cell may be working at its full speed when there are no humans in proximity, and slow down in a safe way when worker presence is required. This enables reaching the maximum available capacity of the system without compromising safety.

In the future, the DSC system will be integrated as a single PC with safety I/O and compatibility with several external sensors for environment monitoring. In addition, more safety device and robot brands will be included in device library to ease commissioning and to attract broader industry. Furthermore, the safety system configuration methodology can be integrated into the configuration software of a robot or a safety device. Existing challenges and future trends on safety can also be found in the literature [9] for further reading.

References

1. ISO/TS 15066:2016, Robots and robotic devices—Collaborative robots
2. ISO 10218-1:2011, Robots and robotic devices—Safety requirements for industrial robots—Part 1: Robot systems and integration
3. ISO 10218-2:2011, Robots and robotic devices—Safety requirements for industrial robots—Part 2: Robots
4. ISO 13850:2015, Safety of machinery—Emergency stop function—Principles for design
5. ISO 13849-1:2015, Safety of machinery—Safety-related parts of control systems—Part 1: General principles for design
6. ISO 13855:2010, Safety of machinery—Positioning of safeguards with respect to the approach speeds of parts of the human body
7. ISO 12100:2010, Safety of machinery—General principles for design—Risk assessment and risk reduction
8. 3HAC048645-001, ABB Robot stopping distances according to ISO 10218-1, product specification reference material
9. L. Wang, R.X. Gao, J. Váncza, J. Krüger, X.V. Wang, S. Makris, G. Chryssolouris, Symbiotic human–robot collaborative assembly. CIRP Ann. Manuf. Technol. **68**(2), 701–726 (2019)

Part III
Dynamic Planning and Monitoring

Chapter 7
Resource Availability and Capability Monitoring

Nikolaos Nikolakis, Kosmas Alexopoulos, and Konstantinos Sipsas

Abstract In a manufacturing system, resources refer to all required items for a production order to be successfully fulfilled. Their effective management can be directly associated with the production supervision and control, affecting the responsiveness, flexibility and thus effectiveness of the entire manufacturing system. As production systems are increasingly incorporating a variety of heterogeneous resources, such as robots operating in the same space with human operators, conventional approaches of production supervision and control are becoming less effective, with an emerging need for smarter approaches and techniques. In particular, production planning, including the allocation of production activities to resources, primarily based on their availability and capabilities, cannot demonstrate increased flexibility and adaptability to changing working conditions, with an impact to the overall system. Nevertheless, the advances in the field of information and communication technologies hold the promise of enabling increased data collection which in turn can create insight over the production operations. Such insight can enable the creation of smart and self-adaptable applications on the shop floor enabling a smart manufacturing system in which humans may operate smoothly with automation systems. In this context, and with focus on a human–robot collaborative manufacturing paradigm, this chapter deals with the effective monitoring of shop floor operating resources via data acquired from their production environment and towards effectively monitoring their availability. This is achieved by means of contextual information generation from raw data which in turn facilitates seamless human–robot collaboration. The concept of a context-aware monitoring system is presented along with a prototype software implementation. Context drives the use of smart services both for monitoring and supervision purposes, as well as for enabling smart control and production planning. Finally, a use case is presented to demonstrate the proposed approach along

N. Nikolakis (✉) · K. Alexopoulos · K. Sipsas
Laboratory for Manufacturing Systems and Automation, University of Patras, Patras, Greece
e-mail: nikolakis@lms.mech.upatras.gr

K. Alexopoulos
e-mail: alexokos@lms.mech.upatras.gr

K. Sipsas
e-mail: sipsas@lms.mech.upatras.gr

L. Wang et al. (eds.), *Advanced Human–Robot Collaboration in Manufacturing*,
https://doi.org/10.1007/978-3-030-69178-3_7

with the implemented prototype, concluding with future research directions as well as identified challenges towards smart monitoring for hybrid and flexible production systems.

7.1 Introduction

Volatile markets, in combination to an increasing demand of personalised products, mandate for flexible and adaptable production systems. Modern automation, as well as software services and cloud computing, is promising technologies towards increasing the adaptability of manufacturing systems [1]. In particular, the value of cloud computing and cloud-based cyber-physical systems in manufacturing was analysed in [2]. Resources, cloud systems, and cloud services constitute the three core components of a cloud manufacturing system [3]. Nevertheless, the human factor remains of major importance due to the cognitive and highly flexible skill set in comparison to existing automation solutions. Hence, it is given particular attention within the modern manufacturing concepts [4], such as cyber-physical production systems [5]. The integration of human cognitive capabilities with an automation system in production runtime has the potential of enabling a manufacturing paradigm of complementary skillsets improving both the flexibility and efficiency of production systems [6]. Towards that end, a context-aware intelligent service system can be used on a manufacturing shop floor to provide the most appropriate information to shop floor personnel according to their status and needs. Several sources of information can be utilised to assist shop floor personnel in performing their tasks.

Manufacturing systems, defined as a combination of humans, machinery and information flow [7], need to achieve rapid response to emerging orders and self-adaptability in low-volume production of medium-to-high variation. In today's production lines human operators interact or even collaborate with automation systems while fulfilling their manufacturing tasks. To decide which production resource, human operator, robot, or machine should be assigned a task or if less or additional resources are needed, towards fully utilising the production capacity and minimising the cost, awareness over the shop floor working conditions needs to be established [8]. Such a smart system monitoring and adjusting the production based on real-time contextual information can demonstrate increased flexibility and adaptability to changing working conditions and production or process requirements [9, 10].

Targeting a hybrid human–robot collaborative cloud-based production environment, this chapter aims to present a web-based, service-oriented, and context-aware system for monitoring the availability of production resources as well as their changing capability for being assigned a production task in production time. Such event-based monitoring is enabled, in this work, through the execution of IEC 61499 compliant function blocks. The connection of the events to a context engine allows for the mapping of shop floor generated information to resource availability and operational status information. Such information is, in turn, valuable for enabling advanced production scheduling and control of the production resources, adapting to changing working conditions and unforeseen events, in a dynamic approach.

The objective of the work presented in this chapter is to monitor the operational status of human and/or robotic production resources in runtime, towards enabling advanced production planning and control approaches, supporting the paradigm of the next generation of production systems. To establish a corresponding perspective, a brief literature review is presented in Sect. 7.2 on the importance of resource supervision or monitoring. In Sect. 7.3 the concept of context and context awareness is briefly highlighted in the context of its importance for manufacturing. Next, the architectural design of a context engine is provided in Sect. 7.4, with the functional analysis of a context-aware resource monitoring system based on IEC 61499 event generation presented in Sect. 7.5. In Sect. 7.6, a prototype implementation is described, and its experimental result with a use case related to the automotive industry is provided in Sect. 7.7. Finally, conclusions and future work are provided in Sect. 7.8.

7.2 Literature Review

The technological breakthroughs in information and communication technologies (ICT) allowed for coupling physical and software components [11, 12] transforming each manufacturing component into a basic element of information [13]. As such, the establishment of a proper data acquisition and modelling mechanism is necessary for enabling advanced features in a production system increasing its flexibility and adaptability.

Modern production systems are nowadays incorporating a variety of production resources and systems, constituting their monitoring and control [14]. A challenge is strengthened by the ever-increasing demand for personalised production of low volumes and high diversity in the product mix. As such, it becomes imperative for a manufacturing system to maintain a smooth and uninterrupted production process when faced with changing conditions of operation. Uncertainty on the shop floor and changing working conditions can have a significant impact. Resource unavailability or breakdown causing deadlocks and production blockages need to be considered, effectively monitored, and dynamically controlled. This is necessary to maintain high levels of operational efficiency and reduce production costs, hence remaining competitive in today's volatile financial landscape.

Regardless of the variety of modelling and control approaches [15], it is widely accepted that any system can be modelled as a discrete event system (DES) [16]. In this context, and considering that flexibility in a manufacturing system is associated to its capability of adapting to new situations, all possible states need to be modelled along with the events triggering a transition from one state to another. Additionally, flexibility can be considered as the tolerance of the state model to external events or unforeseen situations. Considering human–robot collaboration or even interaction in a hybrid assembly system, leader–follower relationships are often established, since the functioning of both types of resources has to be reconciled while sharing the same space [17]. As mentioned in [18], task execution can be partitioned, assigned, stopped, or resumed based on the leading role, collaboration or interaction approaches, or resumed if needed. Real-time information tracking aiming to provide the right information to the right system is of critical importance. As such, a radio-frequency identification (RFID) based system for real-time manufacturing information tracking is proposed in [19]. The resources are equipped with RFID devices, readers, and tags, to create the real-time data capturing mechanism, allowing for tracking production progress, inventory, operational status, etc. In [20], the importance of the adopted data model for information exchange highlighted and of a suitable time-dependent handling strategy, regarding the selection of an appropriate communication method.

To ensure production continuity, failures or changes on working conditions should be directed to a new state for which proper control is established. The establishment of a proper monitoring system will allow for deadlock detection and avoidance if needed, such as a resource becoming unavailable or incapable of carrying out the task assigned to it. Regarding the first part, a resource may become unavailable due to multiple reasons, such as in case of a malfunction or power interruption when considering an automation system, robot, or machine. Unexpected conditions and abnormal situations, such as human medical need or misassembled part preventing the further operations from taking place, fall under the same case. With respect to the aforementioned and towards ensuring smooth production, deadlocks and blockages need to be avoided. This in turn enables cost reduction. As a result, a monitoring system tracking the steps of the production process as well as the operational status of the production resources responsible for the process is of major importance. In particular, a deadlock avoidance approach for supervisory control in manufacturing is presented in [21] adopting the formal tools of automata and Petri nets and proposing a deadlock resolution policy considering multiple unreliable resources. Petri nets are also discussed in [22, 23]. A deadlock detection and recovery approach are presented in the first article based on integer linear programming, while an online diagnosis of discrete events based on labelled Petri nets is discussed in the second. In [24] and concerning the aforementioned recovery strategies are proposed based on the priority of product and availability, used for determining deferent levels of reactivity and recovery strategies. To bridge the gap between theoretical approaches and practical implementations, an approach to reduce the impact of production exceptions in runtime is discussed in [25]. A Petri-net-based workflow is adopted to analyse data acquired from resources enabling resource configuration after a negotiation stage. The implemented strategy is supported to increase the flexibility and effectiveness of assembly systems for complex products. Also, an agent-based approach enabling a semi-hierarchical control solution is presented in [26]. Data are collected in real-time from shop floor devices and resources, integrating real-time data and status information and providing them to a central cloud platform. The central platform then handles the acquired information to enable manufacturing setup and control.

Moreover, virtualisation approaches have been suggested in addition to cloud services to enable software solutions of increased flexibility and scalability. In [27] a resource virtualisation layer along with a virtual shop floor layout and mapping of resources is proposed for achieving greater flexibility and agility to environmental changes. The use of robot operating system (ROS) [28] as a communication layer and its messages for event-driven monitoring and control is discussed in many research articles, such as [29, 30]. More recently, with the advent of artificial intelligence (AI), digital twin models and high processing capabilities, along with AI techniques for activity monitoring via image recognition emerged as discussed in [31]. The actions of human operators are identified through images acquired by optical sensors while hidden Markov models are used to identify deviations from the planned execution sequence or dangerous situations. In addition, an event-driven object-oriented simulation approach is discussed in [32] towards monitoring and coordinating the human–robot collaborative assembly. The proposed digital twin methodology is suggested for holistically integrating operational activities and enabling event-based decision-making and control approaches.

Considering DES, events are used to describe the current state of the system via its monitored components, such as manufacturing resources, systems, and machines. However, as the complexity of a modelled system increases, such as in the case of modern manufacturing paradigms, the state representation and monitoring approaches are becoming challenging. In this context, supervisory control advises the adoption of discrete event models at the highest abstraction level. One potential data model supporting this are the function blocks described by the International Electrotechnical Commission (IEC) 61499 standard facilitating distributed control applications [33]. Moreover, these function blocks can support different levels of granularity from basic function blocks, corresponding to simple operations, up to complex block networks, corresponding to entire processes. In particular, IEC 61499-compliant function blocks are used in [34] to monitor the energy consumption of a robot while proposing a framework for realising human–robot collaboration systems. Also, the notion of context and context awareness may further reduce the complexity of the overall manufacturing system by transforming data into information [35]. In the design of context-aware systems, two important design principles should be considered: (1) a context-aware architecture, and (2) a context model [36].

For the delivery of personalised data and services to manufacturing resources in a production facility, the cognitive domains such as information retrieval, decision-making, and situation awareness are required. For example, the cognitive load, especially of shop floor operators, working under harsh conditions, is not properly addressed during the distribution of information or the operational status of collaborating resources, such as robotic manipulators or automated guided vehicles bringing materials. In this context, data acquisition and processing from the shop floor, leading to the generation of contextual information can effectively be used for monitoring the production process, in terms of data sources or potentially correlated situations. Thus, the availability of a resource, estimated by its proximity to an assembly station, can be retrieved by localisation data from a mobile device or otherwise.

7.3 Concept of Context Awareness

7.3.1 Definition of Context

Currently, the most often cited definition of context is given by Abowd, Dey and Brown in [37]: "Context is any information that can be used to characterise the situation of an entity. An entity is a person, place or object that is considered relevant to the interaction between a user and an application, including the user and applications themselves". They furthermore coined the notion of a *context type* as a piece of information that can be relevant in a particular context. Peter Steggles, in [38], defined context types explicitly as "any description of the world that can be relevant to an application". Abowd, Dey and Brown distinguished *explicit* and *implicit* context types, where the former are directly provided by the user and the latter are sensed unnoticed. The authors also differentiated between *primary* and *secondary* context types. Primary types are location, identity, time and activity, and serve as indices for secondary types such as a few participants of a meeting, or frequency of being at a certain place. Ying and Fu-yuan in [39] furthermore distinguished *direct* from the *indirect* context: while direct context is acquired implicitly via physical (hardware) or virtual (software) sensors or explicitly from the user, indirect context is obtained by interpreting direct context via aggregation and reasoning to higher level context types via so-called *logical sensors* (for sensor types, see [40]). In the literature, the notion of "situation" seems to be equivalent to these kinds of higher level context types.

When defining the notion of context, there is always a threshold between being generic enough to bridge between disciplines (e.g., AI, documentation, cognitive ergonomics, cognitive psychology, business, philosophy, linguistics) and being sufficiently precise to apply the definition in real applications. Thus, it is not surprising that in 2005, Bazire and Brézillon [41] identified at least 150 different definitions of context. From this collection, they raised the question if context can be defined generically at all: Is it a frame for a given object, or the set of influential elements? Is it possible to define it a priori, or only state a posteriori? Is it static or dynamic? When does "context" begin and where does it end? They analysed their set of definitions with statistical methods to find a common denominator: "Context acts as a set of constraints that influence the behaviour of a system (user/computer) embedded in a given task. It varies in structure and nature".

Nevertheless, in both the domain of context modelling as well as its application in context-aware computing, it becomes clear that the definition by Abowd, Dey and Brown today dominates the field as the de-facto starting point for more specialised approaches [42].

7.3.2 *Context Awareness in Manufacturing*

In general, context awareness is the property for the provision of suitable services to a user through the analysis of the user's context. The context can be any information characterising the situation of an entity. The entity can be a person, place, or object relevant to the interaction between the user and a context-aware application. Such applications can react specifically to their current location, time, and other environment attribute and may adapt their behaviour without explicit user intervention and thus, aim at increasing usability and effectiveness. In manufacturing, context-aware applications define a holistic and dynamic context model that considers the context of tools, machines, parts, products, while at the same time, utilises information regarding the planning of manufacturing processes. Context awareness can be used to increase the visibility of operations and their performance in a manufacturing environment. It aims at enabling factory shop floor and office personnel to make a decision, based on a systematic understanding of the system having derived from a real-time sensed context of the factory, instead of a decision based on a fragmentary system view and a limited expert knowledge. In this study, context awareness is perceived as the capability of an information system to provide relevant information and services to the shop floor user. The relevancy depends on the user's task and the means of supporting decisions. These may be taken by shop floor personnel by enabling advance perception and comprehension of tasks (e.g., assembly, maintenance, supervision) allocated to them. Potential errors that may occur during the production process can be avoided through proactive delivery of expert knowledge.

In the design of context-aware systems, two important design principles should be considered: the architecture and context model [43]. The architecture of a context-aware system depends on different requirements, such as the location of sensors, the number of possible users, response times (real-time, soft real-time, pull-driven), or the availability of computing power. More insight into alternative architectures is provided in [43] and [44]. However, in many cases, the layered architecture approach presented in Fig. 7.1 is used, since it separates the detection of the context from its processing.

Apart from the architecture of a context-aware system, the context data modelling approach is also an important aspect. A popular way of classifying context instances is the distinction of the context dimensions to *external* and *internal* [45]. The external

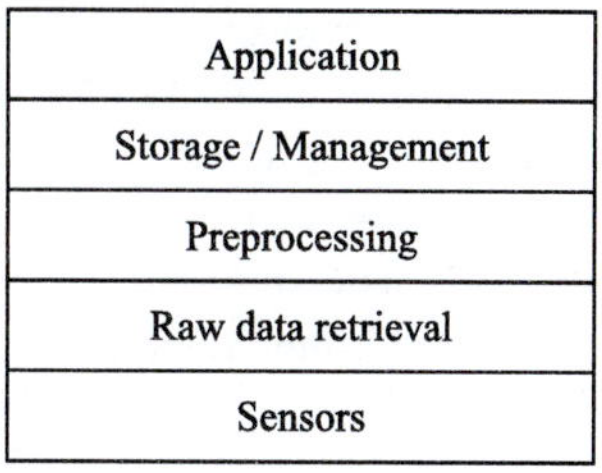

Fig. 7.1 A layered conceptual framework for context-aware systems

dimension refers to the context that can be measured by hardware sensors, i.e., location, light, sound, movement, touch, and temperature or air pressure. In a factory environment, the external content can be subdivided into three entities: places (e.g., workstation, sub-assembly line, assembly line, plant, and offices), people in roles (operators, support personnel, supervisors), and things (machines, tools, conveyors, etc.). External context factors may be sensed through the off-the-shelf sensor technology (e.g., RFID, location technology, machine status). Each of these external entities may be described by various attributes, which can be classified into four categories: *identity* (each entity has a unique identifier), *location* (an operator's position, co-location, proximity, etc.), *value* (or activity, meaning the intrinsic properties of an entity, e.g., machine green/yellow/red status), and *time* (used for timestamps to accurately define the situation, ordering events, etc.). Although the external dimension of the manufacturing context is important and useful for context-aware applications, it should not be the only dimension of context that needs to be captured and used in a manufacturing environment. The internal dimension of context—the user's goals, tasks, work context, business processes, communication, emotional and physical state—also needs to be captured. For example, cognitive load, especially of shop floor operators, working under harsh conditions, should be considered during the information distribution. To provide support through information and communication technologies (ICT) systems in a context-aware manner on a manufacturing shop floor, the cognitive domains such as situation awareness, decision-making capability, and work content are critical and have been given less attention by the scientific literature. However, internal context, such as task and cognitive domain aspects are difficult to be sensed. The work content of an operator could be identified by aggregating different pieces of information from different sources, and information relevant to the external context should be available. Besides defining what sort of entities and data the manufacturing context consists of, a method for modelling the context should also be selected. In [46], context awareness is proposed as an approach to providing additional information in support of the understanding of energy consumption in manufacturing environments.

In [47], a presentation of the most relevant context modelling approaches is given, which are based on the data structures used for representing and exchanging contextual information. Early approaches include *key–value* and *markup models*. Key–value models use simple key–value pairs to define the list of attributes and their values describing context information used by context-aware applications. Markup-based context information models use a variety of markup languages including XML. The main critics of these approaches concern their limited capabilities in (1) capturing a variety of context types, (2) capturing relationships, dependencies, timeliness, and quality of context information, (3) allowing consistency checking, and (4) support reasoning on context, on context uncertainty and higher context abstractions (see [48]). *Graphical and object-oriented models,* such as the Unified Modelling Language (UML), have a strong graphical component (UML diagrams) and are, therefore, also appropriate for modelling the context. The details of context processing are encapsulated at an object level and hence are hidden to other components. Access to contextual information is provided through specified interfaces.

Logic-based models have a high degree of formality. In a logic-based context model, the context is defined as facts, expressions, and rules. A logic-based system is then used for adding, updating, or removing new facts, and thus for updating the context. The inference mechanism can be used to derive new facts, based on the existing rules of the logic-based system. Finally, *ontologies* can be used to represent contextual information due to their high and formal expressiveness and the possibility of applying ontology reasoning. Several context-aware frameworks use ontologies as underlying context models. In [49], an ontology-based approach is used for the development of a context repository for a manufacturing environment. However, the study focuses on enabling self-adaptation of flexible manufacturing processes, based on a context-aware approach and not on information distribution in a manufacturing environment. In a similar manner in [50], an ontology-based context model, also for real-time decision-making, is proposed to optimise the key performance indicators of flexible manufacturing systems (FMS).

In manufacturing, context-aware applications define a holistic and dynamic context model that considers the context of tools, machines, parts, products while at the same utilises information regarding the planning of manufacturing processes. Context awareness can be used to increase the visibility of operations and their performance in a manufacturing environment. It aims to enable the factory shop floor and office personnel to decide, based on a systematic understanding of the system derived from the real-time sensed context of the factory, instead of a decision based on a fragmentary system view and limited expert knowledge.

7.4 Context Engine Development Approach

7.4.1 Context Engine Architecture

The context engine is a module that can extract context and situation information from data sources (e.g., sensors, Manufacturing Execution Systems (MES), Enterprise Resource Planning (ERP)) and then provide functionality that can be used to define the information needed by the user or other systems deployed on the shop floor. The basis for such context-aware services is data gathered from various sources (sensors, inputs of the human operator, etc.) on the shop floor. This information will be used to identify the current context of the services to be provided to the personnel.

To allow context extraction to produce tangible results, a well-defined context model has been adopted. The identified context will be used to support the information services in delivering accurate results based on the actual situation of the shop floor environment. A generic solution for the provision of context-based services to shop floor personnel is provided in Fig. 7.2 and can be applied to different manufacturing systems.

The Context Builder module is responsible for identifying changes in the context and updating the context repository and knowledge. The Context Builder uses all data

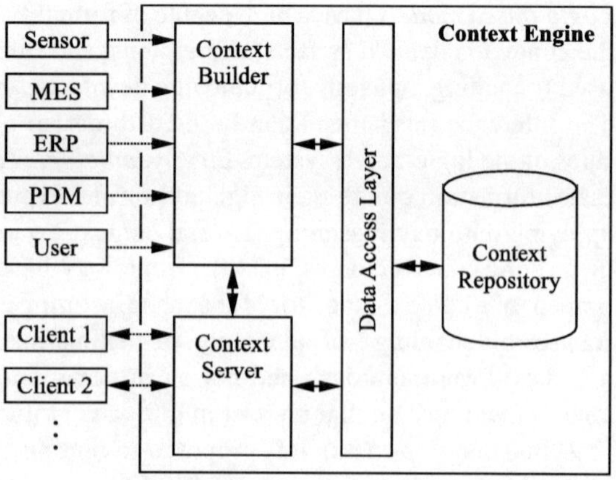

Fig. 7.2 Proposed context engine reference architecture

provided by the data sources to derive the current context. The data from the data sources are evaluated, and then, the context builder proceeds to update the relations between the entities of the context model. In a typical situation, an event that states the arrival of a worker to a workstation will be used to update the relations between the workstation object and the object representing the current location of the user. Apart from the dynamic and continuous context updating the Context Builder, it provides the functionality for static context definition such as workstations and productions units. As shown in Fig. 7.3, the Context Builder consists mainly of two important modules, namely, Context Identification and Context Inference which are described in detail below.

- *Context Identification*: This component is responsible for the identification of the current context, from sensor data, the ontology, and context information stored in the context repository. Context processing can be divided into aggregation and interpretation. The former refers to the composition of raw context information

Fig. 7.3 Context Builder

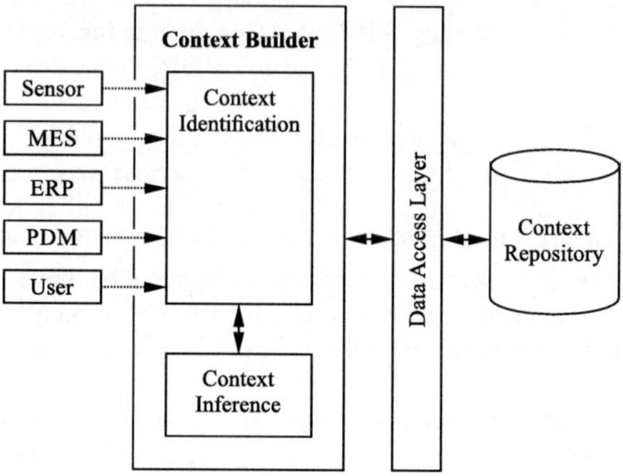

either to gather all context data concerning a specific entity or to build higher level context information.

- *Context Inference*: This component is responsible for generating inferred relations between objects within the context model based on existing relations. For example, when a scheduled task has started at a workstation or area of the shop floor, the activity and status of the corresponding worker are also updated.
- *Context Repository*: Inside the Context Repository, the identified context will be stored for further processing and reuse.

The Context Server provides the means for client applications to access the functionality of the Context Engine. The Context Server may be configured to notify client applications for context changes. Pre-configured changes of entities context can be configured as well as client applications can define statements that "listen" to specific changes in the context.

7.4.2 Context Data Model for Manufacturing

For the integration of context information from different context models, an extended context schema is needed. In this work, the ontology-based approach is proposed for context data modelling. Ontologies allow context modelling at a semantic level, establishing a common understanding of terms and enabling context sharing, logic inference, reasoning, and reuse. Ontology enables high and formal expressiveness of contextual information. Shareable ontologies are a fundamental precondition for knowledge reuse, serving as the means for integrating problem-solving domain representation, and knowledge acquisition modules. Several semantic specification languages such as Resource Description Framework (RDF) and Web Ontology Language (OWL) provide potential solutions for context modelling. These standard formats for data sharing span across applications, enterprise, and community boundaries. Users—both humans and machines—can share and understand the information available on the semantic web.

As shown in Fig. 7.4, the proposed extended ontology schema consists of two parts:

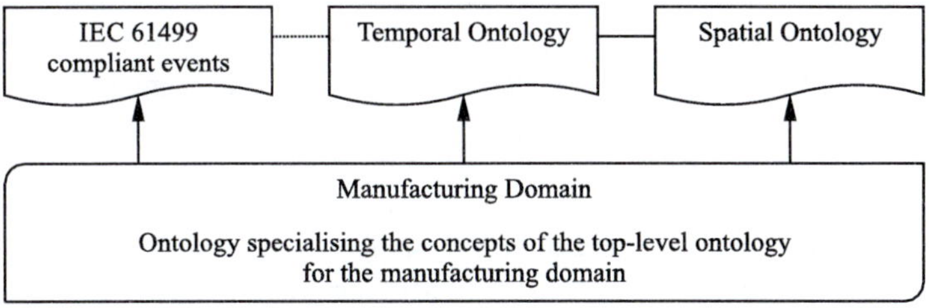

Fig. 7.4 The extended context data model for manufacturing

- *Top-level context ontologies*: This part contains objects describing attributes, such as location, status, or time. The context model consists of two modules:
  - *Temporal Ontology*: Contains entities describing the temporal nature of the context. A property or a relation in the ontology is valid within a time frame. The temporal dimension of a manufacturing context is relevant since in some cases the content to be delivered to the users may depend on historical data and not only the current, active, context. For example, assessing the skills of a specific role to execute a task may depend on the frequency this role has been involved in this task during the previous three months.
  - *Spatial Ontology*: Contains classes describing the spatial/location entities. The definition of a spatial ontology is not in the scope of this chapter and will be elaborated in future releases.
- *Manufacturing domain ontology*: For the status of a current factory to be presented, the ontology model has to include information on the resources (machines, tools, etc.) along with embedded sensors and actuators, the products to be produced, production schedule and orders, the technical and organisational processes. Moreover, it includes events generated by executing IEC 61499 function blocks. It should be noted that while the FBs are used in this study for context generation every other data source could be potentially used in the ontology if meaningful for context generation.

Relations among the ontologies could be possible. For example, the position property of a spatial entity could have a temporal representation and an event may have a location property.

7.4.3 Sensor Input Management

7.4.3.1 Temporal Data Model

Several different approaches have been proposed for representing time in RDF. These can be divided up into three broad classes: (1) versioning approaches that maintain distinct RDF graphs that represent the state of the domain at different points in time, (2) abstract syntax extension approaches that extend the RDF abstract syntax to incorporate notions of time and define corresponding semantics, and (3) conventional RDF approaches that work within the current RDF syntax and semantics and model the changing state of the domain explicitly as triples in the RDF graph, perhaps using N-ary relations. In this work, we are interested in the time dimension of the data mainly for treating them as historical data. The N-ary approach has been selected mainly due to its direct implementation in semantic repositories. In Fig. 7.5, the RDF representation of temporal attributes and object relations is presented using the example of machine status.

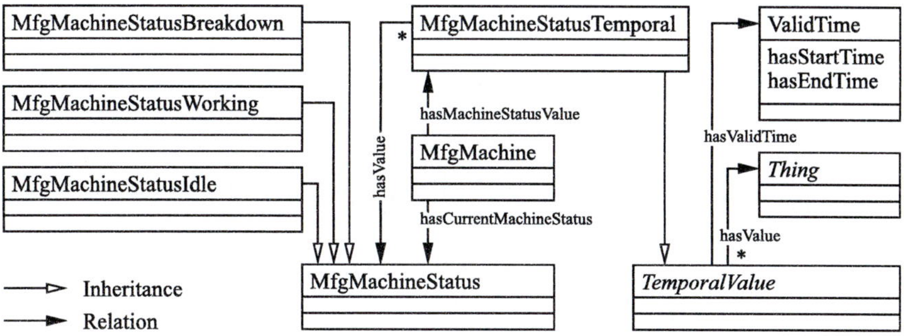

Fig. 7.5 N-ary relational approach representing properties that vary over time

7.4.3.2 Manufacturing Domain Model

The Manufacturing domain model contains the whole real-world objects, such as factory building, shop floor areas, machinery and equipment (tools, fixtures, metrology equipment), logistics, storage areas, transportation, raw materials, semi-finished and purchased parts, and personnel. Moreover, the real-world manufacturing objects are related processes (tasks, work instructions), production orders, to provide information for control tasks. This enables context-aware applications that operate in various areas the using of measures from the real production and the customised delivery of information for specific display and interaction devices. The purpose of the proposed context ontology is to define a fundamental data model for context extraction. The ontology is defined as a *generic device context model* and a *sector-specific context model*. The generic context model, shown in Fig. 7.6, defines concepts such as Product, Part, Resource, Production Schedule, Task, Human Resource, and Machine. Then a sector-specific context model defines concept such as ProductA, PartA1, LoadPartToPallet, and UnloadPartFromPallet.

7.5 Functional Analysis of a Human–Robot Monitoring System in a Hybrid Assembly Station

To define the context of the entities on the shop floor, sensory input from the shop floor is required. In the context of a human–robot collaborative production system, various subsystems are deployed on the shop floor that can produce signals or events useful for the population of the Context Engine model. These signals have been classified in a series of useful events that can be processed by the Context Engine and are presented in Fig. 7.7.

The operation update event represents the status of task execution by signal task start, completion/finished, and failure. During task execution collected data and events invoke a state change. The resource location event indicates the location and status of resources. Resource location is identified via presence in some area (e.g.,

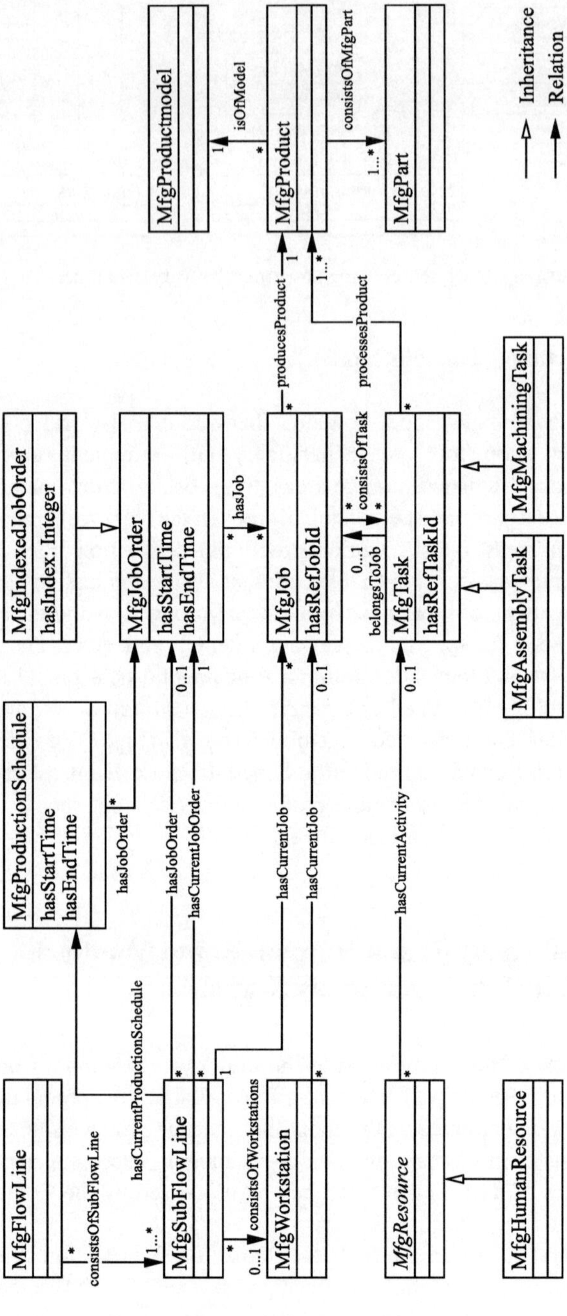

Fig. 7.6 A generic model for capturing context related to production on the manufacturing shop floor

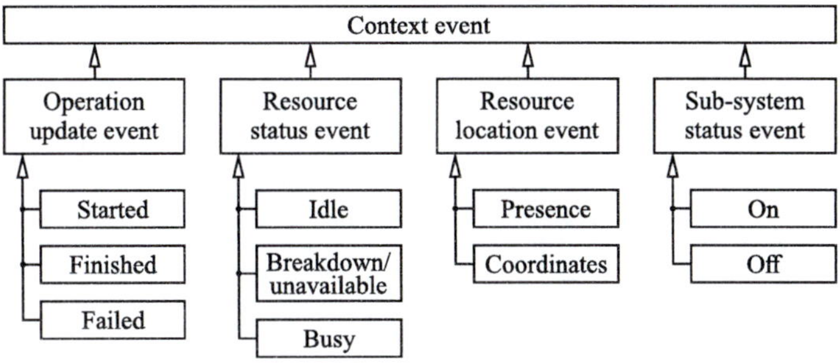

Fig. 7.7 Event model generating contextual information for production monitoring and resource availability

a workstation) or through x-y coordinates which is acquired via a shop floor localisation subsystem. Resource status is characterised by idle when no task is executed, busy when executing a task, and breakdown/unavailable when the resource cannot execute tasks. This event is directly linked to the operation updated status referring to a task execution. Potentially deployed subsystems, such as human operator support, robotic controllers, and safety subsystems, are characterised by on/off status indicating their functional condition and represented by the subsystem status event. The states along with their transitions are illustrated in the state chart in Fig. 7.8 when modelling the monitoring system and considering a rescheduling and resume strategies in supporting the addition of an event handling mechanism.

Rescheduling can be triggered in order not to stop the production process, for example, in case of a resource becoming temporarily unavailable. On the other hand, recovery policies may be triggered in case of a more severe abnormal situation such as power shortages or wrong assembly. Both actions mandate production interruptions, even if automated recovery options are considered, and integrated human intervention is assumed necessary before resuming the production process.

The monitoring component is initialised when the production starts, which in turn is turned on by a production manager or line supervisor. Considering the existence and collaboration not only of human and robot operators but also of supporting subsystems, the proposed monitoring system mechanism is coupled with key software subsystems, each responsible for partially controlling the production, e.g., safety subsystem, robot controller, human controller. The aforesaid software subsystems are being monitored by a "lifebeat" signal, specifying whether or not they are running. At initialisation, the monitoring control component is set to "available" for all workstation-assigned resources, the connected subsystem's status is set to "off", specifying that their status cannot be known in advance and is assumed to be off, and the Supervisory Control is set to an idle mode listening to events and processing them accordingly. Afterwards, the state changing events are the following:

- The subsystem status is updated according to the "lifebeat signal". A lifebeat signal specifies that the component is up and running and can be used during production, enabling the related alternatives for the scheduler, e.g., robot controller

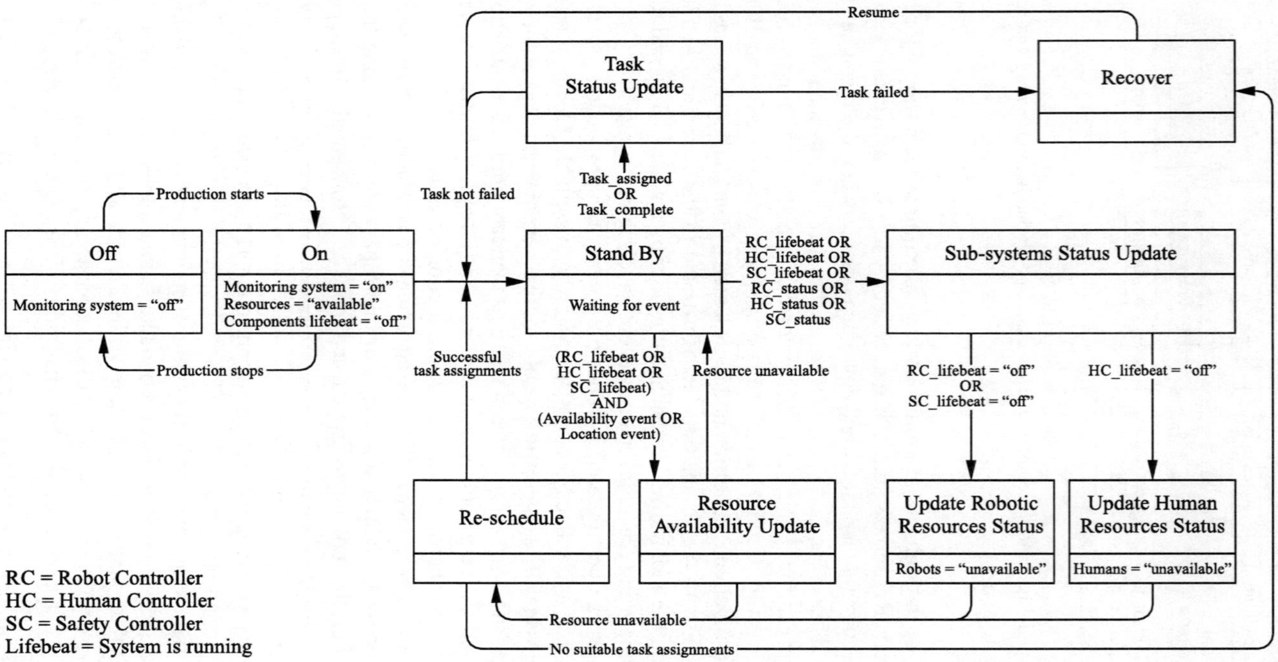

Fig. 7.8 State chart of proposed monitoring system

lifebeat states that the robotic resources can be used for task execution. In case that the subsystem does not provide a lifebeat signal, it is considered off and the schedule is adapted accordingly, updating the status of the resources, i.e., if the safety controller is off then no robot can be safely used since there are also human operators nearby.

- If the subsystems are running, even some of them, the production can begin with the scheduler assigning tasks and the monitoring system waiting for runtime events. An availability event specifying whether a resource is available or not updates the status of the resource and triggers a reschedule with that resource being unavailable. The same status update can be triggered by a location event, suggesting that the resource is leaving the station. In case the availability or location event is not specifying that a resource is unavailable, the state of the monitoring system moves back to the idle state and the production continues as planned.
- IEC 61499 function block generated events are exchanged during task assignments and fulfilment enabling to monitor the execution flow of the production. In the case of an abnormal event stating that the task cannot be fulfilled, recovery actions are required. In that case, a rework task is assigned to a qualified resource, responsible for inspecting the product and/or disassembling it to previous steps, repair it and re-assemble it or discard it.

7.6 Prototype Implementation

7.6.1 Prototype System Architecture

The aforementioned context-aware monitoring methodology based on IEC 61499 function blocks and their generated events has been implemented into a software prototype system concerning a cyber-physical oriented application. The prototype follows a client–server architecture providing a series of web services for data exchange, as presented in Fig. 7.9. Data exchange is facilitated via RESTful APIs and Java Messaging Services (JMS). Raw data from shop floor sensors are either provided via one of the two aforementioned approaches.

More specifically, the prototype implementation supports the following high-level functionalities:

- Web services for accessing the contextual model of the Context Engine. This includes static information such as listing workstations or resources and dynamic information such as resource locations or task execution status.
- Web services for receiving signals from other subsystems such as location events for resources, or task execution updates.
- Messaging mechanisms responsible for notifying clients for specific events, e.g., resource availability or location.

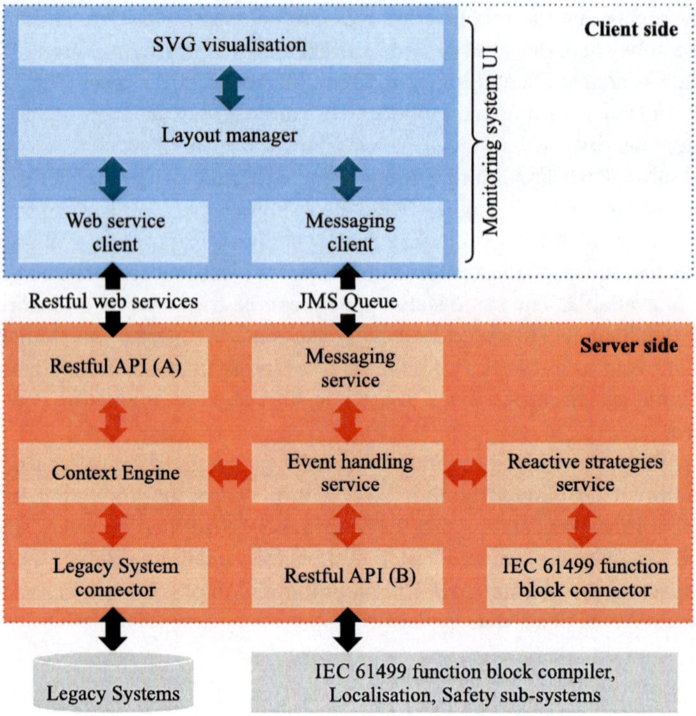

Fig. 7.9 Prototype architecture

- A mechanism for managing incoming signals and transforming them into contextual information that is handled by the Context Engine.
- A service that will handle rescheduling based on the Context Engine situational information on the workstation.

The high-level functionalities presented in the previous section are visualised through the monitoring system's user interface (UI) as shown in Fig. 7.10. Through

Fig. 7.10 Workstation visualisation and monitoring interface

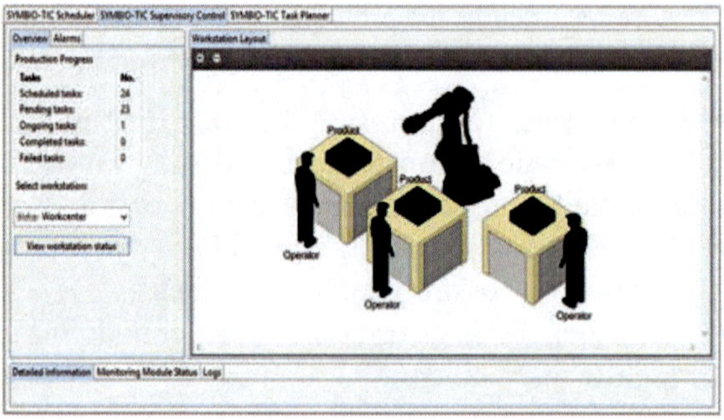

its workstation visualisation interface, it enables a high-level monitoring/supervision of the assembly operations occurring on the shop floor.

7.6.2 Key Enabling Technologies

The client-side component of the monitoring application was developed using Java and the Eclipse Rich Client Platform that enables Eclipse plugin development. The server-side components, while completely decoupled from the client side, were also developed in Java as web applications leveraging the Spring Framework. Moreover, the object-relational mapping framework Hibernate was used to facilitate database persistence and the ActiveMQ implementation of JMS was used for message exchange.

The client-side components come as a standalone application with a pre-packaged Java Runtime Engine (JRE). The only prerequisite for them to run is an active deployment of both the Amorph.pro backend and the server-side components.

The server-side components can be deployed using any servlet container like Apache Tomcat.

7.6.3 Activity Diagrams

Workstation status visualisation or Layout manager: The activity diagram is illustrated in Fig. 7.11; the Layout manager initialises the user interface and populates the

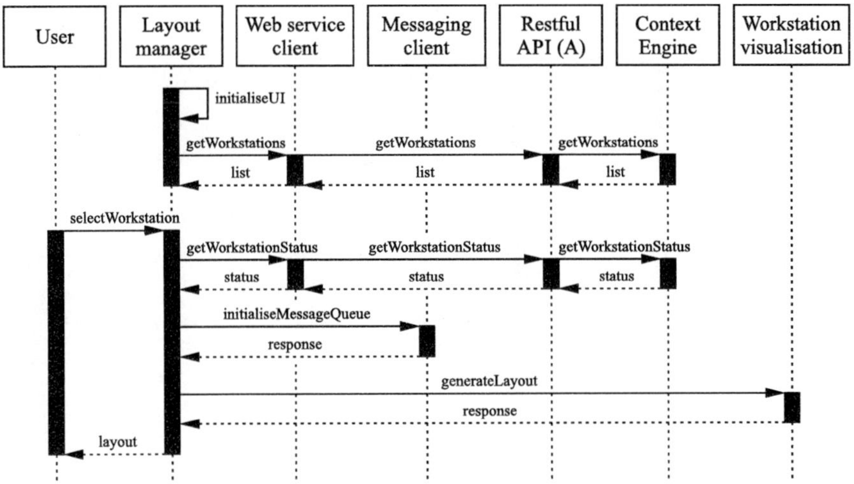

Fig. 7.11 Workstation status visualisation activity diagram

list of workstations. Then, the user selects the workstation to be monitored. After the workstation is selected, the Layout Manager retrieves the status of the workstation from Restful API (A) through the Web service client component. The Restful API (A) retrieves the status information from the context engine and responds with appropriate information. The information is used by the SVG visualisation to produce a layout for the workstation and for populating other UI elements as required. The Layout manager also initialises the message client so that it connects to the appropriate queues in the backend to receive events relevant for the workstation.

Event handling—client side: The messages come from the messaging service of the server side and are fed into the relevant queue monitored by the Messaging client component. When new messages are added, the Layout Manager is notified and proceeds with decoding the message. Then the UI elements are updated including the layout visualisation based on the information provided, see Fig. 7.12.

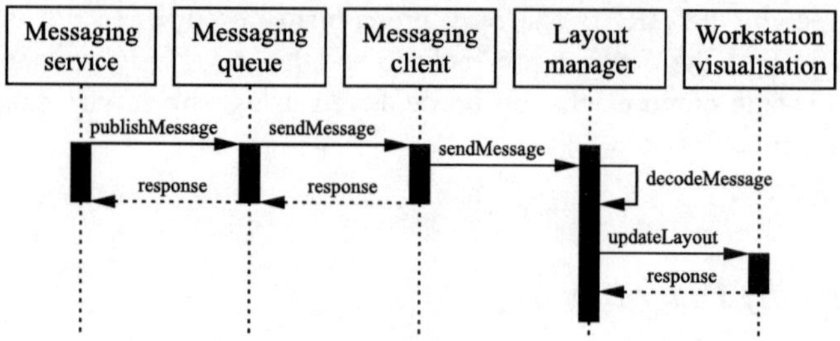

Fig. 7.12 Event handling—client-side activity diagram

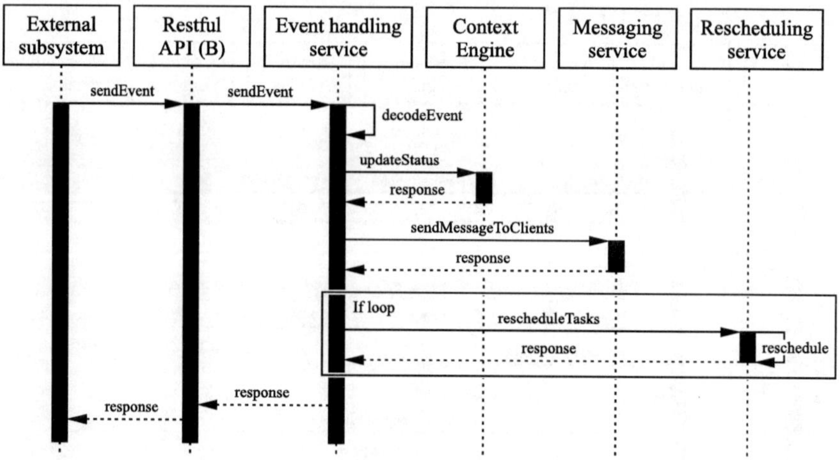

Fig. 7.13 Event handling—server-side activity diagram

Event handling—server side: Figure 7.13 illustrates how events triggered by external systems are handled. It should be noted that external systems include the unit controller, the localisation subsystem, and the safety subsystem.

First, an external system may trigger an event by accessing one of the methods provided by Restful API (B). The details are abstracted here for the sake of page limit. As soon as an event is triggered, the Event handling service is notified. This service is responsible for decoding the event in a format that is appropriate for internal use. Then, the Context Engine is notified to update the state of the context model. Next, the event is transformed into a message and is sent via the Messaging service to the appropriate queue and finally, the Rescheduling service may be triggered depending on the type of the event.

7.7 Experimentation

To test and validate the implemented context-aware monitoring application, an industrially oriented cyber-physical system scenario has been selected including an assembly station where human operators collaborate with industrial robots. The collaboration includes both sharing the same working area as well as collaborating during the task execution. As such, effective and dynamic monitoring in runtime is necessary to enable further control policies that may increase the efficiency and productivity of the assembly station as well as to safeguard the human operators. In particular, the adopted scenario includes the following actions:

- The user selects the workstation. The layout of the workstation is visualised as expected.
- A "presence" event is triggered. The layout and other UI elements are updated correctly.
- A "resource unavailable" is triggered. The UI elements are updated correctly and a new message about schedule generation is received successfully.
- "Task start" events are triggered. The UI elements are updated correctly.
- "Task failed" event is triggered. The user is notified via a pop-up message.

After the selection of the workstation, the layout is visualised, as depicted in Fig. 7.10. The black elements in the layout indicate the positions where resources or products may be located, as soon as they are present in the workstation.

As soon as resources arrive at the workstation, the layout is updated and turns the appropriate elements green, indicating that a resource is present and available for work at the workstation. To trigger the layout change, the application receives messages from the monitoring system backend, which in turn is notified by a localisation system deployed on the shop floor. The updated status of the resources is shown in Fig. 7.14.

In runtime and during the resources' task execution, the operational status of the resources is updated as well. The monitoring system backend, receiving messages from the shop floor subsystems and the function block execution, containing assembly commands or instructions, responds by updating the layout and the section

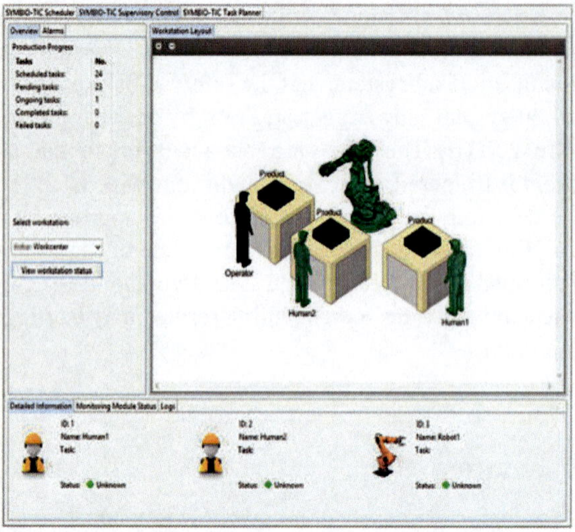

Fig. 7.14 Layout with three resources present

with detailed information about resources. The colour blue is used to signify resources working on a task (Busy) and yellow for the products present at any tables. Figure 7.15 illustrates the situation of one human worker working on a task while the other resources remain idle, facilitating the tracking of production execution as well as resources availability and operational status.

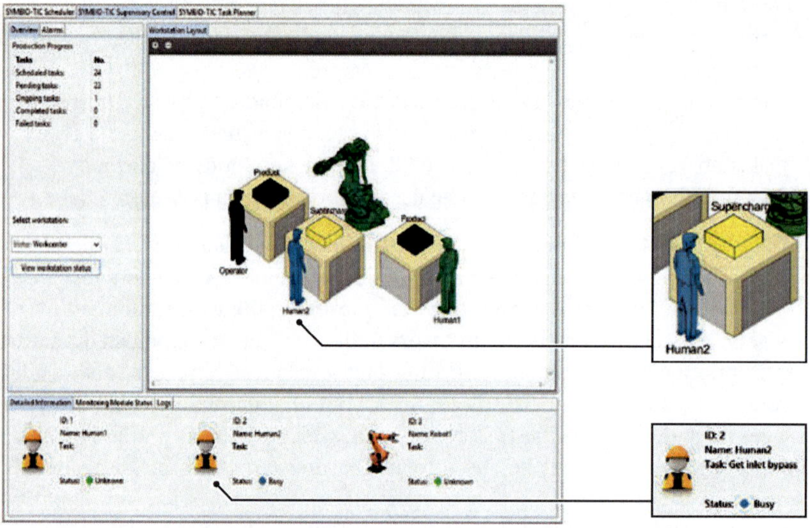

Fig. 7.15 One resource is working on a task

When resources become unavailable, the layout is updated accordingly by changing their colour to red, as indicated for a robot in Fig. 7.16.

Additionally, and considering exception handling and abnormal situations apart from the availability of the resources, proper logging is supported to track the behaviour of the connected systems. Figure 7.17 presents the logs section of the monitoring application where all relevant events that have been received from the backend are listed. The status of relevant subsystems, such as the IEC 61499 function block compiler and the robot safety controller, is presented in Fig. 7.18.

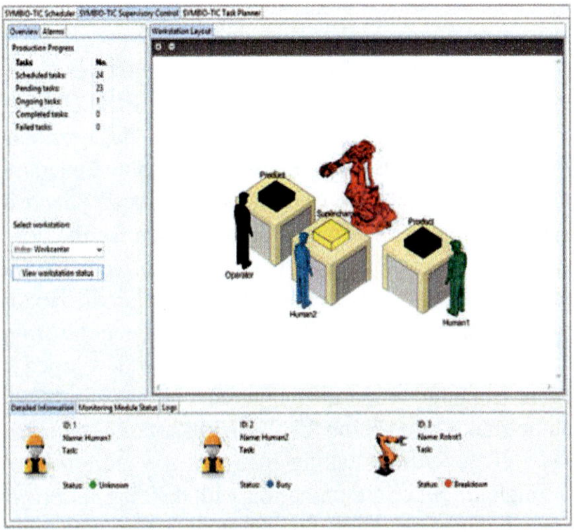

Fig. 7.16 Robot unavailable status update

Detailed Information | Monitoring Module Status | Logs

Event type	Date/Time	Details
ScheduleGenerated	2016.11.10 at 13:12:29	Schedule generated.
ResourcePresence	2016.11.10 at 13:12:37	Human1 arrived
ResourcePresence	2016.11.10 at 13:12:39	Human2 arrived
ResourcePresence	2016.11.10 at 13:12:42	Robot1 arrived
TaskStatusChange	2016.11.10 at 13:12:49	Get inlet bypass started.
ResourceStatusChange	2016.11.10 at 13:13:32	Robot1 Breakdown
ScheduleGenerated	2016.11.10 at 13:13:39	Schedule generated.

Fig. 7.17 Monitoring system logs tab

Detailed Information | Monitoring Module Status | Logs

Name	Status
Safety sub-system	Online
Unit Controller	Online

Fig. 7.18 Monitoring the operational status of connected subsystems required for human–robot collaboration

7.8 Conclusions

This chapter aims to report the conceptual framework of a context-aware information system to support the monitoring of the availability and operational status of a human–robot collaborative manufacturing environment. An RDF-stored knowledge-based approach has been utilised for the collection of contextual entities, from several factory data sources, and represents them through an extended context model for further processing, as detailed under Sect. 7.4.3. The extended manufacturing context model as well as the context management approach provides a common interface for context acquisition, reuse, and updates, which may be utilised by applications in support of shop floor personnel in a manufacturing environment. High-level context is extracted from external data sources such as sensors and MES or ERP systems as well as from the execution of IEC 61499-compliant function blocks. Those function blocks concern simple data exchange, provision of instructions, robotic commands, etc., enabling distributed execution and control of the production system. The proposed framework of a context-aware information distribution system is based on a bottom-up approach for the development of such applications. A practical example from discrete manufacturing industry is used to demonstrate the way that context-dependent information distribution services can be applied to a manufacturing floor, facilitating resources monitoring and status control.

In particular, the monitoring module developed combines characteristics from event-triggered and self-triggered control similar to a holonic system, with the autonomous units being the IEC 61499 standard function blocks. It provides a generic view of the actions taking place on the monitored robotic cyber-physical station, through proper communication with the distributed control components. However, it does not provide a detailed operational view, i.e., visualising a robot movement. It is not capable of simulation rather than high-level monitoring and centralised control.

Future research will evaluate the system's performance when processing a big amount of data. In that perspective, different data technologies such as distributed processing and map-reduce will be considered. Moreover, future research will focus on the development of algorithms that may use the proposed context model and the system architecture for automatic and semi-automatic optimisation of the production system and the deployment of decision support clients on the shop floor. Besides, the ontology-based context model will be improved by future research to allow for the reuse of this model on other types of industries as well as for application domains, mainly maintenance support. Finally, for the consideration of the context in information distribution to be expanded, users' internal context attributes such as cognitive load, should also be considered.

References

1. Y. Cohen, M. Faccio, F. Pilati, X. Yao, Design and management of digital manufacturing and assembly systems in the Industry 4.0 Era. Int. J. Adv. Manuf. Technol. **105**(9), 3565–3577 (2019). https://doi.org/10.1007/s00170-019-04595-0
2. L. Wang, X.V. Wang, L. Wang, X.V. Wang, Latest advancement in CPS and IoT applications. Cloud-Based Cyber-Phys. Syst. Manuf. 33–61 (2018). https://doi.org/10.1007/978-3-319-67693-7_2
3. L. Zhang, Y. Luo, F. Tao, B.H. Li, L. Ren, X. Zhang, H. Guo, Y. Cheng, A. Hu, Y. Liu, Cloud manufacturing: a new manufacturing paradigm. Enterp. Inf. Syst. **8**(2), 167–187 (2014). https://doi.org/10.1080/17517575.2012.683812
4. E. Rauch, C. Linder, P. Dallasega, Anthropocentric perspective of production before and within industry 4.0. Comput. Ind. Eng. 105644 (2019). https://doi.org/10.1016/j.cie.2019.01.018.
5. P. Leitão, A.W. Colombo, S. Karnouskos, Industrial automation based on cyber-physical systems technologies: prototype implementations and challenges. Comput. Ind. **81**, 11–25 (2016). https://doi.org/10.1016/j.compind.2015.08.004
6. J. Krüger, T.K. Lien, A. Verl, Cooperation of human and machines in assembly lines, CIRP Ann. Manuf. Technol. **58**(2), 628–646 (2009). https://doi.org/10.1016/j.cirp.2009.09.009.
7. G. Chryssolouris, *Manufacturing Systems: theory and Practice*, 2nd edn. (Springer, 2006)
8. K. Alexopoulos, S. Makris, V. Xanthakis, K. Sipsas, G. Chryssolouris, A concept for context-aware computing in manufacturing: the white goods case. Int. J. Comput. Integr. Manuf. **29**(8), 839–849 (2016). https://doi.org/10.1080/0951192X.2015.1130257
9. N. Tapoglou, J. Mehnen, A. Vlachou, M. Doukas, N. Milas, D. Mourtzis, Cloud-based platform for optimal machining parameter selection based on function blocks and real-time monitoring. J. Manuf. Sci. Eng. Trans. ASME **137**(4) (2015). https://doi.org/10.1115/1.4029806
10. D. Mourtzis, N. Milas, A. Vlachou, An internet of things-based monitoring system for shop floor control. J. Comput. Inf. Sci. Eng. **18**(2) (2018). https://doi.org/10.1115/1.4039429
11. B. Bordel, R. Alcarria, T. Robles, D. Martín, Cyber-physical systems: extending pervasive sensing from control theory to the internet of things. Pervasive Mob. Comput. **40**, 156–184 (2017). https://doi.org/10.1016/j.pmcj.2017.06.011
12. L. Monostori, B. Kádár, T. Bauernhansl, S. Kondoh, S. Kumara, G. Reinhart, O. Sauer, G. Schuh, W. Sihn, K. Ueda, Cyber-physical systems in manufacturing. CIRP Ann. **65**(2), 621–641 (2016). https://doi.org/10.1016/j.cirp.2016.06.005
13. H. Liu, H. Ning, Q. Mu, Y. Zheng, J. Zeng, L.T. Yang, R. Huang, J. Ma, A review of the smart world. Futur. Gener. Comput. Syst. **96**, 678–691 (2019). https://doi.org/10.1016/j.future.2017.09.010
14. J. Krüger, L. Wang, A. Verl, T. Bauernhansl, E. Carpanzano, S. Makris, J. Fleischer, G. Reinhart, J. Franke, S. Pelegrinelli, Innovative control of assembly systems and lines. CIRP Ann. **66**(2), 707–730 (2017). https://doi.org/10.1016/j.cirp.2017.05.010
15. B. Sencer, C. Okwudire, Control, in *CIRP Encyclopedia of Production Engineering* (Springer, Berlin, Heidelberg, 2014), pp. 281–285
16. G. Maione, D. Naso, Using a discrete-event system formalism for the multi-agent control of manufacturing systems, in *Informatics in Control, Automation and Robotics I* (Dordrecht, Kluwer Academic Publishers, 2006), pp. 125–132
17. X.V. Wang, Zs. Kemény, J. Váncza, L. Wang, Human–robot collaborative assembly in cyber-physical production: classification framework and implementation. CIRP Ann. Manuf. Technol. **66**(1), 5–8 (2017). https://doi.org/10.1016/j.cirp.2017.04.101
18. L. Wang, R. Gao, J. Váncza, J. Krüger, X.V. Wang, S. Makris, G. Chryssolouris, Symbiotic human–robot collaborative assembly. CIRP Ann. **68**(2), 701–726 (2019). https://doi.org/10.1016/j.cirp.2019.05.002
19. Y. Zhang, P. Jiang, G. Huang, T. Qu, G. Zhou, J. Hong, RFID-enabled real-time manufacturing information tracking infrastructure for extended enterprises. J. Intell. Manuf. **23**(6), 2357–2366 (2012). https://doi.org/10.1007/s10845-010-0475-3

20. H. Sun, P. Jiang, Study on manufacturing information sharing and tracking for extended enterprises. Int. J. Adv. Manuf. Technol. **34**(7–8), 790–798 (2007). https://doi.org/10.1007/s00170-006-0637-9
21. H. Yue, K. Xing, H. Hu, W. Wu, H. Su, Resource failure and buffer space allocation control for automated manufacturing systems. Inf. Sci. (Ny) **450**, 392–408 (2018). https://doi.org/10.1016/j.ins.2018.02.043
22. Y. Chen, Z. Li, A. Al-Ahmari, N. Wu, T. Qu, Deadlock recovery for flexible manufacturing systems modeled with Petri nets. Inf. Sci. (Ny) **381**, 290–303 (2017). https://doi.org/10.1016/J.INS.2016.11.011
23. M.P. Cabasino, A. Giua, M. Pocci, C. Seatzu, Discrete event diagnosis using labeled Petri nets. An application to manufacturing systems. Control Eng. Pract. **19**(9), 989–1001 (2011). https://doi.org/10.1016/j.conengprac.2010.12.010
24. P. Berruet, A.K.A. Toguyeni, S. Elkhattabi, E. Craye, Toward an implementation of recovery procedures for flexible manufacturing systems supervision. Comput. Ind. **43**(3), 227–236 (2000). https://doi.org/10.1016/S0166-3615(00)00068-3
25. C. Qian, Y. Zhang, C. Jiang, S. Pan, Y. Rong, A real-time data-driven collaborative mechanism in fixed-position assembly systems for smart manufacturing. Robot. Comput. Integr. Manuf. **61**, 101841 (2020). https://doi.org/10.1016/j.rcim.2019.101841
26. S. Răileanu, F. Anton, T. Borangiu, High availability cloud manufacturing system integrating distributed MES agents, in *Studies in Computational Intelligence*, vol. 694 (Springer, Cham, 2017), pp. 11–23
27. O. Morariu, C. Morariu, T. Borangiu, shop floor resource virtualization layer with private cloud support. J. Intell. Manuf. **27**(2), 447–462 (Apr. 2016). https://doi.org/10.1007/s10845-014-0878-7
28. L. Garber, Robot OS: a new day for robot design. Computer (Long. Beach. Calif) **46**(12), 16–20 (2013). https://doi.org/10.1109/MC.2013.434
29. A. Argyrou, C. Giannoulis, A. Sardelis, P. Karagiannis, G. Michalos, S. Makris, A data fusion system for controlling the execution status in human–robot collaborative cells. Procedia CIRP **76**, 193–198 (2018). https://doi.org/10.1016/j.procir.2018.01.012
30. H. Liu, L. Wang, Remote human–robot collaboration: a cyber-physical system application for hazard manufacturing environment. J. Manuf. Syst. **54**, 24–34 (Jan. 2020). https://doi.org/10.1016/j.jmsy.2019.11.001
31. M. Urgo, M. Tarabini, T. Tolio, A human modelling and monitoring approach to support the execution of manufacturing operations. CIRP Ann. **68**(1), 5–8 (Jan. 2019). https://doi.org/10.1016/j.cirp.2019.04.052
32. A. Bilberg, A.A. Malik, Digital twin driven human–robot collaborative assembly. CIRP Ann. **68**(1), 499–502 (2019). https://doi.org/10.1016/j.cirp.2019.04.011
33. V. Vyatkin, Software engineering in industrial automation: state-of-the-art review. IEEE Trans. Ind. Inform. **9**(3), 1234–1249 (2013). https://doi.org/10.1109/TII.2013.2258165
34. B. Yao, Z. Zhou, L. Wang, W. Xu, J. Yan, Q. Liu, A function block based cyber-physical production system for physical human–robot interaction. J. Manuf. Syst. **48**, 12–23 (2018). https://doi.org/10.1016/j.jmsy.2018.04.010
35. S. Scholze, J. Barata, D. Stokic, Holistic context-sensitivity for run-time optimization of flexible manufacturing systems. Sensors (Switzerland) **17**(3) (2017). https://doi.org/10.3390/s17030455
36. K. Alexopoulos, S. Makris, V. Xanthakis, K. Sipsas, A. Liapis, G. Chryssolouris, Towards a role-centric and context-aware information distribution system for manufacturing. Procedia CIRP **25**(C), 377–384 (2014). https://doi.org/10.1016/j.procir.2014.10.052
37. A.K. Dey, A.K. Dey, G.D. Abowd, Towards a better understanding of context and context-awareness, in *HUC '99 Proceedings of the 1ST International Symposium on Handheld Ubiquitous Computer* (1999), pp. 304–307
38. G.D. Abowd, A.K. Dey, P.J. Brown, N. Davies, M. Smith, P. Steggles, Towards a better understanding of context and context-awareness, in *Lecture Notes in Computer Science (including subseries Lecture Notes in Artificial Intelligence and Lecture Notes in Bioinformatics)*, vol. 1707 (Springer, Berlin, Heidelberg, 1999), pp. 304–307

39. Y. Xu, F.Y. Xu, Research on context modeling based on ontology, in *CIMCA 2006: International Conference on Computational Intelligence for Modelling, Control and Automation, Jointly with IAWTIC 2006: International Conference on Intelligent Agents Web Technologies* (2007), pp. 188–188. https://doi.org/10.1109/CIMCA.2006.181
40. J. Indulska, P. Sutton, C. Johnson, P. Montague, C. Steketee, Context-aware, ambient, pervasive and ubiquitous computing. Comput. Sci. ACSW 2003 **21** (2003)
41. M. Bazire, P. Brézillon, Understanding context before using it, in *Lecture Notes in Computer Science (including subseries Lecture Notes in Artificial Intelligence and Lecture Notes in Bioinformatics)*, vol. 3554 (LNAI, Springer, Berlin, Heidelberg, 2005), pp. 29–40
42. S. van Engelenburg, M. Janssen, B. Klievink, Designing context-aware systems: a method for understanding and analysing context in practice. J. Log. Algebr. Methods Program. **103**, 79–104 (2019). https://doi.org/10.1016/j.jlamp.2018.11.003
43. M. Baldauf, S. Dustdar, F. Rosenberg, A survey on context-aware systems. Int. J. Ad Hoc Ubiquitous Comput. **2**(4), 263–277 (2007). https://doi.org/10.1504/IJAHUC.2007.014070
44. J.-Y. Hong, E.-H. Suh, S.-J. Kim, Context-aware systems: a literature review and classification. Expert Syst. Appl. **36**(4), 8509–8522 (2009). https://doi.org/10.1016/j.eswa.2008.10.071
45. P. Prekop, M. Burnett, Activities, context and ubiquitous computing. Comput. Commun. **26**(11), 1168–1176 (2003). https://doi.org/10.1016/S0140-3664(02)00251-7
46. C. Herrmann, S.H. Suh, G. Bogdanski, A. Zein, J.M. Cha, J. Um, S. Jeong, A. Guzman, Context-aware analysis approach to enhance industrial smart metering, in *Glocalized Solutions for Sustainability in Manufacturing* (Springer, Berlin, Heidelberg, 2011), pp. 323–328
47. T. Strang, C. Linnhoff-Popien, A context modeling survey," Work. Adv. Context Model. Reason. Manag. UbiComp 2004—Sixth Int. Conf. Ubiquitous Comput., vol. Workshop o, no. 4, pp. 1–8, 2004, doi: 10.1.1.2.2060.
48. C. Bettini, O. Brdiczka, K. Henricksen, J. Indulska, D. Nicklas, A. Ranganathan, D. Riboni, A survey of context modelling and reasoning techniques. Pervasive Mob. Comput. **6**(2), 161–180 (2010). https://doi.org/10.1016/j.pmcj.2009.06.002
49. S. Scholze, D. Stokic, J. Barata, C. Decker, Context extraction for self-learning production systems, in *IEEE International Conference on Industrial Informatics (INDIN)* (2012), pp. 809–814. https://doi.org/10.1109/INDIN.2012.6301132
50. M.K. Uddin, J. Puttonen, S. Scholze, A. Dvoryanchikova, J.L. Martinez Lastra, Ontology-based context-sensitive computing for FMS optimization. Assem. Autom. **32**(2), 163–174 (2012). https://doi.org/10.1108/01445151211212316

Chapter 8
Dynamic Assembly Planning and Task Assignment

Konstantinos Sipsas, Nikolaos Nikolakis, and Sotiris Makris

Abstract Human–robot collaboration comes as an evolution to manual production. An increasing number of robots is gradually incorporated in today's production lines. The introduction of automation systems may complement the human factor and skills, and alleviate operators' daily workload. The production process can be executed in collaboration between humans and robotic manipulators achieving improved production efficiency as well as reduced injuries, caused by cumulative stress. Nevertheless, and beyond safety issues that arise from human–robot collaboration, a major challenge remains in how to efficiently break down a manufacturing process to allow for collaborative execution. Different capabilities and constraints need to be considered both in planning the assembly of a product, and its execution. Changes in the operational status of either robots or operators may require modifications to planning as well as the execution steps of a manufacturing process. Such changes supporting human–robot teamwork should be considered and dynamically addressed to maintain efficiency and smooth human–robot collaboration. In this context, this chapter discusses the steps towards semi-automated feature-based assembly planning. A set of evaluation criteria is used to evaluate alternative assembly sequences generated from a product CAD model, which are then mapped onto tasks. The task sequence can then be assigned to production site manufacturing resources for execution. A multi-criteria decision-making model is used for mapping assembly sequences to tasks, while a hierarchical workload model supporting a common level of breakdown is responsible for task assignment to operators, robots or both. A lower level of the model is used for the robot to map a generic description onto commands by the executable robot controller. IEC 61499 function blocks are used as a common data model. The scheduled activities are wrapped into function blocks and dispatched for execution. Runtime events are evaluated and determined if and how to modify

K. Sipsas (✉) · N. Nikolakis · S. Makris
Laboratory for Manufacturing Systems and Automation, University of Patras, Patras, Greece
e-mail: sipsas@lms.mech.upatras.gr

N. Nikolakis
e-mail: nikolakis@lms.mech.upatras.gr

S. Makris
e-mail: makris@lms.mech.upatras.gr

L. Wang et al. (eds.), *Advanced Human–Robot Collaboration in Manufacturing*,
https://doi.org/10.1007/978-3-030-69178-3_8

the existing plan and/or schedule. As a result, dynamic planning and scheduling are facilitated concerning human–robot collaborative manufacturing. The proposed methods have been implemented in software prototypes to validate their feasibility. Finally, this chapter concludes with challenges identified as well as future research directions.

8.1 Introduction

Many efforts have been conducted in the last decades towards making production systems more flexible to changing market demands through smarter production systems and shorter production cycles. While flexibility as a necessity for production systems is not new, as supported by historians such as Scranton [1], it could be conceptualised as an attempt to reduce the firms' risk resulting from an unstable market, as per the potential role of the IT in the transformation of work [1]. However, production dynamics can be enhanced if human skills are combined with modern automation. Towards that direction, the introduction of automation systems improves efficiency in terms of repeatability and operational costs. They are also suitable for hazardous environments while they can relieve employees from ergonomically unfavourable work steps. In particular, industrial robots usually operate in a fenced environment isolated from human operators. The removal of a fence poses several challenges [3]. For example, it could endanger operators in case of a potential collision with an operational industrial robot, either due to malfunction of the robot or in case of an accident. This leaves the production lines with two distinct zones, one of the manual operations and one of the robotic operations. Considering that a job consists of interrelated tasks, routine tasks, being easy to automate, are usually performed by machines, while for more complex problems, skills like intuition, problem-solving, persuasion, creativity, are needed, making them difficult to codify. As a consequence, improved flexibility is pursued in the combination of the strengths of each entity.

Nowadays, numerous new industrial robotic systems are certified for operation in shared working spaces. In these cases, the safety of the human workforce is ensured by—possibly selectively—lowered speed and force of the robot, safety monitored stop and leveraging advanced vision systems to slow operations. This new generation of robots referred to as collaborative robots, designed to be safe for human presence, have significantly reduced operational capabilities when compared to the mainstream of industrial robots which must operate behind fences. The reduced performance of collaborative robots has a significant impact on the production cycle time.

The simultaneous presence and reliance on human and robotic resources in production processes also require coherence in the behaviour of humans and robots sharing the same space, ranging from safe coexistence to coordinated teamwork. Humans and robots taking part in the same production process also need to have an adequate view and plans that enable them to advance the production process towards its goal. The challenge herein is to break down a process to smaller operations that can serve as a common ground for understanding between the two heterogeneous resources. While the human operators may use their cognitive skills to reason and infer from higher-level descriptions, this is not the case for a robot which has to be given an accurate, formal plan to execute.

This chapter discusses the need for planning manufacturing processes accordingly when addressing human–robot collaboration. As such, the different skills should be considered but also the trade-off between assignments to each category. Moreover, in the context of teamwork, changes may be needed in runtime that may affect the planning steps. This cannot be supported by conventional MES systems. As such, an intelligent decision-making framework is proposed to evaluate and assign the tasks to resources, evaluating their competencies and production requirements. Moreover, an assembly feature-based approach bridges the gap between task planning and task execution facilitating the creation of closed-loop control systems between task planning and execution. This is achieved by integrating the feedback from the shop-floor and dynamically adapting the schedule accordingly. The methods adopted for dynamic assembly task planning and task scheduling are discussed while a set of prototype software implementations are provided. Finally, the challenges identified are discussed in the context of collaborative execution, along with potential future research directions.

8.2 Literature Review

Assembly modelling and planning plays an essential role in the manufacturing of products. Assembly parts, each one formed or machined to meet design specifications, are assembled during a procedure aiming to generate the final product with some predefined requirements. An efficient and effective assembly sequence planning, as part of the assembly schedule, is fundamental for the production line as it can not only reduce the production time and cost but also improve the quality of the final product.

In the past years, many efforts and studies have been done in the field of automatic generation of assembly sequences. Some of them use connected graphs to represent the assembly information. In [4], the sequencing of the assembly tasks and components is achieved by an algorithm based on deterministic finite automaton (DFA) rules, engineering decisions and product structure definition. The algorithm runs on an adjacency matrix extracted from the directed graph which was, respectively, created based on the liaison diagram for the assembly parts. In [5], Zhu et al. handled the problem of generating optimal assembly sequences that reduce the complexity of the assembly line. Tasks are presented as nodes of a connected graph and the complexity as weighted edges between them, and thus, the procedure for generating an assembly sequence becomes similar to the travelling salesman problem and is treated with dynamic programming.

Another way to present assemblies and then generate the assembly sequence is through semantic models. In [6], semantic properties are assigned to each semantic individual representing the assembly part and carry information about its temporal, spatial and spatiotemporal position. Moreover, semantic relations between individuals present their real-world relationships. In [7], another semantic approach is presented. The assembly part knowledge is separated into three different levels, referring to the function, structure and feature levels. Then, by reasoning over the ontology, semantic relations are created between parts which can be used for generating the assembly sequence. In [8], the assembly plan semantic representation model contains pre-defined assembly information, such as the geometric constraints and precedence relations between parts along with some dynamic real-time physical constraints, such as the position of the centre of gravity and force strength of each part. In this way, the assembly sequences are modified by the feedback from an interactive virtual environment.

Geometry information of assembly parts alone is inadequate for generating the assembly sequence. In case of a desired efficient assembly sequence planning, additional information is needed. In [9], the assembly connection graph is built by considering both geometric and non-geometric knowledge about assembly parts. The assembly sequencing is generated based on precedence relationships among geometry constraints, special parts and processes. In [9], it is also pointed out that the integration of semantic information in the assembly connection graph could provide more useful information for the designer regarding assembly planning. In [10], a different rule-based approach for generating assembly sequence is presented. In addition to geometrical information, the proposed method introduces machining features, such as aperture and encompassed surfaces that describe components. A model based on these features is created and the assembly sequence is exported based on this model. Finally, in [11], the use of assembly features is introduced. Assembly features are extracted for each product part, stating the connections with other parts. In this way, matching pairs between components are created and the assembly sequence can be achieved by assigning precedence relationships to each assembly feature.

Some of the difficulties related to the automation of assembly sequence generation exist due to the weakness of the inefficient description of assembly knowledge and information. In this study, assembly plans are described through a semantic model. Part geometry and assembly features and precedence relationships are imported to our algorithm, and with the use of a rule-based semantic reasoner in combination with a graph algorithm, the matching and sequencing of the parts are being achieved.

Over the last decade, extensive research has been done in designing flexible production systems with adaptive capabilities, able to react and adapt to the fluctuating market needs. The manufacturing of products in the modern industrial world requires the combined and coordinated efforts of people, machines, and equipment [12]. To address the challenges in a dynamic environment an assembly system should be adaptive. Reconfigurable assembly systems can also provide customised flexibility for a particular product family [13]. One of the main concerns is to find a resolution to the flexibility vs. efficiency dilemma on a small scale, calling for new models and methods of automated assembly planning [14, 15]. Two general approaches are taken to cope with the inherent complexity of assembly sequence planning:

The aggregation that suggests a hierarchical decision scheme separating macro and micro planning [14], and feature-based decomposition that helps to structure domain knowledge around local assembly features

Feature models can be created by recognising features in the geometrical model or by defining the product with features [16]. Assembly features that are derived from the CAD model of the product [17] imply tasks, the use of specific resources, and modes of task execution. Nowadays, due to digital data acquisition, motion capture and visualisation methods, assembly sequence planning are accompanied by virtual evaluation, testing and simulation [17, 18].

However, the simulation of virtual assembly cannot provide complete support for the planning process [18]. Geometric reasoning combined with motion planning should be used for ensuring the feasibility of robotic assembly sequences. Furthermore, recognised assembly features can provide the basis for generating human work instructions as well.

Within this world, robots have gained considerable significance in assembly operations demonstrating great potential [19]. Even with primitive intelligence, robots have revealed their capacity to increase a factory's productivity, efficiency, and quality. The combination of human workers with robots in a manufacturing cell could allow higher flexibility with shorter production cycles leading to a better adaptation to the changes in the market demand profile [14, 15]. A detailed study reviewing human–robot collaboration is presented in [20].

The introduction of robots to manual assembly lines to enhance the competitiveness and efficiency of modern production systems is a research area that has attracted great interest [21]. The modelling of human–robot (H–R) teams using task-based mental models has been researched in [22], inspiring to help robots predict information and resource requirements of their human teammates. H–R task planning methods have been considered as well [20, 21], with the main criterion of reducing the human burden of tasks.

8.3 Dynamic Assembly Planning Overview

The workflow proposed in this chapter starts with the analysis of the CAD product models and the extraction of assembly features from it. This process generates a series of assembly features using a variety of parameters, as detailed in the following sections, such as part geometry and kinematic constraints. Then, the assembly features are fed into assembly sequence planning where possible assembly sequences are generated and evaluated based on a series of criteria. The best sequence is selected and based on user input the final task sequence is generated. This task sequence is used by the scheduling module to produce appropriate allocation of tasks to resources. The task allocation is then sent to the execution and control modules, where task re-allocation (rescheduling) may be requested dynamically based on shop-floor environment changes.

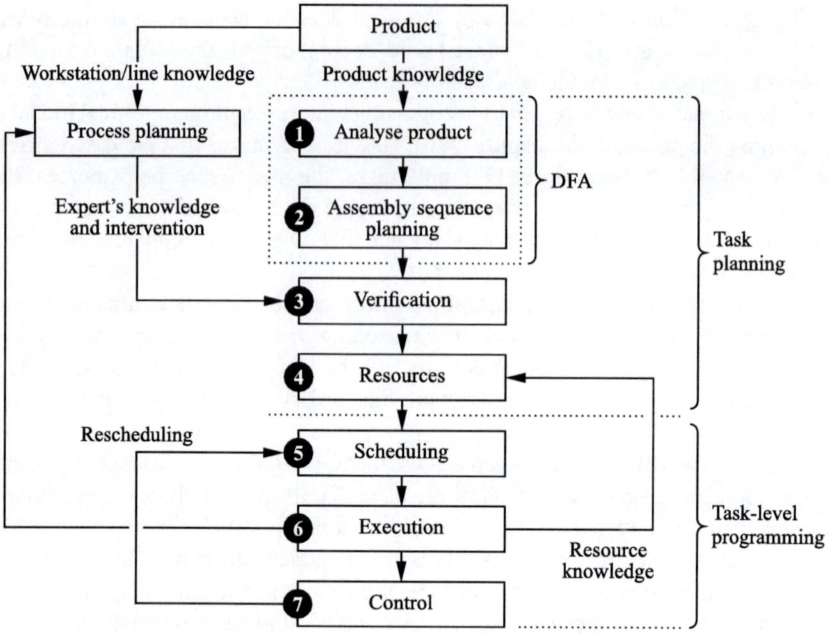

Fig. 8.1 High level representation of the task planner methodology

A high-level overview of the proposed approach towards shared task planning and scheduling is presented in Fig. 8.1.

8.4 Feature-Based Task Planning

An assembly feature contains information regarding the connection of two or more assembly parts. As far as the association of components in an assembly plan is concerned, each assembly task is associated with a specific workstation, resource and feature and each feature is subsequently associated with one or more assembly parts.

An ontology model is created and used to temporarily store and handle the information of the assembly plant. The model consists of five main classes, referring to the assembly tasks, parts, features, resources, and workstations, respectively. The features class contains one subclass, which regards the simple features. The simple assembly features supported by the current development are the placing, insertion and screwing features. Features may be composed of more than one simple feature. Each class of the ontology model has some specified semantic properties that describe it. Assembly tasks, parts, assembly features, resources and workstations that are imported are stated as individuals in the ontology, and the information about them is assigned to them by semantic properties. Then, by reasoning over the ontology, or else by executing semantic rules, the properties of individuals are examined as to which class they belong to, and, in this way, individuals are categorised into the ontology classes.

The algorithm, at first, imports the assembly plan, including the related assembly tasks, parts, features, resources, and workstations. Then, the ontology which is used for handling the assembly entities is created, and semantic rules are executed to extract the connections of the assembly parts. A connected graph is created, presenting these connections. The features precedence is also extracted during the execution of semantic rules, and another connected directed graph is created, declaring the precedence relations of the assembly features. Then, a graph search algorithm runs multiple times on the assembly feature graph and generates feasible assembly sequences. Afterwards, the assembly sequences generated are evaluated and sorted based upon various criteria. In the end, the optimal assembly sequence is exported along with a task list that corresponds to the assembly sequence. The main flow diagram of the algorithm developed is presented in Fig. 8.2.

The algorithm imports the product's assembly parts along with their assembly features. An assembly feature contains information regarding the connection of two or more assembly parts. For example, an assembly feature can be "Screwing Part A to Part B". The whole assembly plan consists of the following components:

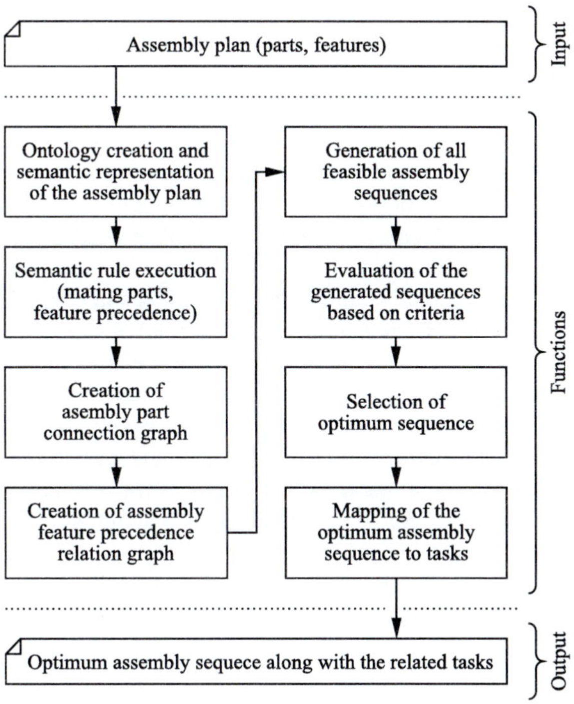

Fig. 8.2 Overview of the proposed algorithm

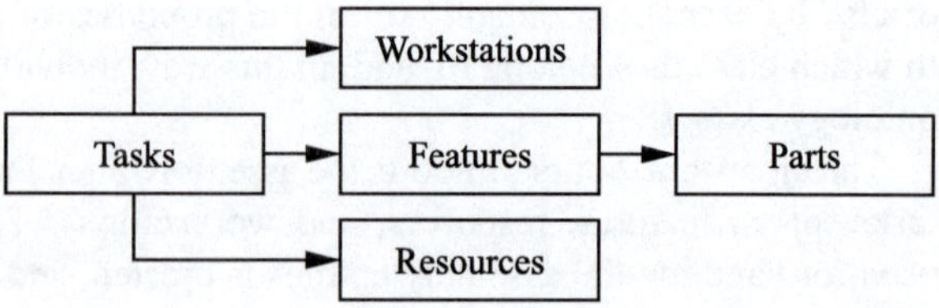

Fig. 8.3 Association of assembly plan components

- Assembly tasks
- Assembly parts
- Features
- Resources
- Workstations

As far as the association of the assembly plan's components is concerned, each assembly task is associated with a specific workstation, resource and feature, and each feature is subsequently associated with one or more assembly parts. The relations of the assembly plan's components are presented in Fig. 8.3. The information of the assembly plan is retrieved from an existing legacy system.

The next step of the algorithm is the creation of an ontology model that is used to temporarily store and handle the information of the assembly plan. The model consists of five main classes, referring to the assembly tasks, parts, features, resources, and workstations, respectively. The features class contains two subclasses, one regards the geometry features of parts and the other regards the assembly features of tasks. Geometry features define the posture or goal of parts in space. Assembly features define the way two or more parts are connected and can be divided into two subclasses, the simple features class and the composite features class. The simple assembly features supported by the current development are placing, insertion, and screwing. Composite features are composed of more than one simple feature. For example, a composite feature can contain both a placing and an insertion simple feature.

Each class of the ontology model has some specified semantic properties that describe it. Assembly tasks, parts, assembly features, geometry, resources and workstations that are imported are stated as individuals in the ontology and the information about them is assigned to them by semantic properties.

The representation of classes and properties of the ontology model is shown as an Entity Relationship Diagram (ERD) diagram in Fig. 8.4. Each box represents an ontology class along with its semantic properties. Property values that are set in bold declare that this property correlates the individuals of this class with other individuals from the class in bold.

After the classification of the ontology individuals and with the use of semantic rules, the matching of the assembly parts and the features precedence relations extraction takes place. To extract the assembly parts connections, each simple feature's information is processed.

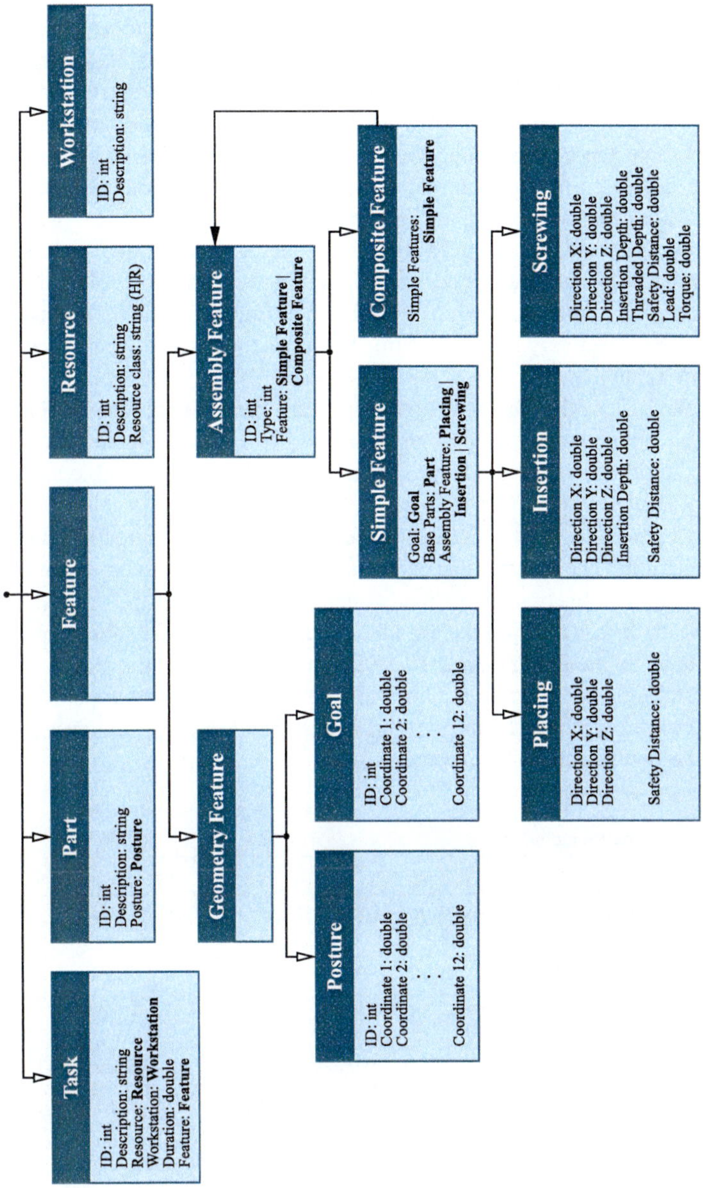

Fig. 8.4 Representation of planning ontology

After all parts connections are found, the connectivity of the parts can be represented by an undirected connected graph, where the vertices will represent the assembly parts and the edges the parts connections.

The last step is the creation of a list of parts, assembly features and assembly feature classes (Table 8.1). It should be noted that the proposed approach for feature-based task planning automates the procedure up to a point and considering the available information. After the generation of the first batch of suggested assembly sequences manual entry from an experienced engineer is required. Furthermore, at the next step about the task planning the same approach is followed. The process is automated up to a point and then manual entry and confirmation are required to proceed to the final step concerning the task scheduling. The base parts, moved parts and assembly feature semantic properties of simple features class individuals declare the parts and the way that they are connected. With the creation of a semantic rule that connects the base parts and the moving parts of a simple feature, the connections of all parts can be effectively found. The rule that does the assembly parts matching is presented in Table 8.2.

A graph search algorithm that runs on the assembly features precedence relations graph has been implemented, to find the total feasible assembly sequences for the assembly of the product. The procedure for the generation of a possible assembly sequence is presented in Fig. 8.5.

The graph vertices represent the features and the connecting directed edges the precedence relations. The features that can be inserted into the sequencing are shown in orange and the features that have already been sequenced in grey.

At first, the algorithm finds all the possible roots of the graph, in other words, the vertices that have no incoming edges. These features declare that the parts that they connect can be joined together at any time independently of the other parts. In this example, there are two roots, features 1 and 3, and the algorithm randomly selects feature 1 as the first feature to be executed.

Table 8.1 Assembly part matching rule

1	2	4	3	3
IF (x is a Simple Feature) AND (a is Base Part of x) AND (b is Moved Part of x) → THEN connect a and b	–	–	–	–

Table 8.2 Assembly features and classes

Part	Name	Feature	Feature name	Class
1	Part 1	1	Placing throttle	Placing
		2	Screwing flange	Screwing
		3	–	Screwing
2	Part 2	4	–	Screwing
		1	–	–
		2	–	–

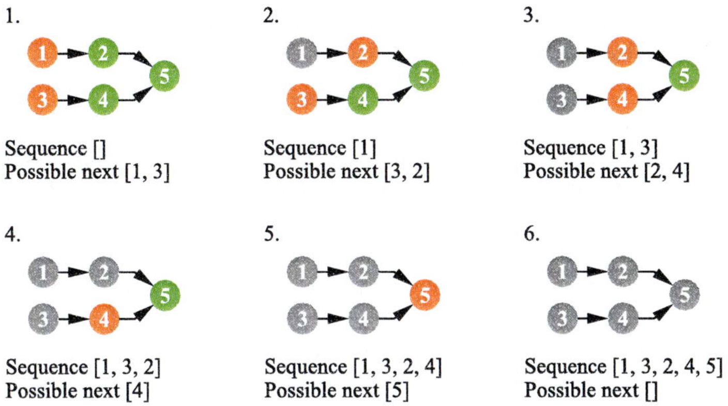

Fig. 8.5 Alternative assembly-feature-based sequences

In the next step, features are sought that can follow feature 1 in an assembly sequence. In the example, these are feature 3 (another root of the graph) and feature 2 (which must be preceded by feature 1). Subsequently, feature 3 is randomly selected from possible next features. The procedure is carried on until all features have been sequenced. In this example, feature 5 will always be last in the feature sequence, as it can only be started after the completion of all other assembly operations.

When all feasible sequences have been generated, they are finally evaluated using the following criteria to identify the most appropriate choice:

N_{SS}: Number of independent stable *subassemblies*

N_{PO}: Number of part re-orientations during assembly

N_{TC}: Number of required tool changes during the assembly

U_{CT}: Consecutive tool usage

Consecutive tool usage is defined as:

$$U_{\mathrm{CT}} \overset{\mathrm{def}}{=} N_{\mathrm{MC}}(1 - P_{\mathrm{LF}} - P_{\mathrm{LC}}), \tag{8.1}$$

where N_{MC} is the maximum number of consecutive assembly features processed using the same tool in an alternative assembly sequence, P_{LF} is the position of last assembly feature in the sequence, and P_{LC} is the position of the last consecutive assembly feature.

Using the aforementioned criteria in an approach weighed by the user, the *assembly sequence auto-evaluation indicator*,I_{ASAE}, expressed in %, is generated as follows:

$$I_{\mathrm{ASAE}} = ((w_1 \cdot N_{\mathrm{SS}} + w_2 \cdot U_{\mathrm{CT}}) - (w_3 \cdot N_{\mathrm{PO}} + w_4 \cdot N_{\mathrm{TC}})) \cdot 100, \tag{8.2}$$

where w_i denotes the weight of each criterion as selected by the tool user/assembly engineer. The default value for w_1, w_2, w_3, and w_4 is 0.25.

Out of this step, alternative assembly sequences are generated for the assembly of a specific product and ranked. The highest-ranking one is suggested to the assembly engineer/user as the most appropriate. This approach can be used iteratively on several parts or products to come up with a set of highest ranked assembly sequences for each part/product.

Following the above-mentioned approach, the assembly engineer will be able to define, in a short time, the optimal approach for assembling one or more products, but not for manufacturing it. This would require knowledge of the specific workstation/line where the operations are executed. Since the approach, up to this point, has not included any knowledge of the specific workstation or assembly line, the user has to refine the appropriate sequence of assembly operations by editing or changing the suggested sequence, or selecting an alternative sequence. As soon as this process is completed a list of tasks has been generated that can be assigned to resources in the shop floor. Section 8.5 discusses in detail how these tasks will be assigned to resources.

Also, it should also be mentioned that as this step is semi-automated and requires user input, the expert may adjust the solution for sequences that don't follow the normal assembly workflow.

The assembly sequence planning core is also supported by a user interface that is responsible for the visualisation of the results, manual editing of the solution, and persistence of the final solution in a database to support task scheduling.

8.5 Dynamic Task Scheduling for Human–Robot Collaboration

The scheduling core is responsible for making assignments of tasks related to the manufacturing of a specific and already analysed product to the available and capable resources of a specific line or workstation. The scheduling procedure requires the definition of a model which can be used for assigning tasks to resources, a decision making algorithm to decide which task shall be assigned to which resource, and the definition of a schedule generation point, or a decision point for the algorithm.

A factory may contain several departments where parts are assembled. A department may include one or more stations. Each station involves many resources, e.g., human operators or/and robots. Similarly, the workload is broken down to the corresponding levels to map the different facility levels (Fig. 8.6).

For human–robot collaboration to be facilitated in a shop-floor environment, the human operator, as well as the robot, could be considered as resources capable of exchanging and/or sharing tasks. As a result, a task could be assigned to one or both of them (Fig. 8.7).

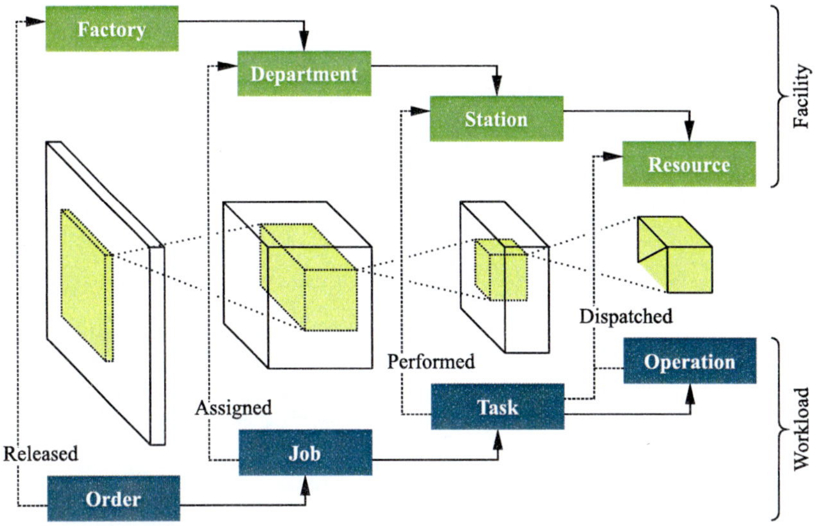

Fig. 8.6 Hierarchical models for facility and workload

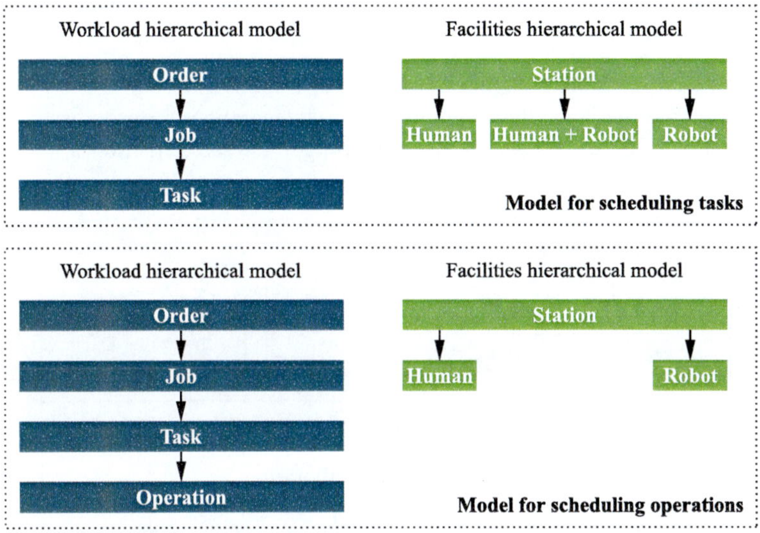

Fig. 8.7 Task and operation models for scheduling human–robot collaborative tasks

A task can be shared in either time or space (Fig. 8.8). *Shared in time* refers to the consecutive work of a human and a robot, in a specific moment, for the fulfilment of the assigned task. *Shared in space*, on the other hand, means that a portion of the task will be carried out by one operator, while the rest can be completed by another. A shared-in-space task may include the serial, the parallel or the mixed execution

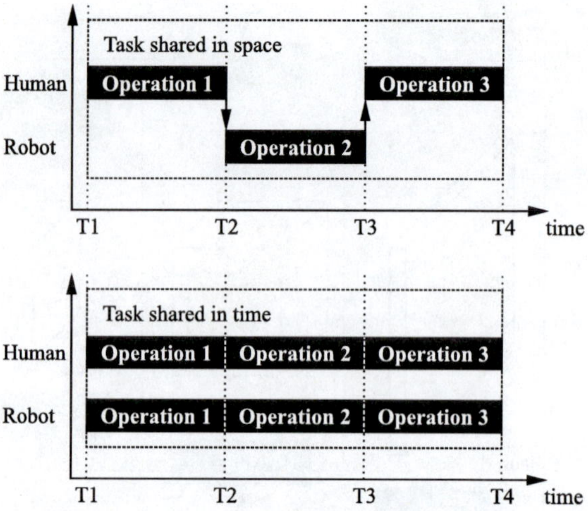

Fig. 8.8 Shared task categories

of smaller task segments. In both cases, it must be assumed that a task describes a combination of smaller sub-tasks, named *operations* throughout this chapter.

The task assignment procedure (Fig. 8.9) is executed at a decision point by a decision-making framework, via the following steps:

- Definition of possible assignments of tasks/operations to resources.
- Selection of criteria for the assignment of tasks/operations to resources.
- Evaluation of generated alternatives according to the selected criteria.

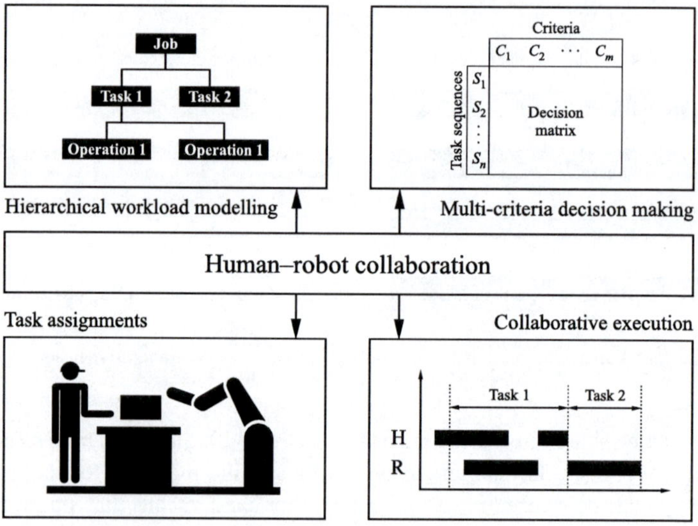

Fig. 8.9 Human–Robot scheduling concept

```
<WORKCENTERS>ID,NAME
<RESOURCES>ID,NAME
<HIGH_LEVEL_TASKS>ID,NAME
<TASKS>ID,NAME
<TASK_PRECEDENCE_CONSTRAINTS>
  <PRE_CONDITION>TASK id
  <POST_CONDITION>TASK id
<RESOURCE_AVAILABILITY>TRUE/FALSE
<TASK_RESOURCE_SUITABILITY>TRUE/FALSE
<START_TIME_OF_TASK>YEAR,MONTH,DAY,HOUR,MINUTES,SECONDS
<END_TIME_OF_TASK>YEAR,MONTH,DAY,HOUR,MINUTES,SECONDS
<DURATION_TIME_OF_TASK>SECONDS
```

Fig. 8.10 Task scheduling input

- Selection of the highest-scoring alternative. After scoring all alternatives in the previous step, the highest-ranking one is selected as the most suitable to match the current objectives.

It is assumed that at least one suitable resource exists for the given task. The suitability of resources is decided upon human capabilities (e.g., flexibility, problem-solving capabilities, complex perception, manipulation) and robot capabilities (repeatability, efficiency, accuracy, high payload capability, etc.).

Additionally, for the suitability algorithm, the availability of the resources and their ability for tool/gripper handling are considered. The input for the decision-making framework includes the resources, the tasks, the task–resource suitability, the task precedence constraints, the duration of tasks, and the time of starting and completing a task (Fig. 8.10).

The proposed decision-making framework selects which resource will be assigned to the execution of a task based on the suitability of the resources.

The alternative solutions that are selected from the decision-making framework are evaluated against multiple criteria to select a good solution in a short time frame. The criteria selection is based on the requirements and specifications of the user. There is no restriction on the kind and number of criteria that can be selected. Some of the criteria implemented up to the time of writing this chapter are summarised below:

Total weight of lifted parts: The criterion refers to the sum of the weight of all the parts lifted by human resources. It is estimated by the following relationship:

$$W_H = \sum_{i=1}^{k} m \cdot w_p, \tag{8.3}$$

where:

w_p the weight of a part lifted by a human resource,

m the iteration n of a part lifted by a human resource, and.
k the total number of parts lifted by a human resource.

The total duration of tasks execution/per cycle time: This criterion is estimated as the sum of the completion time of the n assigned tasks to a resource. The following relationship is used:

$$T_H = \sum_{i=1}^{n} T_c^i, \tag{8.4}$$

where:

T_c^i completion time of task i assigned to a resource,
n the total number of tasks assigned to a resource.

Production or completion rate: The production rate R_P (expressed parts/hour) is inversely proportional to the cycle time T_C of a process (expressed in minutes) and is estimated through the relation:

$$R_P \approx \frac{60}{T_C}. \tag{8.5}$$

This criterion will be used for defining a new cycle time for the shared task scheduling so that the new solution will have a cycle timeless or equal to the estimated cycle time.

Operating cost: The criterion is estimated as the sum of the operating costs of each resource for every task. Two separate formulae are used for estimating the cost of human resources and robots.

$$C_{OH} = \frac{S_D}{T_W} \cdot T_c^i \tag{8.6}$$

$$C_{OR} = C_E \cdot T_c^i \tag{8.7}$$

$$C_O = \sum_{i=1}^{k} C_i, \tag{8.8}$$

where:

C_{OH}, C_{OR} operating costs for human and robot, respectively,
S_D daily salary of the human workforce,
T_W working time of the human workforce,
C_E cost factor for robots,

```
<TASK>ID
<RESOURCE>ID
<YEAR>YEAR
<MONTH>MONTH
<DAY>DAY
<HOURS>HOUR
<MINUTES>MINUTES
<SECONDS>SECONDS
<DURATION_IN_MILLISECONDS>MSECONDS
```

Fig. 8.11 Task scheduling output

C_i the total operating cost per task for robots and humans separately,
C_O the total operating cost per cycle time.

The result of the scheduling method includes the task, the assigned resource per task, the timing factors (start date, month, year, hour, minutes, seconds) and the duration of the task (ms) (Fig. 8.11).

A schedule is initially generated, and the tasks are assigned to the resources after the evaluation procedure described in the previous section. The same schedule is repeated as many times as required to meet the production orders.

During execution time, unexpected events may occur, or changes may be required to match the production targets, e.g., meeting the number of products set for an 8-h shift. In case of such occurrences, the system should be flexible enough to adapt to the newly introduced parameters.

As a consequence, in case of such an event, a rescheduling of the remaining tasks is performed according to a predefined set of rules and constraints. In this case, depending on the nature and stopping point of the task, the manufacturing process may adapt and continue its execution, by re-assigning several operations to a different resource.

The scheduling, as a result of the decision-making algorithm, takes place after an initial triggering event defined by the user before the start of production. This event is the initial decision point for the algorithm. To enable the generation of a new schedule taking into consideration the newly introduced production constraints, a new decision point for the decision-making algorithm has to be created for the event occurred. Afterwards, a new schedule is generated, and tasks are re-assigned to the shop-floor operators.

Additionally, the algorithm takes into consideration the status of the resources in the shop floor at the time of generation of each new decision point. For example, a robot may have malfunctioned or a human resource may have stepped away from their workstation. In such cases, these resources are excluded form task assignment. This information is retrieved from location and status sensing infrastructure deployed in the shop floor.

The scheduling core is supported by a user interface component used for visualising the task assignments to resources using a Gantt chart.

8.6 Task Planning and Scheduling Prototype

8.6.1 System Implementation

The proposed task planning and scheduling system are implemented following a distributed approach which is illustrated hereafter in Fig. 8.12.

The proposed prototype implementation consists of the three major components (1) the user interface, (2) the back-end system, and (3) the shop-floor components.

Shop-floor components

The shop-floor components consist of different software applications deployed on the shop floor. These applications perform various functions such as managing the execution of the tasks and localising of resources. Since these applications exist outside the proposed system, they are not detailed further, however, the need to interface with them has influenced the system design.

Back-end system

The backed-end system is the central component of the proposed system and was developed as a web application using *Java* and the *Spring* framework. It interfaces with the shop-floor applications to intercept events coming from the shop-floor applications, then processes the information received to generate meaningful information for the user interface component. Additionally, it may also support data processing for workflows taking place entirely in the context of the user interface such as the calculation of assembly sequences used for planning for human–robot collaboration.

The development of the back-end system follows a layered architecture as shown in Fig. 8.13.

The three layers that make up the back-end are (1) the Context-aware service layer, (2) the Data processing and context generation layer, and (3) the Shop-floor integration layer.

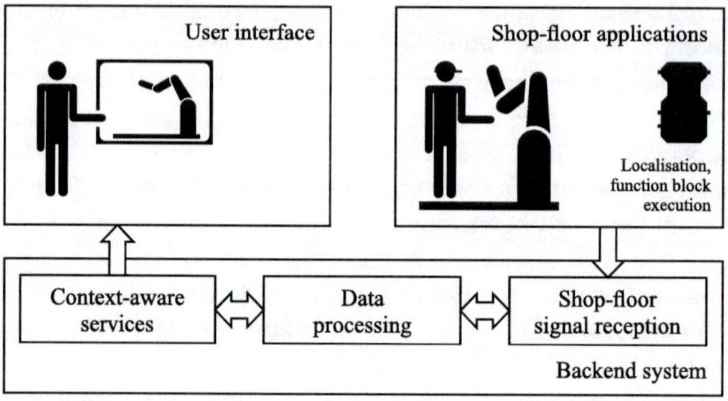

Fig. 8.12 System conceptual design

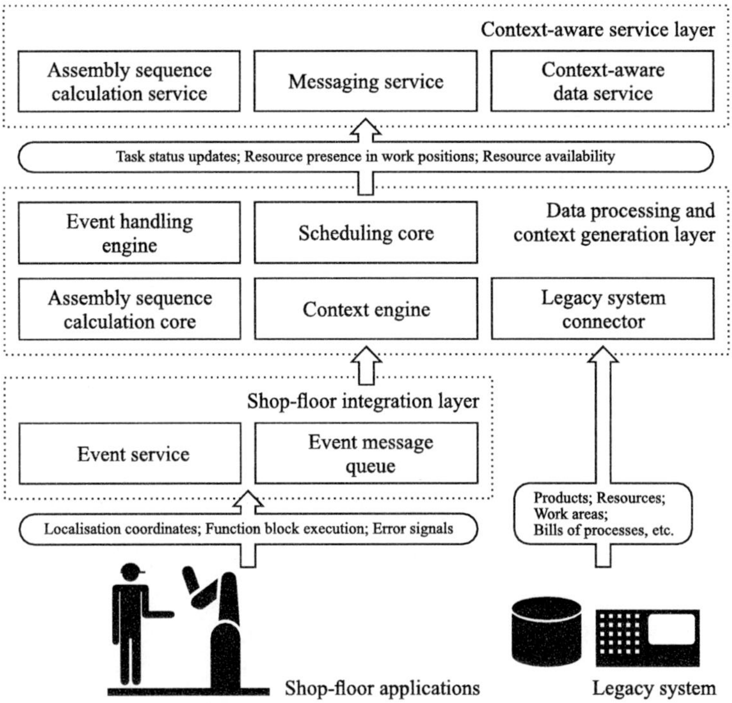

Fig. 8.13 Back-end layered architecture

The Shop-floor integration layer is responsible for receiving signals from the components deployed on the shop floor. To achieve this functionality, it offers two different application programming interfaces (APIs) that the shop-floor applications may use. The first one is a *Restful* service, offering different endpoints for detecting function block execution and error signals. The second is the Messaging Queue implemented using *ZeroMQ* and is used for publishing localisation messages.

The Data processing and context generation layer consists of 5 components: (1) Event handling engine, (2) Context Engine, (3) Legacy system connector, (4) Assembly sequence calculation core, and (5) the Scheduling core. The Event handling engine is responsible for interpreting the shop-floor events to obtain useful information. For example, the localisation coordinates are transformed into resource presence in working position. The Context engine (CE) is responsible for populating the context model, holding a representation of the shop-floor status, with information coming from the Event handling engine and the Legacy system connector. For example, the CE holds the information on where each worker is located, their task at hand, and the progress on the production schedule. The Legacy system connector is responsible for accessing the legacy system of the factory and retrieving information about the resources, the work areas, the products to be produced and so on. The Assembly sequence calculation core is responsible for processing an XML representation of assembly features and identifying the most appropriate assembly sequence using different criteria. The scheduling core is responsible for calculating the task execution schedule based on information about the quantities to be produced, products codes, and available resources.

The Context-aware service layer is responsible for exposing the events and context information to the system's clients (e.g., the user interface). For this purpose, two mechanisms are deployed. The first leverages Restful services to provide information on demand. The second is a messaging service based on ActiveMQ that forwards events from the Event handling engine towards the registered clients.

User interface

The user interface of the system was implemented in Java as an *Eclipse RCP* plug-in, facilitating its integration with an existing commercial Manufacturing Execution System (MES). Its internal structure is illustrated in Fig. 8.14.

The user interface consists of two layers. The Data retrieval layer is responsible for retrieving data from the back-end Restful services and registering for/receiving events from the backend messaging service. The visualisation layer is responsible for visualising the appropriate information and control elements to the user. These elements are used to control assembly sequence calculation, scheduling of tasks, and the status visualisation of the workstations on the shop floor.

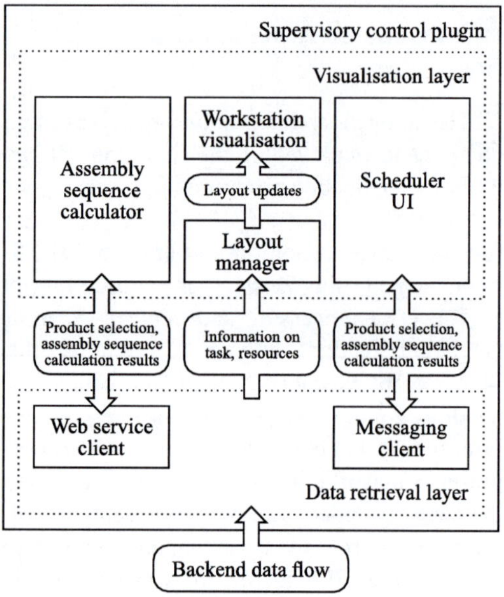

Fig. 8.14 User interface overview

8.6.2 Task Planning and Scheduling

Based on the procedure discussed in Sect. 8.4, the liaison diagram is generated by processing the assembly features. Alternative assembly sequences are generated based on the features of precedence relations. The user selects the desirable validation criteria and selects their weights. Next, the system proposes the assembly sequences that have the highest rank, thus fulfilling the selected criteria in the best way.

The geometrical models of the parts and the relevant elements of the work cell (e.g., tools and fixtures) are encoded into the *Standard Tessellation Language* (STL) format. This triangle mesh representation can be generated by legacy CAD systems and enables efficient collision detection queries.

The presented task planning algorithm has been tested for an assembly plan containing 14 assembly tasks, 12 parts, 8 simple and 2 composite features. The part connections graph as exported from the algorithm is presented in Fig. 8.15.

The suggested assembly sequences generated from the algorithm along with their values of the evaluation criteria printed on screen are presented in Fig. 8.16.

Before being assigned for execution, the selected assembly sequence needs to be mapped to a set of assembly tasks by sequentially processing the assembly features of each part. An assembly feature is mapped to a set of assembly operations or tasks using a set of predefined rules, in combination with some possible features that have been edited in advance and can be manually updated if required.

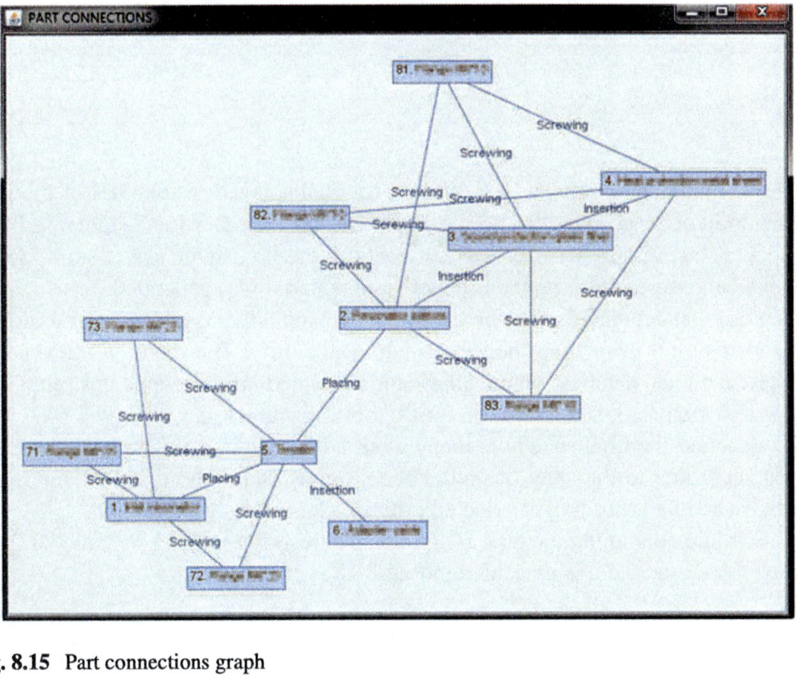

Fig. 8.15 Part connections graph

```
Suggested assembly sequences:
Assembly Sequence 1: [5, 2, 1, 6, 3, 4, 83, 81, 82, 73, 72, 71]
                     Tool changes: 0.25  Stability: 0.667
                     Consecutive tool usage: 1.0  Redirections: 0.571
                     Total score: 1.0

Assembly Sequence 2: [5, 6, 2, 1, 3, 4, 83, 81, 82, 73, 72, 71]
                     Tool changes: 0.25  Stability: 0.667
                     Consecutive tool usage: 1.0  Redirections: 0.714
                     Total score: 0.95
```

Fig. 8.16 Alternative assembly sequences

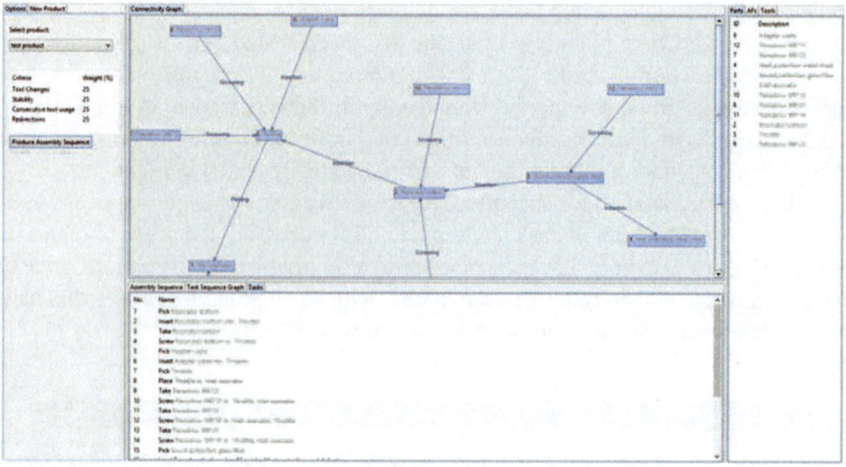

Fig. 8.17 Task planner UI

The automatically generated task sequence is displayed in the GUI of the task planner tool for the production engineer/user who can interact with the tool by editing the sequence (Fig. 8.17). After user interaction, the final sequence is stored in an internal database concerning the product from which it was generated.

The task list generated in the procedure above, including tasks, is being assigned to resources after evaluating their skills and availability. The required information is stored in a local database where all execution related information, constraints and parameters are listed, i.e., execution times, tools, resources, etc.

To generate the final schedule, many alternative assignments are generated and scored according to the asset of predefined criteria. The highest-ranked one is the optimal schedule according to selected criteria, i.e., reducing cycle time.

A schedule containing the task assignments to resources, along with the UI from where it was generated, is presented in Fig. 8.18.

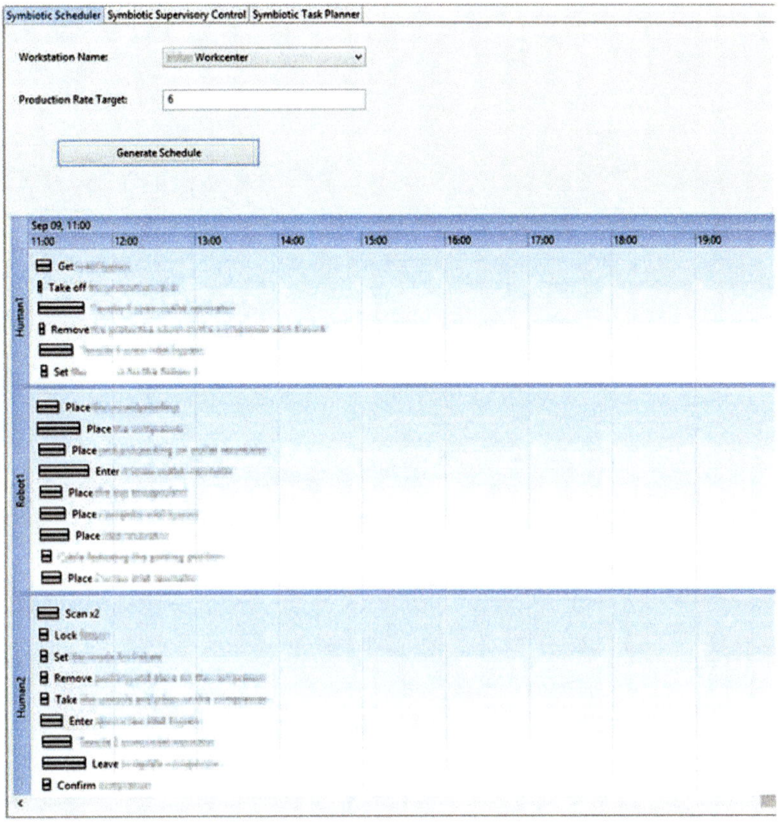

Fig. 8.18 Task scheduling UI

8.6.3 Activity Diagrams

In this section, a series of high-level diagrams will be used to explain how the user interacts with the system and the various components of the system with each other. The sequence diagram shown in Fig. 8.19 depicts the interactions between the various components to produce a task schedule.

In the above diagram, the Scheduler UI components are merged into one (Scheduler UI), and the actions performed by the function block (FB) compiler are omitted. The following actions are described in the diagram:

First, upon UI initialisation, the list of workstations is retrieved from the backend server.

Next, the user selects the relevant workstation and inserts the targeted production rate value.

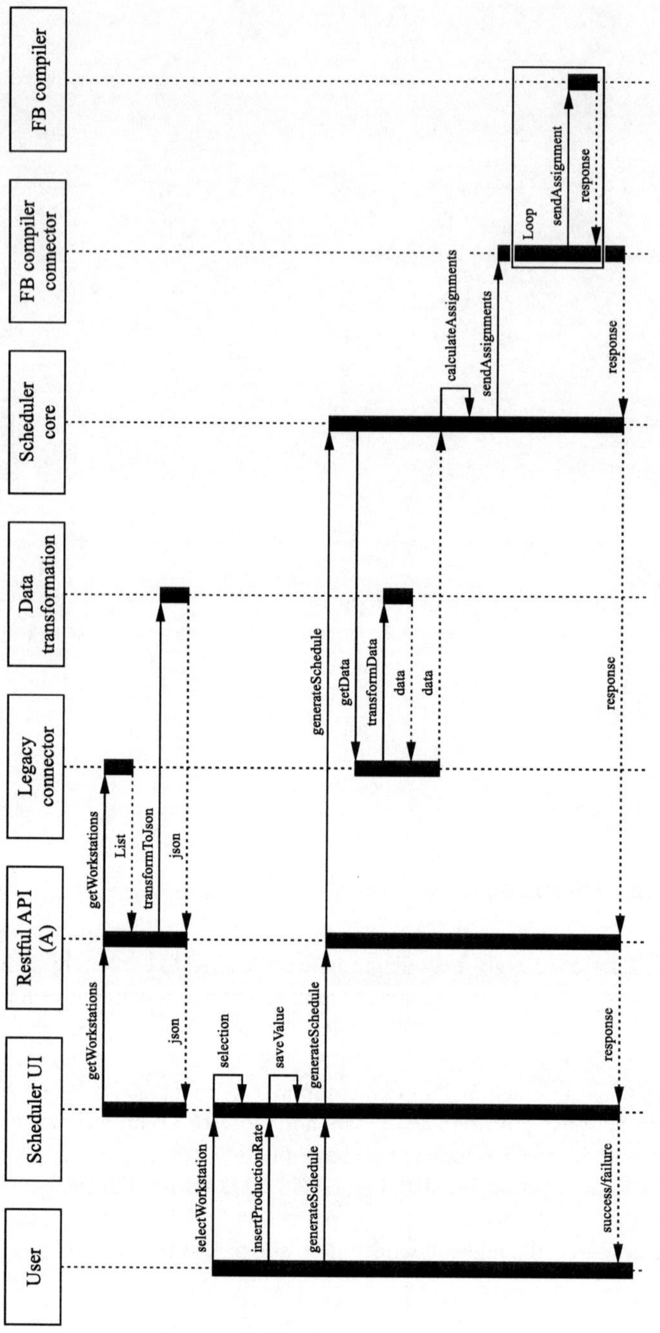

Fig. 8.19 Sequence diagram between components to produce a task schedule

Upon clicking the "generate schedule" button, the Scheduler UI component calls the Restful API methods for schedule generation, in turn, the scheduler core is activated and requests detailed data (e.g., resources, tasks, execution time and constraints) from the legacy system, which then calculates the task assignments and sends them to FB compiler, through the FB compiler connector, for execution.

In parallel, the scheduler core sends a response back to the user through the Restful API and Scheduler UI.

Upon UI initialisation, the list of products is retrieved from the backend server. Next, the user selects the relevant product and modifies the assembly sequence generation criterion values. After clicking the button for the assembly sequence generation, the Layout generator (II) requests the respective XML file of assembly features for the selected product.

Then, the XML is fed into the Assembly sequence processing component (initialise call). Hereafter, the Layout generator (II) requests the connectivity graph and the assembly sequence graph for visualisation.

The workflow continues in Fig. 8.20 where the user edits the assembly graph by adding or deleting features iteratively, then generates the tasks (graph and list) and proceeds to modify the task list by adding tasks.

The workflow concludes when the user selects the "save" option. The Layout generator (II) sends a request to the Restful API (A) through the Service connector (II), the Restful API (A) transforms them into an appropriate format by leveraging the Data transformation component and sends them to the Legacy connector to store them in the legacy system.

8.7 Conclusions

Human–robot teamwork is a domain that received increased attention and research in the latest decade. Complementarities are pursued in the manufacturing resource skillset towards achieving production systems of increased flexibility and adaptability in the context of volatile market demands and emerging financial uncertainty.

Several approaches have been suggested up to now for enabling seamless collaboration between the two different resources. An efficient way is still pending, however, to allow seamless interaction between industrial robots and human operators facilitating shared task execution. Apart from safety issues, appropriate planning and scheduling are required, allowing for this kind of collaboration with much-varying levels of understanding and communication capabilities.

In this chapter, a methodology is discussed supporting human–robot collaboration within a shop-floor environment enabling shared task execution. Also, a generic semi-automated framework for task planning based on the CAD representation of a product has been developed facilitating the analysis of multiple products or one product for multiple workstations. The task planning solution has been complemented by a generic task scheduling solution supporting individual or shared task assignments, using a hierarchical modelling approach. A set of prototype software components

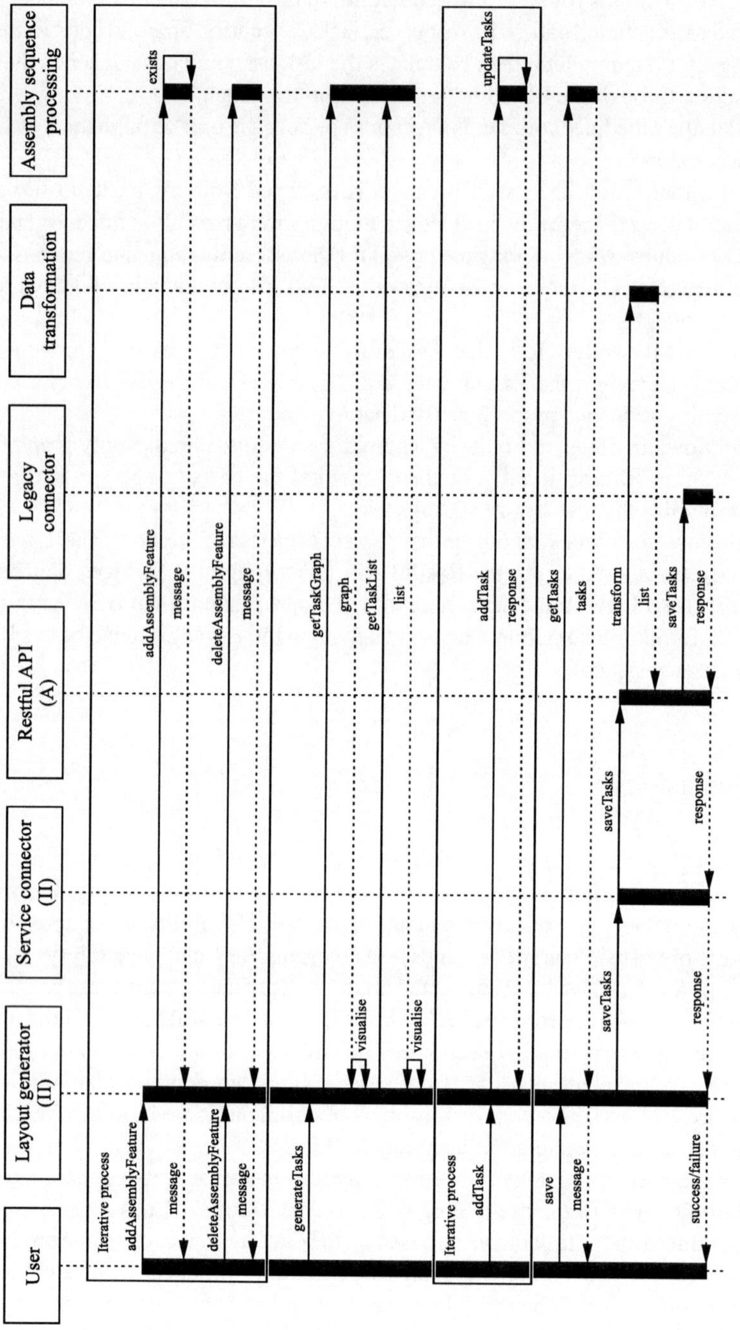

Fig. 8.20 Sequence diagram of task sequence generation and store in legacy system

has been designed and implemented supporting re-planning and re-scheduling based on received events from a monitoring system supervising the shop-floor activities.

As a general conclusion, the proposed approach may offer the capability of generic task planning of multiple products for a specific workstation or for multiple workstations in a semi-automated approach in a short time, and the task scheduling and assignment to specific workstation operators, machines or humans.

The feature-based assembly sequence planner presented here is limited to the most common feature types, i.e., placing, insertion, and screwing. Thus, real-world applications can only be partially supported.

Besides, assembly sequence and task sequence generation require significant user interaction, for editing and validating the generated sequence before submitting it for scheduling and execution.

Finally, readers interested in a human–robot collaborative assembly are referred to [23] for more details.

References

1. P. Scranton, Diversity in Diversity: Flexible Production and American Industrialization, 1880–1930. Bus. Hist. Rev. **65**(1), 27–90 (Apr. 1991). https://doi.org/10.2307/3116904
2. S. J. Winter and S. Lynne Taylor, The Role of IT in the Transformation of Work: A Comparison of Post-industrial, Industrial, and Proto-industrial Organization, Information Systems Research, vol. 7, no. 1. INFORMS Inst. for Operations Res. and the Management Sciences, pp. 5–21, 1996, doi: https://doi.org/10.1287/isre.7.1.5.
3. T. B. Sheridan, Human–Robot Interaction, Human Factors, vol. 58, no. 4. SAGE Publications Inc., pp. 525–532, 2016, doi: https://doi.org/10.1177/0018720816644364.
4. F. Demoly, X.T. Yan, B. Eynard, L. Rivest, S. Gomes, An assembly oriented design framework for product structure engineering and assembly sequence planning. Robot. Comput. Integr. Manuf. **27**(1), 33–46 (Feb. 2011). https://doi.org/10.1016/j.rcim.2010.05.010
5. X. Zhu, S.J. Hu, Y. Koren, N. Huang, A complexity model for sequence planning in mixed-model assembly lines. J. Manuf. Syst. **31**(2), 121–130 (Apr. 2012). https://doi.org/10.1016/j.jmsy.2011.07.006
6. E. Gruhier, F. Demoly, O. Dutartre, S. Abboudi, S. Gomes, A formal ontology-based spatiotemporal mereotopology for integrated product design and assembly sequence planning. Adv. Eng. Informatics **29**(3), 495–512 (Aug. 2015). https://doi.org/10.1016/j.aei.2015.04.004
7. W. Hui, X. Dong, D. Guanghong, Z. Linxuan, Assembly planning based on semantic modeling approach. Comput. Ind. **58**(3), 227–239 (2007). https://doi.org/10.1016/j.compind.2006.05.002
8. H. Lu, H. Zhen, W. Mi, Y. Huang, A physically based approach with human–machine cooperation concept to generate assembly sequences. Comput. Ind. Eng. **89**, 213–225 (Nov. 2015). https://doi.org/10.1016/j.cie.2015.04.032
9. M. Wu, V. Prabhu, X. Li, Knowledge-based approach to assembly sequence planning. J. Algorithms Comput. Technol. **5**(1), 57–70 (Mar. 2011). https://doi.org/10.1260/1748-3018.5.1.57
10. M. Givehchi, A. Haghighi, L. Wang, Generic machining process sequencing through a revised enriched machining feature concept. J. Manuf. Syst. **37**, 564–575 (Oct. 2015). https://doi.org/10.1016/j.jmsy.2015.04.004
11. J. Zhang, Z. Xu, Y. Li, S. Jiang, Framework for the integration of assembly modeling and simulation based on assembly feature pair. Int. J. Adv. Manuf. Technol. **78**(5–8), 765–780 (May 2015). https://doi.org/10.1007/s00170-014-6671-0

12. G. Chryssolouris, *Manufacturing Systems: Theory and Practice*, 2nd edn. (Springer-Verlag, New York, 2006).
13. L. Wang, M. Givehchi, B. Schmidt, G. Adamson, Robotic assembly planning and control with enhanced adaptability. Procedia CIRP **3**(1), 173–178 (2012). https://doi.org/10.1016/j.procir.2012.07.031
14. S. J. Hu, J. Ko, L. Weyand, H.A. ElMaraghy, T.K. Lien, Y. Koren, H. Bley, G. Chryssolouris, N. Nasr, M. Shpitalni, Assembly system design and operations for product variety, CIRP Ann.—Manuf. Technol., vol. 60, no. 2, pp. 715–733, 2011, doi: https://doi.org/10.1016/j.cirp.2011.05.004.
15. L. Wang, S. Keshavarzmanesh, H.Y. Feng, A function block based approach for increasing adaptability of assembly planning and control. Int. J. Prod. Res. **49**(16), 4903–4924 (Aug. 2011). https://doi.org/10.1080/00207543.2010.501827
16. W. van Holland and W. F. Bronsvoort, Assembly features and sequence planning, 1997, pp. 275–284.
17. M.C.Leu, H.A. ElMaraghy, A.Y.C. Nee, S.K. Ong, M. Lanzetta, M. Putz, W. Zhu, A. Bernard, CAD model based virtual assembly simulation, planning and training, *CIRP Ann.—Manuf. Technol.*, vol. 62, no. 2, pp. 799–822, 2013, doi: https://doi.org/10.1016/j.cirp.2013.05.005.
18. Q. Yang, D.L. Wu, H.M. Zhu, J.S. Bao, Z.H. Wei, Assembly operation process planning by mapping a virtual assembly simulation to real operation. Comput. Ind. **64**(7), 869–879 (Sep. 2013). https://doi.org/10.1016/j.compind.2013.06.001
19. S. Angerer, C. Strassmair, M. Staehr, M. Roettenbacher, and N. M. Robertson, Give me a hand—The potential of mobile assistive robots in automotive logistics and assembly applications, in *2012 IEEE Conference on Technologies for Practical Robot Applications, TePRA 2012*, 2012, pp. 111–116, doi: https://doi.org/10.1109/TePRA.2012.6215663.
20. J. Krüger, T. K. Lien, and A. Verl, Cooperation of human and machines in assembly lines, CIRP Ann.—Manuf. Technol., vol. 58, no. 2, pp. 628–646, Jan. 2009, doi: https://doi.org/10.1016/j.cirp.2009.09.009.
21. H. Ding, M. Schipper, B. Matthias, Optimized task distribution for industrial assembly in mixed human–robot environments—Case study on IO module assembly, in *IEEE International Conference on Automation Science and Engineering*, 2014, vol. 2014 January, pp. 19–24, doi: https://doi.org/10.1109/CoASE.2014.6899298.
22. M.A. Goodrich, D. Yi, *Toward Task-Based Mental Models of Human–Robot Teaming: A Bayesian Approach* (Springer, Berlin, Heidelberg, 2013), pp. 267–276
23. L. Wang, R. X. Gao, J. Váncza, J. Krüger, X. V. Wang, S. Makris and G. Chryssolouris, Symbiotic Human–Robot Collaborative Assembly, *CIRP Annals—Manufacturing Technology*, Vol.68, No.2, pp. 701–726, 2019.

Chapter 9
Cockpit: A Portal for Symbiotic Human–Robot Collaborative Assembly

Nikolaos Nikolakis, Konstantinos Sipsas, and Sotiris Makris

Abstract The industrial sector today is experiencing its fourth industrial revolution, powered by technological advances in the fields of robotics, artificial intelligence and the Internet of Things (IoT). Customer demands for mass customisation have increased product variation. Moreover, products are manufactured faster and with higher consistency. In this new context, machines alone do not seem agile enough to keep pace with the growing demand for customised products and shorter cycle times. As digital technology is becoming a pervasive feature of production systems, conventional workstations, equipped with modern embedded systems and ICT, are gradually transforming into smart workstations, in which humans and robots may work and carry out their tasks together. In such a smart environment the human factor is expected to be seamlessly supported by automation systems. To achieve this, several subsystems should be integrated and orchestrated, evaluating key collaboration aspects while ensuring human safety and operational efficiency. However, the increased complexity of such a system of systems requires sophisticated control approaches to maintain high levels of adaptability to unforeseen events that may occur in runtime. For this purpose, a software platform is presented in the following sections, facilitating human–robot collaboration in the context of an assembly workstation. The platform, namely a planning and control cockpit, enables human–robot collaborative assembly, acting as the portal for planning, scheduling, and integration of all required systems to execute, control and adapt the production process to possible events. Key integrated functionalities supporting the collaboration of human operators with industrial manipulators include subsystems in charge of human safety, robot execution control, high-level planning. The platform serves as the main symbiotic orchestrator capable of adaptive assembly planning and execution control of a hybrid assembly station. A prototype implementation is described. The cockpit is

N. Nikolakis · K. Sipsas · S. Makris (✉)
Laboratory for Manufacturing Systems and Automation, University of Patras, Patras, Greece
e-mail: makris@lms.mech.upatras.gr

N. Nikolakis
e-mail: nikolakis@lms.mech.upatras.gr

K. Sipsas
e-mail: sipsas@lms.mech.upatras.gr

L. Wang et al. (eds.), *Advanced Human–Robot Collaboration in Manufacturing*,
https://doi.org/10.1007/978-3-030-69178-3_9

tested and validated in a use case related to the automotive industry and human–robot collaborative assembly in a workstation.

9.1 Introduction

Over the last decade, companies have embraced mass customisation to provide preconfigured customisation value to their customers in an efficient manner and thus increase their competitive advantage and market share [1]. Customer demands for mass customisation have increased the product variation, while the emergence of the digital economy is fundamentally changing the way companies produce goods and access their markets [2]. As a result, a need to increase automation levels in production systems has emerged, responding to factors such as faster business cycles and greater production flexibility tailored to customer demands. Recent technological advances can further improve the flexibility of shop-floor operations, through the adoption of new methods and digital tools [3]. In this context, automation is a vital aspect of the industrial future, enabling accuracy and productivity beyond human ability.

For decades, manufacturing companies have deployed industrial robots to their assembly lines, leaving to their human counterparts less routine and more complex tasks which the robots could not carry out easily. Despite the advance in robotics technology there still are many tasks where humans are much better than robots. Tasks such as gripping, fine assembly, handling textiles or wires remain a challenge for industrial robots.

Further, robots are not good at changing from one task to another. Therefore, there is a need for collaborative work where the strengths of both robots and humans are combined to create a more productive and safe work environment. Moreover, there is a need to design collaborative tasks to ensure the physical efforts of the human worker remain within limits while considering not only production aspects but long-term effects of ergonomy or mental fatigue. This should include not only the design of the specific task but also best fit task allocation. Additionally, the task design must ensure that there will not be a hard collision between the human and the robot.

Towards enabling new production paradigms of increased flexibility and adaptability to internal and external disturbances, production systems are required to make a transition to digital solutions and smart manufacturing [4], adapting dynamically to working conditions. This includes both the collection of runtime events and inputs as well as their interpretation to control signals and their connection to shop-floor actuators, software, hardware or even human operators. As a result, the manufacturing process itself can no longer be treated as a separate black box in the plant control logic but needs to be managed in interconnection to other systems.

This chapter discusses an approach for the planning and control of a hybrid assembly station suitable for human–robot collaboration. A prototype implementation, namely the cockpit, serves as the main platform from which a task planning and scheduling can be performed. A unified data model, the IEC 61499 standard compliant *function blocks* (FBs), is used to wrap high-level tasks. The FBs, as defined in the IEC 61499 automation standard, are used for distributed control of field devices as well as event-triggered supervisory control by monitoring events generated at runtime. The specific FB architecture, merging data and event flows, allows for the creation of a distributed control and execution logic.

9.2 Adaptive Planning and Control for Human–Robot Collaborative Manufacturing

Assembly of small production batch sizes requires frequent changeover. In such cases, flexibility is of major importance, and this is the main reason for preferring manual assembly lines. Flexible robotic systems are considered a prominent solution in such cases. The ROSETTA approach [5] considers three key enabling technologies towards accomplishing this: safety; simplified robot programming and automatic learning/optimisation of the robot program at runtime. Such an environment, where robots would work side by side with human operators and share activities, would need to ensure human safety and provide advanced interaction mechanisms to support a seamless collaboration [6, 7]. Aiming at increasing productivity, it is envisaged that the separation of humans and robots will gradually be lifted in such working environments, with the necessary advanced safety measures made possible by integrated sensors and IT automation. A platform facilitating the creation of applications for human–robot collaboration in an intuitive way is presented in [8]. The software suite, named XRob, provides the main functionalities to create human–robot interactions with applications from inspection to collaborative manufacturing, integrating reasoning, environmental perception and process planning functionalities.

A smart system should adapt autonomously to newly introduced requirements in order to reduce the effort of manual engineering. Even if a production software seems to be adaptable in the sense that a range of products can be manufactured, the variable production processes are statically pre-planned in the software. Towards automating the planning step, engineers tried to model the production process through the functionalities of their controllers resulting in a specific production process. This resulted in the creation of the IEC 61131-3 standard for industrial control. A function block, as described in IEC 61131 and IEC 61499, encapsulates reusability features as well as decoupling. In this context, the IEC 61499 standard can model and be used to develop distributed control systems and applications [9], supporting multiple programming languages. It supports event-triggered communication capabilities [10], supporting the notification of abnormal operation, which is valuable for detecting the loss of critical events. Furthermore, an event-driven model, such as the IEC 61499, can represent any execution control strategy, while the converse is not always true. It provides a consistent model of applications in a system, allowing the reuse of functionalities and data types while offering a standardised way of defining new data types and functionalities.

In addition, an adaptive production plan needs to consider appropriate control of the production equipment at the software and hardware level, and its efficient management. Concerning the aforementioned, it makes sense that manufacturing control systems are moving from centralised forms, where information is stored and being processed in one location, to more distributed approaches. In greater detail, control systems can be classified into three categories: centralised control, decentralised control and hierarchical, based on the overall system structure [11, 12]. A centralised control system uses a centralised controller to monitor and control several subsystems. Sensors capture real-world information and push them up to the controller, which generates command signals to the actuators of the shop floor. To achieve this, the controller should include a decision-making mechanism to analyse the incoming data and issue appropriate responses. A widely used centralised system is the Supervisory Control and Data Acquisition system (SCADA) [13]. SCADA systems have numerous applications in various industries supporting a human–machine interface and one or more Remote Terminal Units (RTUs) connected to the central SCADA server. Considering advanced connectivity such as 5G and concerning multiple devices, the integration of IoT architectures with SCADA can enable interoperability along with optimised resource allocation [14]. However, the complexity and scale of modern industrial systems require advanced control approaches [15]. Considering a computing infrastructure perspective, machine devices have limited capacity in terms of computation and storage. Hence, an infrastructure that can support the storage and processing of large volumes of data is required. In this context, a centralised architecture would require all data to be communicated to a centralised node, which could overload the network infrastructure. Moreover, using a remotely located centralised network/cloud could cause high latency during data transmission, which can be unacceptable. IoT and cloud computing could overcome this bottleneck as argued in [16].

The value of implementing cloud computing in a manufacturing environment was analysed almost a decade ago, in [17]. A fundamental aspect of cloud computing is that everything is a service ranging from the infrastructure to the software. In fact, according to [18], resources, cloud systems and cloud services constitute the three core components of a cloud manufacturing system. A cloud-based system for customised products in the context of smart manufacturing is presented in [19], supporting multiple orders arriving simultaneously with built-in algorithms determining the proper sequence of execution. The importance of controlling manufacturing equipment adaptively in the context of cloud manufacturing is highlighted in [20]. A prototype system is presented for cloud-based control on top of IEC 61499 function blocks. The function blocks, in combination with product manufacturing features, support distributed adaptive control of robotic assembly tasks.

A workspace shared between humans and robots, along with the technologies supporting their collaboration, can be considered a cyber-physical system [21]. The physical assets are "supervised" and controlled by the "cyber" part, which oversees the monitoring, decision-making, planning and adjusting the production process along with human–robot collaboration. In recent years, increasingly more researchers in this field have focused on the Industry 4.0 concept rather than cloud manufacturing. Although these two concepts foster the growth of each other, Industry 4.0 involves a greater emphasis on cyber-physical systems (CPSs) [22]. As manufacturing paradigms are making a shift towards cyber-physical systems [23], new scalable and distributed solutions are required for managing the production process efficiently. Distributed software engineering architectures such as microservices can support scalable and distributed applications. The decentralised concept of (micro) services provides the development teams with more agility and better offering to their clients. An integrated system under the Industry 4.0 concept is presented in [24], which incorporates data gathered from the monitored equipment and alters the maintenance schedule based on the available timeslots. A scheduling algorithm that considers the availability of equipment and technicians, as well as the related processing times, is proposed in [25]. The approach has been validated under a case study but shortcomings in terms of online equipment monitoring as the maintenance activities have already been defined. A service-oriented approach for supporting distributed manufacturing is discussed in [26], where the STEP standard is merged with an XML schema to create a data structure fulfilling the requirements of distributed collaboration for manufacturing.

In the last decade, virtualisation techniques are receiving increased attention. Virtualisation, interoperability and scalability are three attributes of and targets for a cloud manufacturing system. As it regards the increasing importance of service encapsulation and virtualisation techniques for cloud manufacturing, an access model for manufacturing equipment is provided in [27]. The model is supported to provide real-time information for the configuration and scheduling of manufacturing resources in a cloud manufacturing environment. The linking of an additive manufacturing machine to a cloud is presented in [28], enabling users to subsequently upload instruction codes to the cloud for operating the additive manufacturing machine. In [29], a unique digital equipment identifier (DEI) is assigned to each machine to ensure that machines with identical DEIs could be virtually aggregated online. A multi-objective decision-making problem is solved in [30] to map user requests to cloud manufacturing services, thereby maximising the quality of service (QoS) while minimising energy consumption. The use of virtual function blocks (VFBs) is discussed in [31] consisting of wrappers of programs to be deployed via cloud infrastructure and executed at distributed manufacturing nodes. The VFBs support resource monitoring as well as autonomous and synchronous execution. An integration mechanism based on function blocks is presented, integrating various types of manufacturing assets. As a result, cloud technologies and function blocks establish a cross-level integrated manufacturing system.

9.3 Planning and Control Cockpit

9.3.1 Concept

To support seamless human–robot collaboration on a shop floor, as it regards both industrial and collaborative robotic manipulators, critical aspects of the collaboration have to be actively monitored and managed in an effective approach. As discussed in the previous section, in a collaborative workstation, there are three main mechanisms or subsystems that should be considered or established: (1) one related to the robotic behaviour, (2) one for human safety, and (3) one for managing the interaction/communication between the human and the robot. The subsystems grouping various functionalities under the three aforementioned categories are described below:

- *Safety subsystem*: A system, including software and hardware components in charge of overseeing the human–robot shared area and actively safeguarding the human(s). Safety is implemented in terms of adaptive control, adapting the robot behaviour according to the collision risk and possibility to injure an operator who operates in close proximity.
- *Human–Robot Interface*: In order to achieve efficient and seamless collaboration between a human and a machine, a certain level of communication has to be established. Hence, the second main system is the one enabling interaction and communication between a robot and an operator. This can be achieved via gestures, natural language communication or similar.
- *Robot Control subsystem*: All functionalities related to the low-level operational execution and control of a robot.

The components listed above should all be capable of processing information and generating appropriate events as a response in little time, managing individual aspects of the collaboration in execution time.

However, in a production environment, it first must be defined what needs to be produced and in what quantities before manufacturing it. As such, production/assembly planning along with the scheduling of production activities need to complement the components. Such components require more process and decision-making time, in comparison to the aforementioned ones intended for low-level execution. However, all steps are necessary for efficient human–robot collaboration. Different components required for a production environment need to be implemented, integrated and managed efficiently by a central orchestration mechanism. Moreover, all components need to be effectively controlled in a unified way implementing a closed-loop control system, adapting each one to occurring events.

To this end, the purpose of the planning and control cockpit is to facilitate human–robot collaboration in a production environment by serving as the main platform supporting task planning, scheduling and supervision control. Based on the ISA95 model [32], the control cockpit resides on level 3—Manufacturing operations management, integrating and controlling lower-level functionalities. Hence, the planning and control cockpit, as a portal for managing human–robot collaboration in a shared manufacturing environment, should support:

- Software implementation and integration,
- Efficient communication with other modules and devices (if required),
- Monitoring and control functionalities, and
- The implementation of user-friendly interfaces.

The planning and control cockpit is based on the concept presented in [33], extending the use of VFBs discussed in [34] with the functionalities presented in [35].

9.3.2 Objective

The cockpit aims to support human–robot collaboration in task planning and execution, evaluating the feasibility and benefits of the task assignments, in terms of cycle-time, energy consumption, etc. In contradiction to existing software solutions, the robotic resources supported by the cockpit are not limited to collaborative robots. The modelling and architecture of the decision-making algorithms in charge of the task scheduling and re-scheduling are generic enough to capture the operation of collaborative and non-collaborative robots, along with other types of resources operating on the shop floor, such as Automated Guided Vehicles (AGVs) or Unmanned Aerial Vehicles (UAVs).

As such, the platform has been designed according to the following specifications:

- Provide a user-friendly interface for planning and control,
- Integrate task planning, scheduling, and execution control functionalities,
- Support communication interfaces and data exchange to human workers and robotic or similar controllers,
- Monitor and supervise shop-floor operations,
- Control the production in terms of executed operations,
- Enable task-oriented function block design and execution,
- Facilitate energy-efficient robot trajectory planning,
- Support training mode for new workers in production scheduling,
- Support meeting production orders in production scheduling, and
- Connect to a CAD processing module.

Moreover, the cockpit is designed to be adaptable to abnormal situations. Through the supervisory control functionality, awareness of the production condition is created and the production schedule is adapted accordingly for an uninterrupted production process.

Hence, the main objectives of the cockpit are (1) to facilitate the management of collaborative production operations, and (2) to support an increased automation level in the context of a hybrid assembly station where human operators may interact and collaborate with robotic manipulators.

9.3.3 Main Methodologies

Knowledge creation

The planning and control cockpit stores its data in a knowledge structure comprised of five entities that are used to create context over acquired data. The entities are

- *Factory*, providing high-level information about the facility, layout, segmentation of the management structure, etc.,
- *Resource*, referring to the production assets that can execute a manufacturing activity assigned by a scheduling tool and their characteristics
- *Order*, defining the production targets of the factory or a smaller segment/department,
- *Running System*, referring to support systems communicating and interacting with the planning and control cockpit, and
- *Runtime Information*, capturing runtime execution information, such as alarms, operational status, progress, etc.

The presented structure of five entities, as shown in Fig. 9.1, allows the creation of semantic relations which, in turn, enables context modelling at a semantic level. Moreover, reasoning and logic inference are supported, facilitating the creation of awareness of the production environment, based on the analysis of the execution runtime information of adaptive planning and control. Reasoning methods are used to create context out of production domain information, with each incoming batch updating the relations between the entities of the knowledge model.

Task planning and scheduling

Starting from the CAD model of a product, assembly features are extracted as discussed in [36]. The assembly sequence and the task sequence are generated based

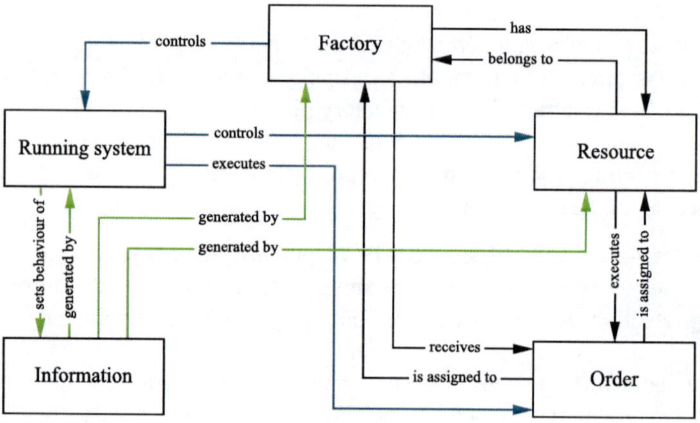

Fig. 9.1 Knowledge structure entities and relations

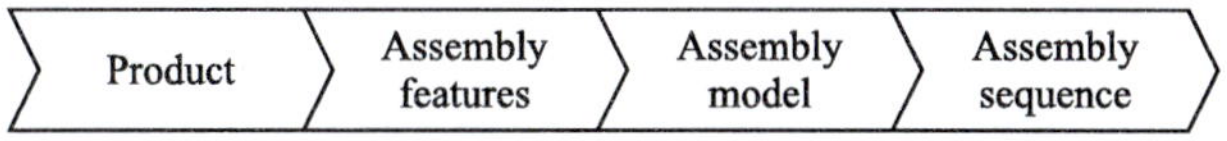

Fig. 9.2 Assembly sequence generation concept

on the evaluation of the feasibility of the connectivity of assembly features in a semi-automated way, following an approach similar to [37]. The analysis of the CAD model includes the following steps, also illustrated in Fig. 9.2:

- Constructing the liaison graph,
- Defining the assembly features, and
- Extracting features parameters.

Alternative assembly sequences are evaluated across a set of criteria, such as tool changes, number of rotations, etc., while the highest ranked one is suggested to the engineer.

Next, tasks are assigned to resources via the decision-making framework. First, a holistic schedule is generated based on the production targets and based on a multi-criteria decision-making framework [38, 39]. Afterwards, runtime data are evaluated and used to modify the scheduling information to generate a new schedule corresponding to the actual working conditions. Finally, tasks are mapped into IEC 61499 function blocks (FBs). The correct sequence of FBs is determined by the scheduler along with the constraints for their distributed execution by the assigned resource.

A high-level representation of the implemented task scheduling concept and its sequential flow is shown in Fig. 9.3, starting from task assignment (1–3), to operational support via the provision of instructions (4) and the need for adaptation in case of occurrence of unexpected events affecting a smooth task execution (5–7).

Supervisory control

IEC 61499 FBs support distributed and event-based execution and control. Events generated and exchanged in runtime can provide insight over the existing working conditions on the shop floor. Moreover, the characterisation of each entity's events and their criticality allows for the creation of suitable execution control logic, which in turn is selected in runtime and determines the appropriate modifications to the existing schedule. Hence, via monitoring the events of the FBs, both monitoring of the production status and supervisory control can be enabled, as discussed in [40]. Also, the connection of the supervisory control component with planning and scheduling facilitates the generation of a schedule complying with the operational conditions on the shop floor. Moreover, abnormal situations as communicated by generated events can be handled with increased flexibility, towards preserving a smooth and uninterrupted manufacturing process.

The supervisory control system oversees the production process and facilitates appropriate correction policies to be applied. As such, the various subsystems that manage a different aspect of production need to be connected to this component. A high-level conceptual diagram is presented in Fig. 9.4, consisting of the main cockpit

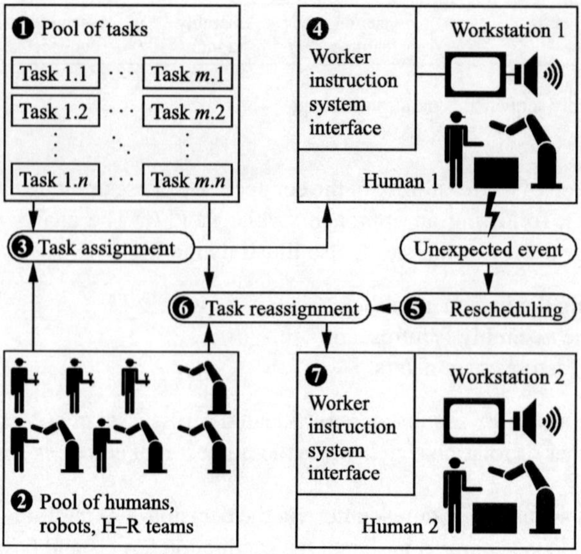

Fig. 9.3 Sequential steps of the implemented adaptive task scheduling approach

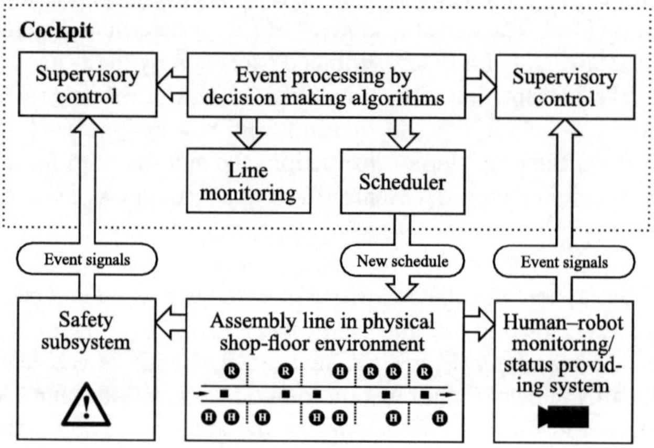

Fig. 9.4 Supervisory control concept

platform and systems in charge of critical functionalities.

9.4 Implementation

9.4.1 Architecture

The implementation of the planning and control cockpit has been carried out to provide an innovative MES system, address the requirements of a human–robot collaborative assembly environment. Its key functionalities include task planning and scheduling and supervisory control, enabling a self-adaptive closed-loop control system.

The cockpit development was carried out in Java and has been based on open-source components, relying on a client–server architecture. The client-side components are developed as plugins to the cockpit platform, whereas the server side is developed as web applications using Java. The server side has been deployed on Apache Tomcat and connected to a MySQL database.

A high-level architecture of the cockpit system is illustrated in Fig. 9.5.

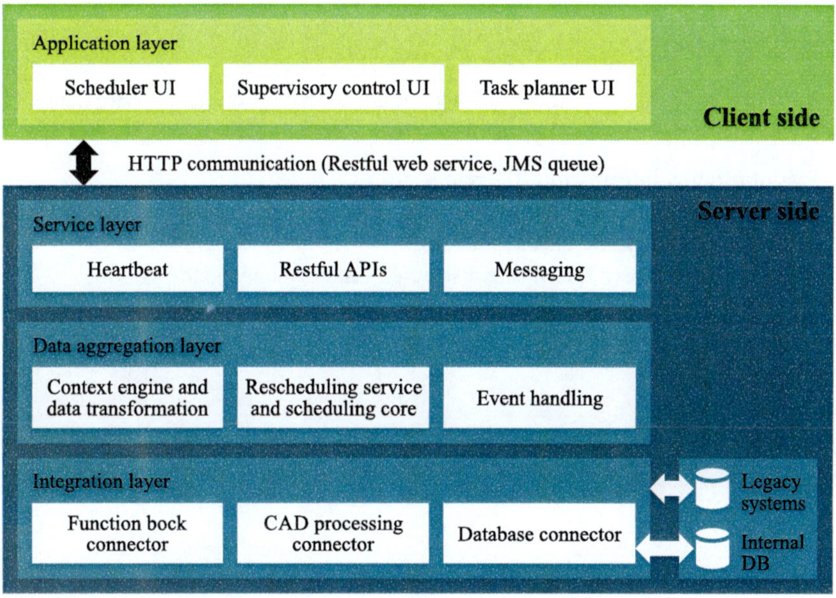

Fig. 9.5 Layered architecture of the planning and control cockpit

9.4.2 Main Components

Task Planner

The task planner component is closely coupled to the scheduling application. The component is used for creating the task sequence for the assembly of a specific product, according to its assembly features. The workflow of the application includes the following steps:

- Assembly features extraction from processing product CAD file,
- Alternative assembly sequence generation, based on a multi-criteria decision-making algorithm where the user may select the weight of the criteria and the preferred sequence,
- Generation of an indicative task sequence which the user may reconfigure and finalise and
- Task sequence dispatching to the central database.

The application targets semi-automated feature-based task planning, through a software component which is lightweight and flexible, in terms of supported user interaction.

The user interface of the task planner (Fig. 9.6) allows the processing of product features to enable the generation of a task sequence. It communicates with the server side via RESTful APIs. It supports the following functionalities:

- List of product information (ID, name, etc.),
- Generate assembly tasks based on the product assembly features,
- Generate assembly and task sequences,
- Interact with the user for editing a selected task sequence, and

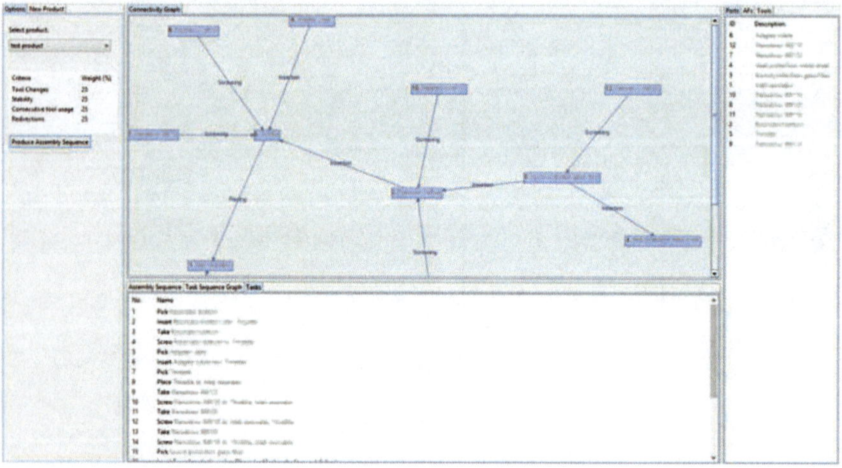

Fig. 9.6 Task planner user interface

- Save task sequence and dispatch for scheduling.

The workflow for generating a task list out of the CAD model of a product begins with a user selecting a station and the CAD whose assembly features will be used for task generation. The evaluation of alternative assembly sequences is performed via a weighted multi-criteria function. The weights of the supported criteria are manually filled in by the user. Afterwards, the conversion to a task sequence takes place. This is performed using a pre-set mapping of assembly features to tasks. Again, the user may edit the task sequence, modify and validate it.

After the validation, the task sequence is stored in the cockpit database and may be used from the scheduling tool. It should be mentioned that the output of the task planned is one of the inputs required by the scheduler. More information is provided in the following subsection.

Scheduler

The scheduler is responsible for making task–resource allocations based on product orders, available resources as well as the status of connected software systems, such as safety subsystems. It is integrated via a public API that offers the following functionalities:

- Schedule or reschedule tasks, based on available resources and production status as inferred by the FB-generated events through the supervisory control component,
- Set the production rate target in parts per hour, and
- Set training mode for newcomers, prioritising the assignments to less experienced operators regardless of the production cycle time.

The scheduler user interface (Fig. 9.7) facilitates access to the features on the server side for scheduling and visualisation of task assignments to resources. It communicates with the server side via RESTful APIs, providing the following functionalities:

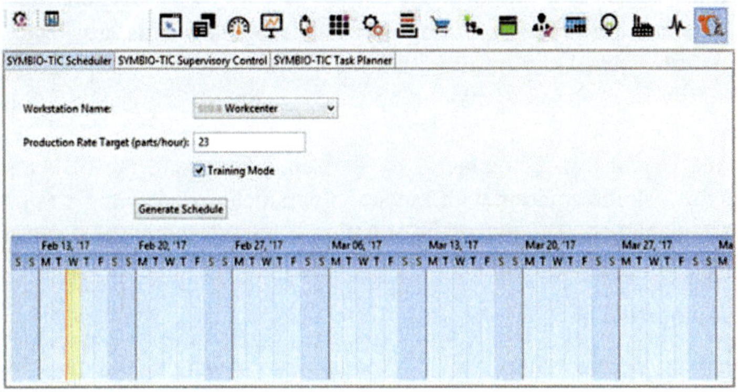

Fig. 9.7 Scheduler user interface

- Gantt chart visualisation,
- List of workstations (e.g., ID, name),
- Selection of product,
- Schedule calculation and assignments, and
- Selection of production mode (training or meeting production rate).

The scheduling application—or scheduler—retrieves the task list and precedence constraints, as generated by the task planner, along with additional information required for the scheduling procedure, such as station equipment and resources, indicative cycle times, resource capabilities, etc. It uses a multi-criteria decision-making framework, implemented in Java, by which alternative task assignments are evaluated and the most appropriate to the user/production-specific requirements are selected and dispatched for execution by the resources in charge. For example, if the selected criterion is the reduction of cycle time, the alternative with the minimum cycle time is selected and related task assignments are dispatched for execution. As a result, the scheduler supports integration with lower level functionalities as mentioned in Sect. 9.3.1.

The task assignments generated from the scheduling component refer to high-level task execution, which is wrapped into IEC 61499 function blocks. It should be mentioned that each high-level task may clear for a human operator to complete, but for a robot requires further conversion to a set of simpler activities which then can be translated into robot commands.

Regarding a user's interaction with the tool, in brief, a user may select a work-station from a drop-down list. Then, optionally, the user may fill in the production rate target and enable or disable the training mode. The production rate signifies the number of products required to be produced for one working period of the day. From the alternative schedules generated, the decision-making framework responsible for the task assignments selects the one best fitting the production requirements, as set by the user, and based on its ranking.

In case the training mode is selected, the target of the schedule is not to meet the predefined number of products but to train the workers. As a consequence, the task assignments are made according to which workers' skills are lower, thus in need of training. At execution time, the parameters used for scheduling, such as the availability of a resource, may be updated, leading to the generation of a new schedule.

After the virtual task assignments by the scheduling tool, the FBs are created wrapping the task information with execution parameters depending on the specific device to be deployed. The related information is also stored in the database of the scheduling tool for manual editing.

Supervisory control

The supervisory control component is responsible for managing the shop-floor operations towards maintaining a smooth, uninterrupted and flexible production. Towards that end, events are communicated to it from all integrated to the cockpit components, and in the context of IEC 61499 standard events. Such events are generated

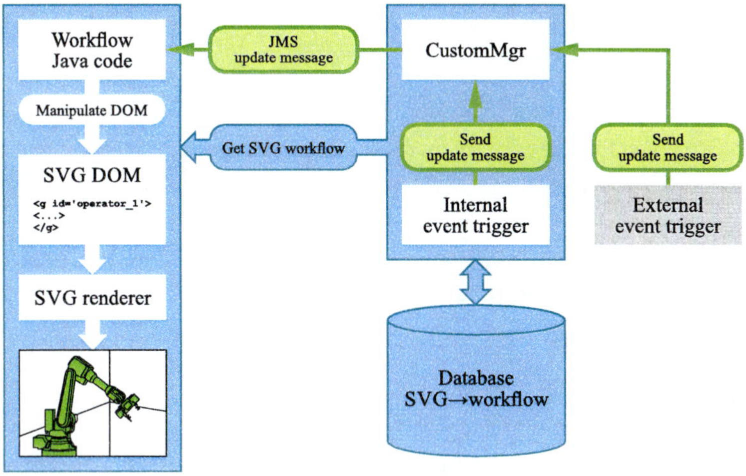

Fig. 9.8 SVG rendering mechanism for supervisory control module visualisation

in runtime and when FBs are executed by the corresponding devices, such as robots. The events may concern resources' status, progress, alarms, etc. Afterwards, the component evaluates the incoming events and "decides" on the best approach to follow to ensure the continuation of the production. Apart from the implemented control logic, the component provides the supervision of the shop-floor operations to its user through the user interface of the client side.

In a similar approach to the scheduler, these components offer public APIs that allow access to their functionalities. More specifically, the supervisory control component adopts a twofold approach for communicating with the server. First, it may access the RESTful APIs that provide access to functionalities such as

- List of workstations (ID, name),
- Workstation layout in SVG format (Fig. 9.8), and
- The initial status of the workstation (resources present and their status, deployed containers) and scheduling information (tasks scheduled, completed, failed and ongoing).

The supervisory control component is responsible for maintaining the state of each workstation, including the presence and status of resources, task execution and issuing alerts about task and resource failure. On the server side, it is not implemented on a single component but is rather realised by the Context engine, Heartbeat monitor service, Messaging service and Event handling service functionalities (Fig. 9.5).

Furthermore, considering that not all components communicate via RESTful APIs, it can subscribe to messaging queues to intercept messages coming from shop-floor deployed systems for monitoring its working conditions, such as

- Status of resources,
- Presence of resources,

- Task execution (completed, failed, and ongoing), and
- Components going offline (e.g., safety system).

More specifically:

The Event handling service offers methods for managing the following events:

- Resource availability,
- Resource presence/location,
- Task status (Start/Finished/Failed), and
- Components going offline (e.g., safety system).

The Heartbeat monitor service offers methods for managing the following events:

- Registering for heartbeat monitoring, and
- Monitoring heartbeat to determine what systems are alive.

The context engine offers the following functionalities:

- Add/remove/access resources (names, status),
- Add/remove/access tasks (names, status),
- Add/remove components for heartbeat monitoring,
- Add/remove components heartbeats,
- Add/remove task schedule, and
- Updating context based on events.

Hence, the supervisory control component supports:

- A centralised event-based monitoring and control tool for shop-floor operations, which in turn enables adaptive schedule generation to the existing working conditions and the operational status of integrated software and hardware components.
- A shop-floor monitoring view containing information about the current production status, operations, resources, etc., all required for an engineer to assess the current status of the production and take additional actions if required.
- Rule-based decision-making algorithms for evaluating the incoming events and ensuring the adaptation of the schedule to the newly introduced requirements of the shop floor, towards meeting the initial user-defined production targets.

9.4.3 Enabling Technologies

The cockpit development was carried out in Java and has been based on open-source components, relying on a client–server architecture. The implemented applications are presented in the following:

- The client-side applications (Supervisory Control, Scheduler and Task Planner) were developed as plugins for the cockpit platform. The technologies involved in

its development were Java and the *Eclipse Rich Client Platform* that enables the development of Eclipse-based plugins.
- The server-side development was also based in Java and took advantage of the features offered by the Spring Framework. A series of open-source components were also used to facilitate the implementation of its features:
- Object Relational Mapping framework *Hibernate* was used to facilitate database persistence
- The *ActiveMQ* implementation of JMS and *ZeroMQ* implementation for Java and *Protocol Buffers for Java* are also supported for message exchange with potential cockpit plugins
- *Apache Jena* enabled the implementation of the context Engine and the Task Planner core.

9.4.4 Hardware and Software Dependencies

The cockpit main platform, as well as the software components that are integrated into it, do not have any particular processing or graphic card requirements. However, depending on the complexity of the production operations and the number of tasks, the processing time required for planning, visualisation and scheduling may be affected by the computer specifications. This is also affected by the integrated plugins.

Neither the client nor the server sides of the cockpit have particular hardware requirements. Both components rely on JRE.

The server side requires a JRE 6 or higher along with deployment of Apache Tomcat v7. It also depends on a connection to a MySQL database (locally or remotely installed). The client side comes packaged with JRE 6 so there is no need for the host computer to have a JRE installed. The client side requires access via the Internet or a local network to the server-side deployment.

9.5 Deployment

The cockpit developments support both local and remote deployment. The cockpit server side can be deployed in the on-premise server room accessing a local network for communication with external components responsible for key human–robot collaborative functionalities such as human–machine interaction, localisation, high-level to low-level translation mechanism, task authoring tools, etc., or in a remote server using the Internet. The client side can also be deployed locally, thus communicating with the server side over the local network or remotely connected to the server side over the Internet. Figure 9.9 illustrates cockpit deployment.

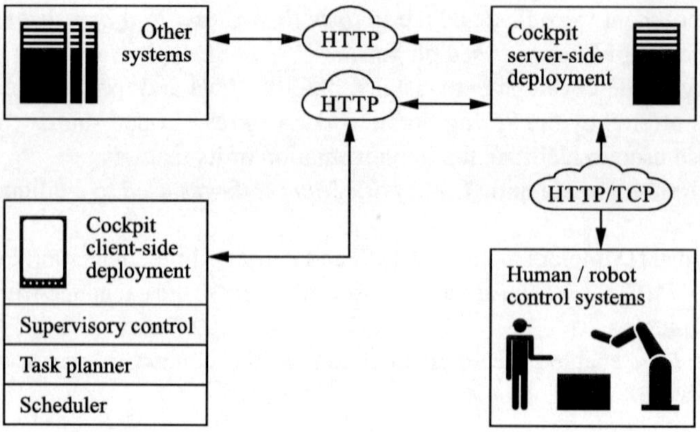

Fig. 9.9 Cockpit deployment

9.6 Conclusions

In an era of extensive digitalisation and increased automation, human competencies and skills may only be enhanced with the adoption of technologies. Still, human dexterity, learning, adaptability and flexibility surpass the current level of automation solutions. Nevertheless, machines can alleviate the workload of human operators and reduce the operational risk and chance of injuries, directly or in the future. As such, human–robot collaboration is widely considered as the intermediate required step towards complete automation, leading to advanced production systems and new opportunities in the social and commercial domains.

The complexity of enabling collaboration between different entities, with different communication and understanding capabilities, is great. Nevertheless, the benefits stemming from bridging the gap and enabling seamless human–robot collaboration in versatile environments are indeed many. As such, a high-level system orchestrating the collaboration towards a common goal is required.

This chapter presents a prototype system for adaptive planning and control of a symbiotic manufacturing system, the planning and control cockpit. The functionalities of the cockpit can be classified based on the time of execution, in planning time and execution time functionalities. The planning time functionalities enable the user of the cockpit to plan the assembly process of a product in a hybrid assembly station up to the start of production. Runtime functionalities enable awareness in the cockpit so that constant evaluation of working conditions and operational systems can be achieved. In turn, this enables supervisory control which facilitates adaptive planning and control of the manufacturing process through a closed-loop control system. As a result, the introduced planning and control cockpit is the next step of an MES targeting collaborative manufacturing environments.

Also, and towards managing all aspects of the collaboration, the cockpit integrates several subsystems, each controlling a different aspect of the collaboration or individual entities, such as safety subsystems, robot controllers, communication, deployment of services, etc.

The cockpit operation implements an event-based control logic depending on asynchronous events generated through the execution of IEC 61499 FBs in distributed software systems or hardware devices. As a result, the events provide insight over the operational status of each component where FBs are executed, enabling different policies to be applied towards adapting the manufacturing process, as well as the behaviour of each entity, to the existing conditions. This allows for increased efficiency, levels of automation as well as seamless collaboration between a human and a robot entity.

The discussed implementation of the planning and control cockpit follows a server–client architecture, supporting to its client side three main functionalities with their user interfaces, the task planner, the scheduler and the supervisory controller.

Nevertheless, while the software structure of the planning and control cockpit relies upon an existing MES system, the planning and control cockpit implementation remain a prototype system, focusing mostly on basic aspects of human–robot collaboration in a production environment, and most importantly, not supporting full integration to potentially existing legacy systems already governing the production. Moreover, the presented prototype considers and has been tested in a single assembly station, with the potential extension to additional stations in the future. The investigations of the latter aspects are required for a future updated cockpit. Besides, the cockpit concerns mostly high-level logic orchestrating production planning and execution subsystems. Virtualisation techniques were investigated for deploying FBs to devices. Such approaches revealed a complementary set of information to the FBs events as well as opportunities to optimise the management and the deployment of high-level tasks to distributed devices. Finally, security aspects are part of the discussed release. Nevertheless, data security is a critical aspect of a safety-critical system.

References

1. D. Mourtzis, Challenges and future perspectives for the life cycle of manufacturing networks in the mass customisation era. Logist. Res. **9**(1), 1–20 (2016). https://doi.org/10.1007/s12159-015-0129-0
2. D. Mourtzis, Design of customised products and manufacturing networks: towards frugal innovation. Int. J. Comput. Integr. Manuf. **31**(12), 1161–1173 (2018). https://doi.org/10.1080/0951192X.2018.1509131
3. K. Jackson, K. Efthymiou, J. Borton, Digital Manufacturing and Flexible Assembly Technologies for Reconfigurable Aerospace Production Systems. Procedia CIRP **52**, 274–279 (2016). https://doi.org/10.1016/j.procir.2016.07.054
4. S. Wang, J. Wan, D. Zhang, D. Li, C. Zhang, Towards smart factory for industry 4.0: A self-organized multi-agent system with big data based feedback and coordination. Comput. Networks **101**, 158–168 (2016). https://doi.org/10.1016/j.comnet.2015.12.017
5. R. Patel, M. Hedelind, and P. Lozan-Villegas, "Enabling robots in small-part assembly lines: The ROSETTA approach—an industrial perspective," Proc. of German Conf.

on Robotics (ROBOTIK), 2012. https://ieeexplore.ieee.org/xpl/articleDetails.jsp?tp=&arnumber=6309522.
6. I. Maurtua, A. Ibarguren, J. Kildal, L. Susperregi, B. Sierra, Human–robot collaboration in industrial applications: Safety, interaction and trust. Int. J. Adv. Robot. Syst. **14**(4), 1–10 (2017). https://doi.org/10.1177/1729881417716010
7. L. Wang, R. X. Gao, J. Váncza, J. Krüger, X. V. Wang, S. Makris and G. Chryssolouris, "Symbiotic Human–Robot Collaborative Assembly," CIRP Annals—Manufacturing Technology, Vol.68, No.2, pp.701–726, 2019.
8. A. Pichler, S. Chandra Akkaladevi, M. Ikeda, M. Hofmann, M. Plasch, C. Wögerer, G. Fritz, "Towards Shared Autonomy for Robotic Tasks in Manufacturing," Procedia Manuf., vol. 11, pp. 72–82, 2017, doi: https://doi.org/10.1016/j.promfg.2017.07.139.
9. V. Vyatkin, IEC 61499 as enabler of distributed and intelligent automation: State-of-the-art review. IEEE Trans. Ind. Informatics **7**(4), 768–781 (2011). https://doi.org/10.1109/TII.2011.2166785
10. S. Wang, C. Zhang, D. Jia, Improvement of type declaration of the IEC 61499 basic function block for developing applications of cyber-physical system. Microprocess. Microsyst. **39**(8), 1255–1261 (2015). https://doi.org/10.1016/j.micpro.2015.07.004
11. T.C. Yang, Networked control system: A brief survey. IEE Proc. Control Theory Appl. **153**(4), 403–412 (2006). https://doi.org/10.1049/ip-cta:20050178
12. A. Sajid, H. Abbas, and K. Saleem, "Cloud-Assisted IoT-Based SCADA Systems Security: A Review of the State of the Art and Future Challenges," IEEE Access, vol. 4. Institute of Electrical and Electronics Engineers Inc., pp. 1375–1384, 2016. https://doi.org/10.1109/ACCESS.2016.2549047.
13. R. Hunzinger, "SCADA fundamentals and applications in the IoT," in Internet of Things and Data Analytics Handbook, Hoboken, NJ, USA: wiley, 2017, pp. 283–293.
14. R. Scattolini, Architectures for distributed and hierarchical Model Predictive Control—A review. J. Process Control **19**(5), 723–731 (2009). https://doi.org/10.1016/j.jprocont.2009.02.003
15. M. Dotoli, A. Fay, M. Miśkowicz, C. Seatzu, Advanced control in factory automation: a survey. Int. J. Prod. Res. **55**(5), 1243–1259 (2017). https://doi.org/10.1080/00207543.2016.1173259
16. C. Wang, Z. Bi, L. Da Xu, IoT and cloud computing in automation of assembly modeling systems. IEEE Trans. Ind. Informatics **10**(2), 1426–1434 (2014). https://doi.org/10.1109/TII.2014.2300346
17. B. Hayes, Cloud Computing. Commun. ACM **51**(7), 9–11 (2008). https://doi.org/10.1145/1364782.1364786
18. L. Zhang, Y. Luo, F. Tao, B.H. Li, L. Ren, X. Zhang, H. Guo, Y. Cheng, A. Hu, Y. Liu, Cloud manufacturing: a new manufacturing paradigm. Enterp. Inf. Syst. **8**(2), 167–187 (2014). https://doi.org/10.1080/17517575.2012.683812
19. X.V. Wang, L. Wang, A. Mohammed, M. Givehchi, Ubiquitous manufacturing system based on Cloud: A robotics application. Robot. Comput. Integr. Manuf. **45**, 116–125 (2017). https://doi.org/10.1016/j.rcim.2016.01.007
20. A.W. Colombo, R. Schoop, R. Neubert, An agent-based intelligent control platform for industrial holonic manufacturing systems. IEEE Trans. Ind. Electron. **53**(1), 322–337 (2006). https://doi.org/10.1109/TIE.2005.862210
21. X. V. Wang, Zs. Kemény, J. Váncza, and L. Wang, "Human–robot collaborative assembly in cyber-physical production: Classification framework and implementation," CIRP Ann.—Manuf. Technol., vol. 66, no. 1, pp. 5–8, 2017, doi: https://doi.org/10.1016/j.cirp.2017.04.101.
22. A.-W. Scheer, "Industrie 4.0 – Wie sehen Produktionsprozesse im Jahr 2020 aus?," IMC Verlag, 2013.
23. P. Zheng, Z. Sang, R.Y. Zhong, Y. Liu, C. Liu, K. Mubarok, S. Yu, X. Xu., "Smart manufacturing systems for Industry 4.0: Conceptual framework, scenarios, and future perspectives," Front. Mech. Eng., vol. 13, no. 2, pp. 137–150, 2018, doi: https://doi.org/10.1007/s11465-018-0499-5.
24. D. Mourtzis, E. Vlachou, A cloud-based cyber-physical system for adaptive shop-floor scheduling and condition-based maintenance. J. Manuf. Syst. **47**, 179–198 (2018). https://doi.org/10.1016/j.jmsy.2018.05.008

25. P. Senra, I. Lopes, J.A. Oliveira, Supporting Maintenance Scheduling: A Case Study. Procedia Manuf. **11**, 2123–2130 (2017). https://doi.org/10.1016/j.promfg.2017.07.342
26. O.F. Valilai, M. Houshmand, A collaborative and integrated platform to support distributed manufacturing system using a service-oriented approach based on cloud computing paradigm. Robot. Comput. Integr. Manuf. **29**(1), 110–127 (2013). https://doi.org/10.1016/j.rcim.2012.07.009
27. Y. Zhang, G. Zhang, Y. Liu, D. Hu, Research on services encapsulation and virtualization access model of machine for cloud manufacturing. J. Intell. Manuf. **28**(5), 1109–1123 (2017). https://doi.org/10.1007/s10845-015-1064-2
28. A. Brant, M.M. Sundaram, A novel system for cloud-based micro additive manufacturing of metal structures. J. Manuf. Process. **20**, 478–484 (2015). https://doi.org/10.1016/j.jmapro.2015.06.020
29. T. Chen, Y.C. Lin, A digital equipment identifier system. J. Intell. Manuf. **28**(5), 1159–1169 (2017). https://doi.org/10.1007/s10845-015-1071-3
30. F. Xiang, Y. Hu, Y. Yu, H. Wu, QoS and energy consumption aware service composition and optimal-selection based on Pareto group leader algorithm in cloud manufacturing system. Cent. Eur. J. Oper. Res. **22**(4), 663–685 (2014). https://doi.org/10.1007/s10100-013-0293-8
31. L. Wang, X.V. Wang, *"Cloud Robotics Towards a CPS Assembly System", in Cloud-Based Cyber-Physical Systems in Manufacturing* (Springer International Publishing, Cham, 2018), pp. 243–259
32. ISA95, Enterprise-Control System Integration—ISA. Retrieved March 30, 2020, from https://www.isa.org/isa95/
33. L. Wang, "Cyber Manufacturing: Research and Applications," TMCE, no. May 2014, pp. 39–49, 2014.
34. X. Wang, X. Xu, Virtualise manufacturing capabilities in the cloud: Requirements, architecture and implementation. Int. J. Manuf. Res. **9**(4), 348–368 (2014). https://doi.org/10.1504/IJMR.2014.066665
35. N. Nikolakis, R. Senington, K. Sipsas, A. Syberfeldt, and S. Makris, "On a containerized approach for the dynamic planning and control of a cyber-physical production system," Robot. Comput. Integr. Manuf., **64**(Aug. 2020), 101919, 2020, doi: https://doi.org/10.1016/j.rcim.2019.101919.
36. Cs. Kardos, A. Kovács, and J. Váncza, "Decomposition approach to optimal feature-based assembly planning," *CIRP Ann.—Manuf. Technol.*, vol. 66, no. 1, pp. 417–420, Jan. 2017, doi: https://doi.org/10.1016/j.cirp.2017.04.002.
37. C. Bikas, A. Argyrou, G. Pintzos, C. Giannoulis, K. Sipsas, N. Papakostas, G. Chryssolouris, An Automated Assembly Process Planning System. Procedia CIRP **44**, 222–227 (2016). https://doi.org/10.1016/j.procir.2016.02.085
38. G. Chryssolouris, K. Dicke, M. Lee, An approach to short interval scheduling for discrete parts manufacturing. Int. J. Comput. Integr. Manuf. **4**(3), 157–168 (1991). https://doi.org/10.1080/09511929108944491
39. K. Efthymiou, A. Pagoropoulos, and D. Mourtzis, "Intelligent scheduling for manufacturing systems: A case study," in *Lecture Notes in Mechanical Engineering*, vol. 7, 2013, pp. 1153–1164.
40. N. Nikolakis, K. Sipsas, S. Makris, A cyber-physical context-aware system for coordinating human–robot collaboration. Procedia CIRP **72**, 27–32 (2018). https://doi.org/10.1016/j.procir.2018.03.033

Chapter 10
Human–Robot Collaborative Workcell Calibration and Control

Gergely Horváth, Gábor Erdős, and Zsolt Kemény

Abstract As in many areas of today's evolving industrial production, collaborative robot cells build much of their flexibility and resilience on an integrated model of the entire workcell that represents the actual physical characteristics of the scene with sufficient fidelity to support efficient assessment, planning, and control up to the workcell level. Nominal models of individual workcell components are mostly available, albeit, they often ignore details of secondary importance. Their exact placement within a unified workcell model, as well as their adjustment to the real production scene, remain to be solved as a part of workcell integration. This matching of nominal component models requires calibration methods which rely on measured data and guarantee the required fidelity of a calibrated virtual representation. The chapter introduces theoretical backgrounds and current technological solutions of workcell calibration, and demonstrates the principles by a case study.

10.1 Introduction

Collaborative robot cells are gaining increasing importance as a new type of work environment where operator and robot work together in a fenceless environment [1]. The flexibility of these cells, however, relies on accurate digitalised knowledge of the environment and the operator, which is the base of reliable and safe supervision systems and advanced control of the cell. Maintenance, repair and overhaul (MRO) activities also require an accurate model of the complex engineering object [2]. With much of the work environment being designed by digital tools, CAD models are

G. Horváth (✉) · G. Erdős · Zs. Kemény
Research Laboratory on Engineering and Management Intelligence, SZTAKI Institute for Computer Science and Control, Eötvös Loránd Research Network, Budapest, Hungary
e-mail: gergely.horvath@sztaki.hu

G. Erdős
e-mail: gabor.erdos@sztaki.hu

Zs. Kemény
e-mail: zsolt.kemeny@sztaki.hu

L. Wang et al. (eds.), *Advanced Human–Robot Collaboration in Manufacturing*,
https://doi.org/10.1007/978-3-030-69178-3_10

usually available, on which a virtual representation can be built. However, one has to ensure that a model relying on nominal designs aligns well with the real environment, i.e., the reference model has to be *calibrated* to reality using measured data.

For every workcell, one can differentiate between *as-designed* and *as-built* models. The term *as-designed* refers to the nominal model of the elements of the workcell, while *as-built* is how the nominal models are realised. The as-built and as-designed models may differ due to various reasons, such as difficult-to-model features (e.g., cables), or minor auxiliaries which may be omitted by a nominal model. Similarly, changes made during the construction process may result in deviations from the intended design. In order to determine the difference between the as-designed and as-built models, they have to be compared in a common reference coordinate frame.

In robotic cells, the robots themselves are usually the mechanical components with the most complex motion capabilities which, in turn, require an accurate alignment of reference frames. Therefore, model calibration in such cells is usually carried out with the robot-related frames being in focus. Robot planning and execution control extends over two main problem domains: one focusing on the spatial mobility—typically, large-scale motion—of the robot, and the other dealing with manipulating the objects in the workspace. The literature of both topics is extensive, making it inevitable to narrow down the scope of examination. Therefore, this chapter deals with the second area only, focusing on robot manipulator calibration.

10.1.1 Workcell Control

Machine and robot control are widely represented in literature. The fundamental knowledge of control is recapitulated in several sources, including the Lagrange formulation, joint space and task space control, PID control, Lyapunov control and many more. Canudas de Wit et al. [3] provide such a comprehensive overview.

Workcell control is a higher-level control problem, and requires a vastly different approach. In association with workcell control, a greater number of workcell objects needs to be operated, and their relations handled. In such cases, the main control challenge does not lie in low-level task control but in system-wide task initiation and handling of success or failure, involving much model-based calculation and discrete logical structures. Chapter 12 [4] demonstrates high-level workcell control—the approach presented there has been utilised in the SYMBIO-TIC project.

A complex set of operations in a workcell can be represented by a graph, with its nodes being the tasks that need to be executed by the individual components of the workcell, and edges representing the interdependencies between tasks. If operations can only be performed following each other in a specific order, the graph will be a single, linear path. If the operation of a machine depends on the outcome of operations of one or more other machines, the situation can be much more complex. In such cases, the graph can have branches or even loops. It is the responsibility of workcell control to follow a path in the graph. When a node is reached and the corresponding task is being carried out, the given component of the workcell needs to be aware of the state of other relevant items in the workcell. In order to achieve this "awareness", either the controller needs to pass the calibrated workcell model to the controlled workcell object, or the object itself needs to be in possession of calibrated workcell data. This is the connection between control and calibration.

10.1.2 Offline and Online Programming

Robots can be programmed by online or offline programming. Both methods have their advantages and shortcomings. *Online programming* is the process when a human operator programs the trajectory of the robot interactively, guiding it manually point-by-point. This is carried out with the robot running, hence the term "online". While online programming usually requires less of a technical—and mathematical—apparatus, it is a time consuming process, demanding manual labour, rarely achieving higher precision due to the very nature of on-site manual adjustment. In the case of *offline programming*, the robot trajectory can be calculated independently of the robot and its actual runtime, provided, adequate knowledge about the robot, the fixture, the workpiece and the workplace is given. Ideally, this calculation can be done automatically, without human intervention upon setting the proper initial conditions.

Online (or manual) programming is usually carried out using a teach pendant [5, 6]. During the process, the operator moves the robot tool centre point into an adequate pose, and then saves the joint settings. This needs to be done for all specific points of a trajectory—in many cases, this can involve a substantial amount of poses to be captured manually. Aside from its labour-intensive nature, there are various problems with this approach. Complex paths cannot be programmed with satisfying precision, and even on flat lines, where the tangent points can be set, the exact normal vectors are hard to capture. This may seriously hinder the accuracy of the manufacturing process. It is possible to extend online programming, by position offsetting [5, 6]. In such cases, key target points are still taught manually by a human operator, however, certain point coordinates are calculated. This approach has the advantage of reducing manual labour demands, however, it still relies on the precision of the robot.

Proper offline programming does not involve runtime use of the robot during the programming phase itself, as it calculates the trajectory based on accurate models. Therefore, much higher precision can be expected than what is achievable by visual or other physical feedback to a human operator. Moreover, offline programming allows fine adjustment of the trajectory in accordance with various metrics. Solutions for offline programming are known to be used in several manufacturing applications, e.g., in thermal spraying [5] or remote laser welding [7].

The idea behind such approaches is to calculate a trajectory based on the accurate model of every element of the workspace, and thereby eliminate the need for actual robot runtime during programming. Modelling has two distinct phases: *physical* and *kinematic*. Physical modelling usually refers to collating the CAD models of the cell. The model of the workpiece is used to generate the robot trajectory, while all the other models are used during collision avoidance calculations. The application of kinematic modelling is twofold. Inverse kinematics is needed when a robot joint trajectory is calculated for a tool trajectory defined in Cartesian workspace. Forward kinematics is used when, e.g., robot motion is simulated for a pre-calculated or pre-recorded joint trajectory. Such simulations can be used to check possible collisions or adherence to velocity and acceleration limits before carrying out the motion in an actual physical environment, or to assess certain properties of the motion plan, such as execution time, estimated performance characteristics, etc. Results of the latter can be fed back to trajectory generation to refine or optimise the tool path according to selected metrics.

10.1.3 Kinematic Modelling

As recapitulated by, e.g., [3], kinematic modelling of a robot is concerned with the description of movement or pose of the robot manipulator with respect to a reference frame. In association with robot kinematic modelling literature usually distinguishes forward and inverse kinematics.

10.1.3.1 Forward kinematics

The kinematic model of a robot consists of n joints, and $n + 1$ links, which is referred to as a *kinematic chain*. The kinematic chain (sometimes represented as a kinematic graph) starts at its *proximal end* with a base link (generally, the ground or the workbench), while its other end (*distal end*) is attached to an *end effector*. With their distal ends not being firmly attached to an artefact in their environment, the robots dealt with in this chapter form an open kinematic chain, and are a serial mechanism without loops in their kinematic graph. Every joint is a representation of the relative motion of the links it connects. The *degrees of freedom* (DOF) are the number of variables that are essential to explicitly describe said motion. These variables are called *driving variables*. Forward kinematics is the following mapping:

$$f(\mathbf{q}) = \mathbf{P}, \tag{10.1}$$

where $\mathbf{q}$ is the vector of all the driving variables q_i, and $\mathbf{P}$ is the representation of the robot pose (position and orientation of the end-effector frame) in Cartesian workspace, with respect to a reference frame. Forward kinematics is used in robot control where the actuators in the joints are the directly controllable parts of a robot. Controlling the robot in joint space is fast and accurate, while the effects of resulting robot movement are usually described and assessed in Cartesian space, requiring a forward kinematic solution for the given joint values.

10.1.3.2 Inverse Kinematics

In the case of inverse kinematics, the following mapping is considered:

$$g(\mathbf{P}) = \mathbf{q}, \tag{10.2}$$

where $\mathbf{q}$ is the vector of all the driving variables, and $\mathbf{P}$ is the representation of the robot pose. It also holds that $f^{-1} = g$, where f^{-1} is an inverse of the forward kinematic mapping f, while g is the inverse kinematic mapping. In this case, a vector $\mathbf{q}$ of joint values is calculated for a pose prescribed in $\mathbf{P}$—this is required, e.g., if a trajectory is defined in Cartesian space, while robot motion is still controlled in joint space. For serial mechanisms, finding an inverse solution is more difficult than calculating forward kinematics, due to a variety of reasons:

- While it is straightforward to generate a closed-form forward kinematic solution of a serial chain with known parameters, it is not guaranteed that a closed-form inverse exists at all. Even if it does, finding the symbolic expressions for a closed-form inverses cannot be fully automated in the general case. Closest to the latter are "mixed-initiative" approaches where alternatives for symbolic expressions are suggested automatically for kinematic pairs, with final selection requiring a human touch. Fortunately, a given closed-form solution can be applied to the same robot in all configurations, therefore, a closed-form inverse planner does not have to consult a human for each point of a trajectory.
- Revolute joints make kinematic mappings of serial mechanisms highly non-linear. This is less of a burden for forward kinematics calculations, but may bear problems in the inverse case. This is especially true for mechanisms where no closed-form inverse exists, and thus, iterative methods are needed. The latter are, namely, known to encounter problems for highly non-linear mappings which possibly have multiple, difficult-to-handle, local extrema.
- While the forward kinematic mapping is always a function (i.e., for a given joint vector $\mathbf{q}$, an unambiguous pose $\mathbf{P}$ can always be determined), the same does not hold for the inverse mapping. In some cases, more than one joint vector $\mathbf{q}$ would satisfy a prescribed $\mathbf{P}$, the inverse solution yielding either a finite number of different joint vectors (typically arranged in branches of *solution pairs*), or an infinite number of possible solutions (e.g., for *kinematically redundant* robots having more DOF than the dimensionality of the workspace for $\mathbf{P}$). In other cases, no inverse solution may exist (e.g., if $\mathbf{P}$ prescribes a pose which the robot is unable to reach).
- Planning a trajectory with inverse kinematics consists in finding a joint space trajectory which realises the desired workspace motion. The calculated joint space trajectory, however, is usually also required to keep to further kinematic requirements, such as a continuous acceleration profile. For this reason, the subsequent joint vectors of a joint space trajectory are preferably selected from the same inverse solution branch. In some cases, the solution branches, however, are intertwined with such complexity that it proves difficult to keep the joint space trajectory on the same solution branch.

10.2 Calibration

Offline planning for the entire workcell—possibly with multiple robots—can fully exploit the capabilities of the robotic devices involved if it can rely on a model that takes into account (1) all relevant components of the cell (2) with adequate precision and fidelity to the actual physical setting.

Robotic cells are complex engineering objects, comprised of several components (robots, other machine tools, conveyors, storage areas, cameras, workpieces, etc.). The latter are, typically, designed separately, and it is assumed that they have separate CAD models available, each having its own reference frame. In order to construct a unified model of the entire workcell, one must be able to relate the aforementioned reference frames to each other, in a way that reflects the relative pose of the components as they are placed in reality. Calibration serves this goal by determining the spatial relations of reference frames, incorporating measured data, and thereby enabling previously isolated component models to be matched to each other, as well as the entire virtual representation to be adjusted to the real production scene.

10.2.1 Expressing Spatial Transformations

Unifying several disjoint models, as well as adjusting them to a real setting, involves the transformation of spatial vectors—be it vertex coordinates or surface normals—into a reference frame which has its origin at a specific location, and has a specific orientation. Linear algebra provides a suitable mathematical apparatus for performing these as automated calculations. A brief summary of some relevant notations in coordinate transformation is given below—for a more in-depth coverage of the topic, the reader is referred to more comprehensive robotics textbooks [3].

10.2.1.1 Rotation Matrices and Expressions of Orientation

When models are transformed between reference frames, vectors such as surface normals are reoriented only, as the placement of their origin does not play a role in calculations. Therefore, the algebraic tools for reorientation can be pictured as switching bases between two right-handed, orthonormal coordinate systems that share the same origin but may have a different orientation. Let us suppose that a vector **p** is expressed in coordinate system (frame) K, using the notation of a column vector:

$$\mathbf{p} := \begin{bmatrix} p_x \\ p_y \\ p_z \end{bmatrix} = p_x\mathbf{x} + p_y\mathbf{y} + p_z\mathbf{z}, \tag{10.3}$$

where **x**, **y**, and **z** are the base vectors of unit length in the designated x, y, and z directions in K. Now, consider another frame K' which has the same origin as K but has a different orientation. The same vector **p** can also be expressed in K':

$$\mathbf{p} := \begin{bmatrix} p'_x \\ p'_y \\ p'_z \end{bmatrix} = p'_x\mathbf{x} + p'_y\mathbf{y} + p'_z\mathbf{z}. \tag{10.4}$$

If **p** is given as expressed in K' with the coordinates $\begin{bmatrix} p'_x & p'_y & p'_z \end{bmatrix}^{\mathrm{T}}$, it can also be expressed in K as:

$$\begin{bmatrix} p_x \\ p_y \\ p_z \end{bmatrix} = \begin{bmatrix} \mathbf{x}'\cdot\mathbf{x} & \mathbf{y}'\cdot\mathbf{x} & \mathbf{z}'\cdot\mathbf{x} \\ \mathbf{x}'\cdot\mathbf{y} & \mathbf{y}'\cdot\mathbf{y} & \mathbf{z}'\cdot\mathbf{y} \\ \mathbf{x}'\cdot\mathbf{z} & \mathbf{y}'\cdot\mathbf{z} & \mathbf{z}'\cdot\mathbf{z} \end{bmatrix} \begin{bmatrix} p'_x \\ p'_y \\ p'_z \end{bmatrix} := \mathbf{R} \begin{bmatrix} p'_x \\ p'_y \\ p'_z \end{bmatrix}, \tag{10.5}$$

where the operator "·" designates the scalar product. **R** is referred to as a *rotation matrix*. Such matrices have the following specific properties which simplify calculations or enable a quick check of numerical correctness:

$$\mathbf{R}^{-1} = \mathbf{R}^{\mathrm{T}} \tag{10.6}$$

$$\det(\mathbf{R}) = 1, \tag{10.7}$$

where $\mathbf{R}^{\mathrm{T}}$ is the transpose of **R**, and $\det(\mathbf{R})$ is the determinant of **R**. A 3×3 rotation matrix has 9 elements, however, these are not fully independent. In fact, an arbitrary rotation can already be specified by 3 independent parameters, and there exist multiple conventions for the correspondence between orientation angles and rotation matrices.

Euler angles and *Tait–Bryan angles* specify an arbitrary orientation by a sequence of rotations around designated axes of a coordinate system. For this purpose, the following "elementary" rotation matrices are used for rotation by α around the z, y, and x axes, respectively:

$$\mathbf{R}(z,\alpha) := \begin{bmatrix} C_\alpha & -S_\alpha & 0 \\ S_\alpha & C_\alpha & 0 \\ 0 & 0 & 1 \end{bmatrix} \tag{10.8}$$

$$\mathbf{R}(y,\alpha) := \begin{bmatrix} C_\alpha & 0 & S_\alpha \\ 0 & 1 & 0 \\ -S_\alpha & 0 & C_\alpha \end{bmatrix} \tag{10.9}$$

$$\mathbf{R} := (x, \alpha) : \begin{bmatrix} 1 & 0 & 0 \\ 0 & C_\alpha & -S_\alpha \\ 0 & S_\alpha & C_\alpha \end{bmatrix}, \tag{10.10}$$

where C_α and S_α stand short for $\cos(\alpha)$ and $\sin(\alpha)$, respectively. The interpretation of these matrices is that they provide a coordinate transformation which is applied to express a vector $\left[\, p_x' \; p_y' \; p_z' \,\right]^{\mathrm{T}}$ of a rotated frame K' in the non-rotated frame K:

$$\begin{bmatrix} p_x \\ p_y \\ p_z \end{bmatrix} = \mathbf{R}(*, \alpha) \begin{bmatrix} p_x' \\ p_y' \\ p_z' \end{bmatrix}, \tag{10.11}$$

where "$*$" stands for any of the above axes. A commonly used type of rotation order for Euler angles is the "z–x–z" sequence:

$$R_{\text{Euler}}(\varphi, \theta, \psi) := R(Z, \varphi) R(x, \theta) R(Z, \psi) = \\ = \begin{bmatrix} C_\varphi C_\psi - S_\varphi C_\theta S_\psi & -C_\varphi S_\psi - S_\varphi C_\theta C_\psi & S_\varphi S_\theta \\ S_\varphi C_\psi + C_\varphi C_\theta S_\psi & -S_\varphi S_\psi + C_\varphi C_\theta C_\psi & -C_\varphi S_\theta \\ S_\theta S_\psi & S_\theta C_\psi & C_\theta \end{bmatrix}. \tag{10.12}$$

It should be noted that "z–x–z" is only one of several possible rotation sequences, and corresponding definitions are recommended to be checked before deploying third-party computational resources.

Another common form of expressing rotations is the *axis–angle representation,* describing a rotation by angle φ around an arbitrary axis $\mathbf{t}$ defined in the original (non-rotated) frame K. Applying *Rodrigues' rotation formula,* the corresponding rotation matrix is:

$$\mathbf{R}(\mathbf{t}, \varphi) := \begin{bmatrix} C_\varphi + (1 - C_\varphi)t_x t_x & (1 - C_\varphi)t_x t_y - S_\varphi t_z & (1 - C_\varphi)t_x t_z + S_\varphi t_y \\ (1 - C_\varphi)t_y t_x + S_\varphi t_z & C_\varphi + (1 - C_\varphi)t_y t_y & (1 - C_\varphi)t_y t_z - S_\varphi t_x \\ (1 - C_\varphi)t_z t_x - S_\varphi t_y & (1 - C_\varphi)t_z t_y + S_\varphi t_x & C_\varphi + (1 - C_\varphi)t_z t_z \end{bmatrix}, \tag{10.13}$$

where t_x, t_y, and t_z are the x, y, and z-coordinates of vector $\mathbf{t}$ in frame K, with $\mathbf{t}$ being of unit length, i.e., $||\mathbf{t}|| = 1$. Due to the latter property, the axis–angle representation is often specified by only three parameters, i.e., the coordinates of $\mathbf{t}$ multiplied by the angle (often specified in radians rather than degrees).

10.2.1.2 Homogeneous Transformations

While rotation matrices are suitable for expressing orientations and re-orienting vectors, they do not introduce any linear offset. While this can be carried out by adding an offset vector separately, the more compact notation of *homogeneous transformations* reduces this to a single multiplication with a 4×4 matrix [3, 8]:

$$\begin{bmatrix} p_x \\ p_y \\ p_z \\ 1 \end{bmatrix} = \begin{bmatrix} r_{11} & r_{12} & r_{13} & d_x \\ r_{21} & r_{22} & r_{23} & d_y \\ r_{31} & r_{32} & r_{33} & d_z \\ 0 & 0 & 0 & 1 \end{bmatrix} \begin{bmatrix} p_x^{'} \\ p_y^{'} \\ p_z^{'} \\ 1 \end{bmatrix}, \tag{10.14}$$

where the elements r_{ij} comprise a rotation matrix $\mathbf{R}$, and d_x, d_y, d_z are elements of the offset vector $\mathbf{d}$ which shifts the origin of the original K into the origin of the shifted and rotated frame $K^{'}$. Homogeneous transformations follow a specific order of rotation and translation:

$$\begin{bmatrix} \mathbf{R} & \mathbf{d} \\ \mathbf{0}^{\mathrm{T}} & 1 \end{bmatrix} = \begin{bmatrix} \mathbf{I} & \mathbf{d} \\ \mathbf{0}^{\mathrm{T}} & 1 \end{bmatrix} \begin{bmatrix} \mathbf{R} & \mathbf{0} \\ \mathbf{0}^{\mathrm{T}} & 1 \end{bmatrix}, \tag{10.15}$$

where $\mathbf{I}$ is a 3×3 identity matrix, and $\mathbf{0}$ is a 3-element column vector with all elements being zeros. Specific properties make it possible to invert homogeneous transformation matrices without actual matrix inversion [3]:

$$\begin{bmatrix} \mathbf{R} & \mathbf{d} \\ \mathbf{0}^{\mathrm{T}} & 1 \end{bmatrix}^{-1} = \begin{bmatrix} \mathbf{R}^{\mathrm{T}} & -\mathbf{R}^{\mathrm{T}}\mathbf{d} \\ \mathbf{0}^{\mathrm{T}} & 1 \end{bmatrix}. \tag{10.16}$$

This makes coordinate transformations with homogeneous matrices suitable even for devices with limited computational power.

10.2.1.3 Describing Robot Joint Transformations: The Denavit–Hartenberg Convention

The typical industrial robot forms a serial mechanism, with joints connecting adjacent links in an open kinematic chain. Each joint represents one kinematic DOF which can either be *prismatic* (allowing linear motion along a specific axis), or *revolute* (allowing rotation about a specific axis). Screw displacements and multiple-DOF motions are usually decomposed to several prismatic-only or revolute-only joints. Being a kinematic chain, the pose of the robot's end-effector frame can be obtained by concatenating the transformations between subsequent links, yielding a transformation with fixed parameters (lengths and angles specific to the given robot) and variables (joint values) in its expressions. The first step in establishing a kinematic model is, therefore, the specification of transformations between adjacent links.

Several conventions exist for a standard description of joint transformations, of which the *Denavit–Hartenberg form* [3, 9] finds most widespread use. This is a relatively compact notation building on the fact that four parameters are sufficient

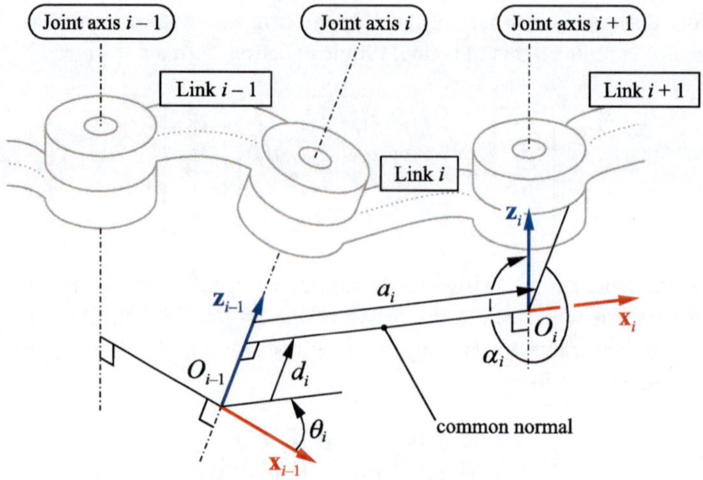

Fig. 10.1 Arrangement of joint axes, frames and parameterised transformations in the Denavit–Hartenberg form

for describing the relative position and orientation of any two lines in 3-dimensional space. To this end, the following numbering and placement conventions are set (see also Fig. 10.1):

- Frame K_{i-1} (with origin O_{i-1}) moves together with link $i-1$.
- The z-axis of K_{i-1} lies on joint axis i.
- The x-axis of K_{i-1} is the continuation of the common normal between joint axes $i-1$ and i. If the common normal is not unique (i.e., the two axes are parallel), the location of O_{i-1} can be freely chosen along joint axis i (in practice, it is preferably chosen so as to reduce the number of non-zero constants).
- The y-axis of K_{i-1} is already determined by the placing the other two axes, as all frames are right-handed (therefore, y-axes are not shown in Fig. 10.1).

With these conventions observed, K_{i-1} is transformed into K_i by a series of transformations in the following order (note that the polarity of rotations and translations follows the right-hand rule and the direction of the frame axes):

1. Rotation around $\mathbf{z}_{i-1}$ by θ_i, so that $\mathbf{x}_{i-1}$ becomes parallel to the common normal between joint axes i and $i+1$. If joint i is revolute, θ_i is the joint variable q_i, otherwise it is a constant parameter.
2. Translation along $\mathbf{z}_{i-1}$ by d_i, so that the rotated image of $\mathbf{x}_{i-1}$ comes to lie on the common normal between joint axes i and $i+1$. If joint i is prismatic, d_i is the joint variable q_i, otherwise it is a constant parameter.
3. Translation along the common normal by a_i, so that the shifted image of O_{i-1} coincides with O_i. The transformed image of $\mathbf{x}_{i-1}$ already coincides with $\mathbf{x}_i$ at this point. In the standard Denavit–Hartenberg convention, a_i is a fixed parameter.

4. Rotation around the common normal ($\mathbf{x}_i$, in fact) by α_i, aligning the image of $\mathbf{z}_{i-1}$ with $\mathbf{z}_i$. In the standard Denavit–Hartenberg convention, α_i is a fixed parameter.

The entire transformation is then:

$$
\begin{aligned}
{}^{i-1}\mathbf{T}_i :=& \mathbf{Rot}(\mathbf{z}_{i-1}, \theta_i)\mathbf{Trans}(\mathbf{z}_{i-1}, d_i)\mathbf{Trans}(\mathbf{x}_i, a_i)\mathbf{Rot}(\mathbf{x}_i, \alpha_i) \\
=& \begin{bmatrix} C_{\theta_i} & -S_{\theta_i}C_{\alpha_i} & S_{\theta_i}S_{\alpha_i} & a_i C_{\theta_i} \\ S_{\theta_i} & C_{\theta_i}C_{\alpha_i} & -C_{\theta_i}S_{\alpha_i} & a_i S_{\theta_i} \\ 0 & S_{\alpha_i} & C_{\alpha_i} & d_i \\ 0 & 0 & 0 & 1 \end{bmatrix},
\end{aligned}
\tag{10.17}
$$

where C and S are short for the cos and sin functions, respectively. The transformation for the entire kinematic chain of n movable links (with link 0 being the base) can then be assembled as:

$$
{}^{0}\mathbf{T}_n = {}^{0}\mathbf{T}_1 {}^{1}\mathbf{T}_2 \cdots {}^{n-2}\mathbf{T}_{n-1} {}^{n-1}\mathbf{T}_n. \tag{10.18}
$$

10.2.2 Classical Calibration Procedure

According to Wu et al. [10], classical calibration methodologies are comprised of four consecutive steps:

1. Modelling,
2. Measurement,
3. Identification,
4. Implementation.

During the *modelling* step, the geometrical model of the robot is to be calibrated. In the model, potential errors are already taken into consideration. Various methods exist for error mapping based on the types of errors to be considered. They can be simple, accounting for arm length or joint angle inaccuracy only, or can be modelled with full-scale transformation matrices [11]. In even more complex cases, the error model can be non-linear [12].

In the *measurement* phase, the pose of the robot end effector is determined. The number of parameters that need to be captured depends on the models used. Usually, six parameters are needed to cover the possible six degrees of freedom of the (Cartesian) workspace: three for position (e.g., with Cartesian coordinates), and three more for orientation (e.g., with Euler angles).

The *identification* step is the fitting of measured data onto the predetermined model. During this step, a variety of curve fitting algorithms can be used, however, their particular choice depends on the underlying robot model. Wu et al. [10] give a summary of frequently applied methods, such as least-squares [13] and damped least-squares (also known as the Levenberg–Marquardt algorithm [14], used for solving non-linear least-squares problems), Kalman filtering [15], or maximum likelihood algorithms [12, 16].

In the *implementation* phase, the calculated error parameters are applied to the robot controller, compensating for and minimising the positioning errors of the robot manipulator.

10.2.3 Mechanical Modelling

The pose accuracy of robots is highly influenced by their geometrical and material structure. As robots are very versatile, capable of complex and flexible movements and can be applied to a multitude of use-cases, the demand emerges to use them in machining, forming or assembly, instead of specialised, often purpose-built, machines [17]. However, robots are less suitable for such tasks, as their open kinematic chain may reduce the stiffness that would be needed for exerting high torques and forces at a precisely positioned TCP. Reduced stiffness, namely, results in elastic deformation in various links of the robot, compromising positioning accuracy when exerting high forces and torques at the end-effector.

Nakamura et al. [18] present a classification of mechanical errors, splitting their sources into geometrical parameters:

- processing and/or assembly errors of robot components,
- thermal expansion of links,

and non-geometrical parameters:

- deflection errors between loaded and unloaded robot links, including backlash,
- transmission errors between gears.

According to the authors, the effect of transmission errors on the robot position is in the order of magnitude of 10 μm, which makes their effect difficult to measure. Instead, computer simulation of the robot postures is applied, with standard random errors, to determine their effect on the TCP pose of the robot.

In addition, the authors carried out experiments to determine the effect of deflection errors and backlash, yielding the following findings [18]:

- "Deflection transferred to angular shift values in each link is nearly the same in all directions.
- In each axis, excluding the driving axis of each link, deflection is proportional to torque.
- Driving axis of each link has a very large angular shift value compared to other axis."

Moreover, it was found that angular deflection is proportional to torque, outside the lost motion range.

Several solutions exist to address such difficulties. A specific option is the modification of the structure of the kinematic chain of the robot. Some possibilities are shorter links which decrease the lever arm; parallelogram loops increasing stiffness on a structural level, or increased actuator power. These techniques can increase the stiffness of the robot structure, decreasing, in turn, the positioning error of the robot.

Another possibility is elasto-geometrical calibration which boils down to the identification of geometrical and elastic models of the structure of the robot [17]. To that end, some authors proposed dynamic models to compensate the errors, by modifying actuator torque on the control side. The difficulty in this approach is that dynamic parameters have to be identified. The authors also mention the need for access to joint-wise control, which is not granted by robot control systems that offer interfaces for TCP motion only.

Another approach is the prediction of the elastic deformation, for which two methods were introduced:

- *Lumped parameter model*—the kinematic chain of the robot is modelled by substituting every link with a set of rigid bodies, which are connected by springs. This model is able to represent the elastic deformation of the links, however, its accuracy greatly depends on the number of springs used for modelling, which proportionally increases the complexity of the model.
- *Finite element method*—applying the Euler–Bernoulli theorem, the robot links are substituted by beam elements. This way, the equivalent stiffness matrix can be calculated, which in consequence makes it possible to calculate the deformation of the structure. Volumetric finite element methods can also be used, however, its high calculation complexity precludes its use in real-time feedback control.

10.2.4 Robot Precision Calibration

Cai et al. [6] introduce a robot cell calibration method using a touch panel. The motivation of their work is the availability of touch panels on smartphones and tablets, and the ever growing sensitivity and resolution of these devices, which can be used for calibration. This approach uses external equipment—a tapping tool—in a two-phase procedure. The goal of the first phase is to determine the orientation of the touch screen, by tapping the touch surface with the touching probe in multiple positions, using translatory movement only. In the second phase, the robot attempts to touch the screen at the same point with different orientations for the same nominal TCP position. The difference between actual and intended touch coordinates can be used to calculate the difference between the robot nominal TCP position and the actual tip location of the touch probe. Applying the captured data to the points measured in the first phase, the actual coordinates of the touch screen origin can be calculated.

Marwan et al. [19] proposed an approach based on an ultrasonic sensor and an Inertial Measurement Unit (IMU), with the former handling position measurement, and the latter measuring orientation. The ultrasonic sensor system is comprised of a transmitter and three receiver units. The transmitter is attached to the end-effector of the robot and sends ultrasonic wave signals to the receivers, which in turn register their distance from the wave source by time-of-flight measurement (TOF). Knowing the accurate location of the receivers, the position of the end-effector can be determined. The setup is sensitive to temperature and other environmental noise, which are compensated by Kalman filtering. The IMU consists of a three-axis accelerometer, two two-axis gyroscopes and one three-axis magnetometer, making it possible to determine the end-effector orientation.

Du et al. [20] proposed an online, image-based calibration solution. The method uses a camera placed on the end-effector of the robot, while a checkboard pattern is attached to a target workpiece. Using the camera image and known geometry of the checkboard pattern, the actual position (and orientation) of the target with respect to the camera can be determined. With the placement of the robot base and the target known beforehand, the transformation representing the TCP pose can be determined. Using multiple measurements and a suitable kinematic model, the necessary parameters in the model can be calibrated using Levenberg–Marquardt algorithm.

For further examples of calibration approaches, the reader is referred to Wu et al. [10], Verl et al. [21], Watanabe et al. [22], and Meng et al. [23].

10.2.5 Workcell Calibration Based on External Sources

In this case, external equipment—usually a 2D (possibly colour) image camera, or 3D depth image camera—is used to compensate the lack of direct pose measurement sensors in today's typical industrial robots. Cameras can also be used for feedback in robot control loops, the latter approach being referred to as *visual servoing* [24].

Fatikov et al. [24] propose a method suitable for automatic integration of multiple microrobots in a sensor system with previously unknown attributes. The approach uses a single camera, mounted under a glass surface on which the microrobots move. The camera tracks the robots by individual LEDs mounted on the bottom of each robot. The tools operated by the robots are tracked by another device—a video microscope—from above the workspace. The tracker algorithm is similar in both cases.

Arai et al. [25] introduce an automatic calibration method for reconfigurable assembly systems. The setup uses two cameras observing robot-mounted LED markers, and applies Direct Linear Transformation (DLT) to restore the 3D spatial coordinates from 2D image coordinates. Having multiple cameras observing the same scene and same objects make it possible to calculate the reference frame transformation between the camera and the robot.

Rückert et al. [26] propose another vision based method for workcell calibration, utilising a reference object of precisely known size and features, observed by a robot-mounted camera. Object recognition algorithms identify the image of the object in the captured scene, and a comparison with the known geometry of the reference object yields its pose relative to the image sensor. Using precise fixtures, the robot pose can be determined with good accuracy, allowing robot-to-cell calibration. The procedure presumes that the self-calibration of the internal sensors of the robot is sufficiently accurate.

Watanabe et al. [27] proposed a method for calibrating virtual models to real scenes in the context of accurately setting workpiece. The method relies on CCD cameras with little a priori requirements regarding their placement or optical camera model—the necessary refinement of parameters is largely based on sensing autonomously generated robot motion.

10.3 Case Study of a Calibration Scenario

This section presents a case study of a method relying on an incomplete point cloud acquired with a fixed 3D camera, with no markers on the calibrated robot, assuming previous knowledge of the CAD geometry of the robot only. To accelerate the process, general purpose GPU programming is utilised. This approach has originally been published in [28].

10.3.1 Problem Statement

Robot cell calibration is the crucial step in determining the as-built model of a robot cell—the section presents an example of this calibration task. The case presented here is built on a collaborative robot cell with a 3D depth sensor mounted at a fixed position, and the virtual scene of the robot cell as a CAD assembly model. The virtual scene is the as-designed model of the robot cell. The depth sensor captures the environment of the robot cell as a point cloud, the latter data considered a partial measurement of the as-built model of the cell. The coordinates of the points are defined in the point cloud reference frame, located at the focal point of the sensor. Figure 10.2 shows a 2D image captured in the actual viewing window of the 3D sensor.

Cell calibration assumes that the depth sensor has a fixed placement in the cell, and no calibration artefacts are used. The fixed location of the depth sensor implies that only partial measurement can be obtained due to occlusion and shadowing. It is also assumed that the virtual scene models only contain relevant objects of the robot cell, i.e., the as-designed and as-built models may differ considerably due to non-modelled artefacts in the robot cell.

The result of the calibration is the alignment transformation that transforms the reference coordinate frame of the as-built model (i.e., the reference frame of the point cloud) to the reference frame of the as-designed model (CAD reference frame of the model), realising the *calibrated workcell model.*

Fig. 10.2 Robot cell as seen in the viewing window of a Kinect 3D sensor

A workflow has been developed (see Fig. 10.3) for calculating the alignment transformation based on the point cloud measurement and the CAD model of the workcell. The basic idea of the algorithm is to locate one or more *frame features* in the measured point cloud and the reference CAD scene, respectively, and calculate the transformation between these two frames. Frame features are geometric features providing means of attaching a local coordinate frame. For example, an Axis Aligned

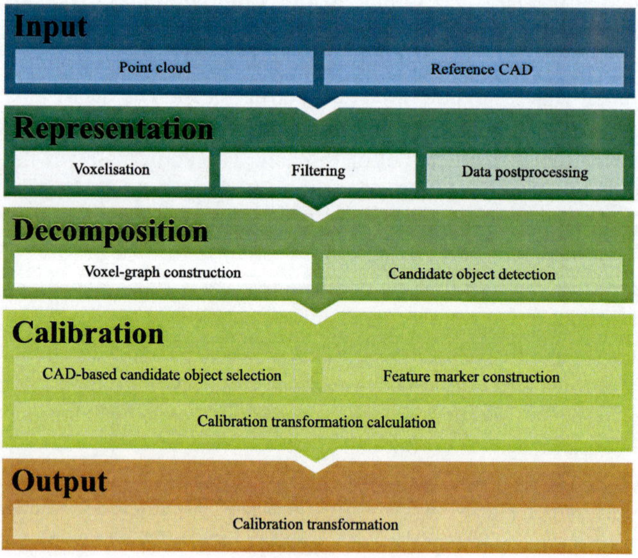

Fig. 10.3 Workflow of the cell calibration process

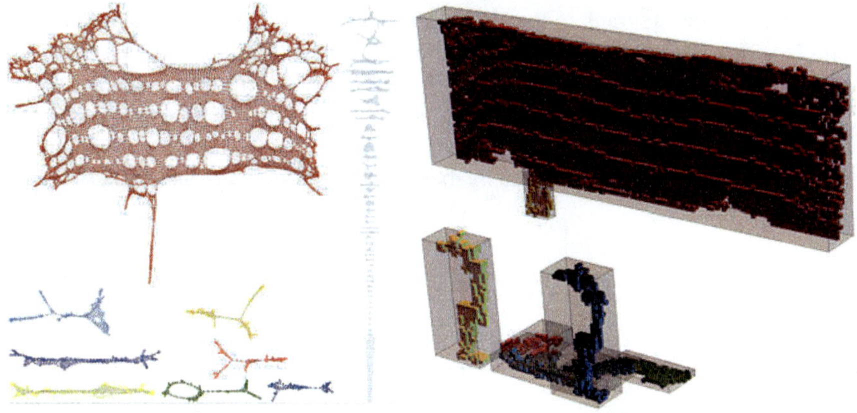

a) b)

Fig. 10.4 Connected voxel subgraphs of the VCG of the robot cell (left) and the corresponding voxel set and AABB rendering (right)

Bounding Box (AABB) is such a frame feature, since a local coordinate frame can be naturally bound to a rectangle (see Fig. 10.5). Since frame feature localisation in a CAD model is straightforward, basically equivalent to sub-assembly selection, only frame feature selection in the point cloud is presented here in detail.

10.3.1.1 Point Cloud Pre-processing

The process starts with *voxelisation*, which takes the point cloud and calculates the position of those voxels—with given dimensions—that contain at least one point. Hereafter, voxels with low point density are filtered out as measurement errors, with the filtering threshold being an input parameter. Next, the connectivity of the neighbouring voxels is calculated as an ordered pair of voxel indices. These edge lists will define the *voxel connectivity graph* (VCG). Figure 10.4 shows a VCG of the robot cell obtained from measured data.

10.3.1.2 Decomposition

The goal of the decomposition phase is to localise a disjoint, connected set of points that may correspond to a given reference candidate CAD component. The underlying assumption here is that the connected voxel subgraphs represent relevant components of the robot cell, which could also be identified in the virtual CAD scene as a component or subassembly.

The connected subgraph of the VCG graph is calculated with the standard *union-find algorithm.* Using the connected subgraphs of the VCG, the disjoint, connected sets of points are calculated together with their AABB (Fig. 10.4). The candidate CAD object is selected as the geometry of the robot model (Fig. 10.5), while the localising feature vector is defined based on the AABB of the geometry, being the normalised AABB dimension given by the sorted (length l, width w, height h) values of a box:

$$\text{reference} \rightarrow \{l, w, h\}. \tag{10.19}$$

The normalised AABB dimensions are also calculated for the indexed disjoint connected sets of points. For the case presented here, these are:

$$\begin{array}{ll} 1 \rightarrow \{357.0, 1307.0, 3997.0\} & 2 \rightarrow \{295.0, 499.1, 988.7\} \\ 3 \rightarrow \{264.0, 470.8, 975.1\} & 4 \rightarrow \{138.4, 469.0, 886.9\} \\ 5 \rightarrow \{52.7, 352.0, 688.0\} & 6 \rightarrow \{112.0, 256.5, 514.3\} \\ 7 \rightarrow \{108.2, 448.0, 476.2\} & 8 \rightarrow \{148.7, 322.5, 535.0\} \end{array}$$

The candidate object detection finds the index of the disjoint connected set of points whose feature vector is closest to the reference feature vector using the chess-board distance function. In case the robot cell contains more than n identical robots, then the first n closest AABB indices are looked for. In the example, these are disjoint connected sets 2 and 3.

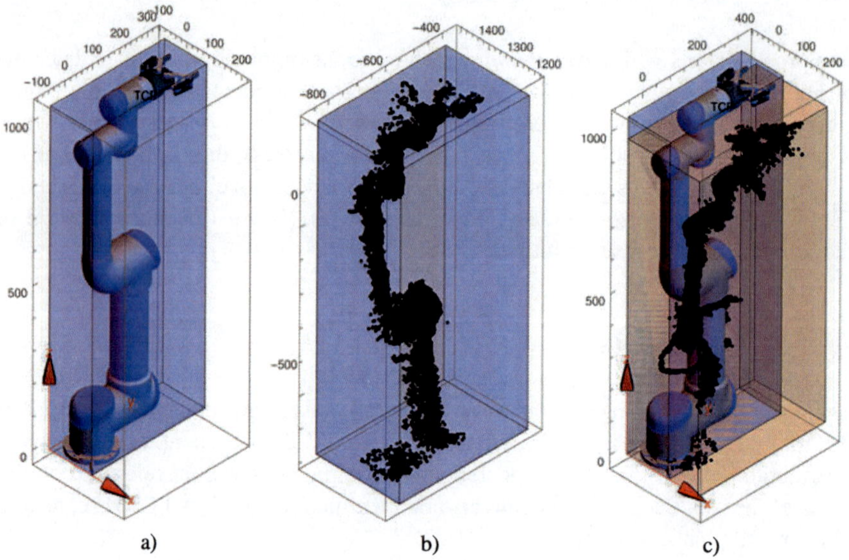

Fig. 10.5 AABB of CAD model (left), AABB and points of a connected subset (middle), and aligned connected subset and CAD model (right)

As the result of the decomposition phase, a pair of one reference CAD object and one disjoint connected set of points is passed to the calibration phase.

10.3.1.3 Calibration

The calibration process aligns the decomposed set of points and the corresponding CAD object. A frame feature is selected on both candidates, then the local coordinate frames are calculated for both frame features, and finally, the alignment transformation between the two frames is calculated. Two types of frame features are defined: the bounding box and the intersecting link cylinders. The calibration process is calculated in both cases, and the results are compared. The feature frame is represented with a homogeneous transformation matrix:

$$\mathbf{f} = \begin{bmatrix} \mathbf{x}_f & \mathbf{y}_f & \mathbf{z}_f & \mathbf{O}_f \\ 0 & 0 & 0 & 1 \end{bmatrix}, \qquad \mathbf{f} \in \{\text{CAD}, \text{PC}\}, \tag{10.20}$$

where $\mathbf{x}_f, \mathbf{y}_f, \mathbf{z}_f$ are the axes of the feature frame, and $\mathbf{O}_f$ expresses the coordinates of the feature frame origin O_f. In the case of the bounding box based form feature, the frame is defined by taking the following assumptions:

1. The origin (O_f) of the frame is placed at the minimum corner point of the AABB.
2. The axes of the feature frame ($\mathbf{x}_f, \mathbf{y}_f, \mathbf{z}_f$) are parallel to the axes of the reference frame.
3. The dimensions along $\mathbf{x}_f$, $\mathbf{y}_f$, and $\mathbf{z}_f$ are calculated for the point cloud frame and the CAD feature frame, respectively. If the $\{\mathbf{x}_f, \mathbf{y}_f, \mathbf{z}_f\}$ axis names are ordered with respect to the AABB dimensions, the sorted list should be equivalent on both feature frames.

Figure 10.5 shows the feature frame placed on the CAD model (Fig. 10.5a) and the point cloud (Fig. 10.5b). The alignment transformation $\mathbf{T}_\text{a}$ is calculated as:

$$\mathbf{T}_\text{a} = \mathbf{F}_\text{CAD}\mathbf{F}_\text{PC}^{-1}, \tag{10.21}$$

where $\mathbf{F}_\text{CAD}$ and $\mathbf{F}_\text{PC}$ are the homogeneous transformations describing the CAD feature frame, and the point cloud frame, respectively. The aligned point cloud and CAD model are shown in Fig. 10.5c. Note that the error of the adjustment is closely correlated with the angular misalignment of the 3D sensor, hence this kind of form feature can only be employed if the sensor orientation is correct.

The frame feature acquisition method applying the intersecting link cylinders exploits the fact that the robot is built up of roughly linear, prism-like segments. Extruded objects, e.g., beams and pipes or various profiles, can be extracted from a given connected set of voxel graphs using the *Branch Connectivity Graph* (BCG) as described in [29]. Using the BCG of the point cloud, cylinders of relatively long

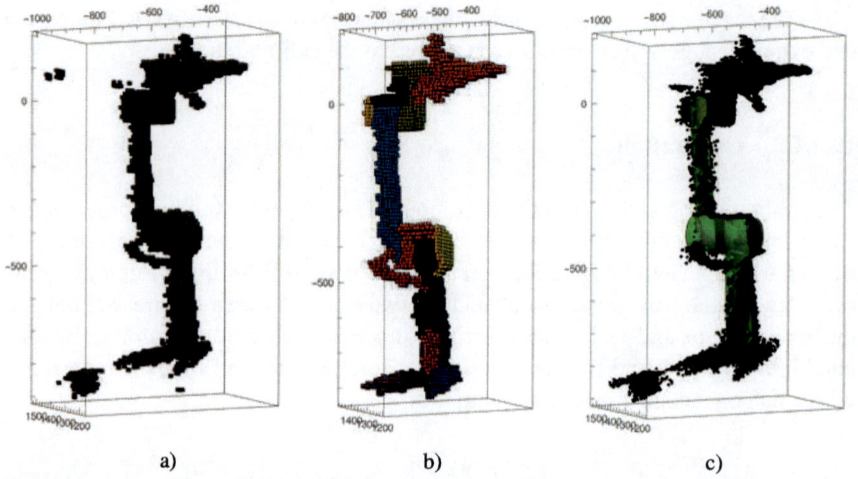

Fig. 10.6 VCG of connected subset No. 3 (left), BCG of connected subset No. 3 (middle), and fitted cylinders on the recognised branches (right)

links (l_1, l_2, l_3) of the robot can be fitted (see Fig. 10.6). These three consecutive intersecting cylinders determine a feature frame in such a way that:

- Origin location $(\mathbf{O}_f)$ of the frame is defined by the intersection of the axes of cylinders representing l_2 and l_3.
- The z-axis $(\mathbf{z}_f)$ is defined by axis of the cylinder representing l_3.
- The y-axis $(\mathbf{y}_f)$ is defined by the intersection (cross-product, in fact) of the axes of the cylinders for l_1 and l_2.

The resulting feature frames are shown in Fig. 10.7a–b, respectively. The alignment transformation $\mathbf{T}_\text{a}$ is calculated as described above, and the aligned point cloud is shown in Fig. 10.7c.

10.3.2 GPU-Based Point Cloud Decomposition and Voxel Graph Construction

First, the point cloud is uploaded to the GPU. The points are represented as a vector of 3 floating point numbers arranged in simple arrays. In most cases, the algorithm is structured such that consecutive threads read continuous blocks of data, achieving *coalesced access* [30].

For efficient use of arrays, every object has to be numbered using a single index. Three different objects are used throughout the application: points, voxels, and voxel edges.

The index of the voxels and edges depends on the size of the bounding box of the point cloud and the size of the voxels. Assume the number of voxels along the x, y and z axes to be d_x, d_y and d_z, respectively. The voxel touching all 3 axes has the coordinates $\{v_x, v_y, v_z\} = \{0, 0, 0\}$, with all voxel coordinates being integer values. For arbitrary voxel coordinates $\{v_x, v_y, v_z\}$, the index v_{idx} of the given voxel is calculated as

$$v_{\text{idx}} = v_x + v_y d_x + v_z d_x d_y. \tag{10.22}$$

The indices of voxel edges are derived from the index of the voxel they are assigned to. An edge connects two voxels for which two out of the three coordinates are equal and the difference of the third one is 1. The edge between them is assigned to the voxel with the smaller voxel coordinate. This way, every voxel has 3 edges and every edge is associated with exactly one voxel, which means the indexing is unambiguous. The index of the edge is $3v_{\text{idx}} + a$ where v_{idx} is the index of the voxel the edge is assigned to, and a is 0 if the edge is parallel to $\mathbf{x}$, 1 if parallel to $\mathbf{y}$ and 2 if parallel to $\mathbf{z}$.

10.3.2.1 Algorithm

The first step is finding the AABB of the point cloud. This is accomplished by finding both the smallest and largest values of the three coordinates. These values are collected separately for every thread block, written back to global memory, after which a specialised form of *parallel reduction* [30] finds the extreme values. The whole point cloud has to be transformed so that its new AABB touches the axes [30]. This is basically a translation.

The following step is the division of the AABB into voxels. The size of the voxels is an input of the algorithm. The coordinates of the containing voxels are calculated for every point and stored in a new array. This is an integer division.

The next step is to choose the relevant voxels, defined as those that have a minimum number of points (*filtering threshold*) inside them. This is done in three steps. First, a *3D histogram* is built for all the voxels where the values will be the numbers of points inside. Histogram building can be done in parallel [30]. Second, every element of this histogram is checked. If its contained point count is above the filtering threshold, the index of the voxel is stored in a new array, otherwise it is assigned the value –1. Third, the new array with the relevant voxel indices is compacted, so that all elements are stored in consecutive positions in the memory. This can also be done on the GPU. For this task, the parallel primitive library CUDPP is used [31].

The last step is to find the edges of the voxel graph. All possible edges are checked whether the corresponding voxels contain more points than the given threshold. This brute force method is still efficient, utilising the huge number of GPU compute threads to generate the edge list. As consecutive threads read consecutive voxel elements from the histogram, the memory read operations can utilise the entire bandwidth of the memory bus, resulting in highly efficient memory transfer operations.

Table 10.1 Runtimes with different options

Number of points	Voxel size [mm]	Filtering threshold	Runtime [s]
100,000	100	5	0.39
100,000	100	100	0.39
100,000	50	5	0.41
1,000,000	50	5	1.86
10,000,000	100	5	14.71
10,000,000	50	5	15.46
10,000,000	50	100	15.15

10.3.2.2 Performance Evaluation of GPU Code

In order to check the added value and speed increase of the GPU implementation, test runs were made with different parameters, on a point cloud with a box size of $2 \times 2 \times 2$ m. The results can be seen in Table 10.1. The specification of the applied hardware is detailed in the next subsection.

10.3.3 Implementation

10.3.3.1 Resources Used

The implemented work cell calibration system uses a Kinect depth sensor via the *Kinect for Windows SDK 2.0* software library. The virtual CAD model is developed in *Wolfram Mathematica v10.0* using the *LinkageDesigner* add-in. The tests have been run on an *Ideum* table computer with Intel i7-4790S 3.2 GHz processor, 16 GB RAM and NVIDIA GeForce GTX 960 graphics card. The voxel graph calculation is implemented in CUDA C, which communicates with Wolfram Mathematica through the *MathLink* protocol. The details of the calibration process for the workcell can be seen in Table 10.2.

The result of point cloud alignment onto the virtual CAD cell is rendered in Fig. 10.8. The same cell is used for gesture control purposes in [32].

10.3.3.2 Precision Evaluation

The accuracy of the fitting is investigated using a test set of points in the vicinity of links 1–3 of the robot (Fig. 10.7). The test set was selected manually by taking every point within a distance of 20 mm from the fitted cylinders (Fig. 10.5c), containing 37,008 points altogether. Next, the shortest distance of every point to the aligned CAD geometry has been calculated. The mean and standard deviation of the distance data are summarised in the last rows of Table 10.2. As it can be seen, intersecting cylinder

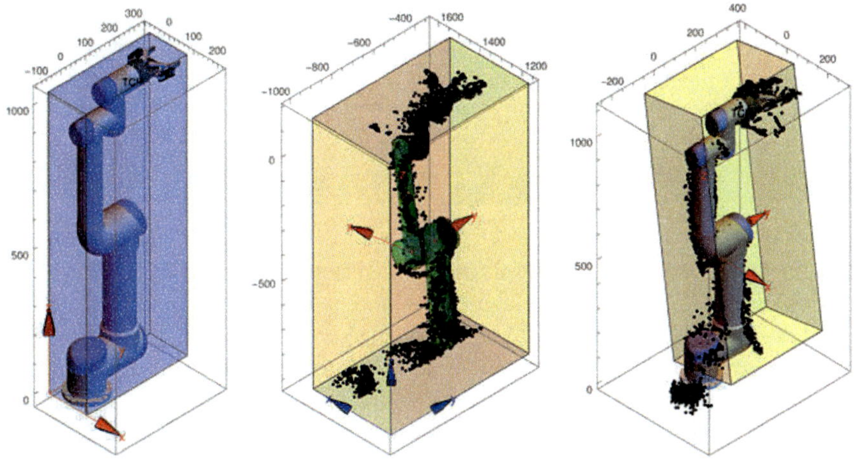

a) b) c)

Fig. 10.7 CAD model and feature frame FrameCAD (left), point cloud, branch cylinders and feature frame F_{PC} (middle), aligned connected subset No. 3 and CAD model (right)

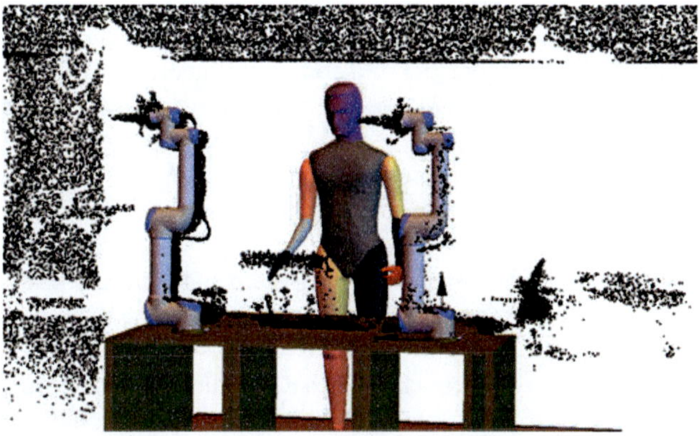

Fig. 10.8 Aligned point cloud and virtual CAD model of the cell

Table 10.2 Parameters of the calibration test run

Number of points in the entire cell	696,975
Data representation time	3.53 s
Bounding box based calibration time	2.30 s
Intersection cylinder based calibration time	15.89 s
Bounding box based mean distance	32.59 mm
Bounding box based standard deviation	24.68 mm
Intersection cylinder based mean distance	10.71 mm
Intersection cylinder based standard deviation	6.85 mm

based fitting produces a 10.71 mm average distance, which is comparable to the 2 mm precision of the Kinect sensor [33]. This result does not take into consideration the inaccuracy of the reference CAD model.

10.4 Conclusion and Outlook

Like many of today's decisive advances of industrial production, human–robot collaboration—both in planning and execution control—relies heavily on model-based operations that require accurate knowledge of relevant elements in the operating environment. Such an integrated model—often amounting to a full-fledged *digital twin*—must reflect and closely follow the physical reality of the production scene. To this end, nominal component models (1) have to be fine-tuned to reflect actual component characteristics, and (2) their exact placement in the scene has to be determined. Both of the aforementioned adaptations rely on measured data which allow the modification and integration of *as-designed* sub-models into an *as-built* model. The process of adapting nominal models to match reality is referred to as *calibration*.

The chapter gave an overview of the theoretical foundation of model calibration, especially regarding robot kinematics and component geometry, following the classical sequence of (1) modelling, (2) measurement, (3) identification, and (4) implementation [10]. An overview of approaches was given, representing several of the most widespread measurement and calibration concepts. As the cases demonstrated, most of today's practically feasible methods rely on contactless time-of-flight or visual measurement principles when the pose of components is to be acquired, while other—possibly contact-based—means of measurement are mostly reserved for calibration of individual robot or machine models. Similarly to classical automated environments, components of collaborative production cells have a possibly complex but certainly structured geometry with characteristic features defined beforehand. This provides reliable guidance during geometrical calibration, as specific features can be identified in an acquired spatial scene, and partial models can be matched accordingly. Much of the power of such calibration approaches, therefore, stems from advanced analysis and identification functionalities.

As a novel contribution, the chapter presented a workflow exemplifying the above mentioned procedure. In the case study, calibration operations relied on a point cloud acquired by 3D depth imaging, which underwent structural decomposition and subsequent matching of relevant features [28].

Due to ongoing fast development of imaging—both 2D and 3D—technologies, it is expected that imaging sensors retain their prevalence in sensing for calibration of workcell models. Evolving in parallel with image acquisition, graphical processing capabilities can be repurposed for model calibration operations, and thereby support the other major stage of producing calibrated virtual representations of structured industrial environments.

References

1. L. Wang, R. Gao, J. Váncza, J. Krüger, X.V. Wang, S. Makris, G. Chryssolouris, Symbiotic human–robot collaborative assembly. CIRP Ann. **68**(2), 701–726 (2019)
2. R. Stark, H. Grosser, P. Müller, Product analysis automation for digital MRO based on intelligent 3D data acquisition. CIRP Ann. **62**(1), 123–126 (2013)
3. C. Canudas de Wit, B. Siciliano, G. Bastin (eds.), *Theory of Robot Control* (Springer, London, London, 1996)
4. Ibid. Chapter 12, S. Ch. Akkaladevil, M. Plasch, M. Hofmann, L. Hiesmair, M. Ikeda, N. Chitturi, A. Pichler, "Programming-free approaches for human–robot collaboration in assembly tasks"
5. S. Deng, Z. Cai, D. Fang, H. Liao, G. Montavon, Application of robot offline programming in thermal spraying. Surf. Coat. Technol. **206**(19), 3875–3882 (2012)
6. Y. Cai, H. Gu, C. Li, H. Liu, Easy industrial robot cell coordinates calibration with touch panel. Robotics and Computer-Integrated Manufacturing **50**, 276–285 (2018)
7. G. Erdős, Zs. Kemény, A. Kovács, J. Váncza, "Planning of Remote Laser Welding Processes," *Procedia CIRP*, **7**, pp. 222–227, 2013
8. J. Cashbaugh, C. Kitts, Automatic Calculation of a Transformation Matrix Between Two Frames. IEEE Access **6**, 9614–9622 (2018)
9. R. S. Hartenberg, J. Denavit, "A kinematic notation for lower pair mechanisms based on matrices," *Transactions of the ASME—Journal of Applied Mechanics*, **22**, pp. 215–221, 1955
10. Y. Wu, A. Klimchik, S. Caro, B. Furet, A. Pashkevich, Geometric calibration of industrial robots using enhanced partial pose measurements and design of experiments. Robotics and Computer-Integrated Manufacturing **35**, 151–168 (2015)
11. S. A. Hayati, "Robot arm geometric link parameter estimation," In *Proceedings of the 22nd IEEE Conference on Decision and Control*, San Antonio TX, USA, 14–16 December 1983, pp. 1477–1483
12. J.-M. Renders, E. Rossignol, M. Becquet, R. Hanus, Kinematic calibration and geometrical parameter identification for robots. IEEE Trans. Robot. Autom. **7**(6), 721–732 (1991)
13. C. F. Gauss, "Theoria Motus Corporum Coelestium in Sectionibus Conicis Solem Ambientium". Cambridge University Press, 2011 (1st published 1809)
14. K. Levenberg, A method for the solution of certain non-linear problems in least squares. Q. Appl. Math. **2**(2), 164–168 (1944)
15. R. E. Kalman, "A New Approach to Linear Filtering and Prediction Problems," *Transactions of the ASME—Journal of Basic Engineering*, **82**(1), pp. 35–45, 1960
16. B.W. Mooring, Z.S. Roth, M.R. Driels, *Fundamentals of manipulator calibration* (Wiley, New York, 1991)
17. S. Marie, E. Courteille, P. Maurine, Elasto-geometrical modeling and calibration of robot manipulators: Application to machining and forming applications. Mech. Mach. Theory **69**, 13–43 (2013)
18. H. Nakamura, T. Itaya, K. Yamamoto, T. Koyama, "Robot autonomous error calibration method for off line programming system," In *Proceedings of the 1995 IEEE International Conference on Robotics and Automation*, Nagoya, Japan, 21–27 May 1995, vol. 2, pp. 1775–1782
19. A. Marwan, M. Simic, F. Imad, Calibration method for articulated industrial robots. Procedia Computer Science **112**, 1601–1610 (2017)
20. G. Du, P. Zhang, Online robot calibration based on vision measurement. Robotics and Computer-Integrated Manufacturing **29**(6), 484–492 (2013)
21. A. Verl, T. Boye, A. Pott, Measurement pose selection and calibration forecast for manipulators with complex kinematic structures. CIRP Ann. **57**(1), 425–428 (2008)
22. A. Watanabe, S. Sakakibara, K. Ban, M. Yamada, G. Shen, T. Arai, A Kinematic Calibration Method for Industrial Robots Using Autonomous Visual Measurement. CIRP Ann. **55**(1), 1–6 (2006)
23. Y. Meng, H. Zhuang, Autonomous robot calibration using vision technology. Robotics and Computer-Integrated Manufacturing **23**(4), 436–446 (2007)

24. S. Fatikow, C. Stolle, C. Diederichs, D. Jasper, Auto-configuration and self-calibration in flexible robot-based micro-/nanohandling cells. IFAC Proceedings Volumes **44**(1), 14630–14635 (2011)
25. T. Arai, Y. Maeda, H. Kikuchi, M. Sugi, Automated Calibration of Robot Coordinates for Reconfigurable Assembly Systems. CIRP Ann. **51**(1), 5–8 (2002)
26. P. Rückert, J. Adam, B. Papenberg, H. Paulus, K. Tracht, Calibration of a modular assembly system for personalized and adaptive human robot collaboration. Procedia CIRP **76**, 199–204 (2018)
27. A. Watanabe, S. Sakakibara, K. Ban, M. Yamada, G. Shen, T. Arai, Autonomous Visual Measurement for Accurate Setting of Workpieces in Robotic Cells. CIRP Ann. **54**(1), 13–18 (2005)
28. G. Horváth, G. Erdős, Point cloud based robot cell calibration. CIRP Ann. **66**(1), 145–148 (2017)
29. G. Erdős, T. Nakano, G. Horváth, Y. Nonaka, J. Váncza, Recognition of complex engineering objects from large-scale point clouds. CIRP Ann. **64**(1), 165–168 (2015)
30. J. Sanders, E. Kandrot, *CUDA by example: an introduction to general-purpose GPU programming* (Addison-Wesley, Upper Saddle River NJ, USA, 2011)
31. H. Nguyen, NVIDIA Corporation (eds.), "GPU gems 3." Addison–Wesley, Upper Saddle River NJ, USA, 2008
32. G. Horváth, G. Erdős, Gesture Control of Cyber Physical Systems. Procedia CIRP **63**, 184–188 (2017)
33. D. Pagliari, L. Pinto, Calibration of Kinect for Xbox One and Comparison between the Two Generations of Microsoft Sensors. Sensors **15**(11), 27569–27589 (2015)

Part IV
Adaptive Robot Control

Chapter 11
Human Motion Recognition and Prediction for Robot Control

Robert X. Gao, Lihui Wang, Peng Wang, Jianjing Zhang, and Hongyi Liu

Abstract The ever-increasing demand for higher productivity, lower cost and improved safety continues to drive the advancement of manufacturing technologies. As one of the key elements, human–robot collaboration (HRC) envisions a workspace where humans and robots can dynamically collaborate for improved operational efficiency while maintaining safety. As the effectiveness of HRC is affected by a robot's ability to sense, understand and forecast the state of the collaborating human worker, human action recognition and motion trajectory prediction have become a crucial part in realising HRC. In this chapter, deep-learning-based methods for accomplishing this goal, based on the in-situ sensing data from the workspace are presented. Specifically, to account for the variability and heterogeneity of human workers during assembly, a context-aware deep convolutional neural network (DCNN) has been developed to identify the task-associated context for inferencing human actions. To improve the accuracy and reliability of human motion trajectory prediction, a functional unit-incorporated recurrent neural network (RNN) has been developed to parse worker's motion patterns and forecast worker's future motion trajectories. Collectively, these techniques allow the robot to answer the question: "which tool or part should be delivered to which location next?", and enable online robot action planning and execution for the collaborative assembly operation. The methods developed are

R. X. Gao (✉) · J. Zhang
Department of Mechanical and Aerospace Engineering, Case Western Reserve University, Cleveland, OH, USA
e-mail: robert.gao@case.edu

J. Zhang
e-mail: jxz170@case.edu

L. Wang · H. Liu
Department of Production Engineering, KTH Royal Institute of Technology, Stockholm, Sweden
e-mail: lihuiw@kth.se

H. Liu
e-mail: hongyil@kth.se

P. Wang
Department of Electrical and Computer Engineering, University of Kentucky, Lexington, KY, USA
e-mail: edward.wang@uky.edu

L. Wang et al. (eds.), *Advanced Human–Robot Collaboration in Manufacturing*,
https://doi.org/10.1007/978-3-030-69178-3_11

experimentally evaluated, with the collaborative assembly of an automotive engine as a case study.

11.1 Introduction

Since their inception in the 1960s, robots have become an integral part of manufacturing automation. Traditionally, robots are pre-programmed to perform repetitive tasks with high precision in environments that are difficult to operate by human workers [1]. Due to safety concerns, robots have been strictly separated from human workers in their operations. As Industry 4.0 is transforming manufacturers into digitised smart factories that are characterised by the seamless integration of the cyber and physical worlds to enable in-situ operation monitoring, process optimisation and adaptive control for informed decision-making, with humans in the centre of the loop, robots are increasingly required to achieve a higher level of communication and dynamic cooperation with human workers as shown in Fig. 11.1, to realise true human–robot collaboration (HRC) [2–4].

HRC is characterised by the dynamic interaction between human worker and robot, based on effective, real-time communication and coordination. A key element for its realisation is that the robot can sense the workspace in real time, analyse the circumstance, and react accordingly. A representative HRC scenario is illustrated in Fig. 11.2, with multimodal communication, planning and collision avoidance [5].

Correspondingly, four functional elements are present in an HRC system: perception, recognition, prediction and action, as shown in Fig. 11.3 [4]. Perception relies on

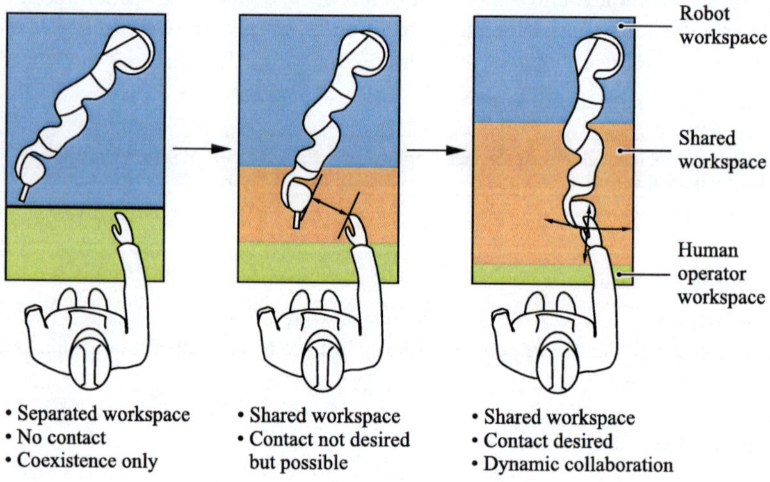

Fig. 11.1 Evolution of workspace, from safety-oriented to collaboration-oriented

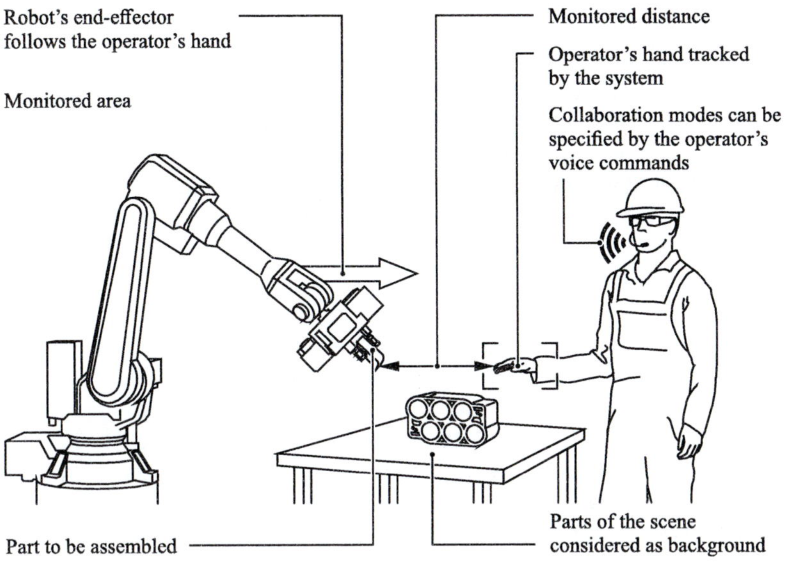

Fig. 11.2 Multimodal symbiotic human–robot collaboration [5]

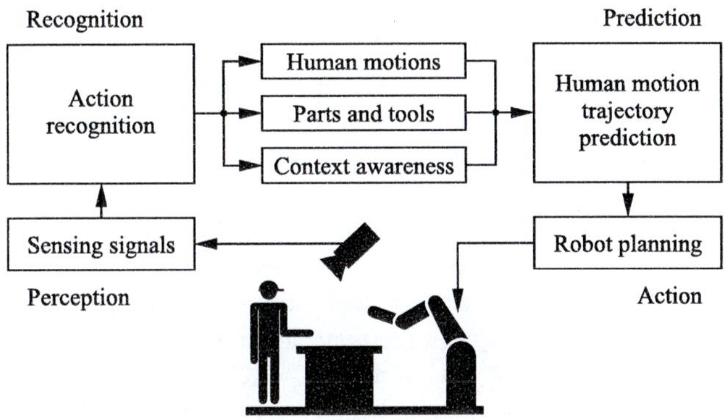

Fig. 11.3 Elements of HRC: perception, recognition, prediction and action

sensors such as vision systems [6] and wearable devices [7], to monitor the manufacturing workspace. They provide in-situ information on the human worker's location and actions, allow them to be *recognised*, and provide the context for the robot to infer the tool and/or part that will be needed by the human work in the next step of the operation [8]. Furthermore, the sequential patterns embedded within the human motion are analysed to establish the basis for *predicting* the trajectory of the human motion in the future, which controls the robot to deliver the tool and/or part to the expected location for handover in a proactive, collaborative manner [9]. Accordingly, human action recognition and motion trajectory prediction have been crucial for establishing an effective HRC system [4].

There have been numerous reported efforts in human action recognition [10–12], where machine learning methods, such as random forest [10], Gaussian mixture models (GMM) [11] and neural networks [12] have been investigated with a recognition accuracy of 70–80% having been achieved. Based on the understanding of the recognised actions, the worker's future action can be inferred for robot planning [13–15]. Hidden Markov model [13], Bayes network [14], Bag-of-words [15], etc., have been used to set up statistical models for this purpose. However, due to the relatively low accuracy in action recognition, the accuracy of future action inference rarely exceeds 80%.

Recently, deep learning, which builds upon artificial neural networks, has emerged as a new machine learning branch with improved capability in discovering and learning complex data patterns [16, 17]. Compared with traditional machine learning techniques, the structure of deep learning contains hierarchically arranged layers to enable multi-level extraction of features embedded in data and allow to amplify the ones that are most relevant to the task (e.g., human action recognition) [16]. In addition, various deep learning architectures have been designed to analyse different data types, such as image and time series. For example, the convolutional neural network (CNN) is inspired by the human visual system and is designed for image processing [16]. Recent reports have shown that deep learning can outperform human experts in various recognition or strategy-related tasks [18, 19] and it has been successfully demonstrated in several application domains, including additive manufacturing and machine fault diagnosis [20, 21]. In HRC, deep learning has been shown to help the robot to improve the recognition accuracy of human commands by analysing the motion of the human worker [22].

One limitation associated with these prior studies on human action recognition is that the collaboration context has not been incorporated. Context awareness involves the identification of the objects the worker is handling, the sequential order of performing actions, and the workspace environment. As significant variability and heterogeneities may exist among workers when performing the same task, the understanding of collaboration context can facilitate action recognition, and is, therefore, of crucial importance for establishing a reliable HRC system.

Human motion trajectory prediction, which forecasts the probable location of the human worker in the next step of the operation, has also attracted much research interest [4]. As a deep learning architecture specialised in sequential pattern analysis [23], recurrent neural networks (RNN) have already demonstrated effectiveness in a variety of manufacturing applications, such as prognosis of structural defect propagation and of tool wear [24, 25]. For human trajectory prediction, an encoder–decoder RNN has been reported in [26], where the encoder analyses the past trajectory and the decoder progressively predicts the future path. The structure was further enhanced in [27] with residual connections that exploit the first-order derivatives to minimise jumps in the predicted trajectory. In [28], a spatial–temporal RNN has been developed to allow simultaneous prediction of a large number of time steps. The training is guided by a long-term temporal objective function, which is designed to learn the variance embedded in human motion. Other researchers have focused on the workspace shared by humans and objects, such as tables and ladders, and allowed their interactions to be explicitly considered in RNN design for improved prediction accuracy [29, 30]. In addition, RNN has also been investigated to simulate realistic human motion [31].

One limitation associated with these studies is that the interactive patterns of human body parts, when carrying out certain actions that can potentially help to improve prediction accuracy, have not been explored. Furthermore, the uncertainty associated with human motion, which is critical to ensure the reliability and safety in HRC, has not been accounted for.

To fill these research gaps, a context-aware deep convolutional neural network (DCNN)-based method for human action recognition and an enhanced RNN model for human motion trajectory prediction have been developed. First, a well-tested DCNN structure, *AlexNet*, has been adapted through transfer learning for the task of human action recognition. In addition, a two-step approach has been designed to allow the network to individually focus on the human action as well as the associated context to improve the recognition performance. Second, leveraging the capability of RNN in sequential pattern analysis, functional units have been incorporated into an RNN-based model to account for the interactions among human body parts for improved motion trajectory prediction. Uncertainty quantification in human motion is carried out via the Monte Carlo (MC) method, which allows to minimise mis-triggering of the robot and enables more reliable and robust collaboration. The methods developed are experimentally evaluated, with the collaborative assembly of an automotive engine as a case study.

11.2 Deep Learning for Action Recognition and Motion Prediction

The developed deep-learning-based methods for action recognition and motion trajectory prediction build upon the theory of conditional probability distribution. Specifically, it refers to the conditional probability distribution of the on-going human action given sensing data, as well as that of the future human body location given its current and past trajectory. The approximation of these two conditional probability distributions using DCNN and RNN constitutes the main task.

11.2.1 DCNN for Human Action Recognition

The effectiveness of DCNN for image processing lies in its ability to distil the relevant information embedded in input image into a high-level feature vector to achieve complex pattern recognition, which constitutes the main motivation for developing DCNN-based human action recognition from video frame images.

11.2.1.1 Fundamentals of DCNN

The main structure of DCNN consists of hierarchically arranged convolutional layers. Given an input image, features are extracted through kernel-based convolution in the convolutional layers, and are organised as feature maps. Multiple kernels can be utilised to extract features of different aspects. Mathematically, to generate N feature maps in the lth convolutional layer, the kernel-based convolution is expressed as

$$\mathbf{x}_j^l = \phi\left(\sum_{i=1}^{M} \mathbf{x}_i^{l-1} * \mathbf{k}_{ij}^l + \mathbf{b}_j^l\right), \quad j = 1, \ldots, N, \tag{11.1}$$

where $\mathbf{x}_j^l$ represents the jth feature map in the lth layer, $\mathbf{x}_i^{l-1}$ represents the ith feature in the preceding, $(l-1)$th layer, $\mathbf{k}_{ij}^l$ denotes the kernel weights corresponding to the jth kernel in the lth layer and the ith feature in the $(l-1)$th layer, $\mathbf{b}_j^l$ is the corresponding bias term, M is the kernel size, $*$ is the convolution operation and ϕ denotes a non-linear activation function. The kernel weights are either trained or pre-defined to detect specific image features. It is seen that the feature extraction process in DCNN at each convolutional layer is based on a weighted sum of features from the preceding layer that passes through a non-linear transformation. As the information of the input image is propagated through the layers, multi-level image representations are extracted, reflected as features from low-level (e.g., curve and edge) to high-level (e.g., domain-specific motif and object), which is critical to achieve accurate image recognition and classification [16].

After each convolutional layer, a pooling layer is commonly appended to fuse the features within local regions of the feature maps by condensing them into corresponding average or maximum values. The pooling process allows to reduce the dimension of the feature maps in the subsequent convolutional layer and improves computational efficiency. Furthermore, pooling alleviates the overfitting issue, which refers to the phenomenon that the performance of the neural network, e.g., human action recognition accuracy, at the training stage far exceeds that at the testing stage. This is mainly caused by the fact that network parameters (e.g., weights) are over-sensitive to small variations in training data (e.g., due to noise) and consequently, cannot effectively capture the underlying data pattern. With pooling, the fused features become less sensitive to these variations since their effects are averaged out [16]. Pooling also allows to analyse the relative locations of features within the image frame instead of the absolute locations, and therefore, making the method robust to image translation.

Taking the highest-level features from the final convolutional layer as inputs, the fully connected layers perform image classification. Specifically, the output of the fully connected layers is a high-level feature vector that represents a conditional probability distribution of the input image class (e.g., human action type in HRC) given the corresponding input image, and the final image classification is carried out by sampling from this distribution. The concept of DCNN-based image classification is illustrated in Fig. 11.4.

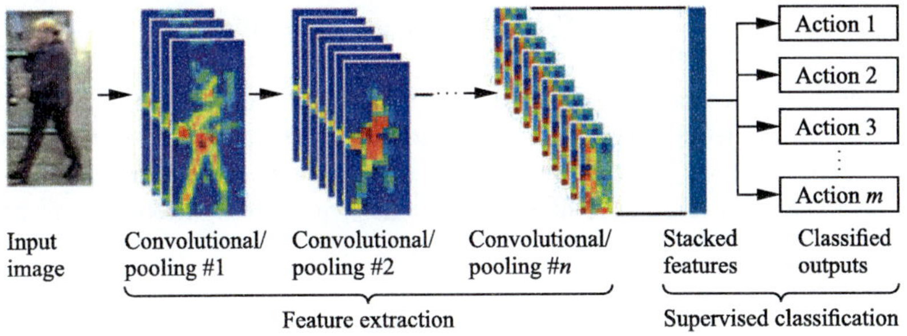

Fig. 11.4 DCNN-based image classification

The structure of DCNN (e.g., number of layers and number of feature maps in each layer) determines the extent to which the input image will be decomposed to analyse the underlying patterns. Therefore, the design of the DCNN structure is critical to ensure network performance. To this aim, rather than designing a network from scratch, a well-tested DCNN structure, AlexNet, is pre-trained as the base network upon which transfer learning is investigated to adapt its feature extraction capability to the domain of human action recognition.

11.2.1.2 AlexNet

AlexNet is developed by Krizhevsky et al. [32] and it has won the ImageNet Large Scale Visual Recognition Challenge (LSVRC) in 2012, which consists of classifying 1.2 million images into 1,000 different classes. Specifically, it has reduced the classification error from 25.8% of the preceding year to 16.4%, representing a 36% performance improvement.

The structure of AlexNet consists of five convolutional layers, three pooling layers and three fully connected layers. In total, there are 650,000 neurons and over 60 million trainable network parameters in the network. This means that a significant amount of computational resources would be needed to retrain this network from scratch for human action recognition. For example, the first convolutional layer contains 96 feature maps with each having a dimension of 55×55, resulting in $55 \times 55 \times 96 = 290{,}400$ neurons. Each neuron is generated from an $11 \times 11 \times 3$ kernel, with "3" corresponding to the number of colour channels in RGB image). This leads to 105,705,600 trainable parameters in the first convolutional layer alone. A more computationally efficient strategy is to leverage the feature extraction capability of a publicly available AlexNet that has been pre-trained on a large-scale dataset (e.g., *ImageNet* [33]) as the base network and adapt it for the task of recognising human action-related images. To this aim, transfer learning is investigated [34].

11.2.1.3 Transfer-Learning-Enabled Network Adaptation

The underlying mechanism that enables the transfer of AlexNet and its adaptation for human action recognition is that the features extracted from the early layers in the neural network are generally more generic as compared to those generated in the later layers, and therefore, are likely to be transferred across different application domains [35]. For example, features extracted earlier are limited to low-level, such as edges and curves, which are shared among images from different domains. On the other hand, when they are assembled into high-level features in the later layers, they become more specific to the categories of a certain domain. In addition, transfer learning further alleviates the issue of overfitting, as the convolutional layers have been well pre-trained and tested on a very large ImageNet dataset.

The transfer of AlexNet involves fine-tuning of all pre-trained weights in convolutional layers to maintain their feature extraction capability and retraining of weights in three fully connected layers to complete the domain adaptation [36]. Specifically, fine-tuning refers to the adjustment of network weights from the values retained at the conclusion of pre-training. As the pre-training dataset ImageNet contains image categories of human activities and tools that are relevant to the task of human action recognition in HRC, the feature extraction capability is expected to be well-established after pre-training and only minor adjustment is expected. Retraining refers to the adjustment of network weights from scratch (i.e., random initialisation). This is due to the fact that the number of output classes varies significantly from ImageNet (1000) to HRC (<10), which requires major changes to the weights in fully connected layers for domain adaptation. The complete transfer process is illustrated in Fig. 11.5 [36].

Fine-tuning of convolutional layers, as well as retraining of fully connected layers, are realised through a backpropagation algorithm [37]. The objective is to minimise human action recognition error $E \stackrel{\text{def}}{=} \left(\mathbf{y}_{\text{predicted}} - \mathbf{y}_{\text{actual}}\right)^2$, where $\mathbf{y}$ denote the classified categories of input images. Guided by this error, network weights are iteratively updated by calculating its gradients w.r.t. the weights:

$$k_{ij,new}^{l} = k_{ij,old}^{l} + \eta\left(\frac{\partial E}{\partial x_j^l}\frac{\partial x_j^l}{\partial k_{ij,old}^l}\right), \tag{11.2}$$

where η denotes learning rate (i.e., step size of each weight update). To prevent oscillation in weight adjustment, a small η is selected for fine-tuning, while a large η is used for retraining. To ensure the convergence of the training process, η is also set to decay gradually. To minimise overfitting, "dropout" is employed to randomly drop neurons at each training iteration [38]. This prevents the network from heavily relying on certain subsets of neurons for action recognition.

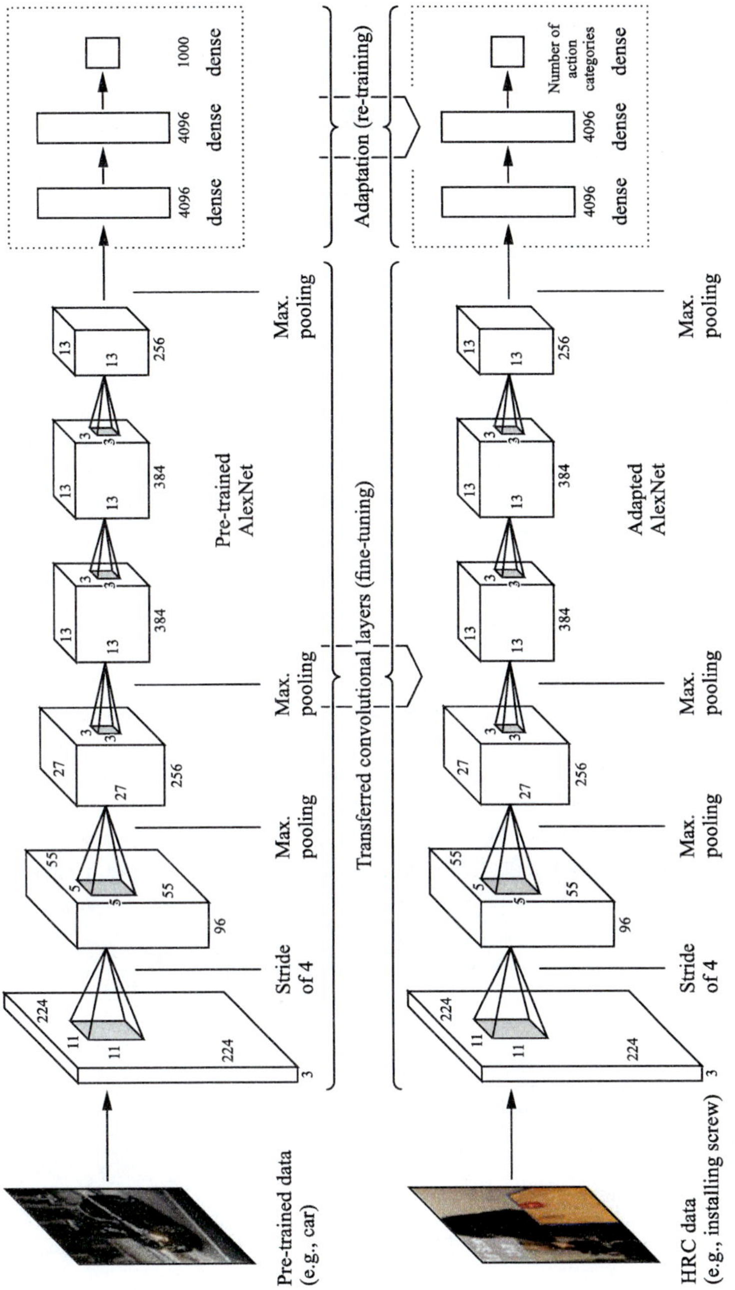

Fig. 11.5 Transfer of AlexNet for human action recognition, adapted from [36]

11.2.2 RNN for Human Motion Trajectory Prediction

While the human action *recognition* answers the question "what is the worker doing now?" and serves as the basis for the robot to infer the tool and/or part to be handled in the next step of operation, human motion trajectory *prediction* answers the question "where will the worker move to next?", which allows the robot to deliver the tool and/or part to the expected location for handover in a proactive manner. The key to accurate motion trajectory prediction is to capture the evolution patterns of human poses that are represented by body joints. RNNs provide the technical foundation for analysing such patterns.

11.2.2.1 RNN for HRC in Assembly

RNNs capture motion evolution patterns by analysing the influence of motion state at each of the time steps on the state at the subsequent step. Motion state can be considered a high-level feature of the motion pattern condensed from raw observations of human poses. The state at time step n, $\mathbf{h}_n$, depends on both the associated observation $\boldsymbol{x}_n$ and the preceding state $\mathbf{h}_{n-1}$, expressed as:

$$\mathbf{h}_n = \varphi(\mathbf{W}_{h-h}\mathbf{h}_{n-1} + \mathbf{W}_{x-h}\mathbf{x}_n + \mathbf{b}), \tag{11.3}$$

where $\mathbf{W}_{h-h}$ denotes the weights representing the influence of the preceding state on the current state, $\mathbf{W}_{x-h}$ represents the weights connecting the observation to the state, $\mathbf{b}$ is the bias, and φ is a non-linear function. As $\mathbf{h}_{n-1}$ also depends on its preceding state, the generation of $\mathbf{h}_n$ makes use of all the available observations up to time n, as shown in Fig. 11.6. The motion evolution pattern is captured via network weights.

Standard RNN structure can be ineffective in motion prediction due to the interaction among different body parts (e.g., arm, leg, and spine). As a solution, two types

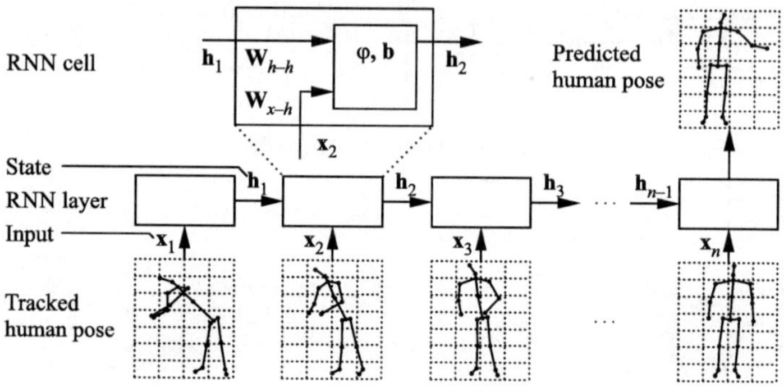

Fig. 11.6 Overview of RNN for motion pattern analysis and prediction [39]

of functional units have been added to the network structure: component and coordination. The component unit analyses the trajectory of a specific body part associated with a certain human pose, and the coordination unit analyses the interaction among body parts. A total of five component (for two arms, two legs and one spine) and four coordination (for arm–arm, arm–spine, leg–leg, leg–spine) units were added to model a complete human body [39].

The network structure of each functional unit is based on the RNN structure in Fig. 11.6, where the input to each component unit is the sequence of past observations of related body joints locations in x–y–z coordinates. For example, for the "arm" unit, the related body joints are wrist, elbow and shoulder. In addition, the output from the coordination units that are associated with the corresponding body parts is also used as input to the component unit. For example, the "arm" unit takes as input the output from the "arm–arm" and "arm–spine" units. Similarly, the input to each coordination unit is the concatenation of body joints locations associated with the interacting body parts. As an example, the input to the "arm–spine" unit is the concatenation of the information on the joints from both the arm and the spine. The RNN structure with functional units is trained in a collective manner by the backpropagation algorithm. In Fig. 11.7, a sample structure of the enhanced RNN is illustrated [39].

Such an enhanced network structure allows for a more accurate prediction of the human motion trajectory in the context of both its own history and the interaction among various body parts, resulting in reduced prediction error. Specifically, the mean deviation between the predicted and actual body joints locations has been reduced by nearly 40% when compared to the RNN without the functional units.

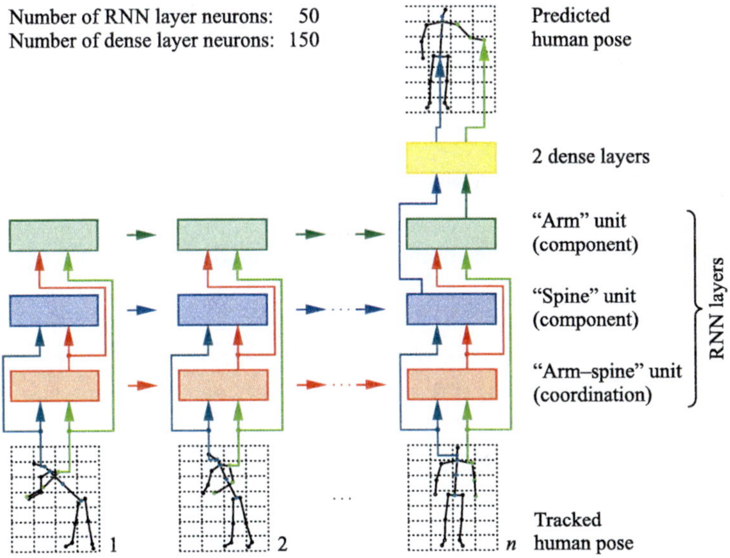

Fig. 11.7 Enhanced RNN for improved accuracy in motion prediction (three functional units shown: arm, spine and arm-spine) [39]

11.2.2.2 Monte Carlo Dropout for Uncertainty Quantification

Human motion exhibits uncertainties. For example, when a worker moves an arm slightly, it is difficult to determine whether the person intends to extend the arm, or it is just the natural movement while maintaining a pose. On the other hand, if the arm moves further away, it becomes less uncertain that the human's intention must be to extend the arm. To account for such uncertainty, probabilistic inference has been incorporated into the motion prediction algorithm, which is formulated as determining the conditional probability distribution $p(\mathbf{y}|\mathbf{X}_n)$, where $\mathbf{X}_n = \mathbf{x}_1, \mathbf{x}_2, \ldots, \mathbf{x}_n$ represents the evolving human pose up to time n, and $\mathbf{y}$ is the future human pose.

To enable varying weight selection for estimating $p(\mathbf{y}|\mathbf{X}_n)$ after network training, the method of MC dropout has been investigated where the network weights were selected randomly using the Bernoulli distribution [40]. Accordingly, $p(\mathbf{y}|\mathbf{X}_n)$ is approximated by performing multiple predictions and averaging the result to arrive at the expected future human pose:

$$\mathbb{E}_{p(\mathbf{y}|\mathbf{X}_n)}(\mathbf{y}) \approx \frac{1}{K}\sum_{k=1}^{K} \hat{\mathbf{y}}(\mathbf{X}_n, \mathbf{W}_{\text{dropout}}), \tag{11.4}$$

where $\hat{\mathbf{y}}$ represents the individual predictions, and the network weights after dropout are expressed as

$$\mathbf{W}_{\text{dropout}} = W \cdot \text{diag}([z]), z \sim \text{Bernoulli}(q), \tag{11.5}$$

where q represents the dropout rate. Statistically, such a probabilistic approach to obtaining $p(\mathbf{y}|\mathbf{X}_n)$ has the advantage of improving the accuracy in human pose prediction and reducing the uncertainty-induced mis-triggering of robot action.

11.3 Experimental Studies

To demonstrate the performance of the developed methods for human action recognition and motion trajectory prediction, a case study of car engine assembly has been conducted. The required parts and tools include (Fig. 11.8) a large cover, a cover cap, two wire collectors, four plugs, a screwdriver, and eleven screws [36, 39].

11.3.1 Human Action Recognition

To evaluate the performance of the transferred AlexNet for human action recognition, an engine assembly process was carried out by a human worker and was recorded by a video camera at 25 frames per second (fps). The assembly is presented by the worker picking up one large cover, four plugs from the table, installing each to the engine

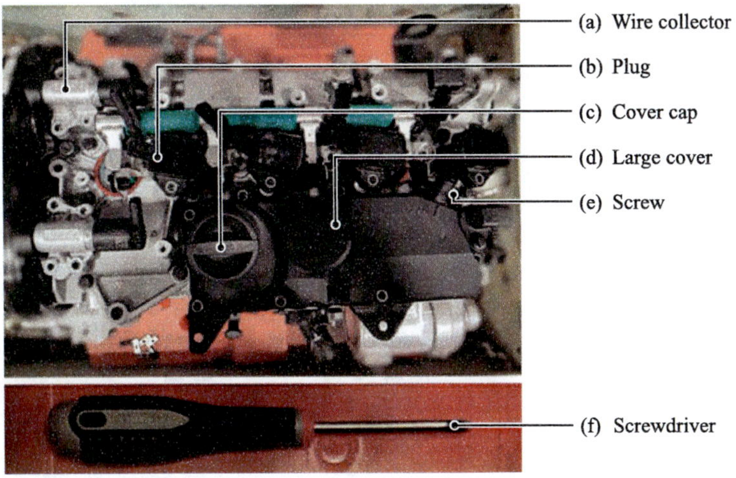

Fig. 11.8 Engine assembly parts and tools [39]

block before picking up a screwdriver to tighten the screws on the installed parts. Considering that human motion associated with certain actions may appear similar, for example, between grasping a part and grasping a tool, the context associated with these actions becomes the key for accurate action recognition. To this aim, a two-step approach has been designed to (1) identify human motion associated with the actions and categorise them into the corresponding categories (i.e., grasping, holding, assembling) by a motion recognition network, and (2) specify from images identified as "holding" which specific tool or part the worker was handling (i.e., context-aware), by a part/tool identification network, as illustrated in Fig. 11.9 [36].

To adapt an AlexNet pre-trained on ImageNet for human motion recognition, images of 9 representative actions in engine assembly were taken to serve as training

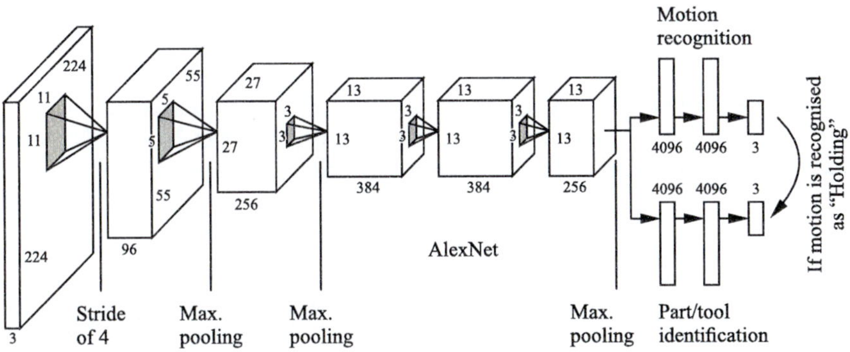

Fig. 11.9 Two-step approach for human action recognition

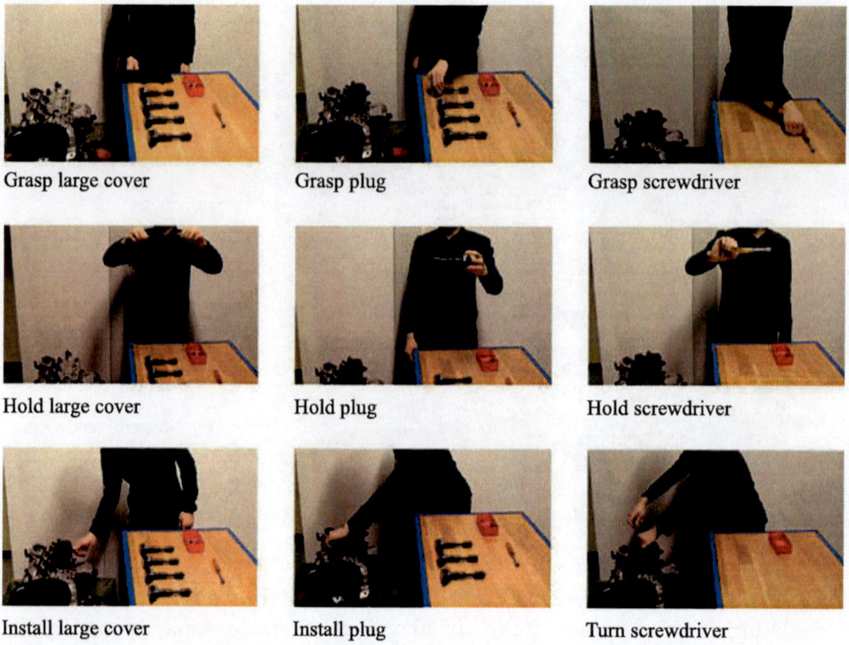

Fig. 11.10 Examples of human action images, adapted from [36]

data, with the sample images illustrated in Fig. 11.10. The action images were classified into three categories: grasping, holding and assembling to train the motion recognition network. For each combination, 10 images were taken. The slight difference in recorded images reflects the variability of human worker in performing the same action. Furthermore, 10 images were taken under different angles for each part and tool, to train the part/tool identification network.

Among all 90 human action images, 80% were used for training and 20% for testing. For the part/tool identification network, all 30 part and tool images (10 images each for large cover, plug and screwdriver) were used for training, while testing images were chosen from those classified as "holding" by the motion recognition network. The selected learning rates during training are shown in Table 11.1. The initial learning rate was set to a relatively small value for the transferred convolutional layers, as these layers have been previously trained on ImageNet dataset and do not require large steps of adjustment as compared to the fully connected layers, which

Table 11.1 Parameter settings related to learning rate

Parameter	Value
Initial learning rate for convolutional and pooling layers	0.001
Initial learning rate for fully connected layers	0.02
Decay rate of learning rate	0.15

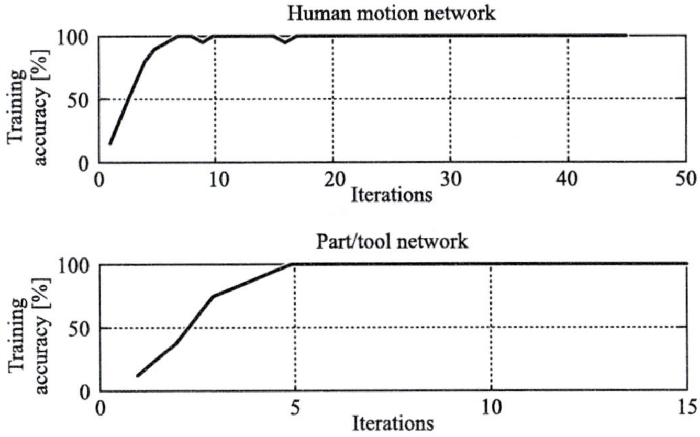

Fig. 11.11 Learning curves of two AlexNet networks [36]

were retrained from scratch. The learning curves of both networks are shown in Fig. 11.11. It is seen that the transferred AlexNet has high learning efficiency with the training accuracy reaching 100% after a few iterations.

Once trained, the two networks were used to process video frame images following the two-step procedure as shown in Fig. 11.9. In total, the recorded video contains 4,485 frames (corresponding to a video length of 180 s). From the available video frame images, one out of every five was used as testing data. It has shown that, for all 897 testing images, only 30 were misclassified, corresponding to an action recognition accuracy of 96.6%.

It is further noted that grasping is typically followed by holding and/or assembling, which is the most logical sequence of performing assembly tasks. This observation establishes the basis for identifying the objects that the worker is handling, such that only the "holding" images are required to be further analysed to identify the object. For example, once the large cover is identified from the video frame as being held by the worker, then it is recognised that the worker works with this cover for the entire period of grasping–holding–assembling. As another example, in a later part of the sequence when both the large cover and all plugs are installed, the recognition of action "holding screwdriver" implies that the worker will turn the screwdriver for the remainder of the assembly operation.

11.3.2 Human Motion Trajectory Prediction

The collaborative assembly workspace for evaluating the performance of RNN-based motion trajectory prediction is shown in Fig. 11.12, where the engine is placed to the right of the worker and a UR5 robot is installed on a workstation to his left. Parts and tools are sorted in colour-coded containers. Human pose is tracked by a

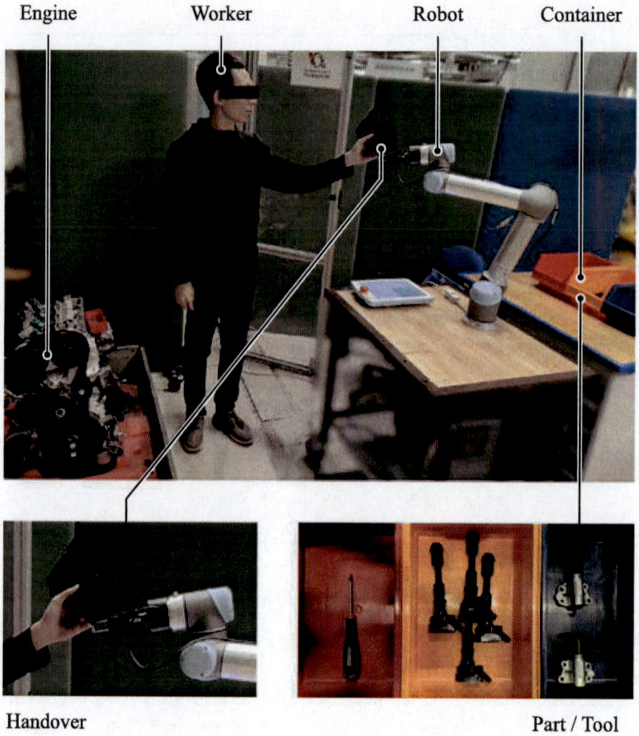

Fig. 11.12 Elements in engine assembly workspace [39]

Kinect sensor at 30 Hz, serving as the basis for real-time motion trajectory prediction. The collaboration between worker and robot is presented by the robot responding to the predicted human motion trajectory, following the human hand in real time as it extends for handover and picking up the relevant part/tool during installation.

The specific objective of predictive modelling of human motion trajectory is to: (1) predict the end location of a motion trajectory at each time step, and (2) evaluate the transition probability to determine if a transition is to occur such that the robot's proactive motion should be triggered. Three pose nodes have been defined: handover (n_1), standing (n_2) and installation (n_3). The entire engine assembly sequence can be represented as a transition graph (Fig. 11.13), where all motion trajectories start from and end at one of these nodes. For example, upon completing handover by taking the screwdriver from the robot, the worker may briefly pause by standing or move directly to installation. For flexibility, the duration of stay at each node is unspecified (i.e., the worker can stay in any node for as long as needed).

To train the RNN network, motion trajectories of the worker from 10 assembly operations were recorded by a Kinect sensor. Relative coordinates are used to account for the variations in the human position relative to the Kinect. For a tracked joint (e.g., wrist), its x–y–z coordinates are evaluated relative to its parent joint (e.g.,

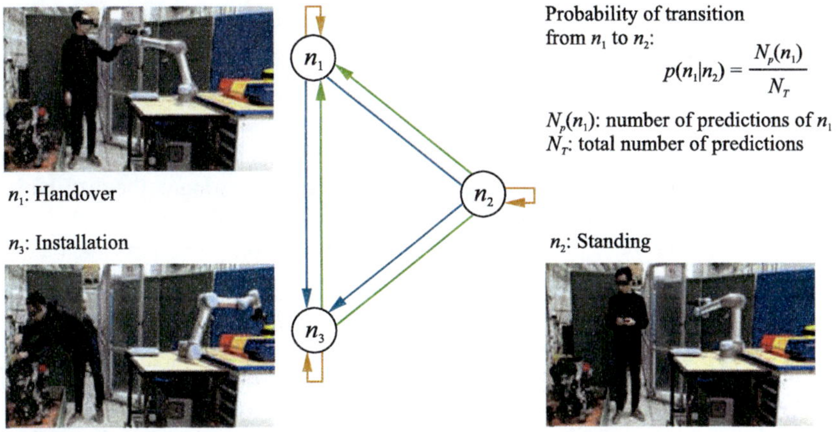

Fig. 11.13 Transition graph between handover, standing and installation [39]

elbow). The network input length (i.e., number of previous time steps to analyse at each frame) is set to 30 frames, corresponding to 1 s of motion. The network output, which represents the end location of the trajectory, is sampled from all poses during the incoming stay at that node. For example, when the worker is moving from handover to installation, the network output is sampled from all poses during the incoming period of installation. This strategy allows the network to explore more possible locations and enhance its robustness. The dataset consists of 8,206 training samples and is randomly split into training and validation sets, with a ratio of 7:3. The mean squared error (MSE) is used as the loss function during training to quantify the mean deviation between the predicted and actual human body joints locations. The network parameters are determined through cross validation.

The benefit of the enhanced RNN structure with two types of additional functional units is comparatively evaluated against an RNN without the functional units and a multi-layer perceptron (MLP). As shown in Fig. 11.14, the enhanced RNN has the

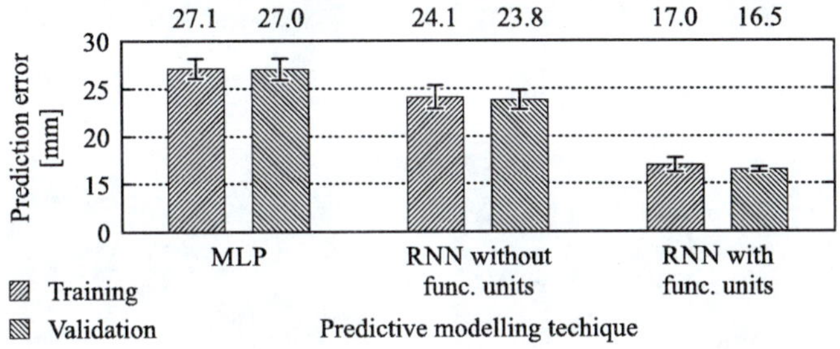

Fig. 11.14 Comparison of prediction error of different network structures [39]

lowest error of 16.5 mm for motion trajectory end location prediction. This corresponds to an error reduction of 44% and 63% from the RNN without the functional units and the MLP, respectively.

The trained network is then evaluated in real-time assembly operations, where the end location of the human body is continuously predicted by the model, as soon as new data from the Kinect sensor is provided. Through MC dropout, the distribution of predicted future locations of the individual body joint is obtained, and the mean value serves as the final predicted location. The transition probability denoted as $p(n_i|n_j)$, $i, j = 1, 2, 3$ and shown in Fig. 11.13, determines if a transition is starting to occur or not. To minimise robot mis-triggering and ensure reliable HRC, the transition is determined to start only when the condition of $p(n_i|n_j) = 1$ is met. Physically, this means that the worker is starting, with 100% certainty, the transition from the action node n_i (e.g., installing) to the action node n_j (e.g., handover). A total of five predictions were made at each time step using MC dropout. The node transition probability is evaluated as the ratio of the number of predictions that belong to the target node to the total number of predictions (i.e., five). The transition is predicted to take place only if all the five results are consistent (i.e., 100% transition probability). Part/tool pickup is triggered when the actual human pose is in the installation node, and handover is triggered when the transition to the handover node occurs from the installation or standing node. As the arm moves and the prediction of the end location of the human hand is updated, the robot also adjusts the target location of its end-effector accordingly.

The collaborative assembly based on motion trajectory prediction is illustrated in Fig. 11.15, using the pickup and handover of a screwdriver as an example. The actual human pose tracked by Kinect is shown in red and its predicted end location in cyan, with the worker's left arm highlighted in thicker lines. The actual and target locations of the robot are in pink and yellow, respectively.

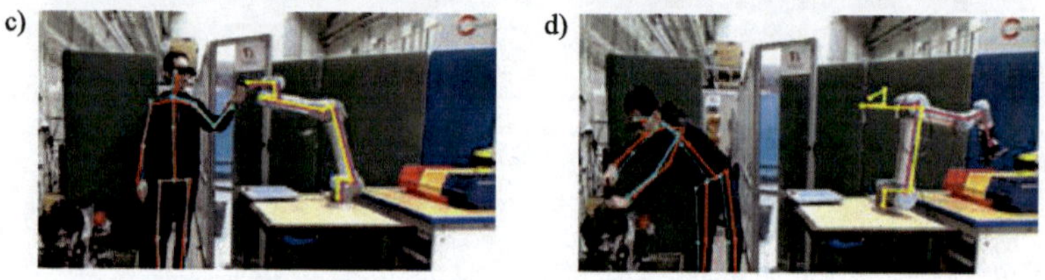

Fig. 11.15 Sample frames from HRC assembly sequence [39]

Fig. 11.16 Uncertainty estimation during a transition sequence [39]

At step (1) the robot is triggered to pick up the screwdriver when the worker is detected as installing the large cover. As the worker completes the installation and starts to extend the left arm in (2), the prediction changes to the handover node. The robot is triggered, and its first target location for handover is generated. As more trajectory information is provided to the network, prediction accuracy improves, as seen in the reduced distance between the prediction and the actual end location of the hand until the arm stops for handover in (3). In (4), the worker returns to the engine and the robot is triggered to pick up the next part.

The effectiveness of uncertainty handling in motion prediction is further evaluated. Figure 11.16 illustrates three nearby time steps around the transition from installation to handover. All five predicted scenarios of the worker's left arm using MC dropout are shown. In (a) the worker is installing a plug and all predictions indicate that he will continue with the installation. In (b), one of the predictions indicates that the worker is moving to handover. With remaining predictions indicating a standing pose, the uncertainty is large (as the transition probability is only 20%), therefore, the robot is not triggered. In (c), all predictions show that the worker is moving for handover, thus the robot will be triggered. It is noted that these results are consistent with human judgement. For example, in (b), the worker is retrieving the arm from the engine. At this moment, it is difficult to determine whether the movement to handover will indeed occur or the worker will stop at a standing position. Correspondingly, large uncertainty is given by the model. As the arm moves further away in (c), the ambiguity is reduced.

It is seen that MC dropout provides an effective measure for uncertainty handling and mis-trigger prevention, allowing for robust HRC. In comparison, if only one prediction is made at each frame without MC dropout, the robot has shown to be mis-triggered one out of four times.

With the two types of functional units added to the RNN structure, it takes ~ 2–3 times as long for the algorithm to execute one complete prediction cycle, from 0.01 s to ~ 0.02–0.03 s. This, however, does not affect the real-time capability when using a Kinect with 30 Hz framerate (or 0.033 s processing time).

11.4 Conclusions and Future Work

This chapter presents deep learning-based techniques for human action recognition and motion trajectory prediction, which are of central importance in enabling effective human–robot collaboration for advanced manufacturing. A context-aware deep convolutional neural network has been developed to account for the variability and heterogeneity of human worker for accurate and robust inference of human actions. A functional unit-incorporated recurrent neural network has been developed to parse the worker's motion pattern and forecast the worker's future motion trajectory. The RNN developed accounts for the coordination between human body parts and allows to quantify motion uncertainty to ensure reliable realisation of HRC.

A case study of automobile engine assembly has been performed to evaluate the performance of the developed methods. For context-aware action recognition, a two-step approach has been designed to allow the DCNN to analyse individually the human action and the associated object that is being handled based on video footages of the workspace. For motion trajectory prediction, human pose information was sequentially analysed by the RNN. Using prediction error as a performance indicator, the methods developed have shown to be able to accurately and reliably determine the on-going human action with a recognition error of 3.4%, and predict worker's future trajectory with a prediction error of 16.5 mm, a 40% improvement over standard RNN.

To further advance the state of the research in HRC, the authors will continue to work in the following areas as future directions:

- Integrated method of human action recognition and prediction, to account for the variability in sequential order when the worker carries out assembly tasks,
- Parallel execution of motion trajectory prediction to enable simultaneous execution of functional units and further improve computational efficiency,
- Adaptive strategy in motion trajectory prediction that allows different functional units to be prioritised for different tasks, since not all human body parts are critical to trigger robot actions during a given assembly operation.

Acknowledgements Gao, Wang and Zhang gratefully acknowledge the funding provided for this research by the US National Science Foundation under the award CMMI-1830295.

References

1. M.R. Pedersen, L. Nalpantidis, R.S. Andersen, C. Schou, S. Bøgh, V. Krüger, O. Madsen, Robot skills for manufacturing: From concept to industrial deployment. Robotics and Computer-Integrated Manufacturing **37**, 282–291 (2016)
2. J. Krüger, T. K. Lien and A. Verl, "Cooperation of human and machines in assembly lines," *CIRP Annals—Manufacturing Technology*, **58**(2), pp. 628–646 (2009)

3. A. Cherubini, R. Passama, A. Crosnier, A. Lasnier, P. Fraisse, Collaborative manufacturing with physical human–robot interaction. Robotics and Computer-Integrated Manufacturing **40**, 1–13 (2016)
4. L. Wang, R. Gao, J. Váncza, J. Krüger, X. V. Wang, S. Makris and G. Chryssolouris, "Symbiotic human–robot collaborative assembly," *CIRP Annals—Manufacturing Technology*, **68**(2), pp. 701–726 (2019)
5. A. Mohammed, B. Schmidt, L. Wang, Active collision avoidance for human–robot collaboration driven by vision sensors. Int. J. Comput. Integr. Manuf. **30**(9), 970–980 (2017)
6. L. Pérez, Í. Rodríguez N. Rodríguez, R. Usamentiaga and D. F. García, "Robot guidance using machine vision techniques in industrial environments: A comparative review," *Sensors (Switzerland)*, **16**(3), p. 335, 2016.
7. P. Kumari, L. Mathew, P. Syal, Increasing trend of wearables and multimodal interface for human activity monitoring: A review. Biosens. Bioelectron. **90**, 298–307 (2017)
8. H. Liu, L. Wang, Human motion prediction for human–robot collaboration. J. Manuf. Syst. **44**, 287–294 (2017)
9. G. Maeda, M. Ewerton, G. Neumann, R. Lioutikov, J. Peters, Phase estimation for fast action recognition and trajectory generation in human–robot collaboration. Int. J. Robot. Res. **36**(13–14), 1579–1594 (2017)
10. A. Roitberg, A. Perzylo, N. Somani, M. Giuliani, M. Rickert and A. Knoll, "Human activity recognition in the context of industrial human–robot interaction," *2014 Asia-Pacific Signal and Information Processing Association Annual Summit and Conference*, 2014.
11. C. Perez-D'Arpino and J. A. Shah, "Fast target prediction of human reaching motion for cooperative human–robot manipulation tasks using time series classification," *Proceedings—IEEE International Conference on Robotics and Automation*, pp. 6175–6182, 2015.
12. H. Hasan, S. Abdul-Kareem, Static hand gesture recognition using neural networks. Artif. Intell. Rev. **41**(2), 147–181 (2014)
13. K. Li, Y. Fu, Prediction of human activity by discovering temporal sequence patterns. IEEE Trans. Pattern Anal. Mach. Intell. **36**(8), 1644–1657 (2014)
14. K. P. Hawkins, N. Vo, S. Bansal and A. F. Bobick, "Probabilistic human action prediction and wait-sensitive planning for responsive human–robot collaboration," *IEEE-RAS International Conference on Humanoid Robots*, 2015.
15. M. S. Ryoo, "Human activity prediction: Early recognition of ongoing activities from streaming videos," *Proceedings of IEEE International Conference on Computer Vision*, pp. 1036–1043, 2011.
16. Y. Lecun, Y. Bengio, G. Hinton, Deep learning. Nature **521**(7553), 436–444 (2015)
17. J. Wang, Y. Ma, L. Zhang, R. Gao, D. Wu, Deep learning for smart manufacturing: Methods and applications. J. Manuf. Syst. **48**, 144–156 (2018)
18. D. Silver, J. Schrittwieser, K. Simonyan, I. Antonoglou, A. Huang, A. Guez, T. Hubert, L. Baker, M. Lai, A. Bolton, Y. Chen, T. Lillicrap, F. Hui, L. Sifre, G. Van Den Driessche, T. Graepel, D. Hassabis, Mastering the game of Go without human knowledge. Nature **550**(7676), 354–359 (2017)
19. Y. Sun, X. Wang and X. Tang, "Deep learning face representation from predicting 10,000 classes," *Proceedings of IEEE Conference on Computer Vision and Pattern Recognition*, pp. 1891–1898, 2014.
20. P. Wang, Ananya, R. Yan, and R. Gao, "Virtualization and deep recognition for system fault classification," *Journal of Manufacturing Systems*, **44**, pp. 310–316, 2017.
21. A. Caggiano, J. Zhang, V. Alfieri, F. Caiazzo, R. Gao and R. Teti, "Machine learning-based image processing for on-line defect recognition in additive manufacturing," *CIRP Annals—Manufacturing Technology*, **68**(1), pp. 451–454, 2019.
22. H. Liu, T. Fang, T. Zhou, L. Wang, Towards Robust Human–Robot Collaborative Manufacturing: Multimodal Fusion. IEEE Access **6**, 74762–74771 (2018)
23. K. Greff, R.K. Srivastava, J. Koutnik, B.R. Steunebrink, J. Schmidhuber, LSTM: A Search Space Odyssey. IEEE Transactions on Neural Networks and Learning Systems **28**(10), 2222–2232 (2017)

24. A. Malhi, R. Yan, R. Gao, Prognosis of defect propagation based on recurrent neural networks. IEEE Trans. Instrum. Meas. **60**(3), 703–711 (2011)
25. P. Wang, Z. Liu, R. Gao and Y. Guo, "Heterogeneous data-driven hybrid machine learning for tool condition prognosis," *CIRP Annals—Manufacturing Technology*, **68**(1), pp. 455–458, 2019.
26. J K. Fragkiadaki, S. Levine, P. Felsen and J. Malik, "Recurrent network models for human dynamics," *Proceedings of IEEE International Conference on Computer Vision*, pp. 4346–4354, 2015.
27. J. Martinez, M. J. Black and J. Romero, "On human motion prediction using recurrent neural networks," *Proceedings of IEEE Conference on Computer Vision and Pattern Recognition*, pp. 4674–4683, 2017.
28. H. Wang, E. S. L. Ho, H. P. H. Shum and Z. Zhu, "Spatio-temporal Manifold Learning for Human Motions via Long-horizon Modeling," *IEEE Transactions on Visualization and Computer Graphics*, pp. 1–1, 2019.
29. A. Jain, A. R. Zamir, S. Savarese and A. Saxena, "Structural-RNN: Deep learning on spatio-temporal graphs," *Proceedings of IEEE Conference on Computer Vision and Pattern Recognition*, pp. 5308–5317, 2016.
30. E. Corona, A. Pumarola, G. Alenyà and F. Moreno-Noguer, "Context-aware Human Motion Prediction". arXiv preprint arXiv:1904.03419., 2019.
31. D. Pavllo, D. Grangier and M. Auli, "QuaterNet: A quaternion-based recurrent model for human motion," *British Machine Vision Conference 2018*, BMVC 2018.
32. A. Krizhevsky, I. Sutskever and G. E. Hinton, "ImageNet Classification with Deep Convolutional Neural Networks," *Advances in Neural Information Processing Systems*, pp. 1097–1105, 2012.
33. L. Fei-Fei, J. Deng, K. Li, ImageNet: Constructing a large-scale image database. J. Vis. **9**(8), 1037–1037 (2010)
34. S.J. Pan, Q. Yang, A survey on transfer learning. IEEE Trans. Knowl. Data Eng. **22**(10), 1345–1359 (2010)
35. M. Oquab, L. Bottou, I. Laptev and J. Sivic, "Learning and transferring mid-level image representations using convolutional neural networks," *Proceedings of IEEE Conference on Computer Vision and Pattern Recognition*, pp. 1717–1724, 2014.
36. P. Wang, H. Liu, L. Wang, and R. Gao, "Deep learning-based human motion recognition for predictive context-aware human–robot collaboration," *CIRP Annals—Manufacturing Technology*, **67**(1), pp. 17–20, 2018.
37. D.E. Rumelhart, G.E. Hinton, R.J. Williams, Learning representations by back-propagating errors. Nature **323**(6088), 533–536 (1986)
38. N. Srivastava, G. Hinton, A. Krizhevsky, I. Sutskever, R. Salakhutdinov, Dropout: A simple way to prevent neural networks from overfitting. J. Mach. Learn. Res. **15**, 1929–1958 (2014)
39. J. Zhang, H. Liu, Q. Chang, L. Wang, and R. Gao, "Recurrent neural network for motion trajectory prediction in human–robot collaborative assembly," *CIRP Annals—Manufacturing Technology*, 2020, Accepted.
40. Y. Gal and Z. Ghahramani, "Dropout as a Bayesian approximation: Representing model uncertainty in deep learning," *33rd International Conference on Machine Learning*, **3**, pp. 1651–1660, 2016.

Chapter 12
Programming-Free Approaches for Human–Robot Collaboration in Assembly Tasks

Sharath Chandra Akkaladevi, Matthias Propst, Michael Hofmann, Leopold Hiesmair, Markus Ikeda, Naresh Chowdary Chitturi, and Andreas Pichler

Abstract Collaborative robots are paving the way for new applications, and are becoming the new frontiers, especially in manufacturing industries. The concept of combining the cognitive flexibility of humans and the physical strength of the robots is rapidly gaining interest and is leading to novel application areas. In this chapter, we introduce various challenges faced by collaborative robotic systems. The focus is on the recent development of programming interfaces to efficiently install collaborative robotic systems in manufacturing applications and the requirement for such interfaces. An extensive review on the state-of-the-art programming approaches is given. Detailed description of novel interfaces that build upon and extend the current state of the art is presented. A qualitative comparison against different criteria required to consider while developing programming interfaces is carried out in detail. The main objective of the organisation and content of the chapter is to introduce different programming interfaces and paradigms that can be exploited for easy programming of robots by non-experts.

12.1 Introduction

The paradigm shift from mass production to mass customisation is driving production systems to handle more product variations, tighter life cycles and smaller batch sizes. Robotics will be one of the main enablers of this transition to the transformable factory of tomorrow. With recent developments, easy-to-use collaborative robots are gaining new foothold and are becoming new frontiers in industrial robotics. Collaborative

S. C. Akkaladevi (✉) · M. Propst · M. Hofmann · L. Hiesmair · M. Ikeda · N. C. Chitturi · A. Pichler
Department of Robotics and Automation Systems, Profactor GmbH, Steyr-Gleink, Austria
e-mail: sharath.akkaladevi@profactor.at

S. C. Akkaladevi
Institute of Networked and Embedded Systems, Alpen-Adria Universität Klagenfurt, Klagenfurt, Austria

L. Wang et al. (eds.), *Advanced Human–Robot Collaboration in Manufacturing*,
https://doi.org/10.1007/978-3-030-69178-3_12

robotic systems allow the combination of the cognitive flexibility of humans with the physical strength and repeatability of robots to achieve high productivity.

However, to achieve efficient human–robot collaboration, several challenges need to be addressed [1], including:

- *User interfaces.* To easily install robots and efficiently collaborate with them, the users should be able to easily interact with them and easily program them for various tasks. When considering manufacturing applications, the users are process experts with limited or no knowledge of robot programming. The interfaces, therefore, should allow the users to easily and intuitively program the robots.
- *Safety.* Safety issues are the primary challenge that must be tackled by any approach implementing collaboration between humans and robots. Indeed, being the intrinsic aim of HRC to allow direct contact between them by eliminating fences, this must be achieved in a safe manner.
- *Intelligence and Autonomy.* Collaborative robots in the industry should not be just regarded as tools that relieve workers of physical fatigue and enhance their capabilities. Therefore, it is very important to endow robots with proper cognitive processing skills and shared autonomy capabilities, so that they can take over some tasks, thus relieving human operators of cognitive effort, especially in complex tasks and scenarios.

Collaborative robotics has been an active field of research due to its potential applications. However, the topic of reducing the programming effort required by an expert using natural modes of communication is still open. In this chapter, the *focus is on presenting different user interfaces* that allow easy programming of complex robotic tasks for non-programming experts.

In literature, different methods for developing agents, which can learn from a human instructor, are described. Common to all methods is the reduction of programming effort. However, in spite of an active interest of the research community, the challenge of easy and intuitive teaching of robots by non-experts is unsolved. Some aspects of this problem are:

- Ease of use for non-programming experts to configure and program robots.
- Ability to handle tasks of high complexity, e.g., industrial assembly.
- Development of multimodal interfaces to capture relevant process knowledge from human users and apply it to the robotic systems.
- The effort required to apply these approaches to teach complex assembly processes (AP) in industrial scenarios is quite high and, in some cases, requires complete remodelling. The main reason is due to the level of abstraction required and the difficulty in adapting to novel situations.
- Moreover, depending on the quality of the task learnt by the robot, the user may decide to re-teach the task or provide feedback to the robot to increase the performance of learning.

This chapter contributes the following:

- Description of different programming paradigms with varying complexity and their applicability to teaching assembly tasks to robots;
- Experimental results of the different programming paradigms that teach complex tasks (assembly) to robots at varying levels of abstraction; and
- Qualitative comparison against state of the art (commercially available) solutions that teach complex tasks to robots.

The objective of this chapter is to describe and compare different programming paradigms for interacting with robots in manufacturing industries. To achieve this objective, a brief review of the state-of-the-art programming approaches is given in Sect. 12.2. In Sect. 12.3, the work of the chapter's authors in developing different programming frameworks and related programming paradigms is explained in detail including the architecture, methodology and experimental results. A qualitative comparison and corresponding evaluation are explained in Sect. 12.4. Finally, in Sect. 12.5, some concluding remarks are provided.

12.2 State of the Art

In the research community, there is a growing interest to solve the challenge of robots learning in complex real-world environments [2–5]. In literature, different methods are described for developing agents, which can learn activities from a human instructor. Common to all methods is the reduction of programming effort. However, the problem of reducing the programming effort required by an expert using natural modes of communication is still an open issue [6]. When viewed from a broader perspective, learning approaches can be subdivided as follows:

- Learning by advice,
- Learning by programming,
- Learning by demonstration, and
- Learning by interaction.

Learning by advice: This is made possible by the use of natural forms of communication. The authors in [7] developed a method whereby advice is given to the learning agent (Reinforcement Learning). In the context of Markov decision-processes (MDPs), advice means that the agent is suggested to execute an action if a particular condition applies. The use of Natural-Language to communicate this advice is particularly effective for non-experts. Such an interface was implemented by [8, 9]. The detection of Natural-Language data has not yet been solved. Therefore, many approaches require that advice be coded in the syntax of scripting or programming languages. This makes access for non-experts more difficult.

Learning by programming: The approaches falling under this category use the task-level programming paradigm for easy and quick reprogramming by non-experts. The task-level programming paradigm [10] is built upon a set of actions, where the actions have the capability to alter the current world state. These set of actions can be seen as formal descriptions of compliant robot motions, and are composed of primitives. These primitives are simple, atomic movements and are combined to form a task [11]. The task could consist of a single action in simple cases and have a multitude of actions in complex scenarios. An example of a primitive is typically a sensory input or a single robot motion described using the Task Frame Formalism (TFF) [12]. From the perspective of the robot operation, any given assembly task is (and could be) broken down into a form of action primitives which the robot can interpret [13]. The main prerequisite for using task frame formalism is to model the world state (in terms of action primitives) and maintain it online. The main drawback of this approach is that the modelling complexity increases exponentially as the complexity of the task increases. To alleviate this problem, use of generic recipes [10] is suggested instead of using such action primitives to build up the assembly task. The terms recipe and skill are used analogously in this chapter. When compared to the action primitives, the recipes are abstracted on a higher level and hence form a bridge between complex tasks and the primitives.

There is an active interest in the research community to use TFF and the approaches differ based on how primitives are combined to describe the task. For example, to support hierarchies and concurrencies in the task, a visual programming tool for defining the flow control is proposed in [14]. In [15], an automated task planning and execution system is shown as a sequence of skills and their parameters, based on the desired goal state and the current state of the world model. To deal with variations in the assembly process, a programming framework that uses knowledge-based components is proposed in [16]. To allow portability across different platforms, the work in [17] details an integrated skill-based framework coupled with task planning.

Learning by demonstration (LBD): LBD is a popular method by which agents learn by physically presenting the task through the human being [18]. The work in [2] provides a detailed overview of relevant methods. A disadvantage of LBD is that complicated tasks can only be presented with difficulty (e.g., if several robotic systems are to solve a problem at the same time). A human-in-the-loop adaptation to correct a batch-learnt policy iteratively and improve accuracy and precision is described in [19]. An approach to enable users to generate skills and robot program primitives for later refinement and reuse, is proposed in [20].

Learning by interaction with the environment is another interesting approach. Recent trends point towards applying Reinforcement learning (RL) [21] for this purpose. In these approaches, learning arises from interaction with the environment. The agent (robot) learns how to behave (which actions to perform when) in order to complete a task in the given environment. The agent learning process (using RL) takes place over discrete time steps by interacting with the environment and gaining experience about the outcome. To reach an optimal policy (the set of actions that lead to the maximum reward), substantial interaction with the environment is required. As a result, the RL approach leads to a memory intensive storage of all state-action pairs in case of complex tasks [22]. Slow convergence towards a satisfactory solution is another drawback of the RL approaches. To overcome these problems, recent approaches use a human teacher in the loop to provide feedback (reward) instead of allowing the agent to "aimlessly" interact with the environment [23–26]. The human teacher in the loop provides feedback, which is used by an RL approach to speed up the convergence time required to reach an optimal policy. Such approaches are termed as interactive reinforcement learning (IRL).

The work in [26] argues that a run-time human feedback as a reward to the IRL approach is beneficial for both the human teacher (to understand the perspective of the robot) and the robot (which learns the optimal policy with the help of the learning algorithm). The user study in this work suggests that the human teachers employ feedback as a single communication channel for various communicative intents (feedback, guidance and motivation). Using this as a basis, the work in [25] conducted a study to realise such IRL approach for efficient real-world robotic systems. In this study, the feedback provided by the user is twofold: first, it includes a reward for preceding actions of the robot, and then, guidance is given for subsequent actions. The results show that such an approach reduces the learning time of the robot and that it is more evident when large state-space (number of interactions possible by the robot in a given environment) is considered. Another approach to accelerate the learning process is given in [27].

To train an agent manually via evaluation reinforcement, a framework that uses real-valued feedback from a human trainer about the agent behaviour is described in [28]. Such feedback allows the trainer to build the agent's policy (interactive shaping). This shaping directly modifies the action policy (the selection mechanism) of the IRL algorithm. Instead of using feedback as an indirect influence, the work in [23] uses it to make a direct statement about the action policy. A framework (TAMER) to apply the approach mentioned in [28] for real-world robotic system is given in [24]. In [29, 30] the robot learns both the desired path and the required amount of force to apply on an object during the interaction.

There are not many approaches which attempt applying IRL to real-world scenarios. This is especially true for robot learning in real-world industrial settings. The work in [6, 11] demonstrates the partial application of IRL to practical demonstrations. Their work demonstrates use-case-specific learning and is limited to the number of actual interactions between the human and the robot. Also, a complete human–robot collaborative assembly process with varying number of objects is not considered.

In contrast to the state of the art, work in this chapter describes different methodologies to programming robots that include:

- Easy intuitive programming framework to learn an assembly process using user interactions,
- Applicability to human–robot collaborative assembly tasks with varying complexity,
- Novel user interfaces for interaction with robotic systems, and
- Programming paradigms that combine different learning approaches (as mentioned above) to achieve short programming cycles.

12.3 Programming-Free Approaches

Developing novel interfaces and interactive frameworks for programming robots is of vital importance to efficiently install and interact with robots in manufacturing industries. This section details different programming methodologies that rely on novel interfaces and programming paradigms to allow non-experts to easily program robot tasks. For each approach, the architecture of the approach is detailed first, followed by the methodology that allows easy programming of robots. This is further followed by the experimental results wherever possible, and the corresponding conclusions.

12.3.1 XRob

The *XRob* software framework enables the creation of complex robot applications within a few minutes. It builds on unique, easy-to-use features that significantly speed up commissioning and make the operation more cost-efficient and flexible than common programming methods. The special software architecture allows easy and intuitive creation of processes and configuration of the components of a robot system via a single user interface. Figure 12.1 provides an overview of the software components within the XRob framework. Each major block represents a component with individual submodules that can be activated depending on the application requirements. A detailed description of the components involved and an extended evaluation of the framework is given in [31], additionally to the brief summary below.

The Perception System software component consists of modules to aggregate data from the current state of the workspace environment. These perception systems track

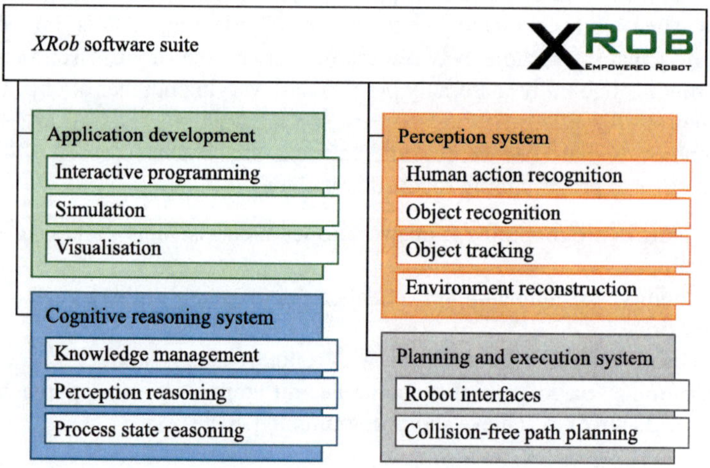

Fig. 12.1 Components and submodules of the XRob software suite [31]

the state of objects and parts, as well as the state of the human worker. Environment reconstruction is used to create a model of the current workspace as input for collision-free robot manipulation planning.

The Planning and Execution system provides modules in order to carry out actions on the robotic target system. XRob supports a variety of robot systems (e.g., UR, KUKA, Stäubli), and provides an open software infrastructure to integrate additional systems. Based on the surrounding environment model, collision-free motion paths are planned online.

Human–Robot Collaboration (HRC), where a human worker and a robotic system perform coordinated actions within the same process, requires a smooth interaction between the robot and the human to stay efficient. This requires a common understanding of the situation and process state between the human and the robots. The component Cognitive Reasoning includes modules (1) to represent process knowledge, (2) to reason on perception data for the derivation of the current state of the environment and (3) to deduce, given a current process state, the next feasible action to be carried out in order to advance in the execution of the process.

The Application Development component is characterised by an intuitive user interface for setting up the sequence of a robotic process. Visualisation of the manipulation paths and objects in the scene is a prerequisite to better understand the robots' "intention". The XRob execution system can be connected to robot vendor specific simulation environments, mimicking a real robot system. This enables early process validation in simulation before real execution.

A cognitive reasoning system is used to enable smooth and intuitive collaboration between robot and human, during coordinated actions within the same process. The subordinate modules provide functionalities [32] to (1) represent and manage domain knowledge, (2) reason about perceived data (e.g., human action, detected objects/parts, robot system state) in order to understand a certain process state and (3) to deduce human, robot, or collaborative actions—given the current process state—in order to drive the process progress further.

The main software components in Fig. 12.1 are tightly integrated and linked together to enable close communication. Depending on the application that needs to be realised, the complexity of the system can be varied by activating or deactivating specific software modules, thus leading to a high scalability of the XRob system.

Previous work in [32, 33] presented examples of how robotic processes in industrial environments can be setup using XRob, considering different levels of shared autonomy among human operator and robot system. The scenarios demonstrate the scalability of the programming framework and a reduced programming effort to setup robotic processes (e.g., robotic coexistence, cooperation and collaboration in [32]), for the process engineer.

12.3.2 Programming-Free Approaches for HRC Assembly Processes Based on Function Blocks

This section presents an architecture that targets programming-free algorithm and application development for safe and efficient HRC in assembly. This development and configuration processes builds on predefined Function Blocks (FBs) according to the industrial standard IEC 61499 [34]. These "building blocks" form the basis for a high-level abstraction of the complete HRC process. Users within this development process make use of tools and defined "building blocks" to tailor application content without involving "classic" programming (i.e., coding).

The initial programming paradigm will be based on a To-do list (task-list) representing the process that has to be executed. Changes in the process, due to internal or external circumstances, are causing rescheduling of the task-list and will further lead to self-reprogramming of the process.

Technologically, this has been realised by the usage of a dynamic FB network, which can be instantiated and changed dynamically at runtime. The FBs are representing a "Container" to carry the content (Commands, Robot Paths, User Messages and others) and represent a FB network of the process, which has to be performed. This network is accessing the predefined content in runtime, to achieve zero-programming convenience for the operator.

Within the project SYMBIO-TIC [35, 36], a development environment for HRC applications was established that consists of a chain of core technology modules. These include (1) IEC 61499 compliant FB design tools and runtime environments, (2) the ABB-RobotStudio robot simulation environment [37], (3) Drag&Bot [38], a web-based graphical programming environment for robotic processes, (4) the SYMBIO-TIC Planning and Control Cockpit, (5) the Unit Controller, an IEC 61499 based execution runtime and (6) the Content Editor to edit information for tasks carried out by human operators. Programming-free development, of course, does include some offline development steps that need to be carried out by system engineers and developers. Figure 12.2 depicts a scheme of the development workflow and involves steps as follows:

1. *Development of building block algorithms (operations)* (FB IDE): This step includes implementation of IEC 61499 compliant function blocks (FBs) through a system engineer. These FBs include functionalities like setting digital I/Os, e.g., of a robot controller, or to perform a specific algorithm required. The resulting FBs are building blocks, for high-level process description using the Drag&Bot tool.
2. *Definition of application specific content*: (Content Editor, Robot Studio, human instructions—HMIC): Process engineers define application specific robot manipulation paths, which are exported as FBs to the Drag&Bot tool. Additionally, the content editor is used to include data content necessary for human worker instructions, and special parameterisation information for robotic and human operations.

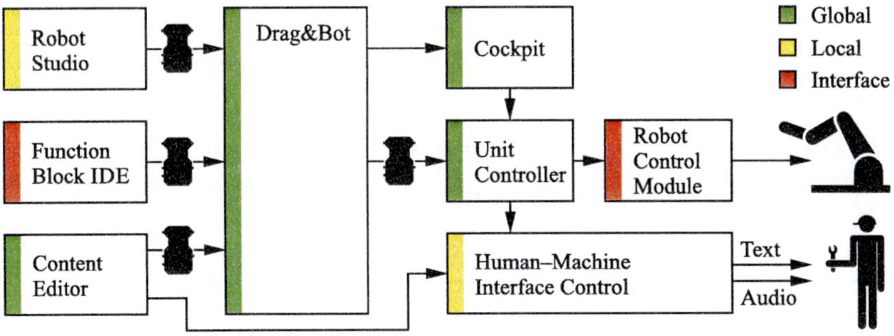

Fig. 12.2 Development Workflow of the SYMBIO-TIC Programming Approach for HRC applications

3. *Composition of tasks, based on single human and robot operations*: Based on defined robot paths, predefined FBs and application specific content, the Drag&Bot web-based programming system can be used by process engineers to compose complex tasks. The distinct single operations are carried out by different SYMBIO-TIC Equipment. Equipment refers to functionally independent hardware and software modules that are capable of executing certain operation types.
4. *(Re-)Scheduling of tasks to be executed to achieve a certain process*: The Cockpit is in charge of scheduling a plan, considering the available human and robot resources. This scheduling procedure is based on the defined tasks of the Drag&Bot platform and the application specific content. In detail, the Cockpit creates instances of task FBs in the Unit Controller and triggers these tasks according to the scheduled sequence. Depending on the current human and robotic resource availability, the Cockpit can be configured to re-schedule plans accordingly.
5. *Execution of the scheduled Process*: Operators act within the SYMBIO-TIC cell and use different modes of interaction to communicate with the Ecosystem modules. The HMIC especially provides means to let the operators confirm executed human tasks or provide feedback in case of process failures. Subordinate software modules, like the Dynamic Safety Control and the Collision Avoidance system (which are not depicted in Fig. 12.2), ensure safety of the operators.

During process execution, all distinct, single operations of task FBs will be executed through SYMBIO-TIC Equipment (e.g., Robot Control Module, Human-Machine Interface Control HMIC). This orchestration of functional equipment is coordinated through the Unit Controller (UC), an execution environment based on the IEC 61499 compliant runtime environment *FORTE* [39]. The UC comes with an interface which can be used by the Cockpit system, to dynamically create task FB instances, and to trigger their execution according to the current schedule. In this fashion, instantiation, parameterisation and wiring of task FBs is feasible, reflecting the whole process sequence.

The evaluation results, generated during the SYMBIO-TIC project [40], show that the realisation of HRC processes, based on the described approach, yields the following benefits compared to traditional assembly process solutions:

- Basic Operations encapsulated in building blocks, yield a reasonable abstraction to the process engineer, thus reducing programming complexity to the composition of operations and tasks.
- A planning and scheduling system (Cockpit) generates additional efficiency and flexibility depending on the existing robotic and human resources.
- The SYMBIO-TIC platform realises "adaptive robot control". This means that the robot control modules can receive information from multiple modalities (e.g., human interfaces, safety subsystems) and generate a meaningful command for the robot (prioritisation of input sources).

Applying the SYMBIO-TIC approach to industrial robots has shown that a safe and trustworthy collaboration among humans and industrial robots is possible.

12.3.3 Intuitive Human–Robot Interaction Based on Capacitive Sensors

Conventional programming approaches involve teach pendants provided by robot manufactures. These teach pendants serve the purpose of robot manipulation, start/stop of applications, programming, robot teaching, etc. However, teaching robots using teach pendants is tedious, time-consuming and is better suited for automated, repetitive tasks. However, for a close collaboration between human and robot, teach-pendant-based programming is not practical, because it does not enable seamless interaction. In the state of the art, there are many solutions available, for example, programming by demonstration, programming by observation, speech-based instructions, to program a collaborative robot [41]. Previous work of the chapter's authors in [42] presents touch-based easy-to-use and configurable user interfaces that are custom to specific collaborative applications. For this purpose, capacitive sensors were developed with ink-jet printing technology and combined with a robot system to enable intuitive HRI.

Figure 12.3 depicts the proposed system architecture for human–robot interaction based on capacitive sensors. A sensor array consists of four capacitive sensors, integrated with a readout-chip (FDC 1004 [43]), providing an I^2C interface. Multiple sensor arrays then act as slaves in an I^2C bus and communicate the sensor values to a central standard microcontroller, e.g., STM 32xx [44].

To enable further processing of data, the master microcontroller provides a TCP interface to enable the configuration of the sensor arrays and to stream sensor data from all connected sensor arrays. The sensor network is linked with a desktop PC

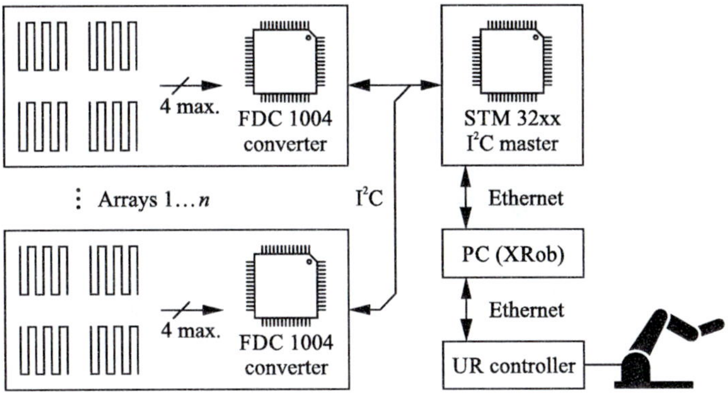

Fig. 12.3 System Architecture Capacitive Interfaces

which runs the robot programming framework XRob (see Sect. 12.3.1). The communication between the robot controller and the XRob runtime is realised using robot vendor specific protocols (e.g., UR "real-time data exchange protocol" [45]).

Figure 12.4 shows the prototype setup of a capacitive sensor array integrated with a UR10 robot, according to the system architecture as described above. The sensor array was ink-jet printed on a PET foil, including printed contacts for the readout chip.

Customisation of the configuration of capacitive sensor arrays is an important issue in realising flexible interaction behaviour. This configuration includes both (1) the interpretation of sensed capacitance values and (2) the triggered interaction behaviour of the robot system. In order to keep the application complexity low on the microcontroller level, the sensor value interpretation and robot response are configured using the XRob framework. The XRob runtime supports customisable "services" that act as background processes. The approach presented here uses a service implementation to perform sensor value interpretation to trigger defined robot system responses upon defined interactions, as described in detail in [42].

Sensor value interpretation is performed based on, e.g., threshold values to determine a "Boolean" on/off state for touch-interaction, or detection of activation patterns (e.g., activating certain sensors of an array in a defined temporal sequence).

System responses include (1) activation of digital outputs (e.g., robot controller), (2) free/hand-guidance mode of the robot, (3) execution of predefined robot programs, or (4) configurable movements of the robot's end-effector.

All of the aforementioned system responses can be configured and triggered using common robot teach pendants as well. The authors of this chapter argue that interaction via capacitive sensors, which are directly applied to the robot system, enables a more seamless interaction, and reduces ergonomic strain. Considering the prototype setup shown in Fig. 12.4, three types of interaction were considered and analysed in [42]: (1) freedrive or hand-guidance mode, (2) fine positioning based on interaction and (3) triggering of robot programs.

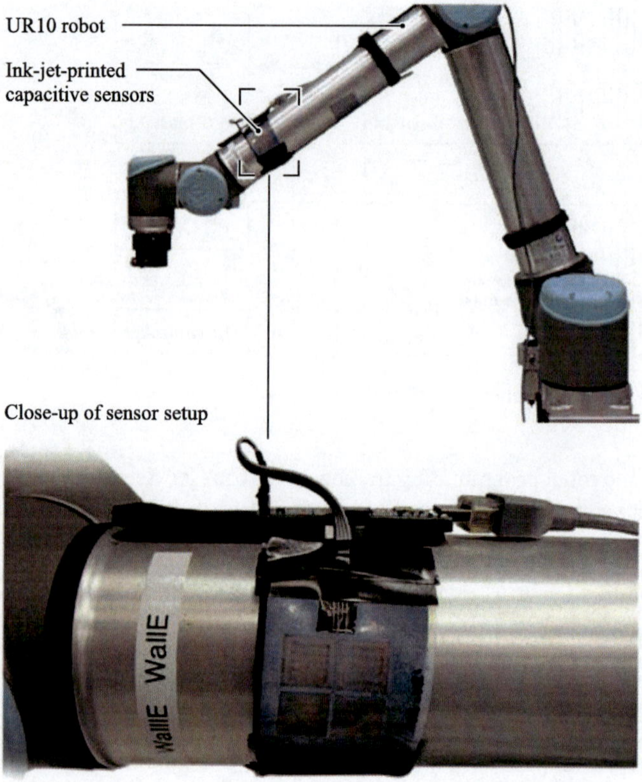

Fig. 12.4 Prototype setup: UR10 robot system equipped with inkjet-printed capacitive sensor array (see [42])

Initial tests showed that the interaction via the capacitive sensors is more natural and intuitive to the user. This is because the interaction can occur close to the robot's end-effector, which is helpful in case of fine-teaching of relevant process positions. Moreover, the amount of "double-focused" attention—where the operator has to watch the robot and the teach-pendant feedback at the same time—is avoided, thus reducing cognitive strain. In the use-case of moving the robot in freedrive mode, the direct interaction can also improve ergonomics. This is because this robot mode typically requires pressing a button at the teach pendant permanently (an example for UR robots is shown in Fig. 12.5).

Lastly, interaction via sensor arrays can also improve productivity, especially in case of process adaptation and reprogramming of process positions. Newly taught positions can be confirmed using another capacitive sensor to, e.g., confirm/save a process point. These functionalities can be tailored based on the configuration of the *Sensor value interpretation* and *System responses* (see explanation above).

Fig. 12.5 Comparison of interaction using teach-pendant (left) to capacitive sensor array (right)

A major future work item includes integration of these concepts with common robot programming and configuration platforms, including, e.g., *UR+* [46]. This aspect also could yield increased industrial relevance and usability.

12.3.4 Multimodal Command Fusion and Interpretation (Multimodal Programming)

Despite increasingly simple programming interfaces that are available for production machinery, a deeper understanding of these systems is needed. For the programming, maintenance and adaption of processes, a highly qualified technician is required in many cases, who is only available to a limited extent. Therefore, the setup time is increased, which consequently reduces flexibility. The goal is to enable an intuitive communication with various modalities so that non-expert users are able to interact with and program production machineries.

In [47], the authors present the conception and development of a multimodal data recognition and fusion system, to interpret and process commands issued through a human actor. The multiple command parts provided by various channels, such as speech, gesture and haptic interaction, are analysed and interpreted in order to generate a valid command statement. This fusion system is intended to be used for human-machine interaction in general. The resulting Command Fusion and Interpretation (CFI) module enables robust control of a robot system based on simple interaction commands. This approach also allows for online adaptation of command structures during runtime, thus providing a high grade of flexibility

The work of the authors targets the development of a rule-based multimodal data fusion and interpretation architecture based on Definite Clause Grammars (DCGs) [48] in order to generate valid commands for human-machine interaction. The classification of data emitted through multimodal channels is not within the scope of this work. As a proof-of-concept demonstration, [47] presents the application of controlling a robot based on commands issued by a human through multimodal channels.

12.3.4.1 System Architecture and Approach

Figure 12.6 depicts a scheme of the architecture of the CFI module. On the left side, different, multimodal input channels are listed. The blocks on top of the figure represent configurability and feedback functionality. A simplified description of the processing sequence of the CFI module is depicted at the bottom. The core functionalities are presented as a block diagram in the image centre. The functional workflow is briefly summarised as follows:

1. The user describes the desired command using different modality channels, captured through recognition systems.
2. All command attributes created are transmitted over several input channels to the CFI module. Arrival of attributes is random and contents can be redundant.
3. Some commands can be executed repeatedly using Heart Beats. During all time, the CFI module filters invalid Heart Beat attributes to ensure a correct interpretation of commands.
4. All attributes are validated by using the *Grammar List* and the *Threshold List*.
5. Each incremental step of validating command structures, the *Hypotheses List* is repeatedly updated by removing invalid hypotheses.
6. If a command is fully received and analysed, the module creates a final statement, and triggers its execution.

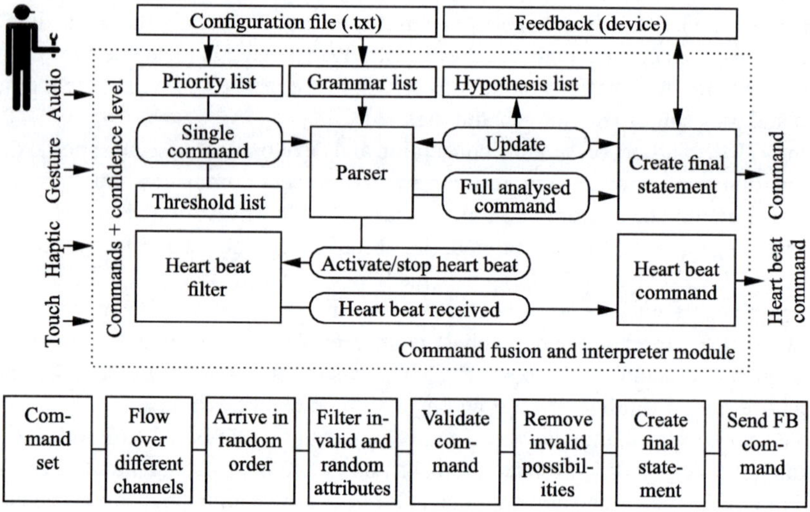

Fig. 12.6 Architecture of the CFI module [47]

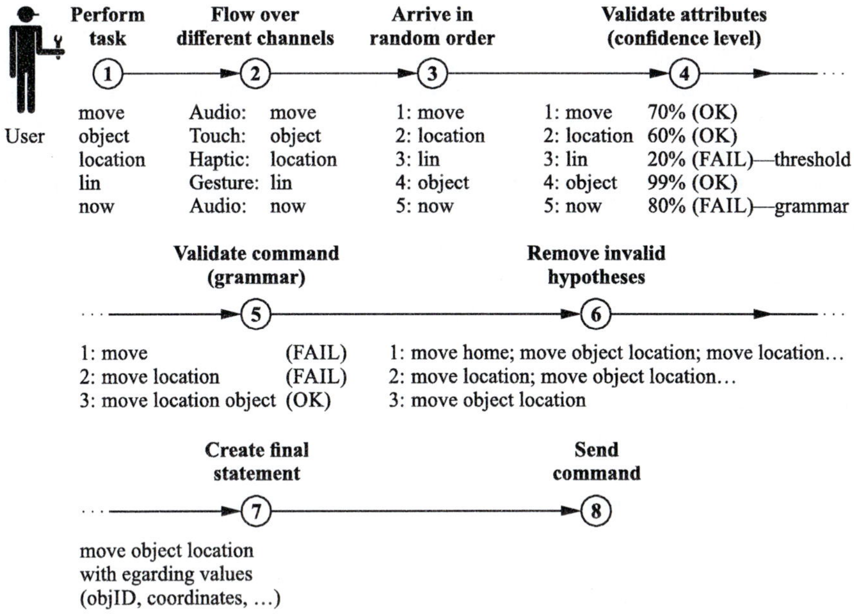

Fig. 12.7 Example of a Command Fusion Execution

For a better understanding of the CFI module, the example in Fig. 12.7 is considered, to explain the individual steps.

Steps 1–3 are covered by individual recognition systems for multimodal channels, and cannot be influenced by the CFI module. Each command attribute arrives at the module in a random order and will be immediately validated in Step 4. In this example, the invalid attribute *now* (not defined in the Grammar List) and the unreliable attribute *lin* (confidence level below threshold) are ignored. The incremental validation of the command takes place in Steps 5 and 6, where each validation in Step 5 creates command hypotheses in Step 6. At every iteration, an attribute is appended to the command structure and hypotheses are generated. After all attributes are appended and validated, only one hypothesis is left and the final statement can be created and sent to the execution unit, as shown in Steps 7 and 8.

The internal modules (according to Fig. 12.6) of the CFI system are designed to comply with the requirements defined in [47]. Simultaneously received commands are handled by the *Priority List* module that assigns a priority to each received command, to avoid ambiguities. Unreliable modalities can cause unintended commands and attributes. Therefore, all invalid commands and attributes are filtered in order to prevent invalid states. The *Grammar List* summarises all valid command structures with the valid combinations of attributes describing a possible command (e.g., “move object location”) using DCGs. Additionally, this list defines the output structure of the final statement. After every validation of a command structure, a pool of hypotheses is created and stored in the *Hypotheses List* in order to reflect all further possibilities. The updated Hypotheses List indicates whether a command is fully received or not.

A *Threshold List* is used to deal with varying quality of the modal input channels. These inputs are received with a corresponding confidence level and will only be accepted over a certain threshold.

The multimodal *Parser* plays a central role in terms of analysing the information gathered. It uses all aforementioned list modules to validate the commands and to generate hypotheses. If a command is fully received, the CFI module creates a final output statement, which follows the defined command structure of the *Grammar List*. A configuration file enables customisation of command structures, priorities and thresholds.

Finally, the feedback mode is enabled and adjusted by the configuration file. If activated, the validation algorithm conducts feedback from the user in order to ensure the correctness of the understood command. The command is illustrated at a device (e.g., screen) where the user is allowed to abort the analysed command within a certain cancellation-timeout.

12.3.4.2 Testing and Discussion

In order to test the CFI module, an environment is needed, including recognition systems as input channels such as speech, haptic, gesture and touch. Additionally, an execution module is needed for testing the impact and robustness of the output statement of the CFI module. The setup established in previous work [47], builds on modalities over four recognition systems, including speech, gesture (full body and hands) and a simple GUI. The output of the CFI module is received by a Universal Robots (UR) controller, operating either in simulation or on real hardware. All modules developed are implemented in C++. An IEC 61499 compliant runtime environment serves as an integration platform in this setup.

Based on this setup, the task of the user was to program a simple pick-and-place application based on the available modalities. In detail, the user had to construct statements to navigate the robot to a workpiece, to grip it, and to place the workpiece at a designated position, as demonstrated in this video source [49].

12.3.4.3 Results

Systems which integrate a CFI module allow non-expert users to use several modalities for programming, controlling or adjusting a machine. Due to the definable grammar, which can be adjusted during runtime, the valid commands can be adapted according to the operator's needs or experience level and create an intuitive interface.

The test use case indicated that the recognition systems used were not accurate enough to gain a high usability of the system. The speech recognition was very sensitive to noise, and therefore, not usable during the time the vacuum gripper is active. The generic command structures enabled programming up to a certain granularity. Thus, high accuracy programming turned out to be challenging. Parameterised commands (e.g., move to coordinates, move distance) are part of future work to increase the usability of the system. Using individually tailored grammar for every user, no expert is needed for programming a machine. This increases the flexibility of a system, because highly qualified technicians are often temporarily unavailable.

12.3.5 Instrumented Tools

In HRC assembly settings, *Learning by Demonstration* (LbD) is of interest in particular because (re)programming takes place more frequently, as these setups are typically applied for smaller batch size production. The authors in [50] introduce a clear taxonomy for high and low level LbD, distinguishing between trajectory-level demonstration and symbolic encoding methods. In [51], two real- world production use cases are presented, differing in process requirements, complexity, required process steps and product geometry or weight. LbD on trajectory level is applied by [52] for a specific application. This work further introduces reusable and task-agnostic motion primitives for assessing the outcome of force-interaction robotic skills.

Trajectory demonstration by hand guidance of a process tool mounted on the robot could be less intuitive than demonstrating it with a hand guided process tool (that is instrumented with sensors to measure process data) because it is not embodied with the human demonstrator. As an intermediate approach between trajectory encoding and symbolic encoding, the authors of this chapter propose to derive the parameterisation of macro-based skills from trajectories demonstrated with an *instrumented tool* rather than from numerical input of process parameters, as introduced in previous work [53].

Figure 12.8 depicts the hardware setup of the instrumented tool. Its main components include a common power tool (e.g., electric screwing tool), a lighthouse tracking sensor of type HTC-Vive [54], and a force–torque sensor (e.g., [55]) that connects the power tool with the U-handle.

As this setup is intended to be integrated into an HRC workcell including a robot system, a hand–eye calibration problem needs to be solved as elaborated in [53]. Additionally, [53] describes an approach to perform an inertia compensation of the externally unloaded instrumented tool. This is required to sense the process forces only. Using the Newton–Euler equations, the forces, torques and wrench of the tool centre point (TCP) can be calculated.

Depending on the application, the retrieved process parameters of the instrumented tool need to be applied to a robotic skill definition, based on a process macro. Considering a screwing process, Fig. 12.9 depicts the scheme of parameterisation necessary to replicate the process using a robot system, where S and E describe the start and end positions of the screwing process, O the pre-position offset. F and T, describe the screwing force and final torque, respectively. Considering a system integration with the skill-based XRob robot programming framework (Sect. 12.3.1),

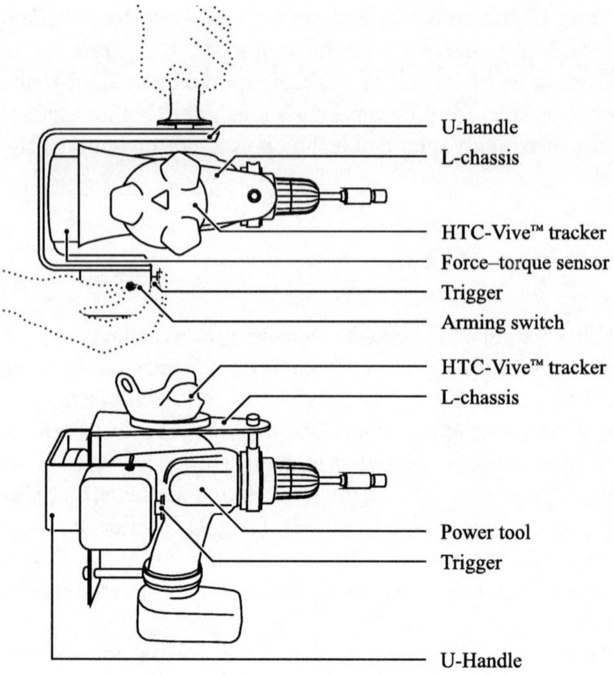

Fig. 12.8 Setup of the instrumented tool, including force/torque and visual tracking sensors [53]

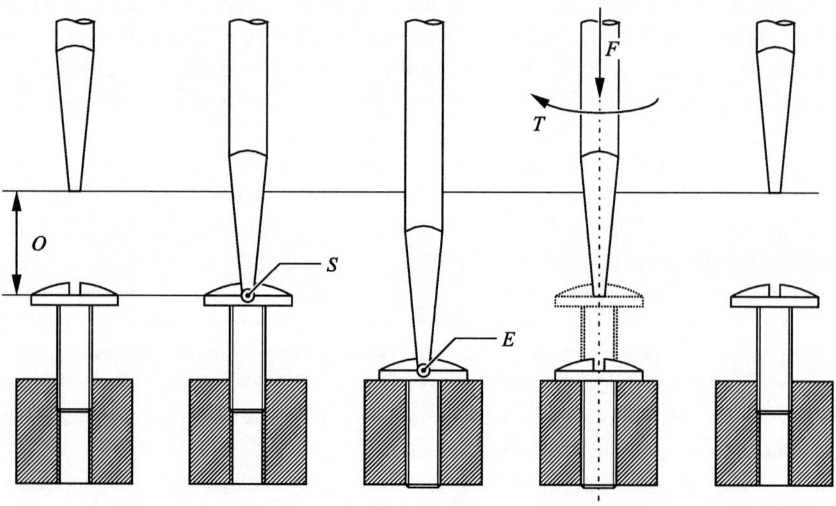

Fig. 12.9 Parameterisation scheme (process macro) for the robotic screwing process [53]

the retrieved process parameters (instrumented tool) need to be provided, e.g., via a REST-API to the programming system.

In order to increase process stability, a multi-step process recipe can be set up with XRob, which includes built-in robot skills for vision-based positioning accuracy compensation. Initial tests show that the described approach can help avoid programming mistakes (e.g., due to mistyping), in case of configuring process values. Moreover, the training time for process operators can be potentially reduced, as less parameterisation expertise is required.

12.3.6 Programming Approach Based on Interactive Demonstration

This approach is based on Interactive Reinforcement Learning that learns a complete collaborative assembly process. The learning approach is done in two steps. The first step consists of modelling simple tasks that compose the assembly process, using task-based formalism. The robotic system then uses these modelled simple tasks and proposes a set of possible tasks to the user at each step of the assembly process. The user is presented with an interactive interface (GUI) to select the suitable action in the given situation. The user selection help the system to relate the current situation in the environment and the chosen action to be performed. The robot performs the chosen action. This is done in an iterative fashion until the assembly is completed, where the completion of the assembly is conveyed by the user (using the interactive GUI). In order to reduce the number of actions proposed, the system considers additional information such as user and robot capabilities and object affordances. These sets of action proposals are further reduced by modelling the proposed actions in a goal-based hierarchy, and by including action prerequisites. The learning framework shows that it is able to learn a complicated human–robot collaborative assembly process and is intuitive to the user. The framework also allows different users to teach different assembly processes to the robot, and demonstrates its capability in dealing with novel situations during the assembly process.

The contributions of this approach to the state of the art include:

- An intuitive programming framework for learning an assembly process, and
- Applicability to human–robot collaborative assembly tasks with novel situations.

To deal with the HRC assembly process, the robotic system should be enabled with cognitive capabilities in order to perceive, reason, plan and execute as shown in Fig. 12.10.

12.3.6.1 Architecture

The architecture of the learning framework is comprised of the following:

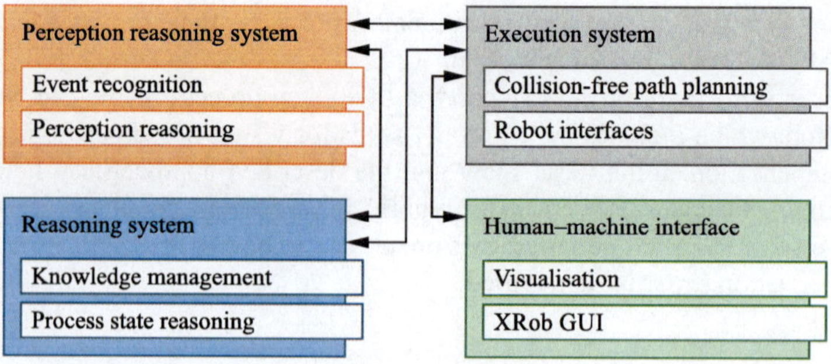

Fig. 12.10 Architecture of the Learning Framework [56]

(1) The *Perception Reasoning System* (PR) understands and interprets the current state of the environment. The PR is equipped with real-time 3D tracking of objects with the help of the Multi-Object Tracker [57, 58], and an action recognition system to recognise the current action performed by the human [59] The object tracker and the action recognition system are combined to recognise the events (actions that interact with objects) using an event detector [31, 32].

(2) The *Reasoning System* represents and abstracts different aspects of the assembly process in the HRC environment. This includes representing the abilities and possible activities of the human and the robot, interplay between human/robot activities and object configurations, current state of the human and the robot with respect to the task, etc. It reasons about the current assembly state by combining information about the current state (given by perception reasoning) and the assembly process knowledge (knowledge management), to make decisions, to plan the next actions accordingly in coordination with the planning and execution system, to behave intelligently, to interact naturally with humans and aid in completing the task. The knowledge management, perception reasoning and the process state reasoning modules together constitute the Cognitive Reasoning System in the architecture.

(3) The *Execution System* module is responsible for plan generation, where the plan consists of actions required to achieve the given task (goals). This includes planning of the task, scheduling of the actions, and also includes planning under uncertainty for an efficient human–robot collaboration. The plans generated are carried out in the real world where the robot plans its path (path/navigation planning) and manipulates the environment accordingly.

(4) The *Human-Machine Interface* is also equipped with communication interfaces that provide a GUI interface for human–robot communication and simulations of the robot's planned behaviour. A detailed description of the architecture is given in [56].

12.3.6.2 Learning Methodology

The solution to teach the robotic system a complex human–robot collaborative assembly process is to first teach the simple tasks (that include one or two agent–object interactions). This is done using task-based formalism [6]. During the second stage, the robotic system can propose a set of possible simple tasks (actions). The reason for choosing the robot to propose a set of actions is to make the learning process more interactive and flexible. The user can then choose the sequence in which these actions should take place, thereby creating a recipe for the complete assembly process.

However, simply proposing a set of actions to the user (given the simple tasks already learnt) is not a valid solution, as there could be many possible actions, and as a result, could overwhelm the user with too many options to choose from. To reduce the state–action space and enable fluent human–robot collaboration, it is assumed that the assembly process starts with an initially known state (known object, human, robot and object configuration in the environment). The system first detects the available objects, the current human state and robot state (with the help of the architecture in Fig. 12.10), and proposes a set of possible actions. The initially generated possibilities are reduced in two steps. First, the user capabilities and the robot capabilities together with object affordances are considered to reduce the action possibilities. Then, these action possibilities are further reduced using action hierarchy and action prerequisites. More details about the learning framework and related nomenclature are described in [60].

Given a reduced list of action possibilities, the user is requested to select a suitable action. Once selected, the action is executed (in case the human chooses to perform the action, the event detection system waits until the action is executed), and then, the resulting state is recorded. The learning framework utilises reinforcement-based learning to record the action policy learnt with the help of user interactions.

12.3.6.3 Experimental Results

The experimental setup consists of a UR 10 robot equipped with a SCHUNK 2-finger parallel gripper as shown in Fig. 12.11. A set of RGBD sensors, as shown in. Figure 12.11a, are used to perform action recognition and object tracking, respectively. The assembly process consists of a set of objects that correspond to a steam cooker. The robotic system observes the user and the environment with the help of the 3D sensors and interacts with the user with the help of an interactive GUI. The interactive GUI is able to visualise the status of the robot in a user-understandable format.

When presented with a set of objects as shown in Fig. 12.11, the robotic system first observes its environment (objects, the user status and its own status) to characterise the status of the assembly process. The goal is to learn the complete assembly process (in this case, the steam cooker assembly). Having recognised the status of

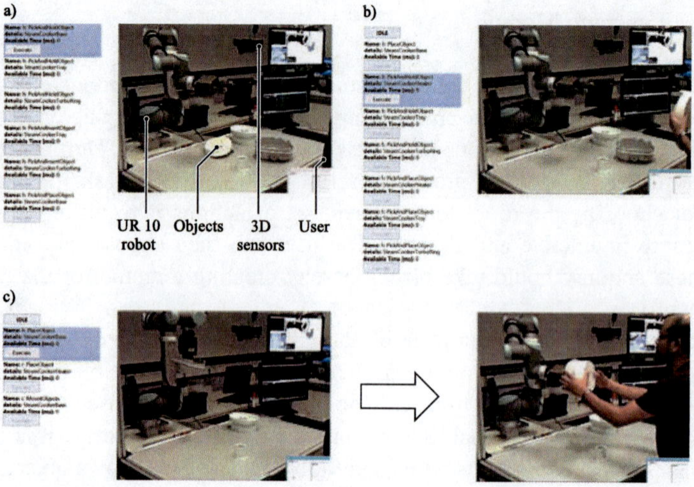

Fig. 12.11 Stages of the approach to learn an assembly process, based on proactive suggestions [56]

the assembly process, the system proactively proposes a set of possible next steps to the user using the interactive GUI. This is also shown in Fig. 12.11a.

All those suggestions are generated by considering the status of the robot, the user, the object configuration and the assembly process. Given the suggestions, the user selects a suitable option for the next action. The robot and/or the human then perform(s) the selected action. Figure 12.11b depicts the example of a suggested action that is executed by the human actor, whereas in Fig. 12.11c a proposed robot action is selected and executed. This results in a change in the object configuration and the environment. The robotics system once again uses the sensory modules to characterise the status of the assembly and propose a set of suitable actions, accordingly. In this fashion, the system learns which action to perform given a status in the AP. Once the AP is learnt, the system, when presented with the set of objects, recalls and remembers the suitable actions, performs them accordingly, and completes the assembly process. A detailed demonstration of the learning framework is given in a video [61] available online.

Once the AP is known, the robotic system can observe the status of the environment and choose a suitable action to perform. However, when considering the dynamic nature of the environment, action execution might fail due to various reasons. Given a new situation where the robotic system characterises the status, chooses an action but fails to execute it, the learning framework is invoked once again. However, the learning framework now only suggests suitable alternatives considering the action previously chosen. For example, as given in Fig. 12.12a, the robotic system chooses to pick up the "heater" object. However, due to external disturbances, the robot fails to grasp the object and fails in the action execution as shown in Fig. 12.12b, c. The robotic system uses the learning framework to suggest a set of possible alternatives

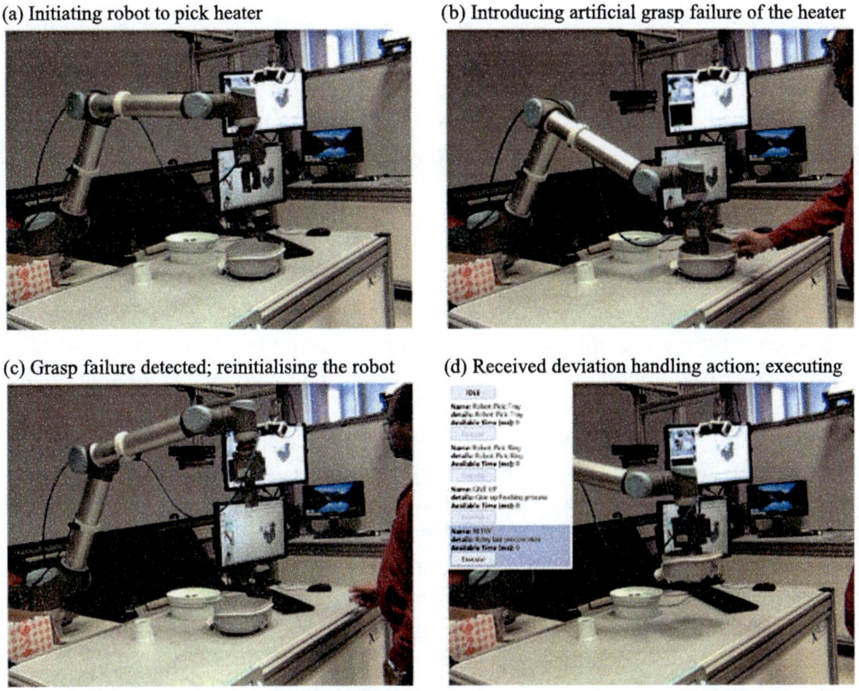

Fig. 12.12 The capability of handling deviations during the execution of an assembly process [62]

that could help in completing the AP. The user chooses, in this case, to retry the same action once again, that is to pick the "heater" object, as shown in Fig. 12.12d. The robot tries the action and succeeds. Thus, the system learns in this case that if it fails to pick a particular object (due to any reason) it can always retry the action, provided the object of interest is still available. A detailed description of handling deviations is given in [62].

12.3.6.4 Comments

The learning framework combines the aspects of Learning by Programming and Learning by Interaction in an intuitive fashion to enable a robotic system to learn complex assembly process. The robotic system is equipped with capabilities to recognise the current situation and perform object manipulations at a task level. The learning framework exploits these capabilities of the robot while teaching the assembly process. Instead of randomly interacting with the environment to find a suitable action policy, the system interacts with the user in an intelligent fashion to learn the action policy. To achieve this, the system observes the environment (the objects, object configuration, the human status) to understand the current situation and proposes a set of possible actions. The user can either select one of the proposed actions to be performed or trigger a new action sequence. This process is done iteratively until the assembly process is completed. In case of deviations, where an action to be performed was not executed (due to disturbances), the system realizes the failure (using proprioception and sensor data). A set of new actions are proposed to recover from the failure, thereby handling the deviation that could occur in the environment.

12.3.7 Programming Approach Using Hybrid Interfaces

This approach aims at combining the aspects of Learning by Demonstration (LbD), Learning by Programming (LbP) and Learning by Interaction (LbI) methodologies in a goal directed assembly process to tackle the problem of easy programming of robotic tasks with the help of a two-phase approach. This work explains the methodology in which the different learning paradigms are planned to be combined to achieve an easy-to-use programming framework. The technology coverage starts with the XRob online workflow editor (Sect. 12.3.1) that allows easy-to-use setup of process workflows and related skill parameters. However, in order to reduce the programming effort, a novel way to demonstrate process trajectories by using instrumented hand guided process tools (Sect. 12.3.5) is presented in this approach. An overview of a promising approach is explained, which allows programming without touching the robot just by demonstrating the process by an expert. The semantic relations between activities executed by the human and robot skills are captured to learn the task sequence of the assembly process. The acquired process knowledge is refined to execute robotic tasks with the help of an interactive graphical user interface (GUI). The system queries the user for feedback, asking for specific information to help the robot complete the task at hand.

12.3.7.1 Architecture and Learning Methodology (Programming by Interactive Demonstrations)

Phase 1: The human operator physically demonstrates the assembly process (AP) to the robotic system by performing a set of activities (interaction with objects). These activities are either demonstrated "hands-free" [63] or using an instrumented tool (Sect. 12.3.5). Each activity corresponds to a task in the AP. From the demonstration (LbD), the system captures [62] the semantic relation between the task performed by the human operator and the corresponding consequences, i.e., effect on the assembly environment [64]. As a result, the knowledge about the sequence of tasks in the assembly process is learnt. This creates a shared understanding between the human and the robot about the AP at the task level. Abstracting the knowledge at the task level allows the system to easily learn complex assembly processes with the help of human demonstrations.

Phase 2: This phase further refines the coarse knowledge obtained in Phase 1, to learn process level parameters required for the actual execution of the task by the

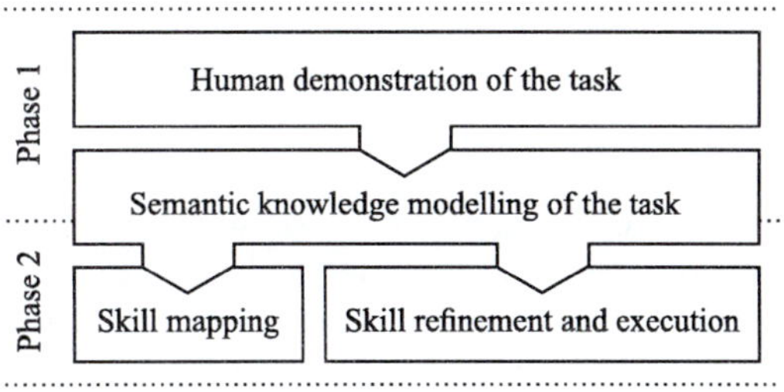

Fig. 12.13 The architecture of the proposed hybrid learning approach [53]

robot. First, the robot maps (LbP) the tasks from the human domain to that of the robot embodiment. The mapping also takes place at a task level, where the robot tasks are formalised using TFF [6]. The robot extracts the required parameters to execute the tasks in a goal directed fashion [65]. In case of missing parameters, the robotic system queries the human with the help of an interactive GUI (as described in 12.3.1 XRob). When the system has to deal with different similar-looking objects, it maps the parameters from previously known objects to the new ones (knowledge transfer [65]) and queries the user for specific modifications. In such cases, the robot queries (LbI) the user only for the parameters that vary, and does not require the user to re-parameterise the complete task. This new parameterisation for similar objects is used in a feedback loop to improve the parameterisation of similar cases in the future [60]. The robotic system in Phase 1 acquires a fast but general representation of the AP. In Phase 2, with the help of intelligent user interactions, the robotic system learns the specific parameters and successfully executes the tasks to complete the AP. This combination could lead to a faster programming phase that is more precise than just demonstrations, and more intuitive than just through a GUI. The system architecture for the proposed learning approach is shown in Fig. 12.13.

For the learning architecture to enable easy teaching of an AP to the robot, there are four modules envisioned:

12.3.7.2 *Human Demonstration of the Task*:

This module deals with the perceptual aspect. This includes representation of agents' (Robot/Human) activities, where activities range from interaction with objects to simple atomic movements. Recognition of such activities including objects in real-time uses state-of-the-art approaches [60], sensor technologies and instrumented tools (as described in Sect. 12.3.5).

12.3.7.3 *Semantic Knowledge Modelling of the Task*:

This involves modelling the knowledge of the assembly process in an action-centric approach for ease of inference [64]. Moreover, it includes learning of the consequences generated by the human agent's activities and their semantic relation to the assembly process. The latter step exploits the learning by demonstration methodology to learn the task sequence of the assembly process and abstract the necessary knowledge.

Learning an assembly process by interactive demonstration requires an abstraction of the knowledge of an assembly process. In order to develop such a knowledge representation, a modelling language combined with a framework to query and reason about the existing data is required. The sub-sections below provide a brief overview of two state-of-the-art approaches to knowledge modelling and processing, and describe the approach of the chapter's authors to represent assembly process knowledge. Furthermore, the approach to deriving the selection and parameterisation of robot skills in order to imitate the human-demonstrated assembly process through the robot is explained. Within the domain of knowledge processing and semantic reasoning, ontology and graph-based techniques [66] are common approaches to realise a knowledge representation. The authors in [67] present a thorough overview of existing ontology languages and their applicability for knowledge modelling and information retrieval. A concrete implementation of an ontology-based knowledge processing system is given by KnowRob [68] especially for the robotics domain. This framework builds on the Web Ontology Language (OWL, especially OWL description logics) and the Resource Description Framework (RDF) to model the domain knowledge and provides mechanisms for ontology reasoning and inference. The *SWI-Prolog* engine, including the Semantic Web Library, enables online querying and adaptation of the knowledge base. Hypergraphs are another example commonly used in knowledge processing and AI to express and reason about domain knowledge. These types of graphs are less restrictive in the definition of edges, as they allow to connect multiple vertices over a single edge [69]. Two examples of hypergraph-based knowledge processing framework implementations (open-source) are *OpenCog* [70] and *GraknAI* [71]. In this work, the GraknAI framework was chosen as an enabler for knowledge modelling and processing.

12.3.7.4 Skill Mapping

The semantic model, targeted in this work, consists of data structures to express an assembly process including a series of *States*, a set of *Events*, and a set of *Relations*. The States represent the individual assembly steps. An Event is capable, dependent on its definition, to advance the assembly process from a specific state to the next. Relations semantically relate Events and States to form transitions between assembly process states, and the corresponding Events to activate the transitions. A more detailed and formal explanation is given in previous work [31]. In addition to these fundamental data structures, the semantic model also describes object types and their configurations (e.g., combinations of objects), human and robot skills, and consequences—generated through events—that lead to changes in the environment. Figure 12.14 depicts a section of the semantic model that describes object types and instances, with given properties and relations.

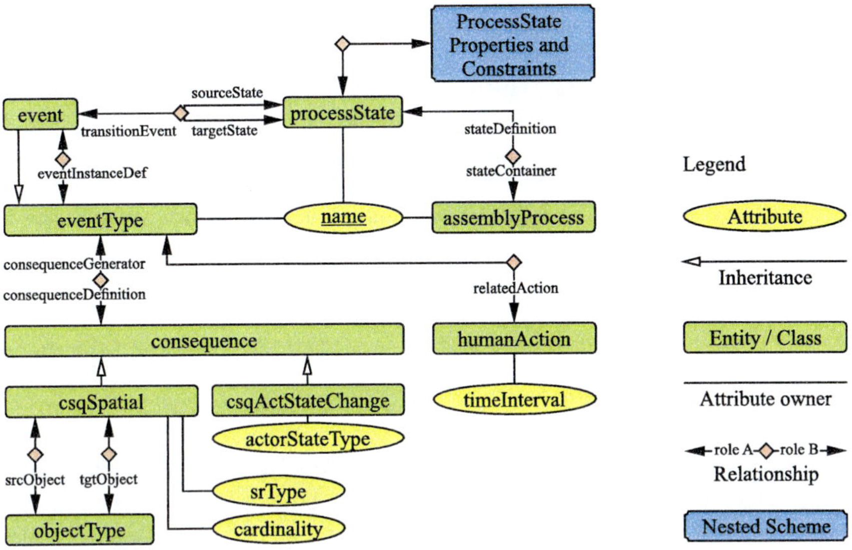

Fig. 12.14 Knowledge representation of object types and instance in GraknAI [72]

All concepts (better known as classes), attributes and relations of the semantic model are defined offline, thus forming the schema for the representation of knowledge. In the offline phase, the knowledge processing framework is populated with relevant process knowledge, including known events and robot skills, along with their causal effects. During the online phase, i.e., learning phase, queries are issued to the knowledge processing framework in order to generate concrete instances of states, events or objects. These queries are parameterised based on the data, which is delivered by the perception systems that observe the environment during the learning phase. Based on the assembly process sequence learnt, the knowledge processing and reasoning system has to answer the question of mapping human skills to robot skills including the parameterisation. This problem is solved by a similar representation of effects caused by both human skills (related to detected events) and robot skills, through so-called consequences. Consequences describe causal changes in the environment, which include (1) changes in the spatial configuration of objects (e.g., combination of two objects), (2) objects disappearing or appearing newly, and (3) displacement of objects.

By applying the concept and relation reasoning functionalities [72] of the knowledge processing framework, direct relations between human skills and robot skills can be established implicitly, by matching the respective causal effects (consequences). This mapping is realised through rule definitions, which are defined in the semantic model and will be evaluated upon write access to the database. In case the parameterisation of an event recorded during the learning phase cannot be mapped to the robot skill, user interaction will be triggered to specify the correct parameterisation.

12.3.7.5 Task Refining and Execution

The module maps the necessary task to execute to match the present embodiment and status of the robot [6]. Each task is further converted into robot skills using the XRob framework (Sect. 12.3.1), and necessary parameters are extracted. An interactive GUI is also developed that enables an intuitive communication between the user and the robot. In case of missing parameters or ambiguities, the robotic system queries the user for feedback to receive the missing parameters with the help of the GUI. However, the system queries the user in an intelligent way posing the questions in a user-centric fashion.

12.3.7.6 Comments

Though learning paradigms for robotic tasks have received extensive attention in the research community, the topic of reducing the programming effort to teach a task to the robot is still open. In literature, different programming paradigms such as Learning by Demonstration (LbD), Learning by Programming (LbP) and Learning by Interaction (LbI) are proposed for teaching/configuring a robotic task. However, these approaches still fall short (in reducing the programming effort) when dealing with complex tasks. This work aims to combine these different approaches in a goal-directed fashion to develop a framework that exploits the advantages of these learning paradigms to alleviate the problem of high programming effort. This chapter first presents a programming framework to easily edit a process workflow (LbP). This is followed by a more detailed description and evaluation about an instrumented tool-based learning by demonstration approach. Finally, a framework is presented that combines these two learning paradigms in an interactive fashion to reduce the programming effort in teaching a complex task to the robot.

12.4 Qualitative Comparison

Robotic solutions for SMEs are comparatively hard to sell/implement since smaller production batches have high product diversity and hence require frequent reprogramming and re-parameterisation of robotic systems. More frequent (shop-floor) re-parameterisation of robots will require new programming paradigms than traditional code-based or CAD/CAM-based robot programming. Therefore, one of the most important challenges to address is to develop intuitive interfaces that allow easy interaction between humans and robots. The existing solutions, however, do not yet allow a non-programming expert to easily interact with robots.

For example, manufacturers of HRC robots usually provide simplified programming UIs for creating robot programs and interfaces to interact with robots. However, specific training and expertise is still expected from users to achieve the task when using such interfaces and the complexity of the programmable task is very limited.

Open source platforms like *ROS* provides a broad range of packages and functionalities for programming robots. However, ROS is programmed code-based and requires a profound knowledge of its system and messaging architecture, as well as advanced software development knowledge, since installed packages (and their versions) are not independent of programs developed and their configurations. Also, ROS/ROS Industrial, lack simplified programming UIs. Therefore, ROS/ROS Industrial are used as prototyping platforms by robot programmers rather than for shop-floor reprogramming or re-parameterisation. To alleviate some of these challenges (especially with UIs), technology providers like Drag&Bot (a software tool) add a simplified UI to ROS Industrial. Complex skills (e.g., 2D template match, 3D object pose recognition) are available as plugins. Other solutions like Artiminds [73] can be described as skill-based offline robot programming tools for multiple robots. While it is a mature proprietary tool also with integration of additional and complex functionality (e.g., cameras + 3D object pose recognition, or force torque sensors and respective functionality such as spiral search for peg in hole operations), it lacks more intuitive interfaces for programming or re-parameterisation of robot programs, and therefore, is an offline programming tool rather than a tool suitable for shop-floor reprogramming.

Table 12.1 compares existing solutions against different criteria that should be considered for programming robots for SMEs and non-experts, namely:

1. Code-based programming approaches require an expert to program robots for the specific task. Such approaches are not suitable for manufacturing domains that require frequent reprogramming of robot tasks, especially when dealing with mass customisation of products and zero lot sizes.
2. CAD/CAM-based methodologies allow users to program/parameterise robot processes with the need for programming knowledge. However, this is less practical because CAD files might not always be available, and trained personnel is required to use the complex CAM tools.
3. The interfaces of the programming tools should be simple and easy-to-use for non-programming experts, so that even workers at the factory floor with process knowledge could program the robots for various tasks.
4. The programming tools provided should be able to deal with complex robotic processes and not just be capable of dealing with simple robotic operations.
5. While parameterisation by hand guidance is an appreciated input modality for interaction, it still is less intuitive for non-programming experts than direct process demonstration with relevant tools.
6. Novel interfaces such as instrumented or tangible interfaces or visual interfaces should be exploited to unobtrusively capture relevant information to program the robot tasks.
7. Process Visualisation is very important to visualise the current status of the robot in the process to the user, so that the user is able to take necessary steps.
8. Ideally, programming interfaces should be applicable to different robot types and should not be limited to a specific robot to allow flexibility.

Table 12.1 Robot programming tool comparison

	Traditional robot SDKs	HRC-robot UIs	ROS Industrial	Drag&Bot	ArtiMinds	XROB™	Octopuz	Robotmaster	SYMBIO-TIC	HoliSafeMRK	Multimodal	Lern4MRK	KoMoProd	ASKRob
Code-based programming	+	(+)	+	(+)	?	×	×	×	×	×	×	×	×	×
Simplified programming approach (e.g., GUI)	×	+	×	+	×	(+)	×	×	+	×	+	+	+	+
CAD/CAM programming	×	×	×	×	×	×	+	+	×	×	×	×	×	×
Programming complex skills/processes	(+)	×	+	(+)	+	+	×	×	+	(+)	×	×	+	+
Hand guidance (kinaesthetic)	×	+	×	+	+	+	×	×	×	+	×	×	×	×
Special interfaces (instrumented interfaces, tangible elements, advanced sensors)	×	×	×	×	×	×	×	×	+	+	+	+	+	+
Process visualisation	+	(+)	(+)	(+)	+	+	+	+	+	+	×	×	×	×
Applicability to different robot types	×	×	+	+	+	+	+	+	(+)	+	(+)	+	+	+
Knowledge portability in case of process variants	×	×	×	×	×	×	×	×	+	×	×	(+)	(+)	+

Legend: +supported(+) possible with some effort × not supported

9. Though current solutions address most of the criteria, the topic of knowledge portability (of robot programs created on one robot setting to another) and handling process variants are still open. Knowledge-based derivation of robot programs could be the answer as it can enable the portability of process knowledge between robotic settings and allow transfer learning in case of process variants.

Table 12.1 compares existing state-of-the-art products available for programming robots for application in the manufacturing domain. The comparison includes technologies provided by traditional robot SDKs, UIs provided by robot manufactures, technology providers like Artiminds [73], Drag&bot [38], Octopuz [74], Robotmaster [75] and compares them against the programming interfaces described in this chapter.

Traditional approaches that use robot-specific SDKs are slowly becoming outdated and are being replaced by novel programming tools. Though the specific SDKs will be relevant, they provide less flexibility and are hard to use for non-experts. The existing state-of-the-art programming tools that can be purchased are better and are paving ways to programming-free approaches to interact with robots. However, the functionality is still limited when considering different robotic settings and process variants. Even in case of slight product variants, the user needs to completely re-program/re-parameterise the robot process when using the state-of-the-art programming tools. In contrast, the methods explained in this chapter provide alternative solutions and point towards possibilities to exploit when dealing with different robotic settings and process variants.

12.5 Conclusion

This chapter deals with the problem of interacting and programming robots for non-programming experts and installing such robotic systems to the manufacturing domain. Specific focus is placed on developing interfaces that allow non-expert users to easily interact with robots. Novel interfaces and the programming paradigms are introduced. Detailed explanation of the methodology is given, followed by the experimental results of the programming paradigm, wherever possible. A comprehensive review of existing state-of-the-art programming approaches is also described. A qualitative comparison of available off-the-shelf programming tools for programming robotic tasks is given with the approaches introduced in this chapter. The comparison considers various criteria that should be considered when developing programming tools and paradigms for easy programming of robots.

Acknowledgements The work presented in this chapter is a cumulative effort from various projects and is jointly funded by projects SYMBIO TIC (EC 637107), Bridge Human–Robot Interaction "TeachBots" (Industrial), HoliSafeMRK (FFG 864872), FELICE (EU GrantID 101017151), DigiManu and Smart Factory Lab (Funded by the State of Upper Austria).

References

1. V. Villani, F. Pini, F. Leali, C. Secchi, Survey on human–robot collaboration in industrial settings: safety, intuitive interfaces and applications. Mechatronics **55**, 248–266 (2018)
2. B.D. Argall, S. Chernova, M. Veloso, B. Browning, A survey of robot learning from demonstration. Robot. Auton. Syst. **57**(5), 469–483 (2009)
3. A. Bauer, D. Wollherr, M. Buss, Human–robot collaboration: a survey. Int. J. Humanoid Robot. **5**, 47–66 (2008)
4. K. Dautenhahn, Methodology and themes of human–robot interaction: a growing research field. Int. J. Adv. Robot. Syst. **4**, 103–108 (2007)
5. M.A. Goodrich, A.C. Schultz, Human–robot interaction: a survey. Found. Trends Hum. Comput. Interact. **1**(3), 203–275 (2007)
6. M.R. Pedersen, L. Nalpantidis, R.S. Andersen, C. Schou, S. Bøgh, V. Krüger, O. Madsen, Robot skills for manufacturing: from concept to industrial deployment. Robot. Comput.-Integr. Manuf. **37**, 282–291 (2016)
7. R. Maclin, J.W. Shavlik, Creating advice-taking reinforcement learners. Mach. Learn. **22**, 251–281 (1996)
8. G. Kuhlmann, P. Stone, R. Mooney, J. Shavlik, Guiding a reinforcement learner with natural language advice: initial results in RoboCup soccer, in *The AAAI-2004 Workshop on Supervisory Control of Learning and Adaptive Systems* (2004)
9. D.L. Moreno, C.V. Regueiro, R. Iglesias, S. Barro, Using prior knowledge to improve reinforcement learning in mobile robotics, in *Proceedings of the Towards Autonomous Robotics Systems (*University of Essex, UK, 2004)
10. M.N. Nicolescu, M.J. Mataric, Natural methods for robot task learning: Instructive demonstrations, generalization and practice, in *Proceedings of The Second International Joint Conference on Autonomous Agents and Multiagent Systems* (2003), pp. 241–248
11. B. Finkemeyer, T. Kröger, F.M. Wahl, Executing assembly tasks specified by manipulation primitive nets. Adv. Robot. **19**, 591–611 (2005)
12. H. Bruyninckx, J. De Schutter, Specification of force-controlled actions in the 'task frame formalism'—a synthesis. IEEE Trans. Robot. Autom. **12**, 581–589 (1996)
13. H. Mosemann, F.M. Wahl, Automatic decomposition of planned assembly sequences into skill primitives. IEEE Trans. Robot. Autom. **17**, 709–718 (2001)
14. F. Steinmetz, R. Weitschat, Skill parametrization approaches and skill architecture for human–robot interaction, in *IEEE International Conference on Automation Science and Engineering (CASE)* (2016), pp. 280–285
15. M.R. Pedersen, V. Krüger, Automated planning of industrial logistics on a skill-equipped robot, in *IROS 2015 workshop Task Planning for Intelligent Robots in Service and Manufacturing* (Hamburg, Germany, 2015)
16. E. Dean-Leon, K. Ramirez-Amaro, F. Bergner, I. Dianov, P. Lanillos, G. Cheng, Robotic technologies for fast deployment of industrial robot systems, in *IECON 2016—42nd Annual Conference of the IEEE Industrial Electronics Society*, vol. 42 (2016), pp. 6900–6907
17. D. Holz, A. Topalidou-Kyniazopoulou, F. Rovida, M. Pedersen, V. Krüger, S. Behnke, A skill-based system for object perception and manipulation for automating kitting tasks, in *IEEE 20th Conference on Emerging Technologies & Factory Automation (ETFA)*, vol. 20 (2015), pp. 1–9
18. A. Billard, S. Calinon, R. Dillmann, S. Schaal, Robot programming by demonstration. Springer Handbook of Robotics **59**, 1371–1394 (2008)
19. W.K.H. Ko, Y. Wu, K.P. Tee, J. Buchli, Towards industrial robot learning from demonstration, in *Proceedings of the 3rd International Conference on Human-Agent Interaction*, vol. 3 (2015), pp. 235–238
20. M. Stenmark, E.A. Topp, From demonstrations to skills for high-level programming of industrial robots, in *AAAI Fall Symposium* (2016)
21. R.S. Sutton, A.G. Barto, *Reinforcement Learning: an Introduction*, vol. 1, no. 1 (MIT press, Cambridge, 1998)

22. U. Kartoun, H. Stern, Y. Edan, A human–robot collaborative reinforcement learning algorithm. J. Intell. Rob. Syst. **60**(2), 217–239 (2010)
23. S. Griffith, K. Subramanian, J. Scholz, C.L. Isbell, A.L. Thomaz, Policy shaping: integrating human feedback with reinforcement learning, in *Advances in Neural Information Processing Systems* (2013), pp. 2625–2633
24. W.B. Knox, P. Stone, C. Breazeal, Training a robot via human feedback: a case study, in *International Conference on Social Robotics* (Springer, 2013), pp. 460–470
25. H.B. Suay, S. Chernova, Effect of human guidance and state space size on interactive reinforcement learning, in *International Conference on Robot & Human Interactive Communication (RO-MAN)*, vol. 25 (2011), pp. 1–6
26. A.L. Thomaz, G. Hoffman, C. Breazeal, Reinforcement learning with human teachers: understanding how people want to teach robots, in *International Conference on Robot & Human Interactive Communication (RO-MAN)*, vol. 25 (2006), pp. 352–357
27. B. Peng, J. MacGlashan, R. Loftin, M.L. Littman, D.L. Roberts, M.E. Taylor, A need for speed: adapting agent action speed to improve task learning from non-expert humans, in *Proceedings of the International Joint Conference on Autonomous Agents and Multiagent Systems* (2016), pp. 957–965
28. W.B. Knox, P. Stone, Interactively shaping agents via human reinforcement: the TAMER framework. Proceedings of the Fifth International Conference on Knowledge Capture **5**, 9–16 (2009)
29. L. Rozo Castañeda, S. Calinon, D. Caldwell, P. Jimenez Schlegl, C. Torras, Learning collaborative impedance-based robot behaviors, in *Twenty-Seventh AAAI Conference on Artificial Intelligence*, vol. 27 (2013), pp. 1422–1428
30. L. Rozo Castañeda, S. Calinon, D. Caldwell, Learning force and position constraints in human–robot cooperative transportation, in *International Symposium on Robot and Human Interactive Communication*, vol. 23 (2014), pp. 619–624
31. S. C. Akkaladevi, M. Plasch, A. Pichler, B. Rinner, Human robot collaboration to reach a common goal in an assembly process, in *Proceedings of the European Starting AI Researcher Symposium (STAIRS)* (2016), pp. 3–14
32. A. Pichler, S.C. Akkaladevi, M. Ikeda, M. Hofmann, M. Plasch, C. Wögerer, G. Fritz, Towards shared autonomy for robotic tasks in manufacturing. Procedia Manuf. **11**, 72–82 (2017)
33. S. Akkaladevi, M. Plasch, A. Pichler, Skill-based learning of an assembly process, e & i Elektrotechnik und Informationstechnik **134**(6), 312–315 (2017)
34. V. Vyatkin, IEC 61499 as enabler of distributed and intelligent automation: state-of-the-art review. IEEE Trans. Industr. Inf. **7**(4), 768–781 (2011)
35. SYMBIO-TIC, Symbiotic human–robot collaborative assembly: technologies, innovations and competitiveness, *H2020-EU.2.1.5.1* (2019). https://www.symbio-tic.eu/, Last Accessed 08 Jan 2020
36. L. Wang, R.X. Gao, J. Váncza, J. Krüger, X.V. Wang, S. Makris, G. Chryssolouris, Symbiotic human–robot collaborative assembly. CIRP Ann. Manuf. Technol. **68**(2), 701–726 (2019)
37. ABB Robotics, Robot studio—the world's most used offline programming tool for robotics (2020). https://new.abb.com/products/robotics/robotstudio
38. Drag&Bot GmbH, Drag&Bot—easy robot operation instead of complicated programming (2020). https://www.dragandbot.com/product/
39. 4DIAC, Framework for distributed industrial automation and control (2020). https://www.eclipse.org/4diac/
40. SYMBIO-TIC Consortium, Demonstrator 3 completion report—automotive component assembly (2019). https://cordis.europa.eu/project/id/637107/results, Last Accessed 15 Jan 2020
41. V. Valeria, P. Fabio, L. Francesco, S. Cristian, Survey on human–robot collaboration in industrial settings: saefty, intuitive interfaces and applications. Mechatronics **55**, 248–266 (2018)
42. N.C. Chitturi, P. Kulha, M. Plasch, M. Ganglbauer, A. Weinbacher, M. Zillich, A. Pichler, Intuitive human–robot interaction with inkjet printed capacitive sensors, in *Fraunhofer Direct Digital Manufacturing Conference 2020* (2020)

43. TexasInstruments, FDC 1004 product description, in *Texas Instruments* (2020). https://www.ti.com/product/FDC1004
44. STMicroelectronics, STM32 32-bit Arm Cortex MCUs (2020). https://www.st.com/en/microcontrollers-microprocessors/stm32-32-bit-arm-cortex-mcus.html
45. UniversalRobots, UNIVERSAL ROBOTS (2020). https://www.universal-robots.com/products/ur10-robot/
46. UniversalRobots, About Universal Robots+ (2020). https://www.universal-robots.com/plus/
47. L. Hiesmair, M. Plasch, H. Nöhmayer, A. Pichler, Intuitive Human Machine Interaction based on multimodal command fusion and interpretation, in *Proceedings of the ARW & OAGM Workshop 2019* (2019). https://doi.org/10.3217/978-3-85125-663-5-02
48. F.C. Pereira, D.H. Warren, Definite clause grammars for language analysis—a survey of the formalism and a comparison with augmented transition networks. Artif. Intell. **13**(3), 231–278 (1980)
49. Profactor GmbH, SYMBIO-TIC multimodal robot programming (2019). https://youtu.be/AbJ8VaxxwzI, Last Accessed 20 Aug 2020
50. A. Billard, S. Calinon, R. Dillmann, S. Schaal, Robot programming by demonstration, in *Springer Handbook of Robotics* (2008), 1371–1394
51. M. Ikeda, S. Maddukuri, M. Hofmann, A. Pichler, X. Zhang, A. Polydoros, J. Piater, K. Winkler, K. Brenner, I. Harton, U. Neugebauer, FlexRoP—flexible, assistive robots for customized production. Austrian Robotics Workshop **2018**, 53–58 (2018)
52. X. Zhang, A. S. Polydoros, J. Piater, Learning movement assessment primitives for force interaction skills (2018). arXiv preprint arXiv:1805.04354
53. S.C. Akkaladevi, A. Pichler, M. Plasch, M. Ikeda, M. Hofmann, Skill-based programming of complex robotic assembly tasks for industrial application. e & i Elektrotechnik und Informationstechnik **136**(7), 326–333 (2019)
54. Vive.com, VIVE tracker (2020). https://www.vive.com/eu/vive-tracker. Last Accessed 13 Jan 2020
55. Ati-ia.com, ATI industrial automation: F/T sensor delta (2020). https://www.ati-ia.com. Last Accessed: 13 Jan 2020
56. S.C. Akkaladevi, M. Plasch, A. Pichler, M. Ikeda, Towards reinforcement based learning of an assembly process for human robot collaboration. Procedia Manuf **38**, 1491–1498 (2019)
57. S. Akkaladevi, M. Ankerl, C. Heindl, A. Pichler, Tracking multiple rigid symmetric and non-symmetric objects in real-time using depth data, in *International Conference on Robotics and Automation (ICRA)* (2016), pp. 5644–5649
58. S.C. Akkaladevi, M. Plasch, C. Eitzinger, B. Rinner, Context enhanced multi-object tracker for human robot collaboration, in *Proceedings of the ACM/IEEE International Conference on Human–Robot Interaction (late breaking reports)* (2017), pp. 61–62
59. S.C. Akkaladevi, C. Heindl, Action recognition for human robot interaction in industrial applications, in *International Conference on Computer Graphics, Vision and Information Security (CGVIS)* (2015), pp. 94–99
60. S.C. Akkaladevi, M. Plasch, S. Maddukuri, C. Eitzinger, A. Pichler, B. Rinner, Toward an interactive reinforcement based learning framework for human robot collaborative assembly processes. Front. Robot. AI **5**(126) (2018). https://doi.org/10.3389/frobt.2018.00126
61. Profactor GmbH, Towards an interactive reinforcement based learning framework for human robot collaboration—video demonstration (2018). https://www.youtube.com/watch?v=hIsWTUIlu3g. Last Accessed 20 Aug 2020
62. S.C. Akkaladevi, M. Plasch, C. Eitzinger, S.C. Maddukuri, B. Rinner, Towards learning to handle deviations using user preferences in a human robot collaboration scenario, in *8th International Conference on Intelligent Human Computer Interaction*, vol. 8, pp. 3–14 (2017)
63. S.C. Akkaladevi, M. Plasch, C. Eitzinger, A. Pichler, B. Rinner, Towards a context enhanced framework for multi object tracking in human robot collaboration, in *IEEE/RSJ International Conference on Intelligent Robots and Systems (IROS)* (Madrid, Spain, 2018) pp. 168–173
64. K. Ramirez-Amaro, M. Beetz, G. Cheng, Transferring skills to humanoid robots by extracting semantic representations from observations of human activities. Artif. Intell. **247**, 95–118 (2017)

65. Y. Mollard, T. Munzer, A. Baisero, M. Toussaint, M. Lopes, Robot programming from demonstration, feedback and transfer, in *IEEE/RSJ International Conference on Intelligent Robots and Systems (IROS)* (Hamburg, 2015), pp. 1825–1831
66. L. Ehrlinger, W. Wöß, Towards a definition of knowledge graphs. SEMANTiCS (Posters, Demos, SuCCESS) **48**, 1–4 (2016)
67. K. Munir, M.S. Anjum, The use of ontologies for effective knowledge modelling and information retrieval. Appl Comput Inform **14**(2), 116–126 (2018)
68. M. Tenorth, M. Beetz, Representations for robot knowledge in the know rob framework. Artif. Intell. **247**, 151–169 (2017)
69. B. Goertzel, Patterns, hypergraphs and embodied general intelligence, in *IEEE International Joint Conference on Neural Network Proceedings* (2006), pp. 451–458
70. D. Hart, B. Goertzel, Opencog: a software framework for integrative artificial general intelligence, in *AGI* (2008), pp. 468–472
71. S. Klarman, Modelling data with hypergraphs—a closer look at GRAKN.AI's hypergraph data model (2017). https://blog.grakn.ai/modelling-data-withhypergraphs-edff1e12edf0. Last Accessed 20 Aug 2020
72. Grakn Labs, GraknAI—A knowledge graph (2020). https://grakn.ai/about. Last Accessed 20 Aug 2020
73. ArtiMinds Robotics GmbH, Artiminds robot programming suite (2013–2020). https://www.artiminds.com/. Last Accessed 20 Aug 2020
75 OCTOPUZ Inc., OCTOPUZ offline robot programming software (2020). https://octopuz.com/. Last Accessed 20 Aug 2020
75. Hypertherm, Inc., Robotmaster—CAD/CAM for robots (2019). https://www.robotmaster.com/en/. Last Accessed 20 Aug 2020

Chapter 13
Sensorless Haptic Control for Physical Human–Robot Interaction

Sichao Liu, Lihui Wang, and Xi Vincent Wang

Abstract Industrial robots can offer fast speed and high accuracy for task execution but are far behind humans in terms of flexibility, adaptability, controllability, and even predictability. Therefore, combining the strength, accuracy, and speed of robots with flexibility, adaptability and controllability can achieve high-quality performance in human–robot interaction. In addition, humans in the manufacturing sector often need to adapt to the pre-planned/pre-defined tasks, dynamically. However, the robots used nowadays are controlled by pre-generated codes that cannot support effective human–robot interaction. Most industrial robots do not have built-in torque/force sensors in the joints and have difficulty in installing external force/torque sensors on the end-effectors. Additionally, the available signals from a robotic system are limited. To address such challenges, this chapter presents an approach to haptically controlling an industrial robot without using any external sensors for effortless and natural human–robot interaction. This sensorless haptic control approach is enabled by the dynamic models of the robot where only joint angles and joint torques are measurable. To have better interaction performance, accurate modelling of the dynamic model of the robot both in the presliding and sliding regimes is developed to estimate the joint torques where the friction model is thoroughly explored. The estimated external force applied to the robot by an operator is transferred into reference position and speed of the robot by an admittance controller, and adaptive admittance parameters are adopted to make human–robot interaction more natural and easier with accurate positioning and control and smooth movement. Finally, the proposed approach is validated by a case study in which a human operator interacts with an industrial robot.

S. Liu · L. Wang (✉) · X. V. Wang
Department of Production Engineering, KTH Royal Institute of Technology, Stockholm, Sweden
e-mail: lihuiw@kth.se

S. Liu
e-mail: sicliu@kth.se

X. V. Wang
e-mail: wangxi@kth.se

L. Wang et al. (eds.), *Advanced Human–Robot Collaboration in Manufacturing*,
https://doi.org/10.1007/978-3-030-69178-3_13

13.1 Introduction

Robots are capable of offering fast, precise, repetitive, and heavy-duty task execution on manufacturing shop floors but they lack the flexibility and adaptability of humans. Therefore, there has been an increasing interest in *human–robot interaction* (HRI) [1], which allows a human operator to work side by side with a robot in close proximity while combining the accuracy and strength of robots with the cognitive ability and flexibility of humans [2]. In such an interactive environment, the human operators and the robots share the same workspace and the same resources. According to the relationship between humans and robots classified by Wang et al. [3], interaction happens if a human and a robot sharing the same workspace are "communicating" with each other. One party interacts with the other through control commands, or any physical contact (either planned or unintended) occurring between them. Both the human and the robot can work on the same task but complete the task step by step in a sequential order. In parallel, a number of factors motivate the wide use of HRI in manufacturing. The combination of complementary strengths of humans and robots leads to increased productivity, flexibility and adaptability, increased robustness, higher degree of resilience, improved ergonomics, and more attractive working conditions [4–7]. In the last decades, research efforts on HRI have been numerous. Many approaches to facilitating multiple interaction modes and adaptive robot control have been reported in the literature [2, 8, 9]. The research of utilising a robot working as a collaborating partner in assembly lines in an unstructured environment has been active, such as in automated assembly systems [3], collaborative manufacturing environments [10], and hybrid manufacturing cells [11]. A robot-assisted but human-guided control solution shows promise in handling highly demanding HRI with enhanced flexibility and adaptability.

However, most industrial robots used today do not have built-in/external sensors facilitating HRI, which makes it difficult for a human operator to interact with them [12]. Additionally, mounting haptic sensors on the robot can be very expensive and often unpractical due to difficulty in mechanical integration. To address this issue, sensorless force detection of physical human–robot contact is viewed as a feasible solution in which external force detection methods are enabled by accurate modelling of the dynamic model of a robot [13, 14]. Thus, the contact force applied by a human operator on the robot can be detected accurately [15]. The detected external force/torque is used as an input for the admittance control that can transfer an external force/torque to the reference position and velocity of the robot [16]. To achieve better compliant behaviour in physical human–robot interaction (pHRI) and haptic applications, an adaptive admittance control strategy is adopted to adjust the admittance parameters online for more natural and easier interaction between a human and a robot [17, 18].

This chapter starts with the definition and applications of pHRI, and then provides detailed treatments on relevant issues with a focus on sensorless pHRI. The remainder of the chapter is organised as follows: Sect. 13.2 gives a short review of HRI, external force/torque estimation, friction model of the robotic system, and control strategies.

The system design of sensorless haptic control for HRI is proposed in Sect. 13.3. Section 13.4 describes the estimation of the contact force and torque for industrial robots. The haptic control strategy for HRI is given in Sect. 13.5. In Sect. 13.6, a case study is reported to demonstrate the efficiency of the proposed approach on robot control performance. Section 13.7 concludes the chapter and discusses future work.

13.2 Literature Review

13.2.1 Human–Robot Interaction

For better definition of HRI, the relationship between humans and robots in the context of manufacturing must be analysed and identified [19]. Wang et al. [20] analysed and summarised the relationship between humans and robots from the following perspectives: (1) workspace, (2) direct contact, (3) working task, (4) simultaneous process, and (5) sequential process. The relationship between a human and a robot is classified as human–robot coexistence, HRI, human–robot cooperation, and human–robot collaboration. HRI involves the physical interaction of humans and robots in a shared working environment with direct contact between humans and robots if necessary. Interaction, by definition, requires communication between robots and humans, which can be driven by voice and gesture commands, haptic instructions, and human thoughts in the form of brainwaves. HRI can certainly occur at the cognitive level with communication between humans and robots through many channels, and one party may guide or control the other locally or remotely, whereas pHRI occurs in the same workspace.

Many research efforts have been focused on the study and characterisation of pHRI [13, 21] that allows a human operator to work/collaborate with a robot in close proximity. In all robotic service and manufacturing tasks, HRI combines the adaptability and flexibility of human operators with the high performance of robots in terms of precision, speed, and payload [1, 22]. In this context, human safety is of utmost importance. Many approaches have been reported in the literature to ensure a safe HRI environment. Before adopting safety control strategies, the causes leading to possible collisions and accidents are to be identified and classified. Therefore, a number of safety strategies have been introduced with a focus on three types of robot safety, and they are classified as crash safety (only "safe"/controlled collisions), active safety (timely detection of imminent collisions), and adaptive safety (intervening in the operation of the hardware equipment and applying corrective actions) [23, 24]. Lasota et al. [25] presented a survey of the potential methods of ensuring safety during HRI and classified them into four major categories: (1) safety through control, (2) motion planning, (3) prediction, and (4) consideration of psychological factors. For constructing a safe HRI environment, the detection of a potential collision between a robot and a human is essential. Seven elementary phases of a collision event pipeline, namely (1) precollision, (2) detection, (3) isolation, (4) identification, (5) classification, (6) reaction, and (7) collision avoidance, are introduced with a focus on collision detection and collision avoidance in a shared environment with humans and robots [26]. Two methods were widely considered for human safety: (1) using a vision system to perform 3-dimensional (D) inspection through 3D models as well as skin colour detection to perform 3D tracking of the human body in a robotic cell [27], and (2) the inertial-sensor-based approach [28] using geometry representation of human operators through a special suit for motion capturing. For example, Schmidt and Wang [29, 30] developed an approach for collision detection between a robot and a human operator [31, 32] based on a Kinect depth camera, and the experimental results show that the collision between depth images and 3D models in an augmented environment can be detected in a timely manner. Fischer and Henrich [33] used multiple 3D depth images for fast collision detection of multiple unknown objects where the depth sensors placed around the work cell were used to observe a common surveilled 3D space. The acquired depth images are used to calculate the minimum distance to any obstacle, and then, the maximum velocity can be limited. Despite the advancement of the technologies developed, the safety standards, such as ISO 10218-1/2 [34, 35], are used to identify collaborative modes and build a safety rule for the interaction between a robot and a human, which are summarised as follows: safety-rated monitored stop, hand guiding, speed and separation monitoring, and power and force limiting.

In recent years, many lightweight collaborative robots offer opportunities for haptic robot control in a shared workspace where a human operator can guide the robot by hands towards a given target position [36–38]. This provides an intuitive control mode for the robot without the specialised expert knowledge of programming, which can avoid the shortages of conventional (re)programming. However, most of the existing industrial robots used in manufacturing are heavy-duty and need a large workspace. This makes the hand-guided control of the industrial robots unfeasible and unreliable. To address such challenges, other communication and control modalities are used for intuitive HRI, such as speech-based commands [39], gesture-based instructions, and brainwave-driven commands [40]. These multimodal intuitive control strategies have been used in many industrial applications of HRI including adaptive collaborative assembly.

Many application-oriented HRI cases show that HRI places the interplay of humans and robots into a cyber-physical environment where human and robotic agents interact in a shared working environment to solve some complex tasks which require the combination of their best, complementing competencies [1]. From the concept of CPS, HRI within a shared working environment can be viewed as an advanced CPS, supported by dynamic control algorithms and deployed devices. The physical and software components are tightly coupled to each other while the physical entities (robots, sensors, monitors, etc.) are controlled by the cyber entities (i.e., algorithms, models, data, events, and so on) [41–43]. Thus, CPS can enable such HRI in areas of dynamic task planning, active collision avoidance, and adaptive robot control. The representative examples of HRI applied in CPS and developed at KTH are summarised as follows: (1) a CPS application of HRI in assembly [44],

(2) a CPS-enabled remote HR system and an AR-based CPS for remote HRI [45], and (3) remote HRI: a CPS application in a hazard manufacturing environment [46]. In [44], humans are the physical part of the CPS, and human instructions (voice commands, gesture instructions, and haptic interactions) are defined as the high-level control commands to control the assembly task execution. For some industrial scenarios that are harmful to human operators, a CPS-enabled remote human–robot system can be applied as a solution for remote interaction between a human operator and a robot. In the physical part, a collaborative robot in a collaborative workstation is controlled by a human operator's haptic lead-through with respect to the AR display of the assembly parts from the cyber representations of a remote workstation. In addition, the industrial applications with pHRI can be found in automated assembly, manual welding [47], automotive manufacturing, cellular manufacturing [48], hybrid manufacturing cells [11], and relevant manufacturing paradigms [5].

13.2.2 External Force/Torque Estimation

Many approaches on the haptic robot control can be classified as sensor-based and sensorless control schemes. The former uses force/torque sensors which are built into the robot joints or mounted at the end-effector to facilitate a human operator to control the robots [49, 50]. Some collaborative robots with built-in torque sensors in the robot joints (Universal Robots [51], KUKA IIWA [52], and ABB YuMi [53]) can allow a human operator to work in a haptic interaction mode in an easy and natural way. However, the cost of these built-in sensors can be up to the total cost of most of the collaborative robots. Most heavy-duty industrial robots do not have any built-in force/torque sensors. A possible solution to haptically control an industrial robot is installing a force/torque sensor at the end-effector where the force applied by a human operator can be sensed and converted into motion control signals for the robot. This control scheme still faces some challenges such as the high cost of sensors and the difficulty in mechanical integration. Therefore, a sensorless haptic control scheme for the heavy-duty industrial robots can be viewed as a promising and attractive solution where the dynamic model of the robot is developed to obtain the control output (the predicted torque), and the detected external force is converted into the reference velocity and the reference position. To enable sensorless haptic control, the external force/torque estimation of the robotic system is the key and a prerequisite.

Many research efforts on external force detection have been reported in the last decades. As reported in the literature, relevant shortages of direct detection with sensors hinder the wide use of external force/torque sensors in robotic systems. Sensorless external force detection, on the other hand, heavily relies on the dynamic modelling of the robotic systems [54, 55]. The accuracy of the dynamic model of the robot can have a direct impact on the accuracy of the external force detection. In some of the external force estimation methods, the external force is defined as the difference between the control output of the dynamic model of the robot and the feedback data from the robot controller where the robot controller used can be accessed through an external interface [12]. An identified inverse dynamic model of a robot is used in the estimation of the external force, in which a least-squares-method-based parameter identification model is required. In addition, adaptive methods used for observing the unknown inertial parameters were also developed to calibrate unknown robot dynamics [56]. These methods can be largely influenced by model uncertainty and the uncertainty of the identified inertial parameters [54]. A dithering-signal-based approach was used to estimate the external force when a robot is not moving [57]. However, this method can cause potential damages to the robot. Most of the contact force estimation methods depend on the residual based on the generalised momentum, while the inertial parameters are required [58]. This method avoids the calculation of joint accelerations by the differential operation of positions and reduces the effect of the noise induced by the differential operation. Furthermore, disturbance observers and adaptive force observers, as well as force estimators, were explored [59 – 61], in which the external force can be viewed as a disturbance. Motor-signal-based solutions are also applied to calculate the contact force in the collaborative assembly [57, 62], however, they require measurable reference currents of the controller. These model-based external force detection methods can easily be affected by unmodelled factors, such as nonlinear characteristics, temperature, and mechanical properties, and cannot produce accurate force estimation.

Recently, neural networks (NN), having the powerful ability of approximation and learning, have been successfully developed to compensate for unknown nonlinear dynamics and unstructured uncertainties. Therefore, one potential solution of the external force estimation that has been actively explored as an alternative way is the NN-based external force estimation method [63]. The strength of NN can be applied in the external force estimation to reduce the drawbacks induced by the model-based methods. Smith et al. [64] proposed two neural-network-based force/torque observers, which are independent of the dynamic model of the robot or haptic devices. In parallel, a model based on Convolutional Neural Networks and Long Short-Term Memory (LSTM) networks represented a potential approach to sensing the interaction forces between surgical instruments and soft tissues [65].

13.2.3 Friction Model for Robot Systems

Friction exists in almost all mechanical systems, which is a complex physical phenomenon affected by a number of factors such as temperature, force/torque levels, position, velocity, acceleration, and lubricant properties [66]. It can lead to tracking errors, and stick and slip motions. For a robotic system, friction in the robot joints has a direct relation to the control performance of the robot. The more accurate the identified friction in the joint space, the better the force-control performance. Friction is a highly nonlinear function of diverse variable parameters, and its behaviour can be divided into two operating regimes: the presliding and sliding regimes [67].

In the presliding regime, the adhesive forces (at asperity contacts) are dominant such that the friction appears to be a function of displacement rather than velocity [68, 69]. This is so because the asperity junctions deform elasto-plastically, thus behaving as nonlinear springs. As the displacement increases, more and more junctions will break, eventually resulting in working in the sliding regime. In the sliding regime, all the asperity junctions are broken apart such that the friction is a function of the velocity [70].

In a mechanical system, the high nonlinearity of friction makes it difficult to obtain models that can describe all the empirical observations. The complex interaction of components, such as gears, bearings, and shafts which are rotating/sliding at different velocities in a robot joint, makes physical modelling difficult [66]. Therefore, friction modelling has been in the research spotlight in manufacturing industries, especially for robotic systems. The study efforts on the friction modelling started from a pure Coulomb model up to complex static and dynamic models where a basic concept of friction is defined as a force that opposes displacement [71]. The models describing friction between two surfaces can be classified into two major groups: static friction models and dynamic friction models [72].

The simplest static friction model is the *Coulomb model* which only considers Coulomb friction and viscous friction [71]. Then, the static friction is taken into account in the friction model. A Stribeck effect of friction is described in which friction has a direct relation with the displacement and velocity [70]. In a *Stribeck model* [73], when the velocity is near zero, the static friction part dominates over others; when the velocity satisfies a certain condition, both the nonlinear and linear parts work; when the velocity is big enough, the friction is mainly linear. Then, the dynamic effect of friction can easily be captured compared with the static friction due to the significant change of the displacement and velocity. Dynamic models [72] (as opposed to static models) make it possible to recreate the processes that occur in a stiction friction regime and at very small sliding velocities, e.g., the hysteresis of presliding friction, frictional lag, stick–slip, and the phenomenon of breakaway force.

The *Dahl model* is one of the earliest dynamic friction models. It has been designed to simulate a symmetrical hysteresis loop observed in bearings subjected to sinusoidal excitations with low amplitudes [74]. A survey of models, analysis tools, and compensation methods for the control of machines with friction was discussed in [75], and a general model structure of friction based on the experimental observation of the friction properties was developed where a switching function between the sliding regime and the presliding regime was proposed for a smooth switch [75]. However, this type of model is difficult to be implemented in a real industrial application. To overcome this drawback, an integrated friction model, known as the *LuGre model*, was formulated for using a set of equations integrating the sliding and presliding regimes without the need of a switching function [76]. Then, a more elaborated friction model developed by Swevers et al., defined as the *Leuven* model [77], included the friction properties of the *LuGre* model and more accurate modelling of the presliding regime using a hysteresis function with nonlocal memory. The *Leuven* model consists of a friction equation and a nonlinear state equation. The *Leuven* model structure allows accurate modelling both in the presliding and the sliding regimes. The model incorporates a hysteresis function with nonlocal memory and arbitrary transition curves. The shortcoming of the integrated model developed by Swevers et al. is the discontinuity in the friction upon closing a hysteresis loop. To tackle this issue, Lampaert et al. [67] gave some modifications to the *Leuven* model. One modification overcomes a detected shortcoming of the original *Leuven* model: a discontinuity in the friction which occurs during certain transitions in presliding. Another modification, using the *generalised Maxwell slip* (GMS) model to implement the hysteresis force, eliminates the problem of stack overflow, which can occur with the implementation of the hysteresis force described in [77]. The GMS model is comprised of several elastoslide elements in parallel, and can better describe the full characteristics of the friction from the presliding regime to the sliding regime [72]. When the force is applied to the component, the asperity junctions deform elasto-plastically. All the elements are sticking and the initial stiffness will be the sum of all the elemental stiffness. When the force on an element reaches the saturation level, the element starts to slip and a new value must be calculated. The total stiffness of the whole system decreases with the spring stiffness of the elements. The whole system is in stick when at least one element is still in stick, the whole system is in slip when all the elements slip.

The model-based friction compensation methods have certain limitations on the identification of the nonlinear part of the friction. To have more accurate modelling of the friction, especially for the nonlinear part, NNs are applied to obtain more accurate compensation of the friction with less computational effort. Neural-network approximation of piecewise continuous functions was used for the friction compensation in the rigid-link robotic systems [78]. Artificial NN-based compensation of friction was developed as generic identifiers to compensate for unmodelled friction effects [79].

13.2.4 Control Strategies for HRI

Robotic control strategies play an important role in the control performance of robotic systems. HRI requires a close interaction between a human and a robot in a shared working environment where a compliance behaviour of the robot is essential to humans. Two main compliant control schemes used in HRI and haptic applications are *impedance control* and *admittance control* [16]. An impedance controller and an admittance controller are the opposite of each other in which the impedance controller accepts the displacement as an input and outputs the force, while the admittance controller can transfer the force into the reference velocity and reference displacement [80]. Impedance control is widely used in the compliant motion of industrial manipulators interacting with the environment. It can build a relationship between a manipulator's end-effector and the force. Robotic systems with impedance control have stable dynamic interaction with stiff environments but have poor accuracy in free space due to friction and unmodelled dynamics. Most of the industrial robots do not have built-in force/torque sensors, thus accurate feedback data of force/torque are difficult to be obtained. Collaborative robots with built-in force/torque sensors offer opportunities for adapting impedance control in robotic motion control. For example, impedance control has been employed in the inner loop torque sensing and control in DLR's lightweight robot [81] and ATR's Humanoid [82]. To have a better performance of pHRI, a variable Cartesian impedance control was deployed to achieve a compliant behaviour of the robot's end-effector in response to forces exerted by the human operator [83]. The performance during pHRI can be enhanced by combining Cartesian impedance modulation and redundancy resolution. The experimental study that is deployed on a 7-degree-of-freedom KUKA LWR4 shows a better performance of variable impedance control in contrast to constant impedance control.

Compared to impedance control, admittance control can provide a high level of accuracy in non-contact tasks but has instability issues when dynamically interacting with a stiff environment. Many research efforts on pHRI with admittance control have been reported. Keemink et al. [84] presented an overview of admittance control as a method of physical interaction control between machines and humans, and an admittance controller framework and elaborate control scheme were also presented for controller design and development. Considering the complementarity of these two control strategies on stability and performance properties, a hybrid system approach that incorporates impedance and admittance control as extreme cases was proposed to enable the continuous switch between them [16]. Kaigom and Roßmann [85] exploited the comprehensive, unique capabilities of the eRobotics methodology to shift adaptable joint admittance control into virtual reality. The advantage is that the targeted joint dynamics are replicated with substantially less effort and at low costs. A control architecture for industrial robotic applications with the compensation of load dynamics, using an admittance control scheme and direct sensing of the human inputs, was developed for HRI, and this control scheme can support the operator of an industrial robot, equipped with a force/torque (F/T) sensor on the end-effector, during collaborative tasks involving heavy payloads carried by the robot [86]. To have an intuitive interaction between a human operator and a robot, a variable admittance control for a 4-degree-of-freedom intelligent assisting device was proposed to improve the system [87]. Meanwhile, adaptive control schemes for industrial robots have been reported in recent years for a better compliant control performance [18]. A new inner-loop/outer-loop robot controller formulation was developed to improve the robustness of pHRI for changes in both robots and human users. Within this control scheme, an inner-loop controller with guaranteed robustness and stability enables a robot to behave like a prescribed admittance model, and an outer-loop controller tunes the admittance model so that the robot system assists humans with varying levels of skills to achieve task-specific objectives [88]. Li et al. [89] proposed a novel control method for series elastic actuator-driven robots used in HRI. Landi et al. [84] presented a strategy for detecting deviations from the nominal behaviour of an admittance-controlled robot and for adapting the parameters of the controller while guaranteeing passivity. For a robot team, Li et al. [90] proposed a novel adaptive control methodology based on the admittance model for multiple manipulators transporting a rigid object cooperatively along a pre-defined desired trajectory.

Lately, NN-based and learning-based admittance controllers were widely designed and developed to achieve effective interaction by modifying the admittance parameters. In [91], an adaptive neural controller using radial basis functions was deployed to guarantee trajectory tracking during interaction with an unknown environment. In [92], a fuzzy learning variable admittance control scheme for a KUKA LWR robot was proposed for human–robot cooperation tasks. In parallel, in [93], a multilayer feedforward neural network was designed using the Cartesian velocity of the robot and the applied force by the operator as its inputs to online modify the virtual damping of the admittance controller.

13.3 System Design

The proposed system allows an operator to haptically interact with a robot without using any haptic device. Instead, it uses a sensorless haptic control strategy to enable the natural interaction between a human operator and a robot with less effort. The system design of the proposed sensorless haptic control for HRI is shown in Fig. 13.1, and the overview of sensorless haptic control for an industrial robot incorporates an HRI and a KRC (KUKA Robot Controller) module, a kinematics and dynamics module, a robotic control module, and a communication module.

In the HRI and KRC modules, a sensorless haptic control strategy for the industrial robot operated by the human workers in a dynamic and unstructured environment is developed to control the robot by a haptic command from the human workers. KRC is responsible for the excitation of the motion and control commands. KRC is not open for robot users, and feedback data available from the robot controller are joint position and joint torque only. The axis-specific actual position of the robot and the current motor torque of an axis are stored in the system variable $AXIS_ACT_MEAS and $TORQUE_AXIS_ACT, respectively. To get access to the feedback data from and send movement commands to the KRC, the communication between the human-machine interface (HMI) and KRC is built based on the open-source cross-platform communication interface *JOpenShowVar* [94] which allows for reading and writing variables and data structures of the controlled manipulators.

The kinematics and dynamics module is responsible for the kinematic and dynamic modelling of the industrial robot. The geometrical parameters of the industrial robot are acquired by the modified DH (Denavit–Hartenberg) coordinate frames. Velocity and acceleration of the joints are obtained by the differential of the available joint positions. The acquired feedback data (joint positions and torques), velocity, and acceleration of the robot are filtered to reduce the effect of the noise induced by the differential. The built-in kinematic model is used for the transfer between the joint positions in the joint space and the end-effector's positions in the Cartesian space, while the dynamic model is used to obtain the control output. The nonlinear Newton–Euler equation of robot dynamics is formulated as a linear identification model with certain inputs (joint positions, velocities, accelerations, and torques). Additionally, the friction in the robot's joints is also considered in the dynamic

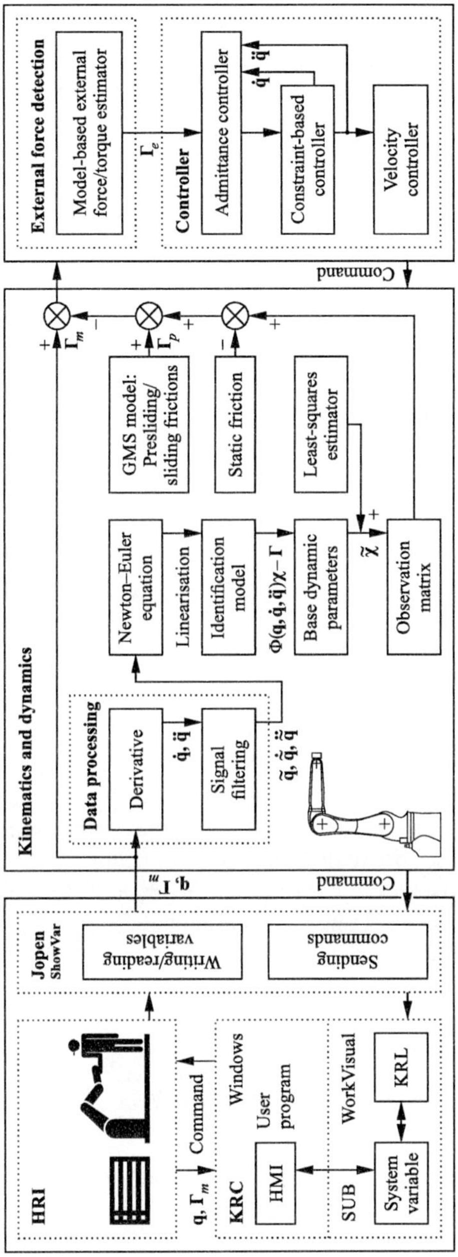

Fig. 13.1 System design of sensorless haptic control for HRI

model for better control performance. The next step for sensorless haptic control is the external force/torque detection which is the key of haptic robot control. The output of the dynamic model of the robot is defined as predicted torques, while the feedback data of the joint torques acquired from the robot controller are the measured torques. If a human operator does not apply any force to the end-effector of the robot, no external force is detected which indicates the predicted torques equal the measured ones. When the human operator applies a force on the robot, the external force induced by this can be detected by the external force detection module where the external force is defined as the difference between the predicted force and measured force. The predicted force can be calculated by the Jacobian matrix based on the acquired joint torques.

The robotic control is to make the robot move when the external force is applied to the robot by a human operator. Then, an adaptive admittance controller is designed to transfer the external force into the reference position and velocity which are defined as the control output. To make the interaction between a robot and a human more natural and easier, adaptive admittance parameters are adopted in the controller. The control output commands are sent to the robot controller for the movement and task execution.

13.4 Estimation of the Contact Force/Torque for Industrial Robots

In this section, the geometric, kinematic, and dynamic models of the robot are presented. The external signals from the robot controllers are obtained via a built-in communication channel when external forces are applied by the human operators. Based on the dynamic model built, the output of the robot system can be defined as a predicted force/torque. The estimation of the external force/torque for the robotic system can be the difference between the predicted force/torque and the force/torque signals measured from the robot controller. In this dynamic model, the friction in the robotic system is considered, and the friction models both in the presliding and sliding regimes are developed which can contribute to good control performance of the robotic system.

13.4.1 Geometric Model and DH Parameters of Serial Robots

This section presents the geometric model of a manipulator, and then DH geometric parameters are identified. Considering the popularity of robots in manufacturing, the geometric model of an industrial robot with 6 degrees of freedom is taken as an example which can show the whole process of modelling the geometry of the industrial robots. The MDH-based coordinate frames of an industrial robot are shown in

Fig. 13.2a, and the geometric model of a real industrial robot (KUKA KR6 R700 sixx [95]) that is used in a case study in the following section is shown in Fig. 13.2b. The corresponding geometric parameters based on the modified DH (MDH) reference coordinate systems built are shown in Table 13.1.

Based on the reference coordinate frames specified for the manipulator, the kinematic model can be developed to obtain the transfer relation between the end-effector's pose in the Cartesian space and the position in the joint space. The transfer function is formulated as follows:

$$\mathbf{X} = f(\mathbf{q}), \tag{13.1}$$

where $\mathbf{X}$ is the end-effector's pose in the workspace, and $\mathbf{q}$ is the robot joint position. The relation between the velocity in the Cartesian space and the joint velocity can be derived by the differential of the transfer function, and is described by

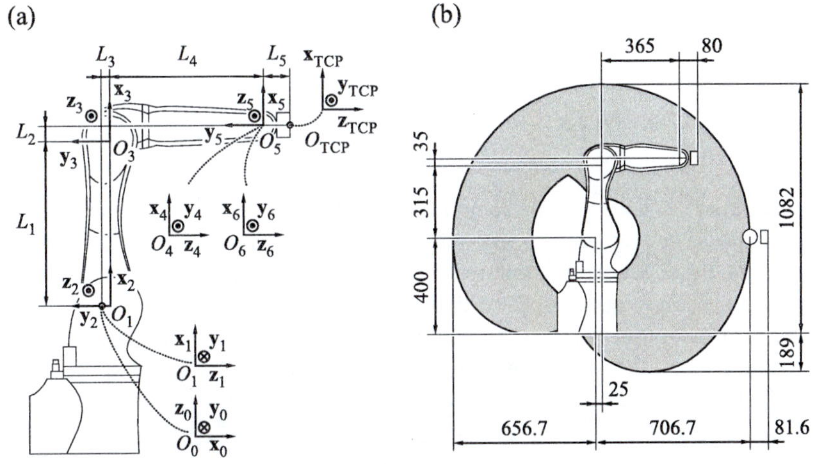

Fig. 13.2 MDH-based coordinate frames of an industrial robot

Table 13.1 Geometric parameters based on the MDH reference coordinate frames built

Joint	α_j	d_j	θ_j	γ_j
1	0	0	q_1	0
2	$\pi/2$	L_3	$\pi/2 + q_2$	0
3	0	L_1	q_3	0
4	$\pi/2$	L_2	q_4	L_4
5	$-\pi/2$	0	q_5	0
6	$\pi/2$	0	q_6	0
TCP	0	0	0	L_5

$$\dot{\mathbf{X}} = \mathbf{J}\dot{\mathbf{q}}, \quad (13.2)$$

where $\dot{\mathbf{X}}$ is the differential of the end-effector's position, and $\dot{\mathbf{q}}$ is the robot joint velocity. $\mathbf{J}$ is a Jacobian matrix. This forward kinematic model can be used to compute the linear and angular velocities of the end-effector in terms of the joint velocities. Then, an inverse kinematic model can be obtained from formula (13.3). It should be pointed out that the inverse of a basic Jacobian matrix is affected by the singularity of the robot, and the inverse of the Jacobian matrix is introduced in [95].

$$\dot{\mathbf{q}} = \mathbf{J}^{-1}\dot{\mathbf{X}} \quad (13.3)$$

13.4.2 Dynamic Model of Serial Robots

The dynamic model of a serial robot with N degrees of freedom can be formulated as follows:

$$\mathbf{M(q)}\ddot{\mathbf{q}} + \mathbf{C}(\mathbf{q}, \dot{\mathbf{q}})\dot{\mathbf{q}} + \mathbf{G(q)} + \boldsymbol{\tau}_f(\dot{\mathbf{q}}) = \boldsymbol{\tau}, \quad (13.4)$$

where $\mathbf{q} \in \mathbb{R}^N$ is the vector of joint variables, $\mathbf{M(q)} \in \mathbb{R}^{N\times N}$ is the mass matrix, which is positive definite, $\mathbf{C}(\mathbf{q}, \dot{\mathbf{q}}) \in \mathbb{R}^{N\times N}$ describes the centrifugal and Coriolis effects, gravity effect can be captured by $\mathbf{G(q)} \in \mathbb{R}^N$, $\boldsymbol{\tau}_f(\dot{\mathbf{q}}) \in \mathbb{R}^N$ denotes friction torques in the joints, and $\boldsymbol{\tau} \in \mathbb{R}^N$ is the joint torque.

The external torques in the robot joints are resulting from the contact force in the Cartesian space of the robot, and a wrench in Cartesian space $\mathbf{f} \in \mathbb{R}^M$ applied to the end-effector and the external torques $\boldsymbol{\tau}_e \in \mathbb{R}^N$ are related by

$$\boldsymbol{\tau}_e = \mathbf{J}^T(\mathbf{q})\mathbf{f}, \quad (13.5)$$

where $\mathbf{J(q)} \in \mathbb{R}^{M\times N}$ is the Jacobian matrix of the robot. According to (13.4), the errors of the torque estimation arise from the inexact compensation of the dynamics, gravity, and friction effects, which derives from unmodelled effects and uncertainties of parameters. The accurate modelling of robot dynamics produces small and ignorable errors from the compensation of the dynamics and gravity. Then, the main error of the dynamic model of the manipulator is induced by the compensation of friction.

13.4.3 Friction Model of the Robot System

The Stribeck model, as the most classical nonlinear friction model, describes many of the friction characteristics. When the velocity is near zero, the static friction part is dominant over others; when the velocity meets a certain threshold range, both the linear and nonlinear parts are remarkable; when the velocity is larger than this threshold, the friction is mainly linear and the nonlinear effect on the friction becomes very small. The Stribeck friction model is formulated by

$$F_f(v) = F_c + (F_s - F_c)e^{-|\frac{v}{v_s}|^{\sigma}} + F_v v, \tag{13.6}$$

where F_c is the Coulomb friction, F_s is the static friction, v is the velocity of the joint, v_s is the Stribeck velocity, σ is the exponent of the Stribeck nonlinearity, and F_v is the viscous friction.

When a single robot joint moves at a constant velocity, the joint torque that derives from the robot movement consists of the gravity and friction torques. Single-joint movement does not produce the centrifugal and Coriolis forces. The dynamic model can be simplified to estimate the joint friction, given by

$$G(q) + \tau_f(\dot{q}) = \tau. \tag{13.7}$$

The estimation of direction-dependent friction is given by

$$\tau_f\left(\tilde{\dot{q}}\right) = (\tau^+ - \tau^-)/2, \tag{13.8}$$

where $\tilde{\dot{q}}$ is the constant joint velocity. τ^+ and τ^- are the torques along the clockwise and counterclockwise directions. Each joint is commanded to move separately with a constant velocity in both directions about a given joint angle. Figure 13.3 shows the experimental excitation trajectory of Stribeck friction estimation of Joint 3 at a velocity of 0.314 rad/s where the friction is obtained using (13.8).

The measurement data of the Stribeck curve for each robot joint with N samples are obtained to identify unknown parameters. The objective function that minimises the residuals of predicted and measured friction torques is formulated as follows:

$$f = \operatorname{argmin} \sum_{k=1}^{N} (\tau_{f,k} - F_{f,k})^2, \tag{13.9}$$

where k is the number of measurement data. The nonlinear solver is adopted to obtain the solutions of the objective function (13.9). Therefore, the sequential quadratic programming (SQP) method, which is characterised by super-linear convergence, is adopted to solve (13.9) with nonlinear characteristics. Here, for simplicity, Joint 3

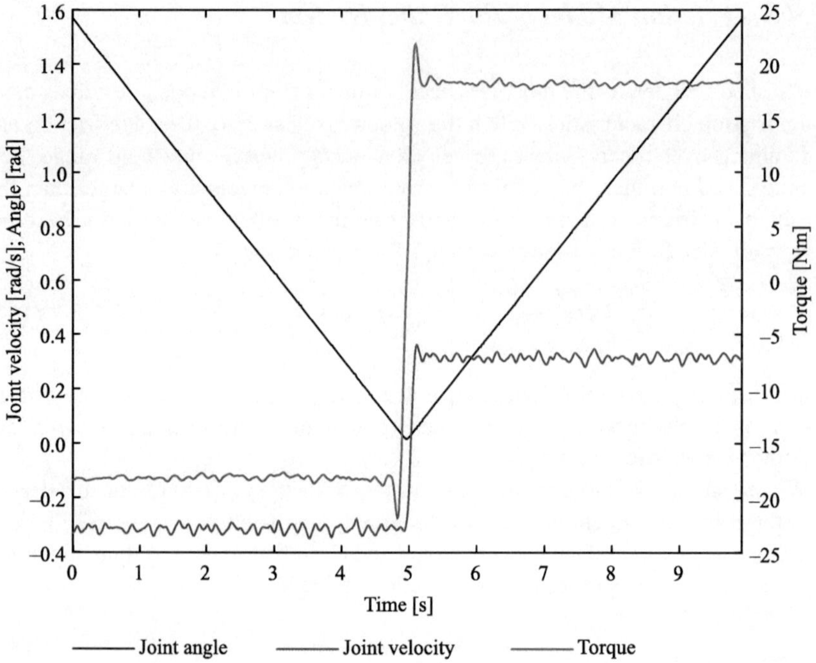

Fig. 13.3 Experimental excitation trajectory of the Stribeck friction estimation of Joint 3 at a velocity of 0.314 rad/s

is taken to show the identification process. The measured friction curves of Joint 3 obtained from the clockwise and counterclockwise motions are shown in Fig. 13.4. The measurement data of Joint 3 (solid blue line) are used to identify the Stribeck parameters, and the dashed red line in Fig. 13.4 is the predicted friction torques. The measured friction shows clear velocity-weakening and velocity-strengthening regimes. Friction shows a nonlinear decrease when the velocity increases from zero until reaching a certain velocity threshold. Then, the friction increases with the increase of the velocity. The identified Stribeck parameters (in Table 13.2) have a capacity of fitting the measured friction of Joint 3.

To have a good control performance of the robot, the friction model in the static state is considered, which is different from the one at a constant velocity. The Stribeck friction model discussed above can identify the friction in the sliding regime in which all the asperity junctions have separated totally. However, the friction in the static state or a presliding regime is a hysteresis function of displacement with nonlocal memory. Therefore, the GMS model is adopted for the accurate modelling of the friction in the presliding regime, and the GMS model consists of a parallel connection of N elastoslide elements. The hysteresis force F_i of each Maxwell-slip element is given as

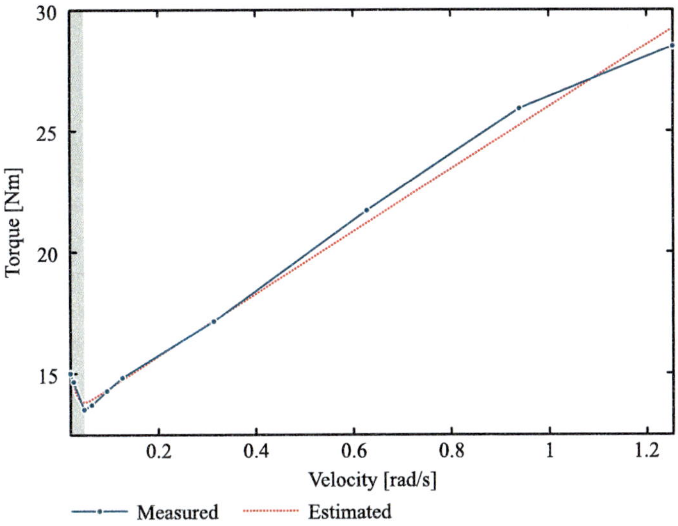

Fig. 13.4 Stribeck curves of measured and predicted joint friction for Joint 3

Table 13.2 Estimated values of Stribeck parameters for Joint 3

Joint	F_c	F_s	F_v	v_s	σ
3	13.157	16.267	12.822	0.024	2

$$\begin{aligned} & If\,|z-\zeta_i| < \tfrac{W_i}{k_i}, \\ & then\begin{cases} F_i = k_i(z-\zeta_i) \\ \zeta_i = const \end{cases} \\ & \text{else}\begin{cases} F_i = sgn(z-\zeta_i)W_i \\ \zeta_i = z - sgn(z-\zeta_i)\tfrac{W_i}{\zeta_i} \end{cases} \end{aligned}, \tag{13.10}$$

where z is the common displacement, ζ_i is the position of element i, k_i is the stiffness, and W_i is the maximum force of the element.

For the GMS model, the friction is defined as the sum of the hysteresis forces of N elastoslide elements, and is given by

$$F_h = \sum_{i=1}^{N} F_i. \tag{13.11}$$

The working procedure of the GMS model can be summarised as follows. The Maxwell-slip element in the initial state is assumed to have zero deflection of the asperity junctions, which means zero relative displacement. When a small force is applied to the element, a deflection of the asperity junctions between the element and the contacting surface is produced while the element keeps stuck. When the applied force increases, the element starts sliding until the force reaches the maximum force of the element. Then, the Maxwell slip system is in the sliding regime when all the elements slip, otherwise, still stick.

To obtain the GMS model, the identification of the unknown parameters of the Maxwell-slip elements is viewed as a nonlinear problem. Then, a periodic sine signal with small amplitudes and low frequencies on the robot is used as the excitation trajectory to estimate the unknown parameters, and the designed trajectory can ensure that the whole process of the excitation stays in the presliding regime. The objective function is formulated as Eq. (13.12), which aims to minimise the sum of squares of the residuals between the measured and predicted presliding friction.

$$F = \sum_{j}^{N} (F_h(j) - \widehat{F}_h(j))^2, \tag{13.12}$$

where $F_h(j)$ is the experimental measurement data of the presliding friction, j is sampling number, N is number of samples, and $\widehat{F}_h(j)$ is the predicted presliding friction.

A sinusoidal signal excitation trajectory for Joint 3 with different amplitudes and frequencies, as shown in Fig. 13.5 (left), is used to identify the unknown parameters in which five Maxwell-slip elements are adopted to approximate the hysteresis curve. The objective function is characterised by the nonlinearity and a nonlinear solver with constraints is adopted. The identified parameter values of the GMS model are shown in Table 13.3. The corresponding mean square quadratic error (MSE) is 0.1571 N^2m^2, which shows a good performance of reducing the tracking errors. The amplitude of the reference trajectory for Joint 3 is 1×10^{-5} rad. In order to reduce the impact of the gravity-induced torque of the robot joints, the excitation trajectory of parameter identification for each joint is operated in the vicinity of the gravity torque-free pose. In this case, the measured torque equals the friction at the

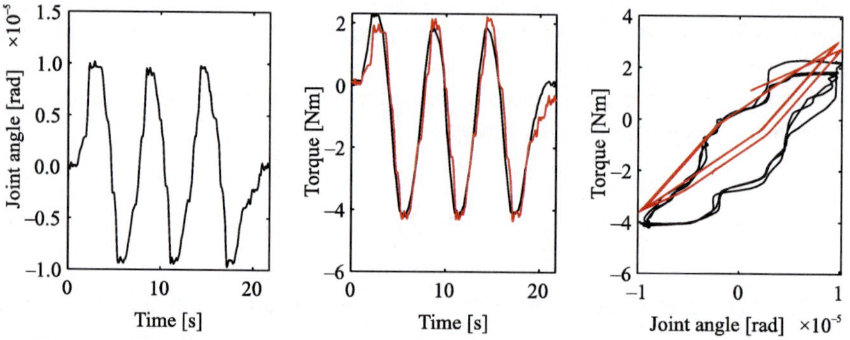

Fig. 13.5 A sinusoidal signal excitation trajectory (left); Measured and predicted friction torques versus time (middle); Measured and predicted friction torques versus relative displacement (right)

Table 13.3 Identified GMS parameters for Joint 3

Para	k_1	W_1	ζ_1	k_2	W_2	ζ_2	k_3
Joint 3	43,138.494	4.617	$9.74 \cdot 10^{-5}$	43,138.616	4.895	$-9.91 \cdot 10^{-4}$	43,138.676
W_3	ζ_3	k_4	W_4	ζ_4	k_5	W_5	ζ_5
4.916	$-9.83 \cdot 10^{-4}$	43,138.193	3.424	$6.90 \cdot 10^{-5}$	43,138.417	3.648	$6.61 \cdot 10^{-5}$

steady state. The experimental sampling frequency is approximately 83 Hz, which is determined by the communication frequency of the KRC. It can be found that all the elements are guaranteed in the presliding regime. Figure 13.5 (middle) shows the measured presliding friction (blue line) and the predicted friction (red line) in the presliding regimes over time. Figure 13.5 (right) shows the hysteresis curve with nonlocal memory in the presliding regime for Joint 3.

Then, the dynamic parameters of the dynamic model need to be identified, which determine the dynamic and gravity effects. The observation model of robot dynamics can be formulated as

$$\boldsymbol{\Phi}(\mathbf{q}, \dot{\mathbf{q}}, \ddot{\mathbf{q}})\boldsymbol{\chi} + \hat{\boldsymbol{\tau}}_f = \boldsymbol{\tau}, \tag{13.13}$$

where $\boldsymbol{\Phi}$ is an $(n \times m)$ observation matrix, n is the number of equations, m is the number of inertial parameters, $\boldsymbol{\chi} \in \mathbb{R}^N$ is the vector of the inertial parameters of the dynamic model, and $\hat{\boldsymbol{\tau}}_f \in \mathbb{R}^N$ is the predicted friction in the presliding and sliding regimes obtained by (13.6) and (13.11). Each joint of the robot has 10 inertial parameters. However, not all of them have an effect on robot dynamics. Therefore, some of these parameters can be grouped, and the parameters that do not affect the behaviour of the dynamic model can be removed. Then, a weighted least-squares method is adopted for parameter identification, and the estimation value of $\boldsymbol{\chi}$ is formulated as

$$\tilde{\boldsymbol{\chi}} = (\mathbf{W}_b^T \boldsymbol{\Sigma}^{-1} \mathbf{W}_b)^{-1} \mathbf{W}_b^T \boldsymbol{\Sigma}^{-1}, \tag{13.14}$$

where $\tilde{\boldsymbol{\chi}}$ represents the base parameters, $\mathbf{W}_b$ is a matrix, $\boldsymbol{\Sigma}$ is the covariance of the prediction error of the motor torque, and b is the number of via points along a specific robot path. To identify basic dynamic parameters, appropriate exciting trajectories need to be designed and Finite Fourier series is chosen as the exciting trajectory to excite the dynamic parameters of the robot.

13.4.4 Estimation of External Force/Torque

In the dynamic model of the robot, the GMS model is adopted to describe the behaviour of friction transition between the presliding and sliding regimes. However, the GMS model presents a function relation of the friction and relevant displacement. The Stribeck friction model describes most of the friction characteristics which are difficult to observe in the transition behaviour of friction in the presliding regimes. To obtain a smooth transition of the presliding regime and the sliding regime, the threshold velocity ϵ is defined to determine the state bound of the presliding and sliding regimes of the robot joint. According to the defined velocity threshold, the dynamic model can be classified by the presliding and sliding regimes and can be reformulated as

$$\mathbf{M}(\mathbf{q})\ddot{\mathbf{q}} + \mathbf{C}(\mathbf{q}, \dot{\mathbf{q}})\dot{\mathbf{q}} + \mathbf{G}(\mathbf{q}) + \mathbf{F}_h = \boldsymbol{\tau}, |\dot{\mathbf{q}}| < \varepsilon \tag{13.15a}$$

$$\mathbf{M}(\mathbf{q})\ddot{\mathbf{q}} + \mathbf{C}(\mathbf{q}, \dot{\mathbf{q}})\dot{\mathbf{q}} + \mathbf{G}(\mathbf{q}) + \mathbf{F}_f(\dot{\mathbf{q}}) = \boldsymbol{\tau}, |\dot{\mathbf{q}}| \geq \boldsymbol{\varepsilon}, \tag{13.15b}$$

where the dynamics of the robot in the presliding regime is described in (13.15a), and (13.15b) is the dynamics of the robot in the sliding regime. Then, the external force estimation can be defined. When an external force is applied to the robot, the detected external torque is the torque residual between the measured joint torques and estimated joint torques, where the measured joint torque can be obtained from the robot controller. The external torques on the joints can be formulated as

$$\boldsymbol{\tau}_e = \boldsymbol{\tau}_m - \boldsymbol{\tau}_p, \tag{13.16}$$

where $\boldsymbol{\tau}_m$ is the measured joint torque that is measured from the robot controller which can either work in "torque mode" or output the current of the motors. $\boldsymbol{\tau}_p$ is the predicted joint torque calculated using (13.15a) and (13.15b).

The transfer relation of the contact force in Cartesian space and the external torques in the joints is described in (13.5). The estimated contact force that derives from the detected external joint torques is formulated by

$$\hat{\mathbf{f}} = (\mathbf{J}^T)^{\dagger}\hat{\boldsymbol{\tau}}_e, \tag{13.17}$$

where $\hat{\mathbf{f}}$ is the estimated wrench, and $(\mathbf{J}^T)^{\dagger} \in \mathbb{R}^{M\times N}$ is the Moore–Penrose inverse of the matrix $\mathbf{J}^T$.

Then, the experiment of the external force detection is performed. The dynamic parameters of the robot are identified using the excitation trajectory of the fifth-order

Finite Fourier series, and a sine trajectory with different amplitudes and velocities implemented on Joint 3 is used for the validation of the identified model by (13.15a) and (13.15b). The appropriate selection of the velocity threshold for the presliding and sliding regimes is important for the accurate estimation of the contact force. A large threshold velocity can lead to the overlap of some of the presliding behaviour in the sliding regime, while if the threshold velocity is small, the presilding friction can be included in the sliding friction regime. Therefore, the threshold velocity for Joint 3 is set as 0.0017 rad/s [2]. For the external force detection in the presliding regime, Joint 3 is commanded to reach the saturation state by executing a movement with a small joint angle at a very low speed along the clockwise direction where the deflection of the asperity junctions of Joint 3 reaches the maximum. Then, the vibration torque induced by the motion of the far-side joint is applied on Joint 3, which can ensure that Joint 3 is in the presliding regime.

The experimental results of Joint 3 moving at a joint velocity lower than the velocity threshold are shown in Fig. 13.6. Figure 13.6a shows the joint velocity of Joint 3, which is lower than 0.0017 rad/s in the whole period. Figure 13.6b, c, d shows the measured friction torque, the predicted friction torque, and their torque residual. From the torque residual with the compensation of the unmodelled friction torque in Fig. 13.6d, the external torques induced by the applied contact forces are detected twice at about 16 s and 24 s, and the threshold of the external torque detection is 2 Nm. It can be found that the torque residual in Fig. 13.6d shows an obvious vibration with a periodic change. The torque residual should be approximately zero when no contact

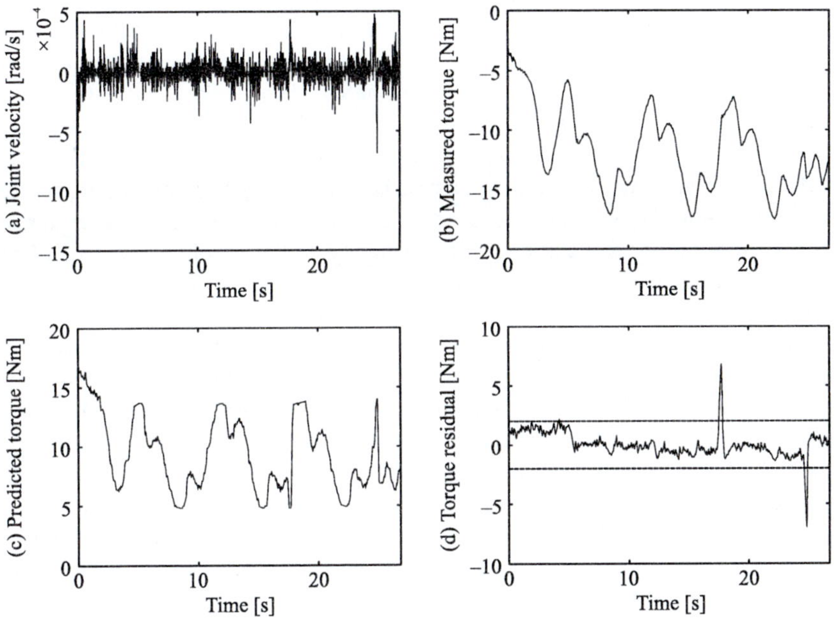

Fig. 13.6 External force detection of Joint 3 in the presliding regime

force is applied to the robot. This vibration is caused by a number of factors: the backlash phenomenon included in the GMS model, environmental elements such as temperature, and unmodelled nonlinear effects. The experimental results show that the proposed approach of the external force detection is effective and acceptable.

13.5 Haptic Control Strategy for HRI

In HRI, a robot interacts with a human operator in a dynamic and unconstructed environment and compliant motion control for the robot is required. Therefore, admittance control is adopted to describe a force-to-motion relationship which can transfer the contact force applied by a human operator into the reference position and the reference velocity. It allows a human operator to haptically control a robot with less effort in a natural way. The formula of admittance control is given by

$$\widehat{f} = M_p \ddot{q} + D_p \dot{q}, \tag{13.18}$$

where M_p is the moment of inertia, and D_p the damping. In the translational and rotational motions, admittance control plays a different role in the transition between the applied contact force and the resulting motion. In rotational motion, admittance control converts $\widehat{f}$ to an angular velocity $\dot{q}$, and an angular displacement q is derived through the integral. For the translational motion in Cartesian space, the external force $\widehat{f}$ is the Cartesian components of the contact force. Admittance control transfers $\widehat{f}$ to a linear velocity reference $\dot{q}$, and derives the displacement q along each Cartesian direction. In velocity control, Laplace transformation of the desired velocity $\dot{q}$ reflects the impact of damping on the steady-state response value whereas the ratio (M_p/D_p) (inertia over damping) influences the dynamics of the robot. It can be concluded that the damping in admittance control can have a large effect on the control performance. For constant admittance control, large admittance parameters require a large contact force to let the robot move at a specified velocity and/or acceleration. This offers opportunities for performing a fine movement with high accuracy in positioning in which the robot is sensible to the force, and the motion can be smooth. Conversely, parameters with low values enable a robot to move at a high velocity and/or acceleration but with poor accuracy. For this drawback, it is important to have an appropriate trade-off between the accuracy of the positioning control and smooth robot motion. Therefore, an adaptive admittance control strategy for the robot is adopted to achieve a natural interaction between a human operator and a robot where the admittance parameters can be adjusted online for better interaction performance. The admittance parameters with high values are expected when performing a fine movement, and in the case of a high-velocity movement, those of lower values are chosen. A discretised desired velocity is therefore obtained by zero-order-hold, which is formulated as

$$\dot{q}_k = \frac{\left(\widehat{f}_k - D_p \dot{q}_{k-1}\right)\Delta t}{M_p} + \dot{q}_{k-1}, \tag{13.19}$$

where $\widehat{f}_k$ is the torque residual at time t_k, $\dot{q}_k$ the operational space linear velocity reference at the time t_k, and Δt the discrete time step. A discretised desired position can also be obtained from Eq. (13.20):

$$q_k = \frac{\left(\widehat{f}_k - D_p \dot{q}_{k-1}\right)\Delta t^2}{2M_p} + \dot{q}_{k-1}\Delta t + q_{k-1}, \tag{13.20}$$

where q_k is the reference angular displacement at time t_k.

Online updating of the admittance parameters is performed to make HRI more natural and with less effort. Haptic interaction with a robot often aims to enable the robot to move and perform desired operations. A typical haptic interaction is described as follows. During an initial state, constant values are assigned to the admittance parameters. Then, a velocity-increase command instructs the robot to move at an increased speed, and a low acceleration is chosen to ensure a safe HRI environment where the admittance parameters are the default values. A fast motion (acceleration) phase is required to reach the target position by means of decreased damping. When approaching the target position, a fine movement with high precision is needed for the accurate positioning that requires high damping. For this purpose, the adaptive updating of the admittance parameters is formulated below.

$$\begin{cases} \widetilde{D}_p = D_p + \alpha|\ddot{q}|sign(\widehat{f}) \\ \widetilde{M}_p = M_p/D_p(1 - \sigma(1 - e^{\delta(\widetilde{D}_p - D_p)}))\widetilde{D}_p \end{cases}, \tag{13.21}$$

where $\widetilde{M}_p$ and $\widetilde{D}_p$ are is the updated inertia and damping, respectively. M_p and D_p are the default ones. The parameter α is to be tuned for preventing the damping from reaching its minimal/maximal values, while σ is used to tune the steady-state ratio (M_p/D_p), and δ is used to control the smoothness according to the ratio variation. To achieve accurate and fine movement, it is required to identify the Cartesian components of the external force exerted by a human on the robot so as to determine the direction of the movement.

13.6 Experiment and Validation

To evaluate the performance of the sensorless haptic control for HRI, a case study is defined and performed on an industrial robot (KR 6 R700 sixx, running KUKA

System Software 8.3 [96]) with a KR C4 controller (KRC in Fig. 13.1). HMI communicates with the robot controller about every 12 ms. Since the movements of the human hand seldom exceed 5–10 Hz, a cut-off frequency of 5 Hz is set for the filter. JOpenShowVar is a middleware between the user program and KRL (see Fig. 13.1) via EtherCAT. The feedback data of the joint position and torque of the robot are collected, and control commands are sent to the robot controller for task execution. The middleware cannot access the robot's data in real time, and the average sampling time is approximately 5 ms. Compared to the control cycle time of 24 ms of the pHRI system, a 5 ms sampling cycle is sufficient for robot control in pHRI.

The experiment of contact force estimation in the Cartesian space is performed on the KUKA robot. A contact force in the negative z-direction of the base coordinate frame is applied to the wrist of the end-effector. The estimated contact force obtained from the proposed model is used to implement the adaptive robot control with a free motion. An admittance controller is used to modify the Cartesian speed reference values based on the detected external force. For the admittance parameters, the default damping is 5 Nm·s/rad, the moment of inertia is 2 kg/m^2, α is 10, and the program override is selected as 10%. The experimental results of Cartesian position reference values, speed reference values, and the estimated contact force in the Cartesian space along the Z-direction are shown in Fig. 13.7. It can be found

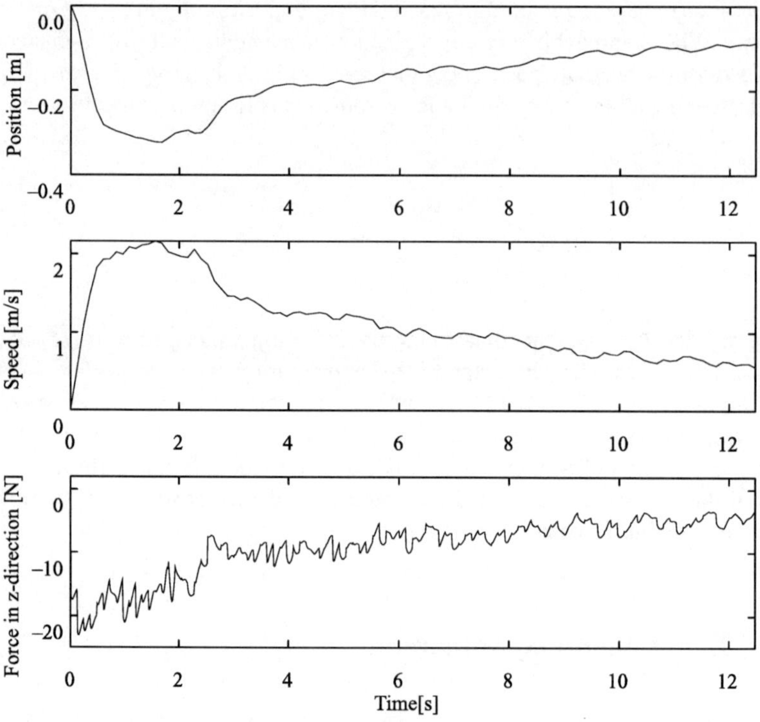

Fig. 13.7 External force detection of Joint 3 in the presliding regime

that the contact force in Cartesian space is detected at $t = 0$ s when the weights that simulate the applied contact force are applied to the end-effector. Then, the weights are continuously removed from the end-effector at a certain time. Accordingly, the contact force is decreased. When the contact force is reduced to 3.5 N, the position and speed reference do not show any change. Therefore, the threshold of the contact force detected in Cartesian space can be determined. Currently, the benchmark test has not been conducted because of the lack of the force sensor, and the threshold of detecting contact force is chosen based on the empirically experimental observation, which is calculated based on (13.18). In addition, the minimum detectable contact force is affected by a number of factors, which include the unmodelled nonlinear effects, the noise of feedback data in the measurement process, and the identification accuracy of the parameters of the dynamic model. The benchmark test will be deployed to evaluate the performance of the proposed method in future work.

The experimental setup of the sensorless haptic control for HRI, as shown in Fig. 13.8, is deployed on an external computer running Windows® 10 operating system and connected with the KRC via an Ethernet cable. The interaction scenario is that a human operator instructs the KUKA robot to move towards a fuel injection tube (marked in blue) which is held by a wooden holder, and then move to the home position of the robot. For simplicity, the proposed control strategy is explained using the third joint (or Joint 3) of the robot.

The experimental results of the HRI process are shown in Fig. 13.9. The default values of M_p and D_p are 20 kg/m^2 and 50 Nm·s/rad. The original configuration of the robot is $\{q_1, \cdots, q_6\} = \{0, -90, 90, 0, 90, 37\}$ in degrees. Then, a human

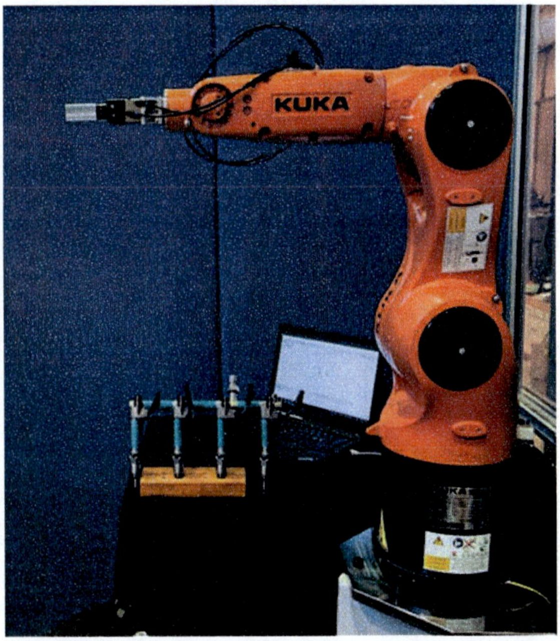

Fig. 13.8 Experimental setup of sensorless haptic control for HRI

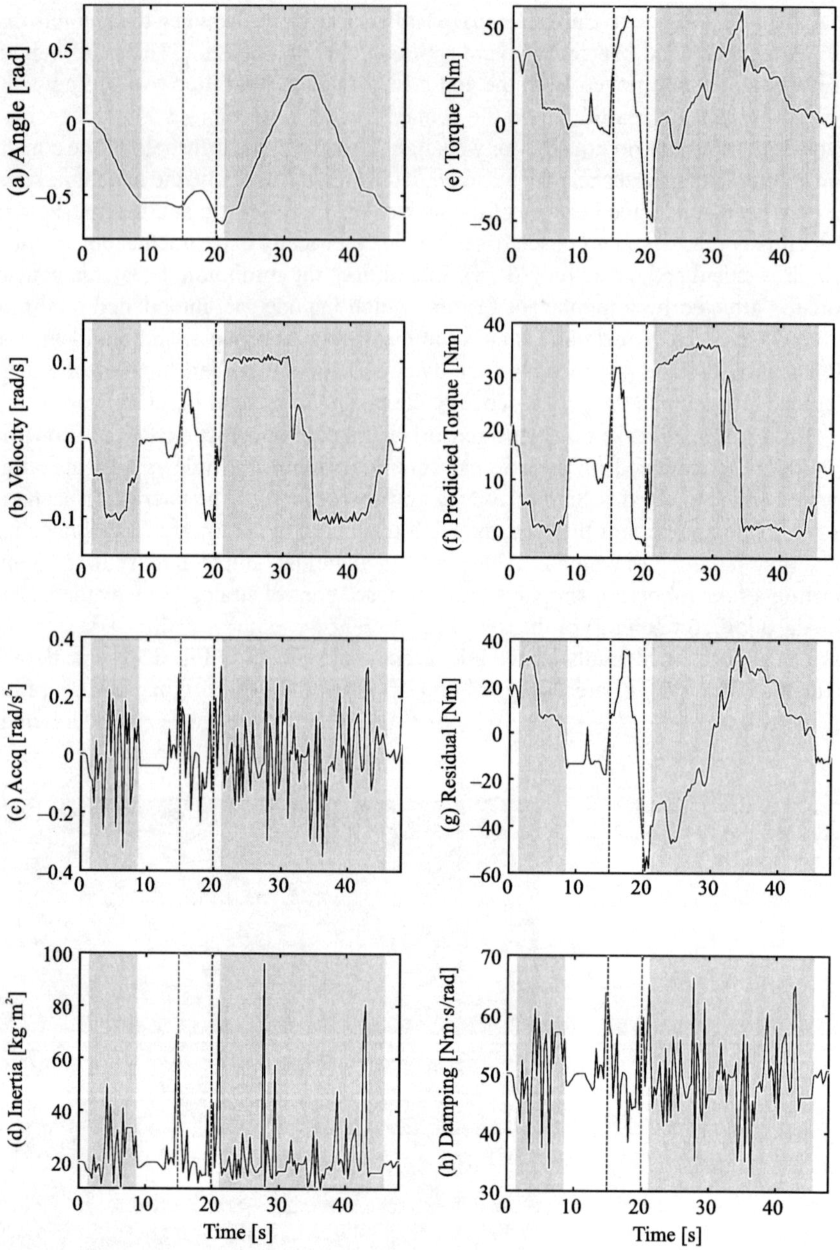

Fig. 13.9 Experimental results of interaction between a human operator and a robot driven by sensorless haptic control

operator applies force to the end-effector and lets the gripper approach the target position. During this process, a fast motion phase with a high velocity is performed when the robot has a large distance with a manipulation object. When close to the target position, a low-velocity instruction is executed for accurate operation, and a distinct admittance parameter adaptation can be reflected. For simplicity, the results of Joint 3 are used to validate the proposed control strategies due to the significant change of Joint 3 in pHRI. The grey areas in Fig. 13.9a describe the motion (angle) of Joint 3 induced by the contact force applied by the human operator. At the initial stage of pHRI, a fast motion of Joint 3 is executed in the time interval 1.4–8.6 s as shown in the grey area of Fig. 13.9a. The generated velocity and acceleration of Joint 3 change accordingly as shown in Fig. 13.9b and c, respectively, and acceleration and deceleration phases of Joint 3 generated by the external force applied are clearly shown in Fig. 13.9b. The measured and predicted torques, and torque residual on Joint 3, also change accordingly, as shown in Fig. 13.9e, f and g, respectively. When the human operator applies the force to the robot, as shown in Fig. 13.9g, the robot can quickly give a response and performs an accelerated movement, and the values of the inertial of moment and damping in Fig. 13.9d and h are decreased such that the velocity and acceleration with large values can be generated. A fine movement is required for smoothly approaching the manipulation object in close proximity. Therefore, the admittance parameters are adjusted with large values to decrease the velocity as shown in Fig. 13.9b. Within this case, the human operator can easily interact with the robot with less effort. It can be observed that both the inertia and damping change accordingly in the proposed adaptive admittance control when the torque residual induced by pHRI is altered. The results obtained in pHRI validate the adaptability and flexibility of the proposed sensorless haptic control method with zero robot programming.

13.7 Conclusions and Future Work

Research presented in this chapter proposes a sensorless haptic control strategy for human–robot interaction without using extra sensing devices. The torques detected based on the robot dynamics are defined as the control commands for the robot's movement and interaction operation. The torque residual is transformed into the reference position and velocity of the robot. Adaptive admittance control is developed to enable fine movement and high precision in position during the interaction. The results of a case study performed on an industrial robot demonstrate and validate the feasibility, flexibility, and adaptability of the proposed approach to assisting the operator during HRI. The proposed approach can be applied in multiple scenarios, e.g., manipulating heavy or long bulky objects, and human–robot collaborative assembly driven by haptic instructions. Our future work will focus on multiple-contact-force-based haptic robot control and precise position identification of external contact force on the robot. This work offers another multimodal support to pHRI, efficiently, in addition to vision and auditory commands.

References

1. L. Wang, R. Gao, J. Váncza, J. Krüger, X.V. Wang, S. Makris, G. Chryssolouris, Symbiotic human–robot collaborative assembly. CIRP Ann.—Manuf. Technol. **68**(2), 701–726 (Jan. 2019)
2. B. Yao, Z. Zhou, L. Wang, W. Xu, Q. Liu, A. Liu, Robotics and computer-integrated manufacturing sensorless and adaptive admittance control of industrial robot in physical human−robot interaction. Robot. Comput. Integr. Manuf. **51**, 158–168 (2018)
3. X.V. Wang, Zs. Keméy, J. Váncza, L. Wang, Human–robot collaborative assembly in cyber-physical production: Classification framework and implementation. CIRP Ann.—Manuf. Technol. **66**(1), 5–8 (2017)
4. EU project: SYMBIO-TIC. https://www.symbio-tic.eu/.
5. L. Wang, From Intelligence Science to Intelligent Manufacturing. Engineering **5**(4), 615–618 (2019)
6. S. Liu, Y. Wang, X.V. Wang, L. Wang, Energy-efficient trajectory planning for an industrial robot using a multi-objective optimisation approach. Procedia Manuf. **25**(August), 517–525 (2018)
7. A.K. Pandey, R. Alami, Mightability maps: A perceptual level decisional framework for co-operative and competitive human–robot interaction. IEEE/RSJ 2010 Int. Conf. Intell. Robot. Syst. IROS 2010—Conf. Proc., pp. 5842–5848 (2010)
8. H. Liu, T. Fang, T. Zhou, L. Wang, Towards Robust Human–Robot Collaborative Manufacturing: Multimodal Fusion. IEEE Access **6**, 74762–74771 (2018)
9. L. Wang, S. Liu, H. Liu, and X. V. Wang. Overview of Human–Robot Collaboration in Manufacturing. InProceedings of 5th International Conference on the Industry 4.0 Model for Advanced Manufacturing 2020, pp. 15–58. Springer, Cham
10. J. Krüger, T.K. Lien, A. Verl, Cooperation of human and machines in assembly lines. CIRP Ann. **58**(2), 628–646 (Jan. 2009)
11. B. Sadrfaridpour, H. Saeidi, Y. Wang, An integrated framework for human–robot collaborative assembly in hybrid manufacturing cells. IEEE Int. Conf. Autom. Sci. Eng. **2016** (November), 462–467 (2016)
12. B. Yao, Z. Zhou, L. Wang, W. Xu, Q. Liu, Sensor-less external force detection for industrial manipulators to facilitate physical human–robot interaction. J. Mech. Sci. Technol. **32**(10), 4909–4923 (2018)
13. A. De Luca, F. Flacco, Integrated control for pHRI: Collision avoidance, detection, reaction and collaboration, Proc. IEEE RAS EMBS Int. Conf. Biomed. Robot. Biomechatronics, pp. 288–295 (2012)
14. A. De Luca, R. Matton, Sensorless robot collision detection and hybrid force/motion control, in *Proceedings of the 2005 IEEE International Conference on Robotics and Automation* (pp. 999–1004). IEEE
15. A. Wahrburg, J. Bös, K.D. Listmann, F. Dai, B. Matthias, H. Ding, Motor-current-based estimation of cartesian contact forces and torques for robotic manipulators and its application to force control. IEEE T. Autom. Sci. Eng. **15**(2), 879–886 (2018)
16. C. Ott, R. Mukherjee, Y. Nakamura, Unified impedance and admittance control, 2010 IEEE Int. Conf. Robot. Autom. pp. 554–561 (2010)
17. G. Kang, H.S. Oh, J.K. Seo, U. Kim, H.R. Choi, Variable admittance control of robot manipulators based on human intention. IEEE/ASME Trans. Mechatronics **24**(3), 1023–1032 (2019)
18. S. Liu, L. Wang, X.V. Wang, Function block-based multimodal control for symbiotic human–robot collaborative assembly, J. Manuf. Sci. Eng., **143**(9), 091001 (2021)
19. J. Schmidtler, V. Knott, C. Hölzel, K. Bengler, Human centered assistance applications for the working environment of the future. Occup. Ergon. **12**(3), 83–95 (2015)
20. X. V. Wang, A. Seira, and L. Wang, Classification, personalised safety framework and strategy for human–robot collaboration, Proc. Int. Conf. Comput. Ind. Eng. CIE, vol. 2018 December, no. December, 2018.

21. A. Bicchi, M.A. Peshkin, J.E. Colgate, Safety for physical human–robot interaction, Springer Handb. Robot., pp. 1335–1348 (2008)
22. S. Liu, L. Wang, X.V. Wang, Symbiotic human–robot collaboration: Multimodal control using function blocks. Procedia CIRP, **93**, 1188–1193 (2020)
23. G. Michalos, S. Makris, P. Tsarouchi, T. Guasch, D. Kontovrakis, G. Chryssolouris, Design Considerations for Safe Human–robot Collaborative Workplaces. Procedia CIRP **37**, 248–253 (2015)
24. G. Michalos, N. Kousi, P. Karagiannis, C. Gkournelos, K. Dimoulas, S. Koukas, K. Mparis, A. Papavasileiou, S. Makris, Seamless human robot collaborative assembly – An automotive case study. Mechatronics **55**, 194–211 (2018)
25. P.A. Lasota, T. Fong, J.A. Shah, A Survey of Methods for Safe Human–Robot Interaction. Found. Trends Robot. **5**(3), 261–349 (2017)
26. S. Haddadin, A. De Luca, A.Albu-Schäffer, Robot collisions: A survey on detection, isolation, and identification, IEEE Trans. Robot. **33**(6), 1292–1312 (2017)
27. J. Krüger, B. Nickolay, P. Heyer, G. Seliger, Image based 3D Surveillance for flexible Man-Robot Cooperation. CIRP Ann. **54**(1), 19–22 (2005)
28. J.A. Corrales, F.A. Candelas, F. Torres, Safe human–robot interaction based on dynamic sphere-swept line bounding volumes. Robot. Comput. Integr. Manuf. **27**(1), 177–185 (Feb. 2011)
29. B. Schmidt, L. Wang, Active collision avoidance for human–robot collaborative manufacturing, in *The 5th International Swedish Production Symposium 6th–8th of November 2012 Linköping*, Sweden (pp. 81–86). The Swedish Production Academy
30. B. Schmidt, L. Wang, Depth camera based collision avoidance via active robot control. J. Manuf. Syst. **33**(4), 711–718 (Oct. 2014)
31. A. Mohammed, B. Schmidt, L. Wang, Active collision avoidance for human–robot collaboration driven by vision sensors. Int. J. Comput. Integr. Manuf. **30**(9), 970–980 (2017)
32. L. Wang, B. Schmidt, A.Y.C. Nee, Vision-guided active collision avoidance for human–robot collaborations. Manuf. Lett. **1**(1), 5–8 (Oct. 2013)
33. M. Fischer, D. Henrich, 3D collision detection for industrial robots and unknown obstacles using multiple depth images. Adv. Robot. Res. Theory, Implementation, Appl. pp. 111–122 (2009)
34. "ISO 10218-1:2011 Robots and robotic devices—Safety requirements for industrial robots—Part 1: Robots."
35. "ISO 10218-2:2011 Robots and robotic devices—Safety requirements for industrial robots—Part 2: Robot systems and integration."
36. J. Krüger, L. Wang, A. Verl, T. Bauernhansl, E. Carpanzano, S. Makris, J. Fleischer, G. Reinhart, J. Franke, S. Pellegrinelli, Innovative control of assembly systems and lines. CIRP Ann. **66**(2), 707–730 (Jan. 2017)
37. A. Casalino, C. Messeri, M. Pozzi, A.M. Zanchettin, P. Rocco, D. Prattichizzo, Operator Awareness in Human–Robot Collaboration Through Wearable Vibrotactile Feedback. IEEE Robot. Autom. Lett. **3**(4), 4289–4296 (2018)
38. S. Zhang, S. Wang, F. Jing, M. Tan, A sensorless hand guiding scheme based on model identification and control for industrial robot. IEEE Trans. Ind. Informatics, **15**(9), 5204–5213 (2019)
39. S. Liu, L. Wang, X.V. Wang, Symbiotic human–robot collaboration: multimodal control using function blocks. Procedia CIR **93**, 1188–1193 (2020)
40. A. Mohammed, L. Wang, Brainwaves driven human–robot collaborative assembly. CIRP Ann. **67**(1), 13–16 (Jan. 2018)
41. L. Wang, A. Haghighi, Combined strength of holons, agents and function blocks in cyber-physical systems. J. Manuf. Syst. **40**, 25–34 (2016)
42. Y. Zhang, S. Liu, Y. Liu, H. Yang, M. Li, D. Huisingh, L. Wang, The 'Internet of Things' enabled real-time scheduling for remanufacturing of automobile engines. J. Clean. Prod. **185**, 562–575 (2018)
43. S. Liu, Y. Zhang, Y. Liu, L. Wang, X.V. Wang, An 'Internet of Things' enabled dynamic optimization method for smart vehicles and logistics tasks. J. Clean. Prod. **215**, 806–820 (2019)

44. L. Wang, M. Törngren, M. Onori, Current status and advancement of cyber-physical systems in manufacturing. J. Manuf. Syst. **37**, 517–527 (2015)
45. H. Liu, L. Wang, An AR-based Worker Support System for Human–Robot Collaboration. Procedia Manuf. **11**, 22–30 (Jan. 2017)
46. H. Liu, L. Wang, Remote human–robot collaboration: A cyber–physical system application for hazard manufacturing environment. J. Manuf. Syst. **54**, 24–34 (2020)
47. M.S. Erden, B. Marić, Assisting manual welding with robot. Robot. Comput. Integr. Manuf. **27**(4), 818–828 (2011)
48. J.T.C. Tan, F. Duan, Y. Zhang, K. Watanabe, R. Kato, T. Arai, Human–robot collaboration in cellular manufacturing: Design and development. IEEE/RSJ Int. Conf. Intell. Robot. Syst. IROS **2009**, 29–34 (2009)
49. S. Liu, L. Wang, X.V. Wang, Sensorless haptic control for human–robot collaborative assembly. CIRP J. Manuf. Sci. Tec., **32**, 132–144 (2021)
50. L.D. Phong, S. Kang, External force estimation using joint torque sensors and its application to impedance control of a robot manipulator. IEEE Int. Conf. Contr. Autom, Syst., pp. 1794–1798 (2013)
51. "Universal Robots. https://www.universal-robots.com/."
52. "KUKA AG, LBR iiwa." [Online]. Available: https://www.kuka.com/en-se/products/robotics-systems/industrial-robots/lbr-iiwa.
53. "ABB Group, IRB 14000 YUMI."
54. W. Khalil, E. Dombre, Modeling, identification and control of robots, Butterworth-Heinemann (2004)
55. J. Swevers, W. Verdonck, J. De Schutter, Dynamic model identification for industrial robots. IEEE Contr. Syst. Mag., **27**(5), 58–71 (2007)
56. C. Yang, Y. Jiang, W. He, J. Na, Z. Li, B. Xu, Adaptive parameter estimation and control design for robot manipulators with finite-time convergence. IEEE Trans. Ind. Electron. **65**(10), 8112–8123 (2018)
57. A. Stolt, A. Robertsson, R. Johansson, Robotic force estimation using dithering to decrease the low velocity friction uncertainties, Proc.—IEEE Int. Conf. Robot. Autom., pp. 3896–3902 (2015)
58. A. Wahrburg, E. Morara, G. Cesari, H. Ding, Cartesian contact force estimation for robotic manipulators using Kalman filters and the generalized momentum. In 2015 IEEE Int. Conf. Autom., pp. 1230–1235 (2015)
59. K.S. Eom, I.H. Suh, W.K. Chung, Disturbance observer based force control of robot manipulator without force sensor. IEEE Int. Conf. Robot. Autom. pp. 3012–3017 (1998)
60. M. Danesh, F. Sheikholeslam, M. Keshmiri, External force disturbance rejection in robotic arms: An adaptive approach, IEICE Trans. Fundam. Electron. Commun. Comput. Sci. **88**(10), 2504–2513 (2005)
61. J. Hu, R. Xiong, Contact Force Estimation for Robot Manipulator Using Semiparametric Model and Disturbance Kalman Filter. IEEE Trans. Ind. Electron. **65**(4), 3365–3375 (2018)
62. A. Stolt, M. Linderoth, A. Robertsson, R. Johansson, Force controlled robotic assembly without a force sensor. IEEE Int. Conf. Robot. Autom. pp. 1538–1543 (2012)
63. A.C. Smith, F. Mobasser, K. Hashtrudi-Zaad, Neural-network-based contact force observers for haptic applications. IEEE Trans. Robot. **22**(6), 1163–1175 (2006)
64. A.C. Smith, K. Hashtrudi-Zaad, Application of neural networks in inverse dynamics based contact force estimation. Proc. 2005 IEEE Conf. Control Appl. 2005. CCA 2005., pp. 1021–1026 (2005)
65. A. Marban, V. Srinivasan, W. Samek, J. Fernández, A. Casals, A recurrent convolutional neural network approach for sensorless force estimation in robotic surgery. Biomed. Signal Process. Control **50**, 134–150 (Apr. 2019)
66. A.C. Bittencourt, S. Gunnarsson, Static Friction in a Robot Joint—Modeling and Identification of Load and Temperature Effects. J. Dyn. Syst. Meas. Control **134**(5), 051013 (2012)
67. V. Lampaert, J. Swevers, F. Al-Bender, Modification of the Leuven integrated friction model structure. IEEE Trans. Automat. Contr. **47**(4), 683–687 (2002)

68. U. Parlitz, A. Hornstein, D. Engster, F. Al-Bender, V. Lampaert, T. Tjahjowidodo, S.D. Fassois, D. Rizos, C.X. Wong, K. Worden, G. Manson, Identification of pre-sliding friction dynamics. Chaos **14**(2), 420–430 (2004)
69. F. Al-Bender, V. Lampaert, J. Swevers, The generalized Maxwell-slip model: A novel model for friction simulation and compensation. IEEE Trans. Automat. Contr. **50**(11), 1883–1887 (2005)
70. D.D. Rizos, S.D. Fassois, Friction identification based upon the LuGre and Maxwell slip models? IFAC Proc. **16**(1), 548–553 (2005)
71. B. Bona, M. Indri, Friction compensation in robotics: An overview, in *Proceedings of the 44th IEEE Conference on Decision and Control, and the European Control Conference*, pp. 4360–4367 (2005)
72. T. Piatkowski, GMS friction model approximation. Mech. Mach. Theory **75**, 1–11 (2014)
73. X. Liu, F. Zhao, S.S. Ge, Y. Wu, X. Mei, End-Effector Force Estimation for Flexible-Joint Robots with Global Friction Approximation Using Neural Networks. IEEE Trans. Ind. Informatics **15**(3), 1730–1741 (2019)
74. P.R. Dahl, A solid friction model. Tech. Report, Aerosp. Corp. El Segundo, **158** (1968)
75. B. Armstrong-Hélouvry, P. Dupont, C.C. De Wit, A survey of models, analysis tools and compensation methods for the control of machines with friction. Automatica **30**(7), 1083–1138 (1994)
76. C.C. de Wit, H. Olsson, K.J. Astrom, P. Lischinsky, A New Model of Control Systems with Friction. IEEE Trans. Automat. Contr. **40**(3), 419–425 (1995)
77. J. Swevers, F. Al-Bender, C.G. Ganseman, T. Prajogo, An integrated friction model structure with improved presliding behavior for accurate friction compensation. IEEE Trans. Automat. Contr. **45**(4), 675–686 (2000)
78. R.R. Selmic, F.L. Lewis, Neural-network approximation of piecewise continuous functions: Application to friction compensation. IEEE Trans. Neural Networks **13**(3), 745–751 (2002)
79. M.K. Ciliz, Adaptive control of robot manipulators with neural network based compensation of frictional uncertainties. Robotica **23**(2), 159–167 (2005)
80. G. Ferretti, G. Magnani, P. Rocco, Impedance control for elastic joints industrial manipulators. IEEE Trans. Robot. Autom., **20**(3), 488–498 (2005)
81. C. Ott, O. Eiberger, W. Friedl, B. Bauml, U. Hillenbrand, C. Borst, A. Albu-Schaffer, B. Brunner, H. Hirschmuller, S. Kielhofer, R. Konietschke, A humanoid two-arm system for dexterous manipulation, Proc. 2006 6th IEEE-RAS Int. Conf. Humanoid Robot. HUMANOIDS, pp. 276–283 (2006)
82. G. Cheng, S.H. Hyon, J. Morimoto, A. Ude, J.G. Hale, G. Colvin, W. Scroggin, S.C. Jacobsen, CB: A humanoid research platform for exploring neuroscience. Adv. Robot. **21**(10), 1097–1114 (2007)
83. F. Ficuciello, L. Villani, B. Siciliano, Variable Impedance Control of Redundant Manipulators for Intuitive Human–Robot Physical Interaction. IEEE Trans. Robot. **31**(4), 850–863 (2015)
84. C.T. Landi, F. Ferraguti, L. Sabattini, C. Secchi, C. Fantuzzi, Admittance control parameter adaptation for physical human–robot interaction. IEEE Int. Conf. Robot. Autom. pp. 2911–2916 (2017)
85. E. G. Kaigom, A new eRobotics approach : Simulation of adaptable joint admittance control, 2013 IEEE Int. Conf. Mechatronics. Autom., pp. 550–555, 2013.
86. S. Farsoni, C.T. Landi, F. Ferraguti, C. Secchi, M. Bonfe, Compensation of Load Dynamics for Admittance Controlled Interactive Industrial Robots Using a Quaternion-Based Kalman Filter. IEEE Robot. Autom. Lett. **2**(2), 672–679 (2017)
87. A. Lecours, B. Mayer-St-Onge, C. Gosselin, Variable admittance control of a four-degree-of-freedom intelligent assist device. In 2012 IEEE Int. Conf. Robot. Autom. pp. 3903–3908 (2012)
88. I. Ranatunga, S. Member, F.L. Lewis, D.O. Popa, S.M. Tousif, S. Member, Adaptive admittance control for human–robot interaction using model reference design and adaptive inverse filtering. IEEE T. Cont. Syst. T., **25**(1), 278–285 (2017)

89. X. Li, Y. Pan, G. Chen, H. Yu, Adaptive human–robot interaction control for robots driven by series elastic actuators. IEEE Trans. Robot., **33**(1), 169–182 (2017)
90. Y. Li, C. Yang, W. Yan, R. Cui, A. Annamalai, Admittance-based adaptive cooperative control for multiple manipulators with output constraint. IEEE Trans. Neural Networks Learn. Syst., **30**(12), 3621–3632 (2019)
91. C. Yang, S. Member, G. Peng, Y. Li, Neural networks enhanced adaptive admittance control of optimized robot–environment interaction. IEEE Trans. Cybern. **49**(7), 2568–2579 (2018)
92. F. Dimeas, N. Aspragathos, Fuzzy learning variable admittance control for human–robot cooperation. In 2014 IEEE/RSJ Int. Conf. Intell. Robot. Syst. pp. 4770–4775 (2014)
93. A.N. Sharkawy, P.N. Koustournpardis, N. Aspragathos, Variable admittance control for human–robot collaboration based on online neural network training. In 2018 IEEE/RSJ Int. Conf. Intell. Robot. Syst., pp. 1334–1339 (2018)
94. F. Sanfilippo, L.I. Hatledal, H. Zhang, M. Fago, K.Y. Pettersen, Controlling Kuka Industrial Robots: Flexible Communication Interface JOpenShowVar. IEEE Robot. Autom. Mag. **22**(4), 96–109 (2015)
95. "KUKA KR6 R700 sixx." [Online]. Available: https://www.kuka.com/en-de/products/robot-systems/industrial-robots/kr-agilus.
96. KUKA. System Software. https://www.kuka.com/en-se/products/robotics-systems/software/system-software/kuka_systemsoftware.

Chapter 14
Intelligent Human–Robot Assembly Enabled by Brain EEG

Abdullah Mohammed and Lihui Wang

Abstract This chapter reports a framework that can facilitate the interactions between a human's EEG (electroencephalography) signals and an industrial robot. This can be achieved by using an EEG headset that captures the brain signals of the human and send it via Bluetooth to a local workstation for signal processing, feature extraction and classification. The system developed provides the ability for a shop-floor operator to control the robot using own brain signals. The system can cooperate with other channels of communications (gesture, voice, etc.) to strengthen the collaboration between the human and the robot during shared assembly operations. Such a collaboration aims to fuse the high accuracy of the robot with the high versatility of the human. Therefore, the aim is to exploit the strength of both sides and enhance the quality and adaptability of human–robot collaborative assembly operations. This approach is applicable to other types of robots as well, for example ones used for assisting people with severe disability.

14.1 Introduction

During the history of the industrial robotics, several developments were reported to enhance the collaboration between industrial robots and shop-floor operators. These collaborations were explored using different methods of communication. Most of these communication channels were focused on the possibility of establishing an intuitive and robust way of interaction. This interaction should provide the possibility of jogging a robot, teaching it the intended path and run the deployed program. In addition, communication between the human and robot should provide the ability for the human to stop the robot at any point using emergency stop signal to avoid any potential consequences if the robot is moving in a dangerous way.

A. Mohammed · L. Wang (✉)
Department of Production Engineering, KTH Royal Institute of Technology, Stockholm, Sweden
e-mail: lihuiw@kth.se

A. Mohammed
e-mail: agmo@kth.se

L. Wang et al. (eds.), *Advanced Human–Robot Collaboration in Manufacturing*,
https://doi.org/10.1007/978-3-030-69178-3_14

14.2 Literature Review

During the history of industrial robotics, several developments were reported to enhance the collaboration between industrial robots and shop-floor operators. These types of collaboration were explored using different methods of communication. First, the methods were mainly using a teach pendant to connect to a robot controller. Due to that, several researches such as [1–3] worked on improving the design of the pendants, others such as [4, 5] worked on developing more versatile interaction between the teach pendant and the robot controller. Meanwhile, other researches such as [6–8] developed smartphone software to be utilised instead of a conventional teach pendant. This showed that the touch screen and voice commands using the smartphone provide intuitive ways of interaction between the human and the robot and shorten the learning curve for the operator. After that, the focus moved toward building a GUI (Graphical User Interface) to control the robot using a local workstation [9–12].

Despite the efforts, establishing natural and intuitive channels of communication between the human and the robot is still a significant challenge and it requires further investigation. Using brainwaves to communicate with the robot was one of the directions that have a potential for future implementations. A number of researchers such as [13–23] worked on controlling humanoid robots by analysing the human's EEG signals and translating them into a number of commands. Additionally, [24–36] focused on using brainwaves to navigate a mobile robot in an unknown environment.

Several approaches were reported in recent years to control a wide range of robots. For instance, [37] used brainwaves to control a planar robot. In addition, [38] developed an approach to controlling a space robot working in remote locations. Furthermore, [39] developed an approach to navigating a research robot using an open-source EEG-headset. Moreover, [40, 41] explored the possibility of translating the human brainwaves to meaningful control commands for a chair robot navigating in indoor environment. Meanwhile, [42–44] used brain signals to control an industrial robot and perform a defined set of tasks. Similar approaches were developed and reported in [45–48] to control robot arms using brain commands. In addition, [49] introduced an approach to utilising brainwaves combined with RFID (Radio Frequency Identification) for controlling a robot used in medical applications. Furthermore, [50] focused on controlling a vacuum cleaning robot navigating an indoor environment.

In addition, several researchers used different techniques to analyse the signals captured from human brains. For example, the work reported in [51] used commercial software to develop a GUI to train the human commands and translate them into a set of control signals to the robot. Other examples are the works presented in [52, 53] which utilise genetic algorithm techniques to analyse brainwaves and control a robot accordingly. By looking at Fig. 14.1, it is clear to see a growing interest in using EEG in the robotic field as it has a high potential of applications.

In parallel with advances in human–robot interfaces, collaboration itself has also undergone considerable development in recent years. Readers interested in human–robot collaboration in general are referred to [54].

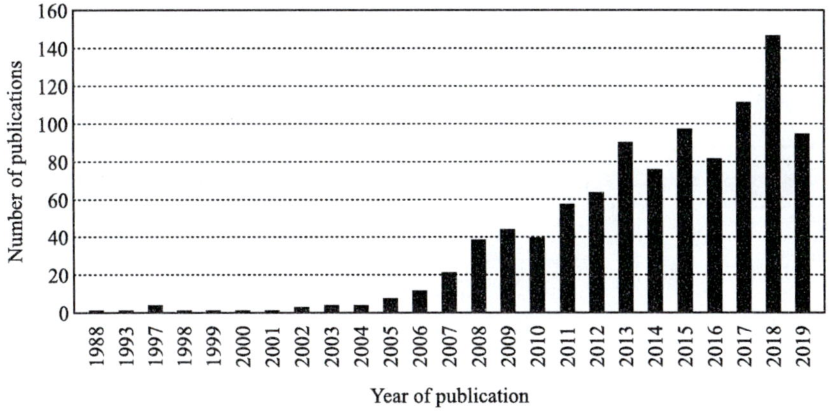

Fig. 14.1 Distribution of EEG research in robotics over years

14.3 Concept and System Design

In this research, a *g.Nautilus* EEG headset, as shown in Fig. 14.2, is used. The device has been selected as the EEG measuring headset to record brain activities. The headset uses a set of 16 electrodes attached to the human scalp in multiple locations. The actual locations are defined using the 10–20 system method and connected to a Bluetooth transmitter, where 10 and 20 refer to the fact that the actual distances between adjacent electrodes are either 10% or 20% of the total front–back or right–left distance of the skull. Figure 14.3 illustrates the locations of these electrodes along the scalp.

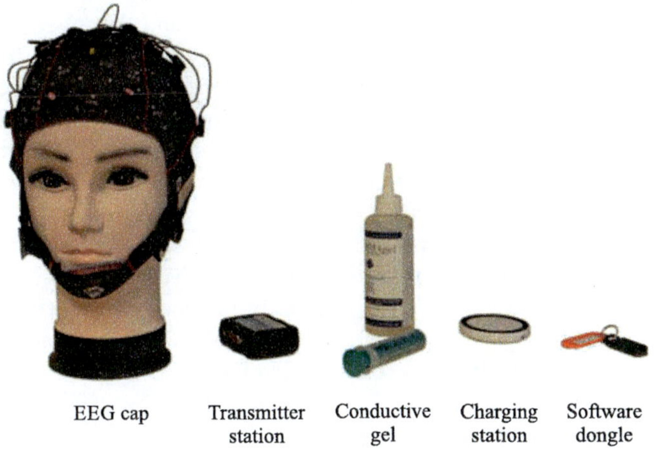

Fig. 14.2 The hardware used in the experiment

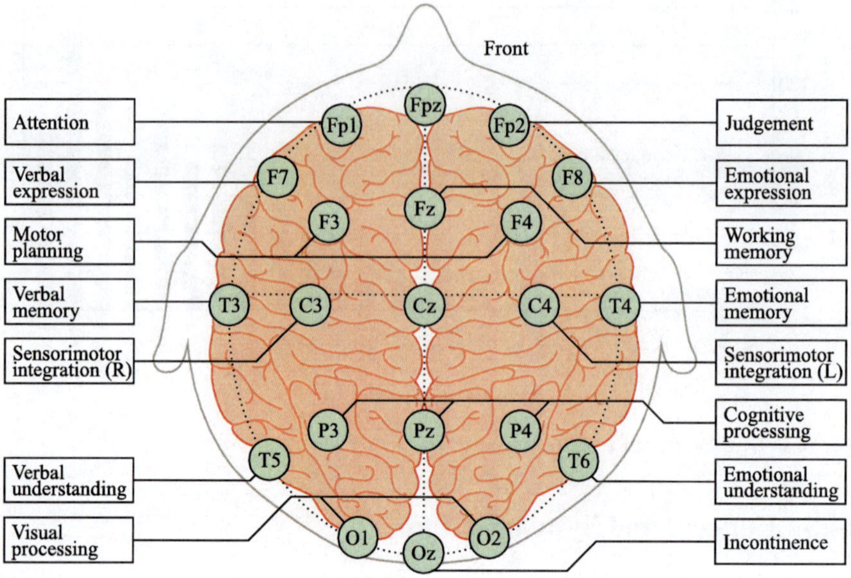

Fig. 14.3 The 10–20 system's electrode locations based on the brain's functions

As shown in Fig. 14.4, the process consists of four stages. The first one (1) is removing the artefacts that come from the captured EEG signals using different types of techniques explained in the Materials and methods section. The second one (2) is analysing the EEG signals and extracting their features to be used in the next stage, the third stage (3) classifies the EEG signals based on their features to a number of classes, and the last stage (4) translates the classes of the EEG signals into the corresponding robot commands, which is achieved by using IEC 61499 function blocks. The communication between system modules and the robot controller is performed using Java language.

An experimental setup is established to perform the training and the evaluation of the system. As shown in Fig. 14.5, the main section of the experimental setup is a central server which is responsible for executing the machine learning algorithms. The components of the system are connected using a local router installed in the laboratory. An industrial robot is used to perform the experiment.

The captured EEG signals satisfied the desired quality of the approach. A sample of the raw input EEG signals is depicted in Fig. 14.6. The artefacts in the signals can be clearly seen in the figure and in different frequencies. A common type of artefact is the one caused by eye blinking of the human. This type of artefact can be clearly seen in the signal. However, other types of artefacts require more advanced techniques to detect.

The EEG signals were captured using the *BCILAB* plugin in *Matlab*® as a development environment.

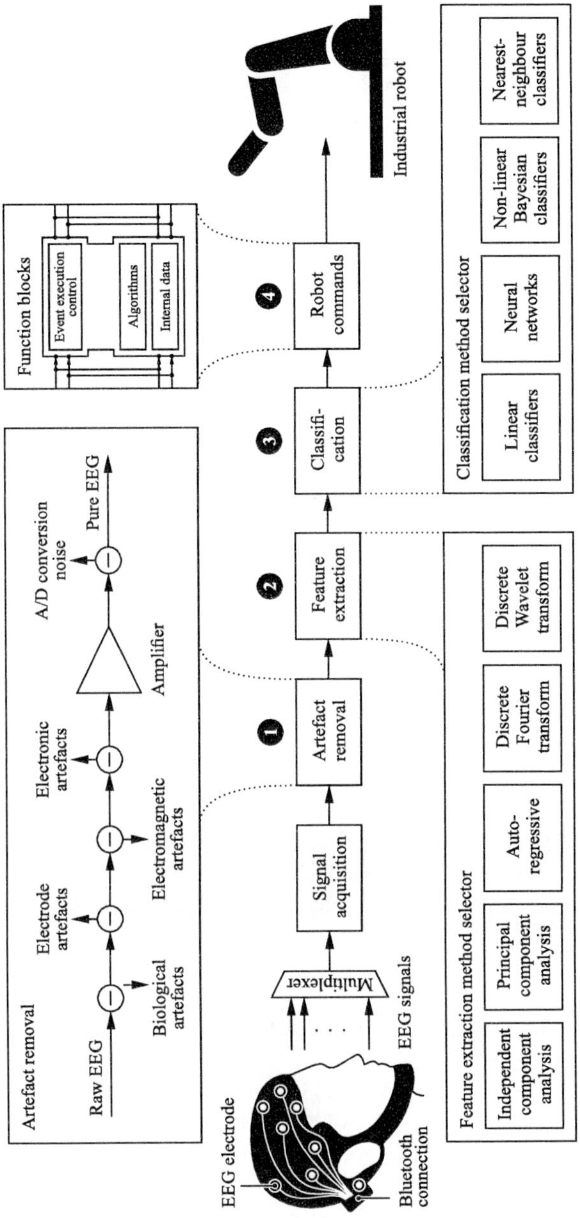

Fig. 14.4 Brain–robot interface framework

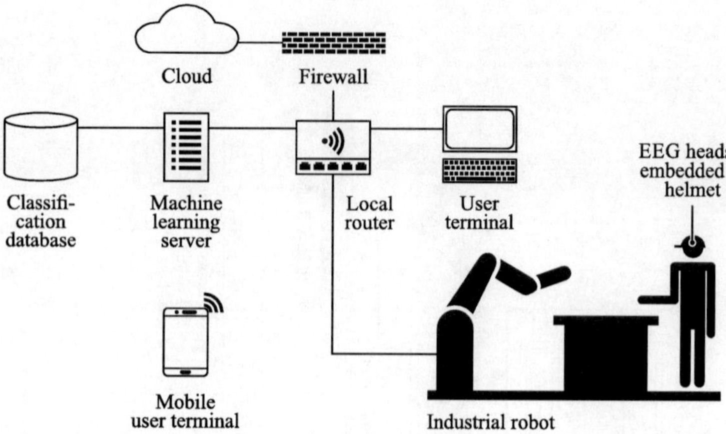

Fig. 14.5 Network infrastructure for brain–robot interface

After removing the artefacts, the EEG signals were then fused together to be analysed and to understand their topography. Figure 14.7 shows an example of the EEG power topography during a training session.

14.4 Methodology

14.4.1 Feature Extraction

The features of EEG play a major role in identifying the suitable classifier that can be considered during the design of BCI. For instance, the features of signals can be extracted from: amplitude, BP (Band Powers), PSD (Power Spectral Density), AR (Auto-regressive), AAR (Adaptive Auto-regressive), time frequency features and inverse model-based features. Moreover, the following properties of the EEG affect the features of the signals:

- *Outliers and artefacts*: one of the main challenges in extracting the features of the EEG is that they have a poor signal-to-noise ratio. This means that the signal power is much lower than the noise power in the captured waves.
- *High dimensional feature vector*: this is due to the fact that the EEG signals have multiple features which are usually extracted from multiple channels and during multiple time segments.
- *Time-based activities*: the activities of the brain heavily rely on at which time they are captured.
- *Non-stationary*: the EEG signals are highly unstable and they can vary significantly over time and during the training sessions.

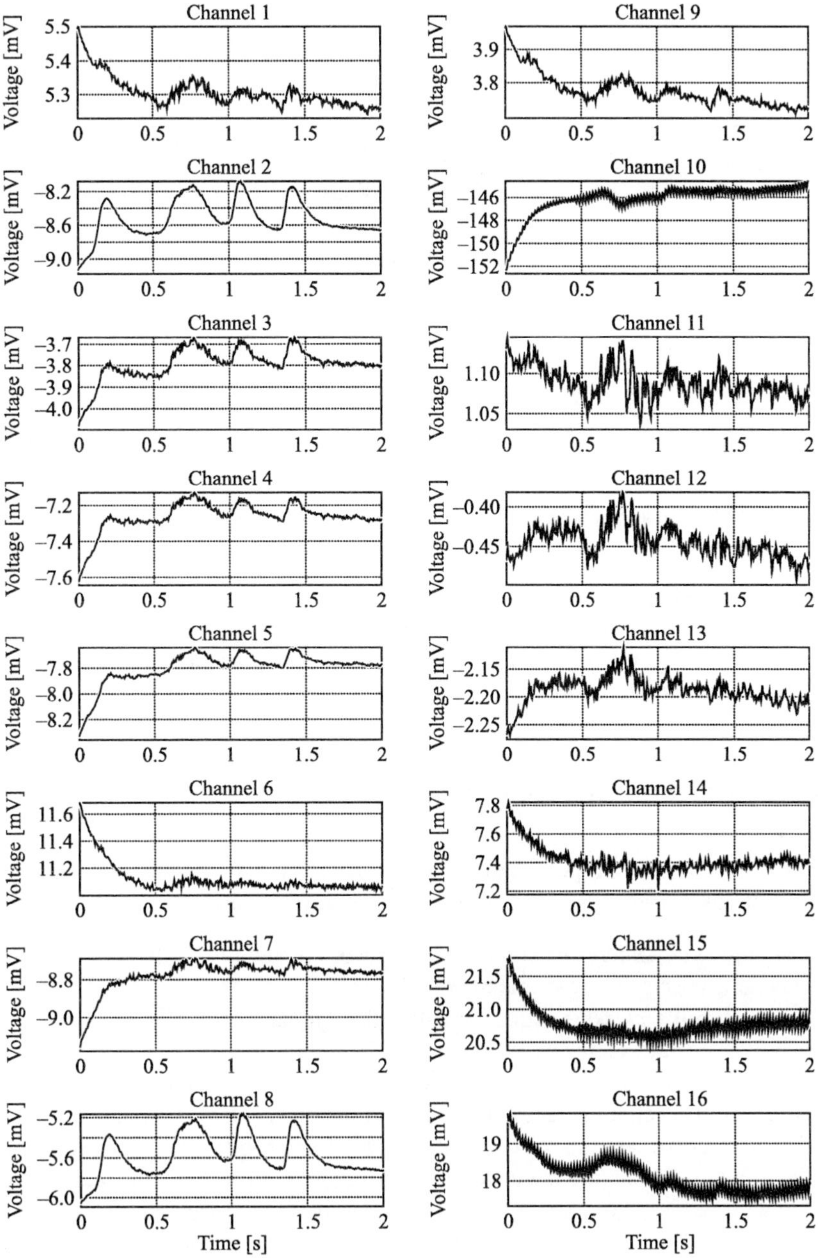

Fig. 14.6 Samples for EEG raw signals

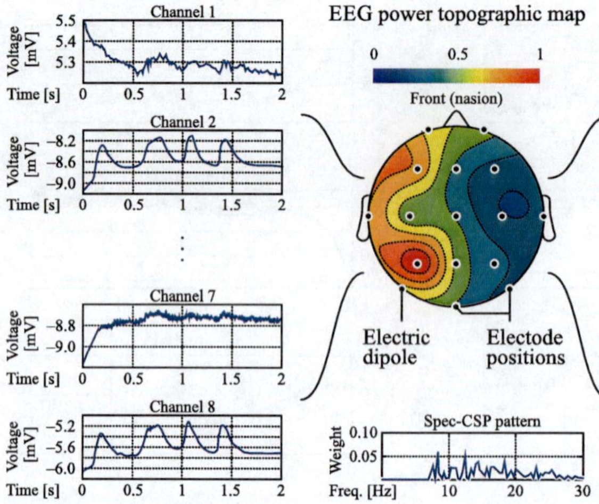

Fig. 14.7 EEG power topography

- *Limited training set*: the training of the BRI (Brain–Robot Interface) system is time-consuming, and therefore the training set is small. However, the training set in the medical field could be sufficiently enough since it can be recorded by patients for weeks or months.

14.4.2 Classification Algorithms

In order to select the proper classifier for the BRI, we first need to understand the specifications of each type of classifiers that are currently used:

- *Generative classifier*: this classifier consists of computing the likelihood of each class and selects the most suitable one.
- *Discriminative classifier*: this classifier discriminates the classes in order to classify a feature vector directly.
- *Static classifier*: this type classifies a single feature vector; therefore, it is not able to take into account the temporary information during classification.
- *Dynamic classifier*: this type is able to classify a sequence of feature vector; therefore, it can capture the temporary information during classification.
- *Stable classifier*: this classifier is robust against the small variations in the training set and these variations usually will not affect the performance of the classifier.
- *Unstable classifier*: this classifier has a major drawback since small variation in the training set may lead to significant changes in the performance.
- *Regularised classifier*: this type of classifier is carefully regulated to prevent overtraining; therefore, the classifier can be highly robust to outliers.

14.4.3 Classification Problems

There are usually two major problems that exist in the field of BRI, these problems are:

- *The curse of dimensionality*: there is an exponential relationship between the data needed for the classifier and the dimensions of the feature vectors. Therefore, there is a high probability that the classifier will give poor results.
- *The bias–variance trade-off*: there are three possible sources of classification errors: the first is the noise that is usually captured in the BRI system and is an irreducible error, the second is the bias which determines the divergence between the estimated mapping and the best mapping, the third is the variance which represents the sensitivity of the training set.

In order to minimise the classification errors, the bias and the variance need to be low. However, in reality it is highly difficult to achieve. Therefore, a trade-off needs to be considered.

The MSE (Mean Square Error) of the BRI system is mainly affected by the three sources of errors mentioned previously, and it can be explained using Eq. (14.1).

$$\begin{aligned} MSE &= E[(l^* - f^*(v))^2] + [(f^* - E[f(v)]^2] \\ &= N^2 + B(f(v))^2 + V(f(v)), \end{aligned} \tag{14.1}$$

where l represents the label that the classifier found, v is the feature vector and f is the mapping used in the classifier.

The experimental activities showed that the selected software is well-structured with good maintenance quality and sufficient documentation. In addition, the software has a number of video tutorials that were useful in knowing how to use the software as well as some theoretical knowledge needed for processing the EEG signals. However, a few points might need to be improved to enhance the software: (1) the software is developed using Matlab® which has a performance limitation compared with other development environments, and (2) the software had a few bugs that have been discovered during development; after fixing them, the system is working properly.

14.5 Implementation and Results

14.5.1 System Training

The first stage of the implementation for the proposed system is to conduct a certain number of training sessions by recording the EEG signals from a number of participants. The training was achieved by applying the following criteria:

- Providing a relaxing environment for the participant to have a clear mind and the EEG signal for the brain idle state is recorded as a reference.
- The participant is asked to think about a robot control command and the EEG signals are recorded for 10 min.

The system is then trained based on the recorded signals from the participants.

14.5.2 *Artefact Removal*

There are a few techniques that can be implemented to eliminate the artefacts captured in the EEG signals. Some artefacts have a certain frequency nature; therefore, some methods use low-pass filters to remove those artefacts. Figure 14.8 shows an example of the EEG signal processing during which noises can be detected and removed.

The following is a brief description about the techniques used in artefact removal for the captured EEG signals:

Linear Regression Methods: Due to its simplicity, these techniques were commonly used in removing artefacts from EEG signals in the 90 s. Nowadays, these methods are still used in removing EOG (Electrooculography) artefacts. To fulfil the objective of the method, the linear regression uses a linear model explained in Eq. (14.2) to determine the relationship between the EOG artefacts and the EEG signals. Then, the method subtracts the EOG from the captured EEG as shown in Eq. (14.3).

$$E_o = E_r + \alpha O_v + \beta O_h \tag{14.2}$$

$$\hat{E}_r = \hat{E} + \hat{\alpha} O_v + \beta O_h, \tag{14.3}$$

where E_o is the observed EEG, E_r is the real EEG, O_v and O_h are the vertical and horizontal EOG artefacts, α and β are the EOG coefficients and β indicates that the variable is approximated.

- *Filters*: Filters are usually not common in removing the artefacts from the EEG signals. The most commonly used filter for removing the artefacts are: adaptive filters, Weiner filters and Bayes filters.
- *BSS (Blind Source Separator)*: There are two common BSS used to eliminate artefacts from the EEG signals:

    (a) *PCA (Principal Component Analysis)*: PCA is used to remove the artefacts of eye movement EOG from the EEG signals. It is achieved by generating a random vector **x** consisting of a number of elements $x(1), x(2),..., x(T)$. The first stage is to determine the first principal component of **x** using Eq. (14.4).

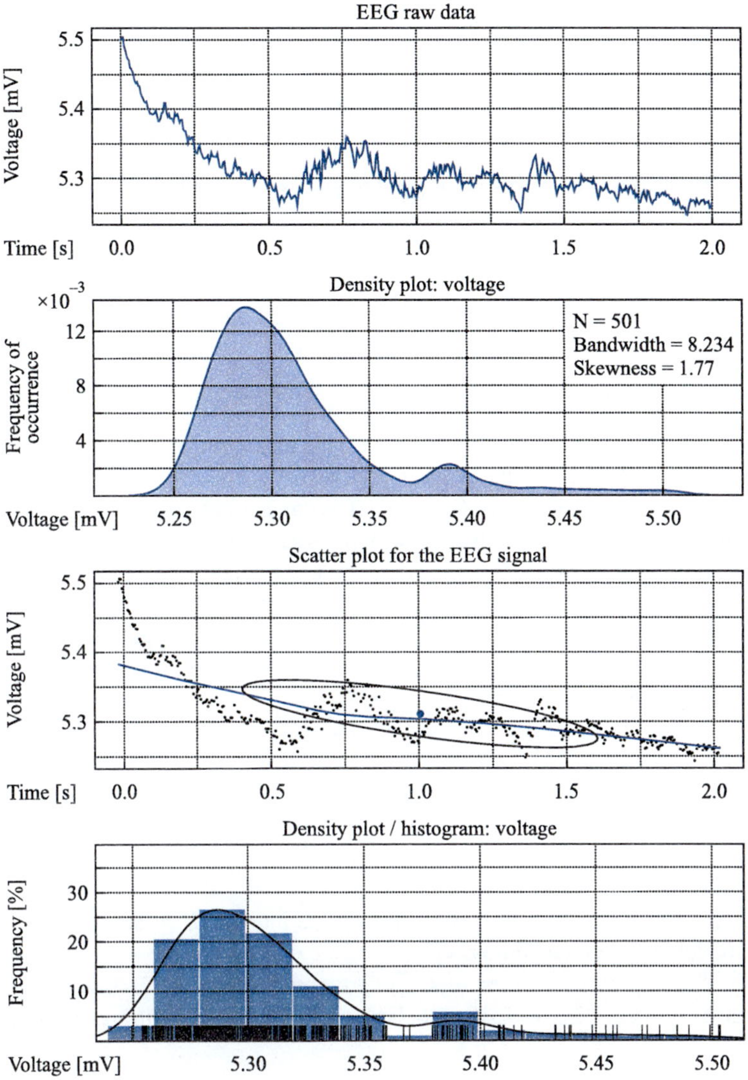

Fig. 14.8 Example of EEG signal processing

$$y_1 = \sum_{k}^{n} w_{k1} x_k = \mathbf{w}_1^{\mathrm{T}} \mathbf{x}, \tag{14.4}$$

where y_1 is the first principal component of **x**, and w is the weight for the analysis.

(b) *ICA (Independent Component Analysis)*: ICA tries to divide the EEG signals into statistically independent sources and then omit the artefacts.

Equation (14.5) explains how to calculate the independent components.

$$y = B^T x = Q^T s = \sum_i q_i s_i, \tag{14.5}$$

where y is the independent component, **Q** is the coefficient matrix and **s** is the source.

14.5.3 Feature Extraction

For the sake of brevity, only the commonly used methods are listed below.

- *ICA (Independent Component Analysis)*: This technique is a supervised learning technique due to the fact that it produces a set of independent component vectors.
- *PCA (Principal Component Analysis)*: The main usage of this technique is to captured EEG data into a subspace with low-dimension.
- *DWT (Discrete Wavelet Transform)*: It is a non-stationary time-scale analysis technique that is suitable to be utilised to detect the features of the EEG signals.
- *AR (Auto-Regressive)*: As described in Eq. (14.6), this technique can be described as a linear regression of a current observation for a series versus one or more prior observations of the series.

$$y(t) = \sum_{i=1}^{p} \Phi_i y(t-i) + \varepsilon_t, \tag{14.6}$$

where $y(t)$ represents the time-series that will be modelled, Φ_i are the coefficients of the model and ε_t is the white noise.

DFT (Discrete Fourier Transform): As presented in Eq. (14.7), this technique is achieved by transforming the EEG signals between the time domain and the frequency domain.

$$x_k = \sum_{n=0}^{N-1} x_n . e^{-\frac{i2\pi}{N} kn}, \tag{14.7}$$

where x_k is the DFT of the sequence $\{x_n\}$.

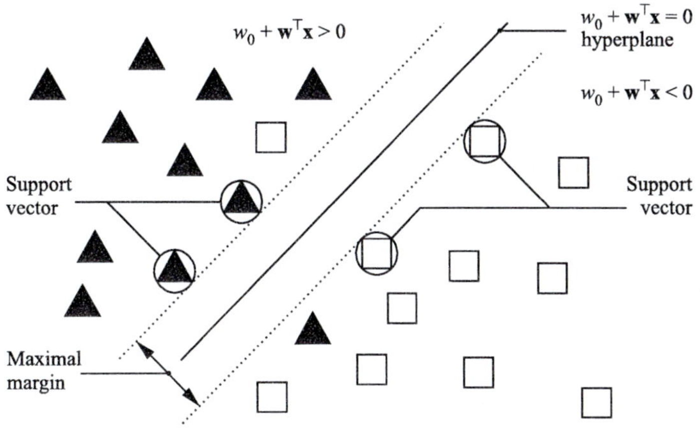

Fig. 14.9 Using hyperplane to identify two classes

14.5.4 Classification

The classification algorithms that are usually used in BRI can be categorised into four sections: Linear classifiers, Neural networks, Non-linear Bayesian classifiers and Nearest-neighbour classifiers. Figure 14.9 illustrates one example.

Below is an explanation of classification algorithms that are investigated, and the features of each.

(1) *Linear classifier*: This classifier is basically a discriminant algorithm which utilises linear functions to distinguish between classes.

(a) *LDA (Linear Discriminant Analysis)*: also known as Fisher's LDA is a method used to find a linear combination of features to construct a hyperplane that can divide the data into multiple classes. This classifier depends on the Bayes theorem which is shown in Eq. (14.8).

$$P(M|N) = \frac{P(N|M)P(M)}{P(N)}, \tag{14.8}$$

where M and N are events and $P(N) \neq 0$.
Based on the theorem, the probability of N used by each class can be related to the input by taking on the value of N in each class.

(b) *Linear Support Vector Machine*: this classifier chooses an optimal hyperplane based on the fact that it maximises the margin around the hyperplane, see Fig. 14.9.
A "kernel trick" shown in Eq. (14.9) is used to determine the optimal hyperplane.

$$K(x, y) = \exp\left(\frac{-(x-y)^2}{2\sigma^2}\right), \tag{14.9}$$

where x and y are the data points.

(2) *NN (Neural Networks)*: The following is a summary for the neural networks that have been investigated:

(a) *MLP (Multilayer Perceptron)*: As presented in Eq. (14.10), this network generates a single output based on the input set of values and by adjusting the input weights.

$$y = \varphi\left(\sum_{i=1}^{n} w_i x_i + b\right) = \varphi\left(\mathbf{w}^T \mathbf{x} + b\right), \tag{14.10}$$

where $\mathbf{w}$ represents the weight vector, $\mathbf{x}$ represents the input vector, and b is the bias.

(b) *LVQ (Learning Vector Quantisation)*: It is a NN algorithm that has the objective of selecting a prototype vector that should have the minimum distance to the input vector. The distance d is calculated using the Euclidean distance explained in Eq. (14.11).

$$d(\mathbf{x}, \mathbf{c}) = \sum_{i=1}^{n} (x_i - c_i)^2, \tag{14.11}$$

where n represents the number of attributes, $\mathbf{x}$ represents the input vector, and $\mathbf{c}$ represents a given prototype vector.

(c) *DNN (Dynamic Neural Network)*: The main feature for this type of NN is that it has a dynamic relationship between the input and the output of the network. Equation (14.12) defines the mathematical model of this NN.

$$y_i(t) = g_i(x_i(t)), \tag{14.12}$$

where y_i represents the output of the i th DNU (Dynamic Neural Unit) and g_i is the neural output function.

(d) *RBFNN (Radial Basis Function Neural Networks)*: This type of network fits a sum of radially symmetrical functions as shown in Eq. (14.13):

$$f(\mathbf{x}) = \sum_{p=1}^{N} w_p \varphi\left(\mathbf{x} - \mathbf{x}^p\right), \tag{14.13}$$

where φ is the basis function, w_p are elements of the weight vector $\mathbf{w}$, $\mathbf{x}$ is the input vector, and N represents the number of the basis functions defined.

(e) *BLRNN (Bayesian Logistic Regression Neural Network)*: The relationship between the input and the output of the network can be explained by Eq. (14.14):

$$\begin{aligned} y[n] = & {}_{(i=1)}^{(N-1)} a_i y[n-i] \\ & + {}_{(j=1)}^{(N-1)} b_i j y[n-j] x[n-i] \\ & + {}_{(i=0)}^{(N-1)} c_i x[n-i], \end{aligned} \tag{14.14}$$

where $x[i]$ is the input, $y[i]$ is the output, N is the order of the recursion, b_i is the weight for the bilinear portion and c_i is the weight for the feed-forward portion.

(f) *PeGNC (Probability Estimating Guarded Neural Network Classifier)*: The architecture of the network is based on GNC (Guarded Neural Classifier). The output of this network can be described in Eq. (14.15):

$$y_j = \sum_{i=1}^{k} q_{ij} c_i,\ j = 1, .., N, \tag{14.15}$$

where y_j is the output component, q_{ij} is the weight and c_i is the classification unit.

(3) *Non-linear Bayesian classifiers*: Two types of Bayesian classifiers are summarised below:

(a) *Bayes quadratic*: The classifier can be formulated using Eq. 14.16.

$$\hat{j} = \arg\max_j \log \pi_j - \frac{1}{2} \log \left| \sum_j - \frac{1}{2} (x - \mu_j)^T \sum_j^{-1} (x - \mu_j) \right|, \tag{14.16}$$

where $\mathbf{x}_j$ is the input, π_j is the prior probability of $\mathbf{x}_j$, and $\boldsymbol{\mu}_j$ is the mean of $\mathbf{x}_j$.

(b) *HMM (Hidden Markov Model)*: This classifier uses Markov probability distribution which is explained in Eq. (14.17).

$$P(S_{1:T}, Y_{1:T}) = P(S_1) P(Y_1|S_1) \prod_{t=2}^{T} P(S_t|S_{t-1}) P(Y_t|S_t), \tag{14.17}$$

where S_t is the current state and Y_t is the current output.

(4) *Nearest neighbour classifiers*: For this type of classification, two common classifiers are investigated:

(a) *kNN (k Nearest Neighbours)*: This classifier uses the Euclidean distance which is defined in Eq. (14.18) to decide a dominant class with respect to the training set.

$$d(\mathbf{x}, \mathbf{y}) = d(\mathbf{y}, \mathbf{x}) = \sqrt{\sum_{i=1}^{n} (y_i - x_i)^2}, \tag{14.18}$$

where $\mathbf{x} = (x_1, x_2, \ldots, x_n)$ and $\mathbf{y} = (y_1, y_2, \ldots, y_n)$ are two points in Euclidean n-space.

(b) *Mahalanobis-distance-based classifier*: Unlike the previous classifier, this one uses the *Mahalanobis distance* calculated in Eq. (14.19) to determine the dominant class based on the training set:

$$d_c(x) = \sqrt{(x - \mu_c)^T M_c^{-1} (x - \mu_c)}, \tag{14.19}$$

where $d_c(\mathrm{x})$ is the square of the Mahalanobis distance, x is the distance of the vector from the mean, the vector of the observation, $\boldsymbol{\mu}_c$ is the vector of mean values of independent variables, $\boldsymbol{\mu}_c^{-1}$ is the inverse covariance matrix of independent variables.

14.6 Discussion

The human brain is a highly advanced organ that works with the nervous system to observe the human surroundings, analyse the observed information and then coordinate the other parts of the body accordingly. A thought associated with a command leads to the activation of different parts of the human brain, which significantly relies on the brain's interpretation of the word and how the brain has evolved in the early stages of a human's life. The activity of the brain can be recorded and considered a signature for the given command. As an example, Fig. 14.10 illustrates how different parts of the brain are triggered when a human is thinking about moving the robot.

14.7 Conclusions

The main challenges of the utilisation of human brain signals in the industrial environments are:

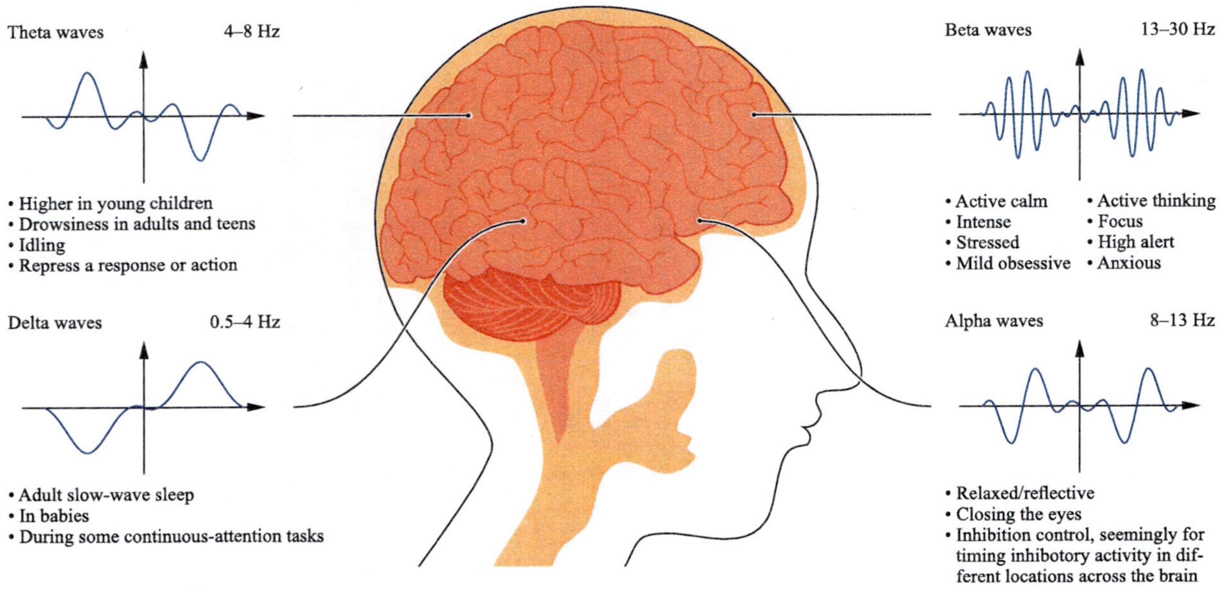

Fig. 14.10 Characteristics of main types of EEG signals

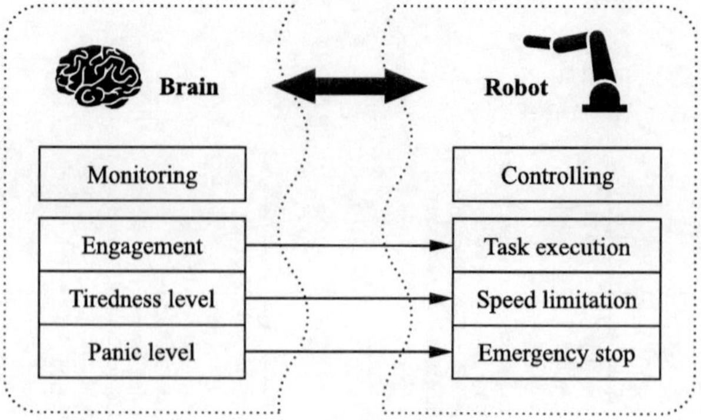

Fig. 14.11 Major research aspects of the brain–robot interface

1. The current modules used for data acquisition are still expensive and inconvenient to wear for a long period of time. To solve the problem, companies are developing implantable and adaptable sensors together with low power consumption devices that can process the captured data locally. In addition, a new generation of low-cost wireless EEG headsets starts to appear and has a significant impact on improving the quality of the captured EEG signals.
2. The processing of the captured data and the reduction of the external interference and noise is a must, especially for the wireless wearable devices.
3. The quality of captured EEG signals relies heavily on the position of the electrodes on the human scalp.

There are variations in the captured EEG signals from one individual to another because each brain is developed in a different circumstance. To have a reliable module of communication, the current approaches require relatively long sessions of training, calibration and adaptation. The technique continues to be evaluated with respect to other comparable channels of communications such as facial expressions and head movements. This can be solved by combining BCI with other more mature channels of communications to enhance the reliability and the intuitiveness of the approach. A further effort needs to be implemented to improve the flexibility of the approach in handling a wider range of commands. This can be done by improving the classification methods.

References

1. K. S. Kwon, "The ergonomic design of robot teach pendants for human–robot interaction", *Proceedings of the 35th SICE Annual Conference. International Session Papers (IEEE, 1996)*, pp. 1233–1236
2. E. Garcia, S. Vergara, F. Reyes, M. Vargas, A. Palomino, G. Villegas, "Development of a Teach Pendant for an ABB Robot ", *2009 International Conference on Electrical, Communications, and Computers (IEEE, 2009)*, pp. 256–259

3. M. Gao, L. Ye, Y. Yan, "Design and Implementation of Teach Pendant for Six Degrees of Freedom Industrial Robot", *2015 Fifth International Conference on Instrumentation and Measurement, Computer, Communication and Control (IMCCC) (IEEE, 2015)*, pp. 1929–1933
4. H. Fukui, S. Yonejima, M. Yamano, M. Dohi, M. Yamada, T. Nishiki, "Development of teaching pendant optimized for robot application", *2009 IEEE Workshop on Advanced Robotics and its Social Impacts (IEEE, 2009)*, pp. 72–77
5. G. Lisi, M. Hamaya, T. Noda, J. Morimoto, "Dry-wireless EEG and asynchronous adaptive feature extraction towards a plug-and-play co-adaptive brain robot interface", *2016 IEEE International Conference on Robotics and Automation (ICRA) (IEEE, 2016)*, pp. 959–966
6. S. M. Abbas, S. Hassan, J. Yun, "Augmented reality based teaching pendant for industrial robot", *2012 12th International Conference on Control, Automation and Systems (IEEE, 2012)*, pp. 2210–2213
7. Y. Jan, S. Hassan, S. Pyo, J. Yoon, "Smartphone based control architecture of teaching pendant for industrial manipulators", *2013 4th International Conference on Intelligent Systems, Modelling and Simulation (IEEE, 2013)*, pp. 370–375
8. M.-J. Oh, S.-M. Lee, T.-W. Kim, K.-Y. Lee, J. Kim, Design of a teaching pendant program for a mobile shipbuilding welding robot using a PDA. Comput. Aided Des. **42**, 173 (2010)
9. L. Wang, M. Givehchi, G. Adamson, M. Holm, A sensor-driven 3D model-based approach to remote real-time monitoring. CIRP Ann. **60**, 493 (2011)
10. L. Wang, R. Sams, M. Verner, F. Xi, Integrating Java 3D model and sensor data for remote monitoring and control. Robotics and Computer-Integrated Manufacturing **19**, 13 (2003)
11. L. Wang, A. Mohammed, M. Onori, Remote robotic assembly guided by 3D models linking to a real robot. CIRP Ann. **63**, 1 (2014)
12. Y. Gao, Y. Su, W. Dong, W. Wang, Z. Du, X. Gao, Y. Mu, "U-Pendant: A universal teach pendant for serial robots based on ROS", *2014 IEEE International Conference on Robotics and Biomimetics (ROBIO 2014) (IEEE, 2014)*, pp. 2529–2534
13. N. Prakaksita, C.-Y. Kuo, C.-H. Kuo, "Development of a motor imagery based brain–computer interface for humanoid robot control applications", *2016 IEEE International Conference on Industrial Technology (ICIT) (IEEE, 2016)*, pp. 1607–1613
14. L. Zhang, M. Jiang, D. Farid, M.A. Hossain, Intelligent facial emotion recognition and semantic-based topic detection for a humanoid robot. Expert Syst. Appl. **40**, 5160 (2013)
15. W. Li, C. Jaramillo, Y. Li, "Development of mind control system for humanoid robot through a brain computer interface", *2012 Second International Conference on Intelligent System Design and Engineering Application (IEEE, 2012)*, pp. 679–682
16. J. Tang, J. Jiang, Y. Yu, Z. Zhou, "Humanoid robot operation by a brain–computer interface", *2015 7th International Conference on Information Technology in Medicine and Education (ITME) (IEEE, 2015)*, pp. 476–479
17. N. Zhang, J. Jiang, J. Tang, Z. Zhou, D. Hu, "A Novel Steady-State Visually Evoked Potential-Based Brain–Computer Interface Paradigm to Steer a Humanoid Robot", *2014 Fourth International Conference on Instrumentation and Measurement, Computer, Communication and Control (IEEE, 2014)*, pp. 774–778
18. M. Bryan, J. Green, M. Chung, L. Chang, R. Scherer, J. Smith, R. PN Rao, "An adaptive brain–computer interface for humanoid robot control", *2011 11th IEEE-RAS International Conference on Humanoid Robots (IEEE, 2011)*, pp. 199–204
19. P. Gergondet, S. Druon, A. Kheddar, C. Hintermuller, C. Guger, M. Slater, "Using brain–computer interface to steer a humanoid robot", *2011 IEEE International Conference on Robotics and Biomimetics (IEEE, 2011)*, pp. 192–197
20. K. Naveed, J. Iqbal, H. ur Rehman, "Brain controlled human robot interface", *2012 International Conference of Robotics and Artificial Intelligence (IEEE, 2012)*, pp. 55–60
21. N. Prataksita, Y.-T. Lin, H.-C. Chou, C.-H. Kuo, "Brain–robot control interface: Development and application", *2014 IEEE International Symposium on Bioelectronics and Bioinformatics (IEEE ISBB 2014) (IEEE, 2014)*, pp. 1–4
22. A. Finke, A. Knoblauch, H. Koesling, H. Ritter, "A hybrid brain interface for a humanoid robot assistant", *2011 Annual International Conference of the IEEE Engineering in Medicine and Biology Society (IEEE, 2011)*, pp. 7421–7424

23. K. Takahashi, S. Maekawa, M. Hashimoto, "Remarks on fuzzy reasoning-based brain activity recognition with a compact near infrared spectroscopy device and its application to robot control interface", *2014 International Conference on Control, Decision and Information Technologies (CoDIT) (IEEE, 2014)*, pp. 615–620
24. Y.-C. Yu, A. Nawroj, S. Wang, L. Gabel, "Mobile robot navigation through a brain computer interface", *2012 IEEE Signal Processing in Medicine and Biology Symposium (SPMB) (IEEE, 2012)*, pp. 1–5
25. T. Geng, J. Q. Gan, "Motor prediction in brain–computer interfaces for controlling mobile robots", *2008 30th Annual International Conference of the IEEE Engineering in Medicine and Biology Society (IEEE, 2008)*, pp. 634–637
26. X. Fan, L. Bi, Y. Li, K. Jie, H. Ding, "A P300 brain–computer interface for controlling a mobile robot by issuing a motion command", *2013 ICME International Conference on Complex Medical Engineering (IEEE, 2013)*, pp. 707–710
27. A. O. Barbosa, D. R. Achanccaray, M. A. Meggiolaro, "Activation of a mobile robot through a brain computer interface", *2010 IEEE International Conference on Robotics and Automation (IEEE, 2010)*, pp. 4815–4821
28. C. Farmaki, G. Christodoulakis, V. Sakkalis, "Applicability of SSVEP-based brain–computer interfaces for robot navigation in real environments", *2016 38th Annual International Conference of the IEEE Engineering in Medicine and Biology Society (EMBC) (IEEE, 2016)*, pp. 2768– 2771
29. Y. Chae, S. Jo, J. Jeong, "Brain-actuated humanoid robot navigation control using asynchronous brain–computer interface", *2011 5th International IEEE/EMBS Conference on Neural Engineering (IEEE, 2011)*, pp. 519–524
30. X. Perrin, R. Chavarriaga, F. Colas, R. Siegwart, J. d. R. Millan, "Brain-coupled interaction for semi-autonomous navigation of an assistive robot", *Robotics and Autonomous Systems* 58, 1246 (2010)
31. A. I. Nawroj, S. Wang, Y.-C. Yu, L. Gabel, "A brain–computer interface for robot navigation", *2012 38th Annual Northeast Bioengineering Conference (NEBEC) (IEEE, 2012)*, pp. 15–16
32. P. Rousche, R. Clement, C. Visser, J. Cruz, J. Si, "Neurabotics: an integration of neurons, rats and robots for advanced studies of brain–computer interface systems", *First International IEEE EMBS Conference on Neural Engineering, 2003. Conference Proceedings. (IEEE, 2003)*, pp. 462–465
33. K. R. Lyons, S. S. Joshi, "Paralyzed subject controls telepresence mobile robot using novel sEMG brain–computer interface: Case study", *2013 IEEE 13th International Conference on Rehabilitation Robotics (ICORR) (IEEE, 2013)*, pp. 1–6
34. W. Jia, H. Pu, X. Luo, X. Chen, D. Huang, O. Bai, "Reliable control of a quadruped walking robot in uneven terrain environment based on noninvasive brain–computer interface", *2012 IEEE International Conference on Mechatronics and Automation (IEEE, 2012)*, pp. 1608–1614
35. R. Upadhyay, P. Kankar, P. Padhy, V. Gupta, "Robot motion control using brain computer interface", *2013 International Conference on Control, Automation, Robotics and Embedded Systems (CARE) (IEEE, 2013)*, pp. 1–5
36. A. Vourvopoulos, F. Liarokapis, "Robot navigation using brain–computer interfaces", *2012 IEEE 11th International Conference on Trust, Security and Privacy in Computing and Communications (IEEE, 2012)*, pp. 1785–1792
37. A. Ubeda, E. Ianez, J. Badesa, R. Morales, J. Azorin, N. Garcia, "Control strategies of an assistive robot using a Brain–Machine Interface", *2012 IEEE/RSJ International Conference on Intelligent Robots and Systems (IEEE, 2012)*, pp. 3553–3558
38. Y. Bian, X. Wang, L. Zhao, S. Cui, "A new control system for space robot based on brain–computer interfaces", *2016 IEEE Advanced Information Management, Communicates, Electronic and Automation Control Conference (IMCEC) (IEEE, 2016)*, pp. 232–236
39. P. Gargava, K. Sindwani, S. Soman, "Controlling an arduino robot using Brain Computer Interface", *Proceedings of 3rd International Conference on Reliability, Infocom Technologies and Optimization (IEEE, 2014)*, pp. 1–5

40. C. Hema, M. Paulraj, S. Yaacob, A. Adom, R. Nagarajan, "Robot chair control using an asynchronous brain machine interface", *2010 6th International Colloquium on Signal Processing & its Applications (IEEE, 2010)*, pp. 1–4
41. F. Galan, M. Nuttin, E. Lew, P. Ferrez, G. Vanacker, J. Philips, J. Millan, A brain-actuated wheelchair: asynchronous and non-invasive brain–computer interfaces for continuous control of robots. Clin. Neurophysiol. **119**, 2159 (2008)
42. E. Hortal, D. Planelles, C. Costa, E. Ianez, U. Eduardo, J. Azorin, E. Fernandez, SVM-based Brain–Machine Interface for controlling a robot arm through four mental tasks. Neurocomputing **151**, 116 (2015)
43. A. Ubeda, E. Ianez, J.M. Azorin, Shared control architecture based on RFID to control a robot arm using a spontaneous brain–machine interface. Robotics and Autonomous Systems **61**, 768 (2013)
44. E. Ianez, J.M. Azorın, A. Ubeda, J.M. Ferrandez, E. Fernandez, Mental tasks-based brain–robot interface. Robotics and Autonomous Systems **58**, 1238 (2010)
45. J. Wang, Y. Liu, J. Tang, "Fast robot arm control based on brain–computer interface", 2016 IEEE Information Technology, Networking, Electronic and Automation Control Conference (IEEE, 2016), pp. 571–575
46. H. Wang, X. Dong, Z. Chen, B. E. Shi, "Hybrid gaze/EEG brain computer interface for robot arm control on a pick and place task", *2015 37th Annual International Conference of the IEEE Engineering in Medicine and Biology Society (EMBC) (IEEE, 2015)*, pp. 1476–1479
47. E. A. Pohlmeyer, B. Mahmoudi, S. Geng, N. Prins, J. C. Sanchez, "Brain–machine interface control of a robot arm using actor-critic re-enforcement learning", *2012 Annual International Conference of the IEEE Engineering in Medicine and Biology Society (IEEE, 2012)*, pp. 4108–4111
48. N. Waytowich, A. Henderson, D. Krusienski, D. Cox, "Robot application of a brain computer interface to staubli tx40 robots—early stages", *2010 World Automation Congress (IEEE, 2010)*, pp. 1–6
49. A. Eleni, "Control of medical robotics and neurorobotic prosthetics by noninvasive Brain–Robot Interfaces via EEG and RFID technology", *2008 8th IEEE International Conference on BioInformatics and BioEngineering (IEEE, 2008)*, pp. 1–4
50. M. Hoenen, K.T. Lübke, B.M. Pause, Non-anthropomorphic robots as social entities on a neurophysiological level. Comput. Hum. Behav. **57**, 182 (2016)
51. M.A.A. Kasim, C.Y. Low, M.A. Ayub, N.A.C. Zakaria, M.H.M. Salleh, K. Johar, H. Hamli, User-friendly labview gui for prosthetic hand control using emotiv eeg headset. Procedia Computer Science **105**, 276 (2017)
52. R. Roy, M. Mahadevappa, C. Kumar, Trajectory path planning of EEG controlled robotic arm using GA. Procedia Computer Science **84**, 147 (2016)
53. V. Azimirad, M. Alimohammadi, A. Joudi, A. Eslami, M. Farhoudi, Analysis of PSO, AIS and GA-based optimal wavelet-neural network classifier in brain–robot interface. IRBM **36**, 240 (2015)
54. L. Wang, R. X. Gao, J. Váncza, J. Krüger, X. V. Wang, S. Makris and G. Chryssolouris, "Symbiotic Human–Robot Collaborative Assembly," *CIRP Annals—Manufacturing Technology*, Vol.68, No.2, pp. 701–726, 2019

Part V
Multimodal Decision Support

Chapter 15
Worker Assistance Suite for Efficient Human–Robot Collaboration

Gergely Horváth, Csaba Kardos, András Kovács, Balázs Pataki, and Sándor Tóth

Abstract Providing efficient worker assistance is of key importance in successful human–robot collaboration as it can reduce the cognitive load of the workers. The human and the robot together form an environment where bidirectional information exchange and context awareness play a crucial role. Using multiple channels of communication allows fitting this context to the worker's preferences, which helps increase efficiency and the quality of production. This chapter introduces a novel worker assistance suite, which supports multimodal communication. The structure of the system and its interfaces are discussed and the implementation is detailed.

15.1 Introduction

This chapter introduces the aspects of worker assistance related to Human–Robot Collaboration (HRC) [1]. The increased flexibility, which is demanded from modern production systems, especially in the case of HRC, requires advancement on the worker assistance side as well. This chapter gives an overview of related work and formulates the problem as providing multimodal context-dependent assistance

G. Horváth · Cs. Kardos (✉) · A. Kovács
Research Laboratory on Engineering and Management Intelligence, SZTAKI Institute for Computer Science and Control, Eötvös Loránd Research Network, Budapest, Hungary
e-mail: csaba.kardos@sztaki.hu

G. Horváth
e-mail: gergely.horvath@sztaki.hu

A. Kovács
e-mail: andras.kovacs@sztaki.hu

B. Pataki
Department of Distributed Systems, SZTAKI Institute for Computer Science and Control, Eötvös Loránd Research Network, Budapest, Hungary
e-mail: balazs.pataki@sztaki.hu

S. Tóth
SANXO-Systems Ltd., Budapest, Hungary
e-mail: sandor.toth@sanxo.eu

L. Wang et al. (eds.), *Advanced Human–Robot Collaboration in Manufacturing*,
https://doi.org/10.1007/978-3-030-69178-3_15

for human workers. This is followed by a detailed walk-through of the definition, development, and implementation of an assistance system. The concepts and results presented here are based on papers by Kardos et al. [2, 3].

15.2 Related Work

Proper work instructions are of key importance in production systems as they contribute to the quality of the resulting products. Work instructions describe the operations, parts, equipment, resources, safety regulations, and commands applicable during the execution of the production tasks. Paper-based documentation is still the most widely applied form of delivering work instructions, however, its static structure makes it difficult to maintain and update over time. With shortening product life cycles and growing product diversity, there is a need to replace paper-based instructions with enhanced (digital) Worker Instruction Systems (WIS) [4, 5].

The main advantage of using WIS over paper-based solutions is their ability to be linked to multimedia instruction databases or other information sources, allowing dynamically changeable content. This suits the requirements of quickly changing manufacturing environments. Not only many companies have developed their own WIS solutions, but also several commercial frameworks are available, with various options to link to instruction databases. Most of them offer interfaces to import product and process data from Enterprise Resource Planning (ERP), Product Lifecycle Management (PLM), and Manufacturing Execution Systems (MES), supporting the definition of instruction templates, which the imported information can be linked to. Application of WIS allows maintaining consistent and up-to-date work instructions.

The modalities offered by commercial WISs are mostly visual—e.g., text, figure, video, and 3D content—in some cases extended with audio instructions or commentary. The required user interaction is, in most cases, provided by contact-based means, e.g., a button or a touchscreen, however, more advanced cases can also rely on Near Field Communication (NFC) or Radio Frequency Identification (RFID) technologies.

Commercial WIS applications mostly focus on maintaining the consistency of work instructions, however, with the increasing complexity of production systems, there is a need for enhanced support to the human operator as well. On the one hand, assembly systems offer an efficient answer to handling the growing number of product variants and the need for customised products, while, on the other hand, adapting to this variety puts additional cognitive load on skilled workforce [4, 6]. More information to be handled by the operator means more decisions to be made, which can easily lead to errors in these decisions [7]. In order to handle this increased cognitive burden, improvements in WISs have to be implemented. With the prevailing technologies offered by cyber-physical devices, new research focuses on ways of utilising them in worker assistance systems.

In [4], authors discussed how digital models can be used to generate virtual assembly for simulation, planning, and training purposes. The various modalities available in modern production systems and WISs are also introduced (see Fig. 15.1).

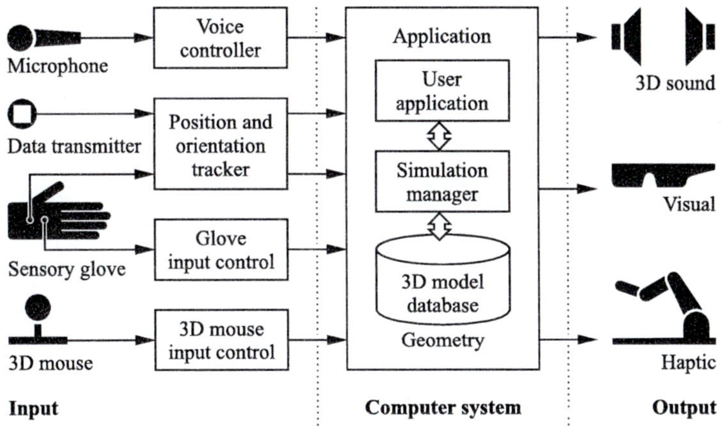

Fig. 15.1 Input and output interfaces in a VR-based WIS, based on [4]

In [8], as an early contribution to the field, three important questions (aspects) were proposed regarding worker information systems: what (content), how (carrier), and when (timing) to present. Based on the work of Alexopoulos et al. [9], Claeys et al. [10] focused on the content and the timing of information in order to define a generic model for context-aware information systems (Fig. 15.2). Context-awareness and tracking of assembly process execution are also the focus of research by Bader and Aehnelt [11], and Bannat et al. [12].

A special field of interest can be quality check or maintenance operations, with an even higher variety of tasks—Fiorentino et al. [13] suggested addressing this challenge using augmented reality instead of paper-based instructions to increase productivity, while Erkoyuncu et al. [14] proposed improving efficiency by context-aware adaptive authoring in augmented reality.

Lušić et al. [6] aimed to provide a methodology for selecting WIS for assembly, and they identify the following key characteristics:

- The applied modalities of the communication,
- The degree of mobility, i.e., the application of stationary or mobile devices,
- Dynamic or static provision of the information, where dynamic provision requires a suitable representation of the sequence of operations and their tracking,
- Flexibility of information delivered: context-aware or predefined information,
- Source of information delivered: local or distributed (web-based) data flow.

The impact of WIS on work performance is analysed in [15], and a causal model and KPIs are proposed to discover the relationship between work instructions and production (see Fig. 15.3). They propose the following classification of the functions of a WIS:

- Functions of perception, such as documentation, monitoring, and context recognition,
- Functions of instructions for known processes, such as initiation of processes, information output, learning, and practicing,

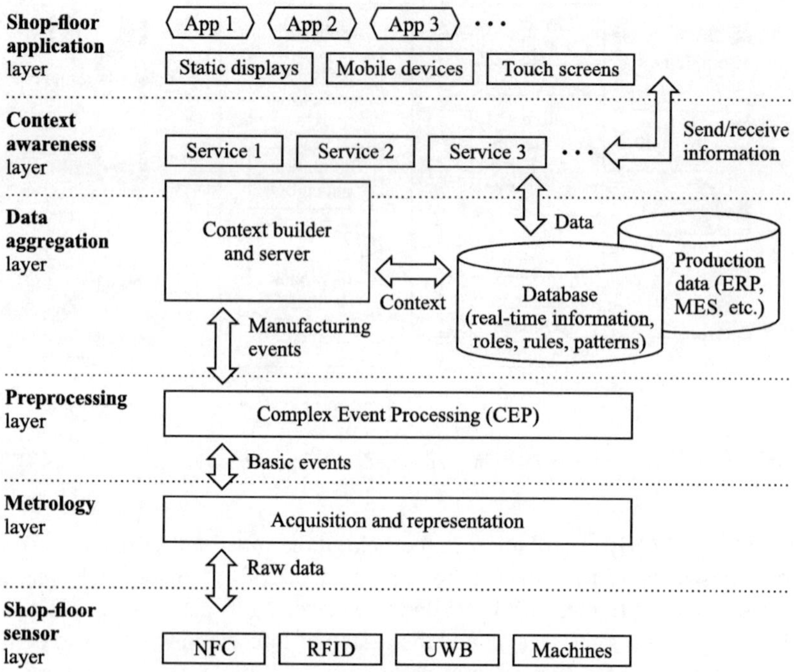

Fig. 15.2 A generic model for context-aware information systems [10]

- Functions of instructions for unknown processes, such as decision-making support and communication.

Morioka and Sakakibara [16] identified that WISs (by providing skill-dependent and multimedia content) have a key impact on shortening the learning and training phase for workers. They proposed a system for extracting expert knowledge and transferring it as multimodal learning materials to novice workers (see Fig. 15.4.). Vernim and Reinhart [17] also confirmed that better performance can be achieved by carefully selecting the media to deliver the content.

This review of currently available software solutions and works on WIS solutions shows that—besides recent development of commercial systems (connectivity, handling various contents, etc.)—technologies delivered by cyber-physical devices offer new ways to further improve the capabilities of WIS solutions [17]. The overview also shows that WIS for assembly can provide the necessary support to human workers in flexible production systems. The emerging field of collaborative robotics aims to even further increase the flexibility offered [18]. However, as the availability of (skilled) human workforce is often the bottleneck in production systems, successful collaborative applications require efficiency from both human and robot actors [16].

Nevertheless, recent trends in assembly systems (e.g., large product diversity and, especially, the prevalence of collaborative applications) make it necessary to increase the efficiency of bidirectional communication by providing processing context-related information and multimodal interfaces.

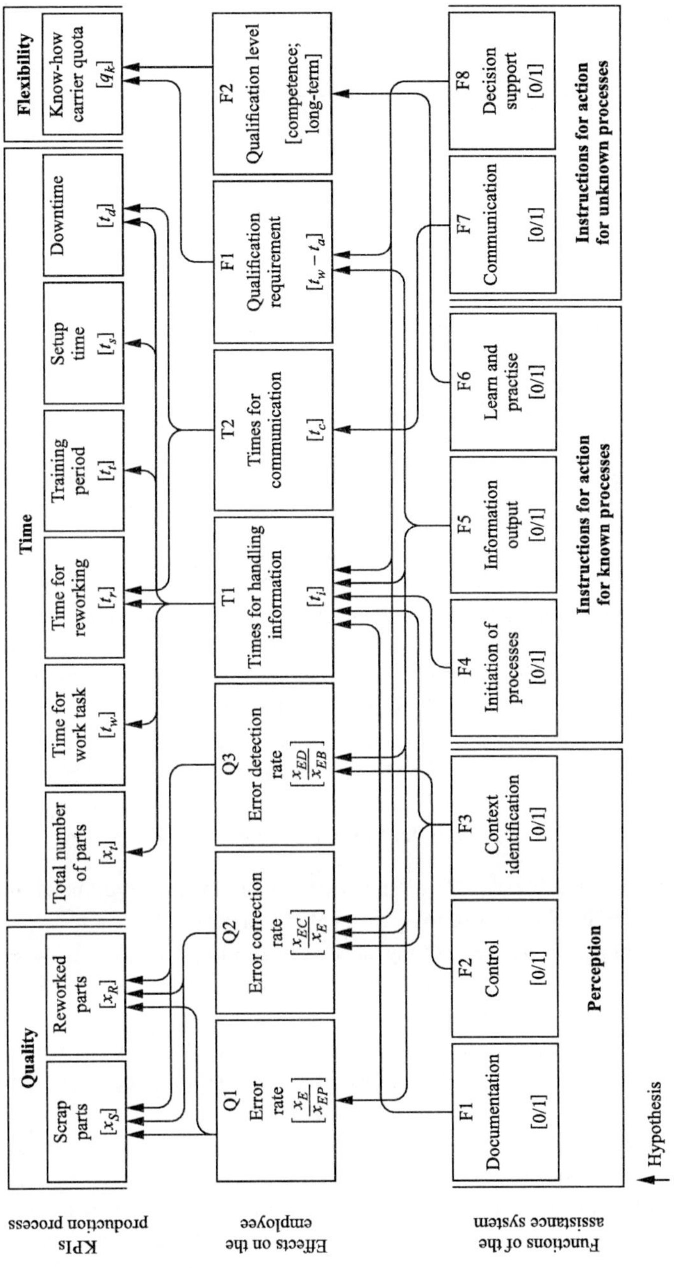

Fig. 15.3 A causal model for analysing the effects of worker assistance on production processes, based on [15]

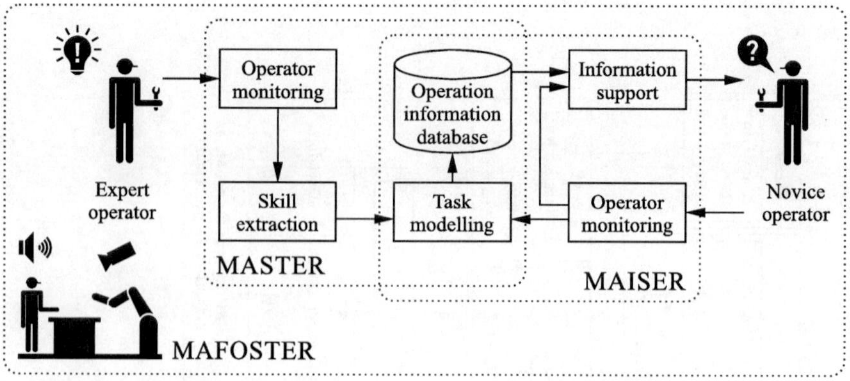

Fig. 15.4 Extraction–storage–delivery model for multimodal instruction, based on [16]

15.3 Problem Statement

Assuming that human–robot collaboration (HRC) takes place in a workcell equipped with a single robot, and a sequence of tasks to be executed is given, with each step assigned to either human workers or to the robot, a WIS is required to provide bidirectional interfaces for the human workers. Based on the literature review, the expectations and requirements towards such an assistance system can be formulated, and the problem can be stated as the definition of a worker assistance system capable of processing and storing input data coming from connected systems which defines the context of the worker's activity and the process subject to execution, and Providing multimodal bidirectional communication to and from the human worker, with the main goal of delivering content according to the specifications of process engineering.

It is also assumed that HRC assistance is provided not by a standalone system but as a part of an HRC ecosystem, the latter being composed of the following:

- A Unit Controller (UC) controlling and tracking the task execution within the workcell and also implementing robot controller functions,
- A Cockpit responsible for scheduling and rescheduling the tasks between the available resources (e.g., robot or human worker) and monitoring the shop floor, and
- A Mobile Worker Identification (MWI) system for locating human workers across the shop floor and within the workcell, also transmitting this information towards the Cockpit.

The operation of a collaborative WIS is (at least) twofold: content is created and the interfaces of the system are defined *offline* (i.e., not during production), while content delivery and the interactions with the system take place during production, when the system is *online*. These two basic use cases define two basic user types (roles) in a collaborative WIS: the *process engineer* (PE) who authors the content offline, and the *operator* who interacts with the system, typically when it is online.

15.4 Data Model

The general concept of providing the worker with a wearable mobile assistance suite is guided by the recognition that wearable components do not necessarily need to implement autonomous or smart behaviour. The most important function of the worker assistance system is to provide multimodal and bidirectional communication through multiple devices between the worker and the Unit Controller (UC). As the UC is a single, dedicated instance in every workcell, this mode of operation is best supported by a server–client architecture where the majority of the business logic is implemented on the server side. Placing the logic into a central component is advantageous as it increases the compatibility offered towards the connected devices which, therefore, are able to connect and communicate via standard interfaces and protocols (e.g., HTML5). The server–client-based human-machine interface system implemented in accordance with the aforementioned requirements is named Human-Machine Interface Controller (HMIC). Figure 15.5 shows its major structure and connections to other workcell components. The HMIC consists of two main parts. (1) The *backend* is responsible for implementing the business logic (i.e., the instruction data model and database) required for the content handling, and the interfaces for communication towards the Cockpit and the UC. (2) The *frontend* provides the necessary interfaces towards the devices of the wearable assistance suite.

The HMIC backend is coupled with a database (HMIC DB; see Fig. 15.6.) that stores the context data and all the relevant information delivered to the worker through the assistance suite. These data are queried during execution-time content generation by the backend before delivery, in order to have the delivered information fit (1) the context defined by the process execution, (2) the devices available at the worker, and (3) the skills of the worker using these devices. The delivery of the instructions is triggered by the HMIC input interfaces. The most important fields of the database are detailed below.

Similar to the representation applied in the Cockpit and the UC, the *task* is a fundamental element of the HMIC representation, as it links operations to workstations. It is the smallest element tracked by the Cockpit during execution.

An *operation* is the smallest element of task execution in the HMIC DB. A task is built up by a sequence of operations.

Instructions cover one or more operations, which are delivered to the worker in the format defined by the instruction. For each instruction, there is a skill level specified, which allows having different skill-level-dependent instructions for the same operation(s). A typical approach here, which was used in the presented implementation as well, is to specify low granularity instructions for the beginner workers and to combine multiple instructions covering several operations into one instruction for the more skilled workforce.

The *worker* table contains the defined task skills, information regarding the devices available for the worker, and the identifier of the workstation where the worker is located. This field is updated as a result of worker identification.

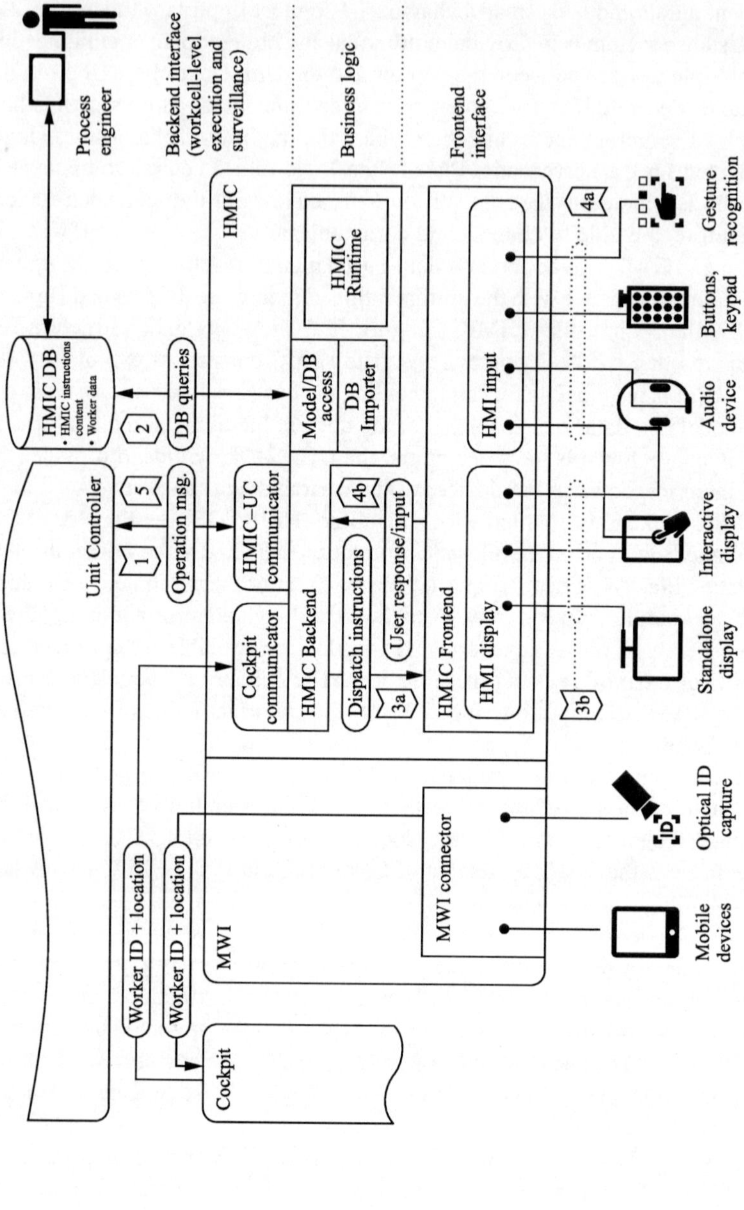

Fig. 15.5 Overview of the HMIC system

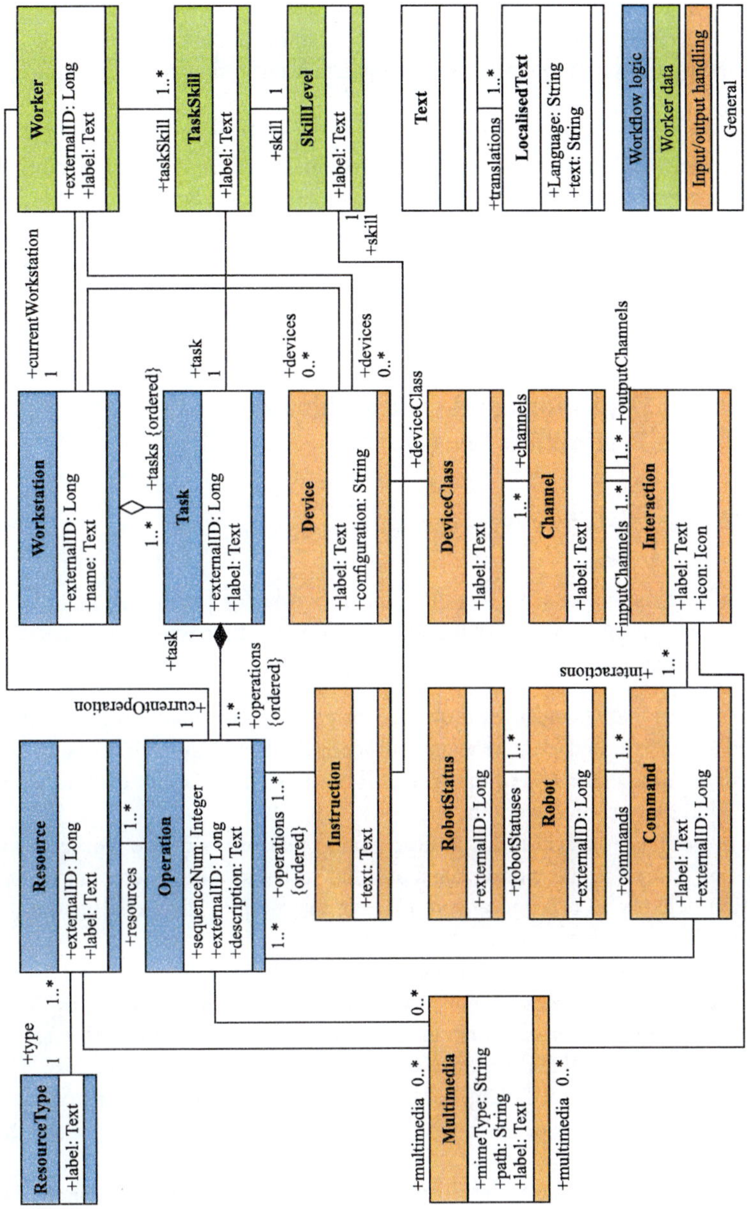

Fig. 15.6 Data model of the HMIC WIS

The *task skills* assigned to each worker contain records that describe the skill level of the worker for a given task. Therefore, instead of having one skill level for every activity, it is possible for a worker to have different skill levels for different tasks, which is useful, for example, when the worker is transferred into a new role. (By default, there are three levels available: beginner, trained, and expert.)

Input and output devices available and assigned to each worker are stored in *device* tables.

Each device is an instance of a *device class* object, which specifies the input and output channels available for a certain device. As the device class can have multiple channels, it is possible to define multimedia devices, e.g., a smartphone device can be defined to provide channels for audio input/output, screen, and buttons.

Using *interactions*, it is possible to define how specific input channels are allowed to receive commands from the worker.

15.5 Context Definitions and Communication with the Production Ecosystem

15.5.1 Contexts of Process Execution

In order to deliver instructions in a way that suits the worker and the situation the best, the worker assistance system must obtain information about the following aspects of the working context:

- *Process context*: tasks and operations are originally specified externally, then managed and distributed to the relevant workcells by the Cockpit. Upon execution of a predefined task, the UC of the workcell notifies HMIC about the task to be executed and the assigned worker.
- *Worker properties context*: skills, abilities, and preferences of the worker. This information, stored in the worker database, has to be taken into account during instruction delivery in order to ensure that the instructions displayed match the requirements of the given worker.
- *Worker location context*: the current location of each individual worker within the plant detected by the MWI systems. MWI notifies the Cockpit about significant changes in worker location, which in turn forwards this information to all relevant parties, including the HMIC instance belonging to the affected workcells. The location within the workcell also enables delivering location-aware instructions (e.g., increasing font size or volume when the worker is further away from the devices).
- Worker activity context: actions performed by the workers that encode commands to the robot (or other assembly system components) are captured by sensors of various types placed in a wearable suit or deployed at the workstation. These devices are managed directly by HMIC, which also interprets the commands

(acknowledgement of a task, failure, etc.), and forwards the commands to the UC.

15.5.2 Information Flow

When an operation is to be performed in the workcell, the UC sends a request for displaying the adequate instructions to the worker. The content, the delivery mode, and the formatting of the instruction is selected in accordance with the current process execution context, including the operation and worker identities (process context), worker skill level and individual preferences (worker properties context), and the precise location of the worker within the workcell (worker location context). Then, the instructions are delivered to the selected HMI display device. Upon termination of the operation, the worker responds with one of the predefined status responses for the task, typically, a success or failure response. This action is matched with the messages valid in the given context, and the selected message is passed on to the UC. The UC then determines the next action to perform, in view of the outcome of the preceding operation.

Figure 15.7 shows a series of detailed work instructions displayed to a novice worker. The status responses available for the operation are displayed at the bottom of the screen.

Means of ad-hoc communication towards the worker are also provided in order to be able to communicate outside the above-described normal process flow of displaying operation-related instructions. This is typically needed when exceptions occur in the execution, requiring human intervention for correction (removal of obstacles, repositioning of the workpiece to resume normal process flow, etc.), or recovery (removal of faulty workpiece or taking the cell offline for tool replacement, repair, etc.). In such cases, this information is displayed as a message to the worker, and a confirmation message is returned to the UC. This essentially corresponds to the information flow of normal process execution, however, the context and nature of messages and actions are set by the detected exception.

Similar to the ad-hoc messaging, the worker needs to be notified about changes in robot/cell status (e.g., when the robot starts moving). Unlike the previous cases, this information does not need to be confirmed by the worker. A confirmation is sent to the UC instead, however, this only means that the message was successfully displayed.

Similarly, ad-hoc commands can be issued by the worker to intervene with the normal process flow when needed (e.g., slow down, halt robot, retract robot, and resume operation). The UC can specify the set of valid commands or messages for a given operation, which are then displayed along with the robot status as a palette of command options, using the same information flow as explained in the previous paragraph: the worker can issue one of the displayed ad-hoc commands through the input devices available in the given context. Once recognised, the command is sent to the UC.

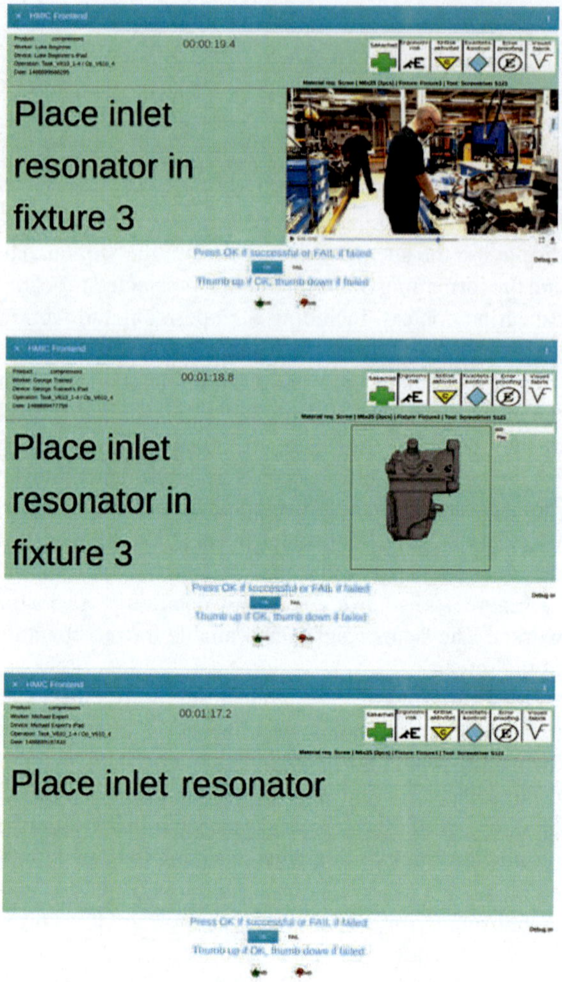

Fig. 15.7 Examples of an instruction generated for novice (top), trained (middle), and expert (bottom) workers

15.6 Multimodal Interaction

In order to present work instruction to the human worker with the modality (or multiple modalities) that suits the given situation the best, visual (text, image, video, or 3D animation) and audio modalities are used. In addition, multimodal human-to-machine commands allow the human worker to submit commands to the robot and to the assembly system in various novel contactless (e.g., hand or body gestures, or audio commands) and traditional contact-based modalities, for which the implementation is discussed below.

15.6.1 Hand Gestures

Hand gesture recognition is an effective and easy way for human workers to interact with a WIS. Hand gesture communication is fast, natural, and easy to learn.

The wearable gesture recognition device is a glove with sensors that is capable of measuring the momentary state of the worker's hand, recognises previously recorded gestures, and sends the corresponding command that is paired with it. The Gesture Glove interface is mobile: while the worker wears it, there is no need to be close to a button or touch screen. It is ideal to operate in noisy environments where voice communication would not work. It also works in occlusion or disadvantageous lighting conditions which are not ideal for visual communication.

The gestures produced by the glove and interpreted by the gesture interpreter can be used to confirm the end of tasks (e.g., when a task is successfully finished by the worker), or can be used to control robots either to start/stop the robot or instruct the robot's movements.

The Gesture Recognition device has three main parts: a glove with sensors, a microcontroller unit, and a client application. The Gesture Glove uses 12 inertial measurement units (IMU): two on each finger, one on the hand, and one on the wrist. These IMUs contain three 3-axis microelectromechanical sensors (MEMS): a 3-axis gyroscope, a 3-axis accelerometer, and a 3-axis magnetometer. The gyroscope measures rotation, but the signal integration causes a drift which must be corrected. For drift correction, the accelerometer and magnetometer are used. The accelerometer measures gravity, and the magnetometer measures the Earth's magnetic field. Because the accelerometer and magnetometer are slow and noisy, we cannot rely on them alone, so all three sensors are needed for correct spatial orientation measurement.

15.6.2 Body Gestures

Gesture recognition using body gestures is a convenient way of communication in situations when buttons or other more sophisticated interfaces are not available. It has the advantage that the operator needs no separate equipment to wear or operate, as the necessary sensors are all part of the working environment.

To have an effective body gesture recognition system in place, there must be a way to constantly observe the whole body of the operator. It may not be possible all the time because of occlusion, be it workbenches, other workers, robot movements, or just the self-hiding nature of some movements. It is, therefore, better to define redundant and unambiguous gestures. Although recognition of gestures is one crucial aspect, it is also essential that these gestures should not look akin to gestures that are regularly used by the worker during the assembly operation. Failure to fulfil the latter requirement may lead to the misdetection of gestures, which is unacceptable.

Example gestures used during the implementation are standing in an upright position, facing a sensor, and holding both hands straight to the side. This way the operator signals that the operation is finished, basically sending the command OK to the system. Standing upright and keeping one arm to the side and another high up in the air may signal that the operator had some trouble during the current operation, and thus sends the FAIL command.

In the example presented here, a Microsoft Kinect v2 was selected as a depth camera. For depth measurement, the Kinect v2 camera utilises time-of-flight sensors, but the camera also features RGB sensors. During implementation, the Microsoft Kinect SDK, version 1409, was used. This SDK is the reference implementation used in association with the Kinect sensor, however, it is only available on Microsoft Windows platform. The recognition software had a dictionary-like configuration file, containing gesture names and the associated commands that are to be sent upon gesture recognition. For gesture definition, the official Microsoft Visual Gesture Builder was used. Although there were other gesture proposals besides the ones mentioned above, many had to be dismissed because either the SDK could not handle them firmly, and therefore recognition quality was poor, or they were easily confused with regular movements of the operator.

15.6.3 Voice Commands

Voice-based interaction can be useful in cases when workers operate in a moderately noisy environment, and it is easier for them to say commands out loud rather than use hand gestures or touch screens.

Voice recognition works using a hot word (e.g., "Symbiotic"). When the voice-to-text engine recognises the hot word, it starts to interpret the subsequent words as commands to the system.

For each work instruction, one can define voice commands, i.e., the words expected to be said out loud by the worker for the purpose of communication with the robot. These voice commands, once interpreted, are converted to actual commands that are also used in other interaction modes to send information back from the worker to the system.

For example, when the worker is finished with the operation, he/she can finalise the work by responding with an OK command or signal, whereas a FAILED command can signalise a failure. In case of voice interaction, the worker can say the hot word + OK (e.g., "Symbiotic OK") to generate an OK command, or the hot word + FAIL (e.g., "Symbiotic Fail") to generate a FAIL command.

In the same way, when the worker controls the robot in a specific situation, saying the hot word + START ROBOT (e.g., "Symbiotic Start robot") lets the robot do its operations and send a ROBOT_START command, or it can be stopped with a hot word + STOP ROBOT (e.g., "Symbiotic Stop robot") to send a ROBOT_STOP command.

For text recognition, a possible solution is to use Google's text-to-speech engine available in the Chrome browser, which makes voice recognition available on any device capable of running a recent Chrome browser. Google's text to speech is of high quality and allows voice recognition without any training by the speaker and allows voice commands in practically any language.

It was found that voice recognition works best if the worker uses a headset with a good quality microphone. This can also enhance voice recognition when working in a noisy or crowded environment.

15.7 Mobile Worker Identification and Tracking

An important component of the proposed worker assistance suite, which provides the worker location context for the HMIC, is Mobile Worker Identification (MWI). Worker identification, localisation, and tracking methods based on sensor information are crucial for efficient HRC. Nevertheless, the adequate technology and the required functionalities strongly depend on the given application environment, such as the area to cover (complete shop floor or single workcell) or the desired accuracy.

In the current application, the goal was to cover a limited area on the shop floor close to the robot and the conveyor system, with an accuracy requirement below 0.5 m. Additional orientation information (the direction the human worker faces) was also desired. Finally, the MWI system developed had to be easy-to-integrate and easy-to-set up, cost-effective, and scalable, with a possibility to add, remove, or rearrange the sensors in case of modifying the area covered. These requirements could be satisfied with an area-scan camera system that tracks workers' motion with the help of QR codes attached to the workers' clothing. Since working with an industrial robot requires wearing a protective helmet, attaching the QR codes to the helmet is proposed; see Fig. 15.8. QR codes ensure fast readability and greater storage capacity compared to standard barcodes. The reading of QR codes can be robust, also when codes move. A part of the robustness of QR codes in the physical environment is their possibility of error correction—if this option is utilised, they continue to be correctly readable even when a part of the QR code image is obscured, defaced, or removed.

The requirement of covering a large area implies that a multi-camera system, consisting of low-cost, easy-to-integrate camera nodes, is needed, where each node is composed of local computing devices and cameras with suitable short-range optics. For the particular application, Raspberry Pi 3 single-board computers with IDS XS2 cameras (range: 0.5–2.2 m) and INTEL NUC i3 computers with UI3540LE cameras and TAMRON M12VM412 optics (range: 1.3–3 m) were used (a comparable node is shown in Fig. 15.9). Positions, orientations, and view angles of the camera nodes are established at the time of calibrating the system. Figure 15.10 illustrates one experimental setup with 7 smart camera nodes for MWI in a robotic workcell.

The captured images are scanned for QR codes (where multiple QR codes in a single image are allowed), and the code positions in the image plane are determined.

Fig. 15.8 Workers identified by MWI based on QR codes attached to their helmets

Fig. 15.9 Example of a smart camera node with a Raspberry Pi 3 and a camera attached

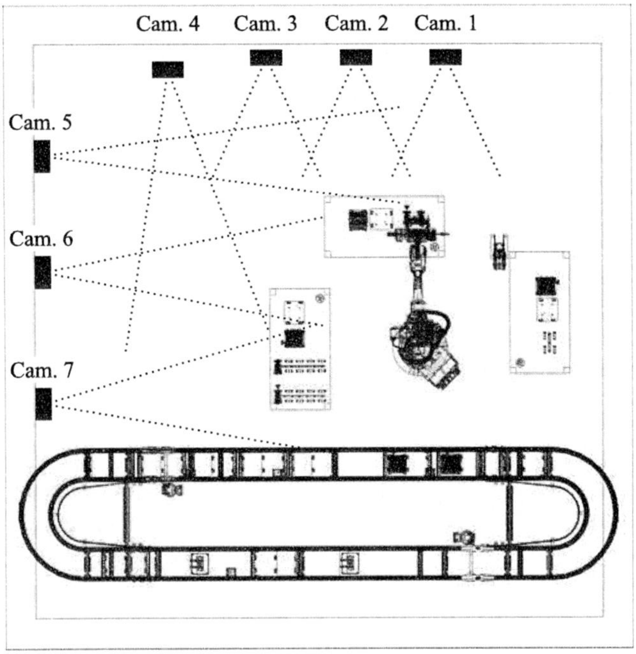

Fig. 15.10 Experimental setup with 7 smart camera nodes for MWI

Possible perspective distortions of the QR code are corrected, which is required for the robust reading of QR codes. The exposure time of the cameras is continuously adjusted according to the contrast and intensity parameters of the QR codes. The identified QR codes and their positions are subsequently submitted to a server application via Wi-Fi and TCP/IP communication, which performs the localisation of the human operators using one of the two possible calculation methods: when only one camera sees a given QR code, the position of the code is calculated by the size of the QR code and its position in the image plane. More accurate positioning can be achieved by triangulation when multiple cameras capture the same QR code within a given time window.

With the above physical architecture and computation methods, MWI enables delivering dedicated content to the authenticated workers corresponding to their skill levels, as well as selecting the appropriate display device depending on their position and head orientation in the HMIC.

15.8 Implementation

The HMIC backend was implemented as a *Spring Framework*-based Java web application providing 3 communication interfaces:

- *ZeroMQ* messaging between the HMIC and UC. For this, the HMIC provides a server port to receive messages from the UC and connect back to a server port provided by the UC to send messages. This way, messages between HMIC and UC can be sent asynchronously.
- *REST* interface for accessing data managed internally by the HMIC, such as the descriptive data of the Workstations, Workers, Operations, and Instructions.
- *WebSocket* interface using STOMP messages via *SocketJS* to handle the delivery of instructions and user input.

The HMIC frontend and other device handlers, such as the gesture glove and Kinect-based gesture controllers, connect to the HMIC via this interface, receive instructions, and send back commands as necessary. The HMIC frontend is the main client to the HMIC system providing a display of instructions on various devices including monitors, tablets, mobile phones, and Augmented Reality (AR) glasses. It also handles user interactions in the form of button-based inputs. The HMIC frontend is implemented as a *NodeJS* application with *React*-based user interface.

The implemented HMIC system was demonstrated in the laboratory simulation of an automotive assembly use case, where 29 parts were assembled in 19 tasks (for details, see [19]). The content was prepared and delivered to the system via XML files structured according to the HMIC DB schema. In the laboratory environment, the limited number of test users did not allow a comprehensive formal evaluation of every aspect of the implemented system, however, further tests are planned along the principles outlined in [20]. In order to test the multimodal interfaces, the following devices were connected to the system: a large-screen tablet device, a smartphone, a gesture glove, a Kinect device, and a pair of AR glasses. The devices with HTML5 web browsers were registered directly in the frontend of the system, while those without web browsers were connected via their client applications developed, which provide the necessary interfaces according to the specifications. During the tests, the system was able to handle all devices in parallel, thus providing a wide variety of simultaneous communication channels. The assembly tasks and the worker identification signals were issued by using mock interfaces of a UC and a Cockpit. The system was able to handle inputs from and to deliver content to multiple devices with an adequate response time.

15.9 Conclusions

Human–robot collaboration in industrial production relies heavily on efficient and dependable communication between human and robotic resources. This is not only due to the fact that the interlinking of robots and humans with variable cognitive capabilities and skill levels needs appropriate information exchange, but also due to the high operation diversity in production areas where collaborative settings have the highest expected benefits (i.e., cases with large product variability and operations that often need a "human touch"). Trends and expectations outline the need for a worker assistance system that goes beyond today's prevailing practice of push-buttons and paper-based work instruction sheets by (1) enabling dynamic changes to the instruction content, (2) extending static content with more expressive multimedia elements (video, audio, virtual, or augmented reality), (3) integrating instruction delivery and machine-to-human signalling with human-to-machine communication, and (4) integrating the context-awareness of production control (including worker identification) with adaptation to individual user characteristics such as skill level, current mental/physical state, or preferences in the mode of communication.

The chapter explained a possible solution for a worker assistance system by presenting a pilot implementation integrated into a proposed multi-level production control architecture, and offering possibilities of adaptive content generation, selection of multiple communication modes, and recognition of current working context. The chapter presented selected technologies supporting communication over a variety of channels establishing a multimodal character of the suite, and discussed advantages and drawbacks of each principle. In addition, a possible architecture of a worker assistance suite was presented, which was validated in a pilot implementation representing an assembly scenario from the automotive industry.

The development of underlying technologies and changes in industrial production practice make it probable that the use of integrated worker assistance suites experiences considerable growth, especially in cases where current instruction and feedback practices are known to be a hindrance to work efficiency. Worker assistance systems have close dependencies to both process planning (preparing content offline) and process execution (deploying content with production being online). Therefore, the maturing of the multimodal communication paradigm is anticipated to bring about supporting technologies enabling routine coupling with existing solutions and technologies in production planning and control.

Acknowledgements Research for work presented in this chapter has been in part funded by the "National Lab for Autonomous Systems", Ministry for Innovation and Technology, and the National Research, Development and Innovation Office, Hungary.

References

1. L. Wang, R.X. Gao, J. Váncza, J. Krüger, X.V. Wang, S. Makris and G. Chryssolouris, "Symbiotic Human–Robot Collaborative Assembly," *CIRP Annals—Manufacturing Technology*, 68(2), pp. 701–726, 2019.
2. Cs. Kardos, Zs. Kemény, A. Kovács, B.E. Pataki, and J. Váncza, "Context-dependent multimodal communication in human–robot collaboration," *Procedia CIRP*, vol. 72, pp. 15–20, Jan. 2018.
3. Cs. Kardos, A. Kovács, B.E. Pataki, and J. Váncza, "Generating Human Work Instructions from Assembly Plans", in UISP, Proceedings of the 2nd Workshop on User Interfaces and Scheduling and Planning. Delft, the Netherlands **2018**, 31–40 (2018)
4. M.C. Leu, H.A. ElMaraghy, A.Y. Nee, S.K. Ong, M. Lanzetta, M. Putz, W. Zhu, A. Bernard, CAD model based virtual assembly simulation, planning and training. CIRP Ann. **62**(2), 799–822 (2013)

5. M. Lušić, K. Schmutzer Braz, S. Wittmann, C. Fischer, R. Hornfeck, and J. Franke, "Worker Information Systems Including Dynamic Visualisation: A Perspective for Minimising the Conflict of Objectives between a Resource-Efficient Use of Inspection Equipment and the Cognitive Load of the Worker," *Advanced Materials Research*, vol. 1018, pp. 23–30, Sep. 2014.
6. M. Lušić, C. Fischer, J. Bönig, R. Hornfeck, and J. Franke, "Worker Information Systems: State of the Art and Guideline for Selection under Consideration of Company Specific Boundary Conditions," *Procedia CIRP*, vol. 41, no. Supplement C, pp. 1113–1118, Jan. 2016.
7. Å. Fast-Berglund, T. Fässberg, F. Hellman, A. Davidsson, J. Stahre, Relations between complexity, quality and cognitive automation in mixed-model assembly. J. Manuf. Syst. **32**(3), 449–455 (2013)
8. E. Hollnagel, Information and reasoning in intelligent decision support systems. Int. J. Man Mach. Stud. **27**(5), 665–678 (Nov. 1987)
9. K. Alexopoulos, S. Makris, V. Xanthakis, K. Sipsas, A. Liapis, and G. Chryssolouris, "Towards a Role-centric and Context-aware Information Distribution System for Manufacturing," *Procedia CIRP*, vol. 25, no. Supplement C, pp. 377–384, Jan. 2014.
10. A. Claeys, S. Hoedt, H.V. Landeghem, J. Cottyn, Generic Model for Managing Context-Aware Assembly Instructions. IFAC-PapersOnLine **49**(12), 1181–1186 (Jan. 2016)
11. S. Bader and M. Aehnelt, "Tracking Assembly Processes and Providing Assistance in Smart Factories," in *Proceedings of the 6th International Conference on Agents and Artificial Intelligence (ICAART)*, Angers, France, 2014, pp. 161–168.
12. A. Bannat, F. Wallhoff, G. Rigoll, F. Friesdorf, H. Bubb, S. Stork, H.J. Müller, A. Schubö, M. Wiesbeck, M.F. Zäh, "Towards optimal worker assistance: a framework for adaptive selection and presentation of assembly instructions," in *Proceedings of the 1st international workshop on cognition for technical systems, Cotesys*, 2008.
13. M. Fiorentino, A.E. Uva, M. Gattullo, S. Debernardis, G. Monno, Augmented reality on large screen for interactive maintenance instructions. Comput. Ind. **65**(2), 270–278 (Feb. 2014)
14. J.A. Erkoyuncu, I.F. del Amo, M. Dalle Mura, R. Roy, and G. Dini, "Improving efficiency of industrial maintenance with context aware adaptive authoring in augmented reality," *CIRP Annals*, vol. 66, no. 1, pp. 465–468, 2017.
15. T. Keller, C. Bayer, P. Bausch, J. Metternich, Benefit evaluation of digital assistance systems for assembly workstations. Procedia CIRP **81**, 441–446 (2019)
16. M. Morioka, S. Sakakibara, A new cell production assembly system with human–robot cooperation. CIRP Ann. **59**(1), 9–12 (Jan. 2010)
17. S. Vernim and G. Reinhart, "Usage Frequency and User-Friendliness of Mobile Devices in Assembly," *Procedia CIRP*, vol. 57, no. Supplement C, pp. 510–515, Jan. 2016.
18. G. Michalos, S. Makris, J. Spiliotopoulos, I. Misios, P. Tsarouchi, and G. Chryssolouris, "ROBO-PARTNER: Seamless Human–Robot Cooperation for Intelligent, Flexible and Safe Operations in the Assembly Factories of the Future," *Procedia CIRP*, vol. 23, no. Supplement C, pp. 71–76, Jan. 2014.
19. Cs. Kardos, A. Kovács, and J. Váncza, "Decomposition approach to optimal feature-based assembly planning," *CIRP Annals*, vol. 66, no. 1, pp. 417–420, 2017.
20. P. Hold, S. Erol, G. Reisinger, W. Sihn, Planning and Evaluation of Digital Assistance Systems. Procedia Manufacturing **9**, 143–150 (2017)

Chapter 16
Augmented Reality Enabled Human–Robot Collaboration

Xi Vincent Wang and Lihui Wang

Abstract Augmented Reality (AR) provides an interactive approach to combining the physical manufacturing environment with computer-generated virtuality. However, it is still challenging to establish an efficient AR system due to the complexity of the actual manufacturing environment, and the difficulties in human–robot collaboration (HRC). In this research, a feasible AR system is developed with a novel closed-loop structure. A feasible AR-based human–robot interaction method is also developed based on the HoloLens device, together with an advanced compensation mechanism. Compared with the previous research, the proposed work aims to accurate robot control based on AR. The proposed system is then implemented and validated through case studies, and quantifiable performance assessment on system response time, compensation accuracy, and other characteristics.

16.1 Introduction

In recent years, the manufacturing industry calls for a new generation of paradigms, systems and technologies, and new initiatives have been launched worldwide. Despite different terminologies initiated in different countries, the common spirit of the ongoing efforts is to strengthen manufacturing with the help of new automation, information and communication technologies. Among the novel technologies, Augmented Reality (AR) is considered a promising approach as it provides fast and efficient interaction methods for operators by combining the physical manufacturing environment with virtuality in cyberspace [1]. It is also aligned with the objectives of Cyber-Physical Systems, that is, the deep-intertwined hardware and software supporting the interaction with the physical context. In parallel, the Human–Robot Collaboration (HRC) approach also aims to offer a friendly and safe working environment

X. V. Wang (✉) · L. Wang
Department of Production Engineering, KTH Royal Institute of Technology, Stockholm, Sweden
e-mail: wangxi@kth.se

L. Wang
e-mail: lihuiw@kth.se

L. Wang et al. (eds.), *Advanced Human–Robot Collaboration in Manufacturing*,
https://doi.org/10.1007/978-3-030-69178-3_16

where an operator and an industrial robot can work in a shared space at the same time [2, 3]. HRC provides efficient support to both manufacturing and disassembly operations [4].

To achieve a more flexible and interactive manufacturing solution, it is logical thinking to combine the AR approach with the HRC application. Thorough system design is required to efficiently integrate multiple technology components. Meanwhile, the manufacturing tasks require high precision during execution, for example, welding, assembly, or machining. It is essential to develop an AR-based HRC approach with high accuracy. Therefore, this research is conducted to answer the research questions of (1) how to establish a comprehensive AR-based HRC system integrating essential technology modules efficiently, and (2) how to deploy a compensation mechanism in the AR-based HRC system. The research contributions in this chapter include a novel AR system based on wearable devices and a closed-loop mechanism that is capable of compensating potential execution deviations towards accurate HRC processes and operations. It tackles the accuracy challenge of the HRC with AR devices in the previous research. During system implementation and evaluation, the performance of the proposed system is validated in terms of response speed, and the result of the compensation mechanism is evaluated via accuracy measurement.

16.2 Related Work

In 2008, Reinhart et al. [1] introduced a programming system based on Mixed Reality (MR). Interactive drag-and-drop functions were realised for robot programming [4]. Similarly, an MR approach was proposed for collaborative design [5, 6]. In this approach, the virtual prototype can be manipulated and its process sequence can be decided. Research was later extended to an AR-based interaction system from design to virtual assembly training [7]. Ong et al. [8, 9] combined the AR technology with a finite element analysis for easy interpretation of the data and the location of critical regions of the product. Recently, a teleoperation approach is proposed to manipulate industrial robots by using a gamepad [10].

For the planning phase, Akan et al. [11] introduced an AR approach based on a multimodal language. The planning focus is shifted from a traditional coordinate-based to intuitively object-based approach. Jiang et al. [12] presented a facility layout planning system based on augmented reality. An optimisation algorithm based on Analytical Hierarchy Process Genetic Algorithm (AHP-GA) was integrated with the AR approach, towards automatic planning suggestions.

At the process execution stage, operator-support methods were proposed to provide virtual instructions during assembly [13, 14]. The AR approach is utilised for direct communication for the control of both mobile and industrial robots [15, 16]. The production and process information is provided to the user to enhance the operator's integration with the assembly process.

Kardos et al. [17] proposed a touch-based solution developed to control industrial robots. Similarly, a touchscreen-based approach was introduced to manipulate the mobile robot remotely [18]. The approach provides intuitive operation via the third-person view through the touchscreen.

Lambrecht and Kruger [19] developed a hand-held device and a motion tracking system. Both static and dynamic gestures can be captured to define poses, trajectories, and tasks. Hofmann et al. [20] also proposed a Go & See system on a smartphone-enabled AR environment to identify the bottlenecks in production lines. Finally, Maly et al. [21] compared the AR environment on the touchscreen-based smartphone and in-air gesture-based glasses. These two approaches are evaluated from four main perspectives, that is, marking limitation, view limitation, visualisation techniques, and interaction techniques.

For the phase after task execution, Erkoyuncu et al. [22] introduced an AR system supporting context-aware maintenance. The proposed system can provide overlaid data animation, AR content, and context awareness in the virtuality.

To recapitulate, despite the significant development of AR-based robotic approaches in recent years, there is still a lack of a comprehensive structure design for AR-based industrial robot systems. Previous research work is based on open-loop systems and there is no compensation mechanism proposed in the literature. Thus in this research, a novel AR-based robot system is proposed with a detailed and feasible system structure design. A closed-loop compensation mechanism is also developed for accurate HRC operations.

16.3 Proposed System

As shown in Fig. 16.1, the proposed system consists of four major layers, that is, AR Layer, Virtual Environment, Robot Operating System (ROS) with robot commands, and the Robot Controller layer.

16.3.1 AR Layer

The *AR Layer* is the main interaction cockpit for the user. All the user commands and inputs are read and interpreted from here. The interaction logic of the AR Layer is summarised in Fig. 16.2. After the AR system is launched, the *Robot Recognition Module* first identifies the geometry and location of the physical robot and deploys the Virtual Robot 1 (master model) right over the physical robot in the AR vision. Then, the *User Interaction Module* recognises and analyses the operator's gesture. According to the user's hand gesture inputs, for example, tap, gaze, or grip, the user interactions are interpreted into moving commands to drive the Virtual Robot 1 (master model), whose posture will be moved, or rotated accordingly. The master model represents the final location and posture of the industrial robot. Conventionally,

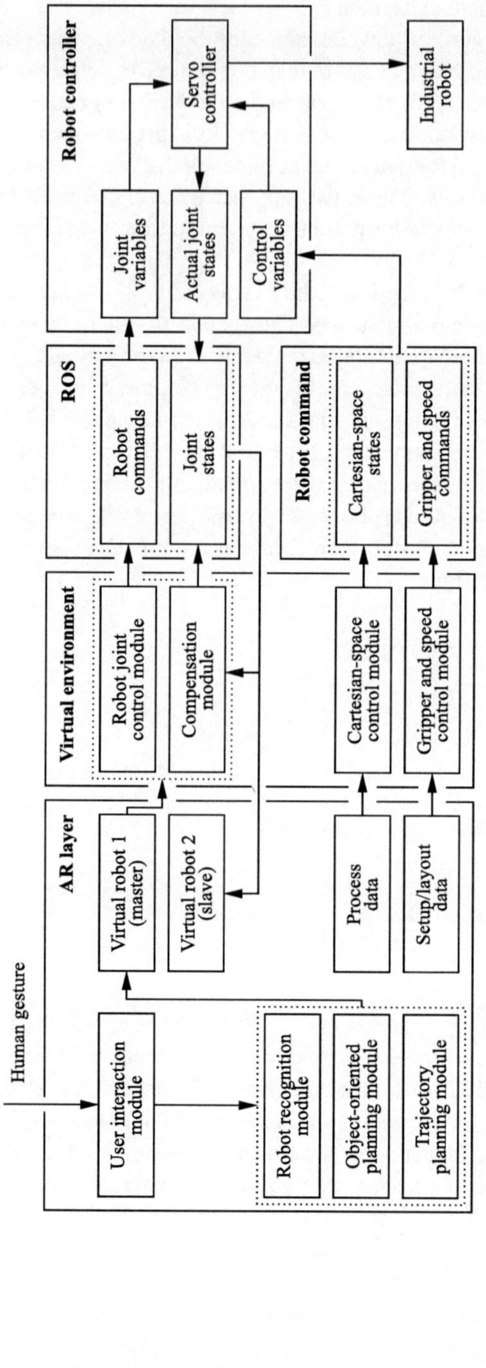

Fig. 16.1 System structure

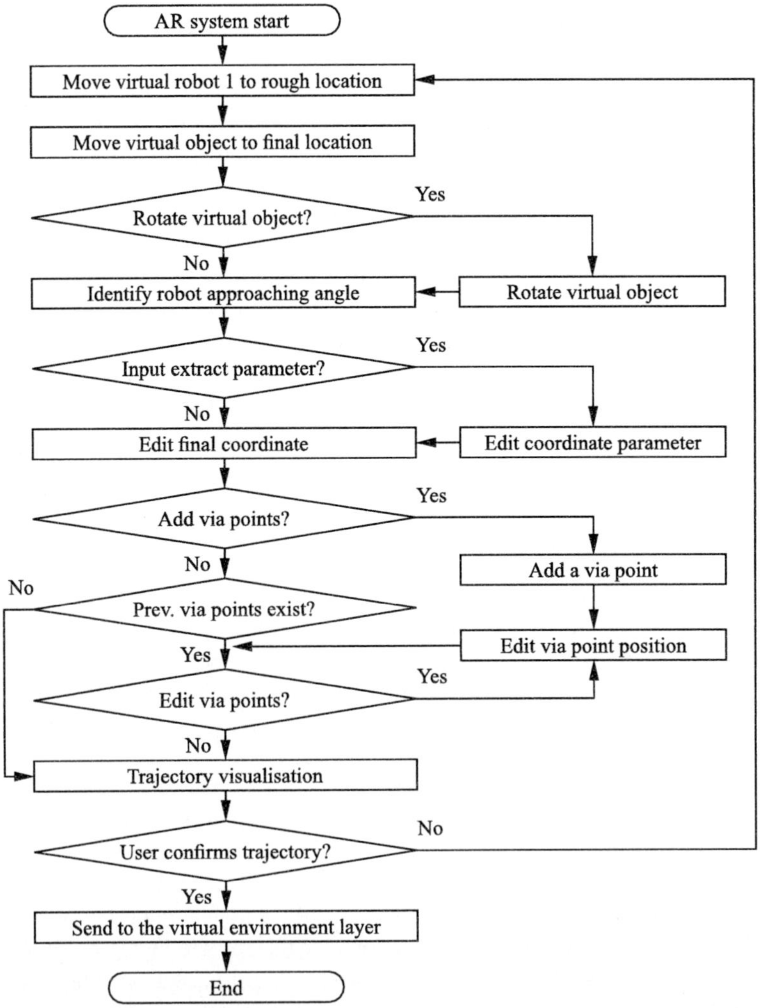

Fig. 16.2 Interaction logic of the AR Layer

it is difficult to identify detailed and accurate robot coordinates and posture in the AR space. Hence in this research, the *Object-oriented Planning Module* is also developed. This enables the user to manipulate a generalised virtual object to precisely define the desired final position and angle of the end-effector. The user can also input exact parameters of the coordinates when the physical objects' position and orientation are available before the collaborative process.

The *Trajectory Planning Module* is also developed at the AR device layer. The operator can set multiple via points in the virtual space in the AR, to address the feasible best trajectory of the robot, for example, avoiding potential collisions, or passing through narrow areas. Then, the robot processes through the via points step by step toward the final destination, according to the user's definition. The planned path and via points can be edited by the user as long as needed. During the interactive review phase, the determined trajectory is dynamically visualised by Virtual Robot 1 to provide an intuitive result to the user. After the trajectory plan is confirmed by the user, the result is sent to the virtual environment layer and the interaction module is terminated.

In the proposed system, two virtual robot models are actually established in the AR. During the trajectory execution, while the Virtual Robot 1 (master model) presents the final posture of the robot, Virtual Robot 2 (slave model) reflects the dynamic posture of the robot in real-time, according to the current joint states fed back from the robot controller. The joint position, speed, torque data can also be visualised in the AR space, thus forming the first closed-loop in the system. If there are any potential issues identified by the user, the robot can be stopped, and the user modifies the plan and improve the trajectory plan accordingly.

16.3.2 Virtual Environment

The *Virtual Environment* layer includes the *Robot Joint Control Module*, which reads the trajectory plan, as soon as it is confirmed by the end-user at the AR Layer (Fig. 16.1). The robot plan is converted from the trajectory plan into robot commands for an industrial robot. At this stage, there are two types of controls of the robot and two control modules are developed in the proposed system, respectively.

The Robot Joint Control Module manipulates the robot based on the combination of every joint, and the *Cartesian-space Control Module* utilises the final position of the end-effector in the Cartesian space. In practice, the manufacturing data (e.g., process plan, robotic cell layout, fixture geometry), is static knowledge that is maintained before the operation planning. This type of static data is normally stored in the Cartesian space. Thus, the system developed in this research provides two mechanisms to integrate dynamic planning and static control knowledge via both joint combination and Cartesian space. It also offers more potentials for interfacing with different AR devices and robot systems in the future. The Denavit–Hartenberg form is utilised to establish the coordinate transformation,

$$^{i-1}\mathbf{T}_i = \mathbf{Rot}(\mathbf{z}_{i-1}, \theta_i)\mathbf{Trans}(\mathbf{z}_{i-1}, d_i)\mathbf{Trans}(\mathbf{x}_i, a_i)\mathbf{Rot}(\mathbf{x}_i, \alpha_i), \tag{16.1}$$

where $^{i-1}\mathbf{T}_i$ is the transformation locating link i with respect to link $i-1$, θ_i is the joint angle, d_i is the link offset, a_i is the link length, and α_i is the link twist (see Chap. 10 [23] for a more detailed description of transformations and the Denavit–Hartenberg form). Eventually, the robot command is transferred to the next layer, together with the results of the Gripper & Speed Control Module.

16.3.3 ROS, Robot Command and Controller Layers

In this research, the *Robot Operating System* (ROS) is utilised to process the robot commands, and the proprietary robot data file is responsible for maintaining the Cartesian space states respectively. The robot commands are eventually sent to the control variable interfaces within the industrial robot controller.

As mentioned above, the manufacturing scenario has a high requirement for robot operations. From the perspective of low-level control, there are classic approaches like Proportional–Derivative (PD) and Proportional–Integral–Derivative (PID) controllers. However, the dynamic approach might affect the robot's stability and linearity. Hence in this research, a post-compensation mechanism is developed instead without altering the original control cycle. It needs to be noted that the actual position of the robot is sent back to ROS only at the last phase of the robot operation execution. The Compensation Module automatically compares the expected position variables against the actual robot positions. An error threshold defined before the execution based on the nature and requirement of the task. When the difference is larger than the threshold, the compensation algorithm is initiated accordingly and generates a compensatory movement command for the robot controller towards the planned position. Then, the robot advances again until the full completion of the task. Thus, the control accuracy of the system is guaranteed.

In practice, it is feasible to deploy the compensation mechanism at either the ROS level or the virtual level. In the proposed system, it is deployed at the *Unity PC* layer to decrease the computation task on the robot side and to provide the potential of more complex compensation mechanisms in the future, while ROS is utilised as the computer–robot interface only. As a result, the actual robot position is also dynamically looped back to the AR device, which drives the virtual model representing the actual robot posture constantly. Thus it achieves a *closed-loop* control mechanism for the AR-based human–robot collaboration with high accuracy throughout the system.

Figure 16.3 shows the control logic in the proposed work. When the control loop is started, the virtual robot pose is set the same as the physical robot, based on the exact join variables received from the controller. The system has five main control modules, that is, 6-axis control, C-space control, speed control, gripper control and compensation request. For better safety and clearer control logic, every module needs to be manually confirmed by users before the physical robot is driven by commands. For example, when the real robot is under 6-axis control mode, the user needs to confirm the new pose for the virtual robot in the AR environment, and then the data can be sent to initiate the execution. After the physical robot moves to the defined position, the user can define the next position or choose other control modules. The only functionality that does not have a safety confirmation is the Compensation Request, which can compensate for minor deviations between the expected position and the physical location. Because the data is collected from the controller and compared automatically, and the compensational movement is in a very limited range, users can initiate the robot movement without safety concerns.

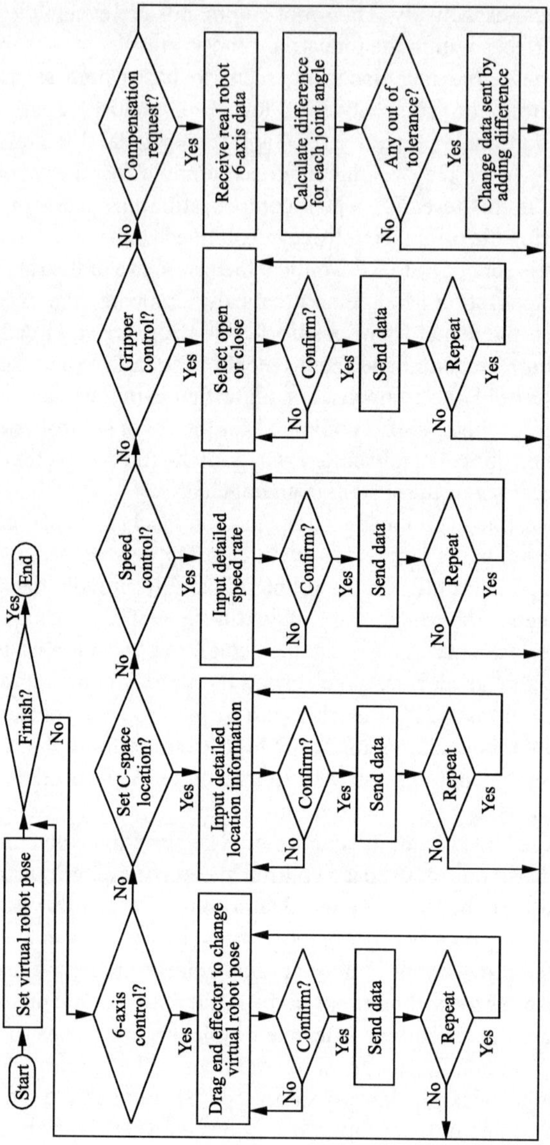

Fig. 16.3 Flow chart of the control logic

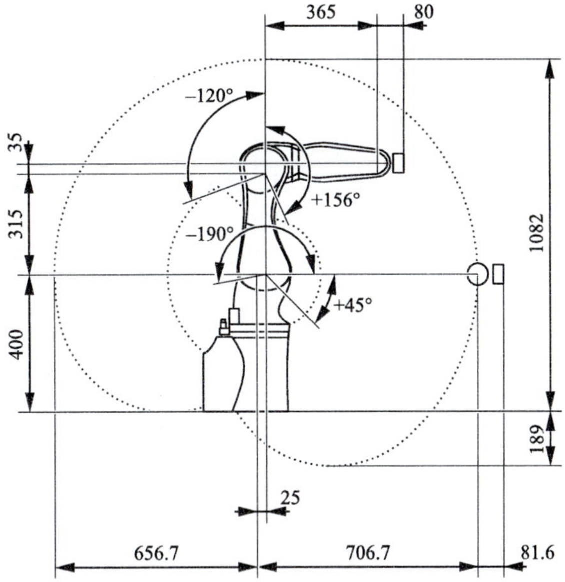

Fig. 16.4 Key parameters of KUKA KR6 R700 Sixx [24]

16.4 Deployment Structure, Implementation and Performance Evaluation

16.4.1 Test Robot

During the implementation, the industrial robot utilised is KUKA KR6 R700 Sixx. It is a compact 6-axis robot from the KR AGILUS series. Its energy supply system and controller are integrated within the KR C4 controller. A Windows PC is also utilised as a platform to design and test the graphic interface and also to serve as a broker computing node connecting the AR environment and the robot. The hardware is connected using a TCP/IP connection through Ethernet ports.

Unfortunately, KUKA does not support complete 3D models of industrial robots. Thus the virtual robot arm was re-created in SolidWorks based on the parameters from KUKA technical file as shown in Fig. 16.4. As Unity only supports FBX data format during importing, the assembly SolidWorks data is first imported to 3D Max for conversion into FBX format, and then transferred to Unity.

16.4.2 Trajectory Planning

The proposed system is implemented in a real industrial robot cell to evaluate the feasibility and performance. The deployment structure of the proposed system is

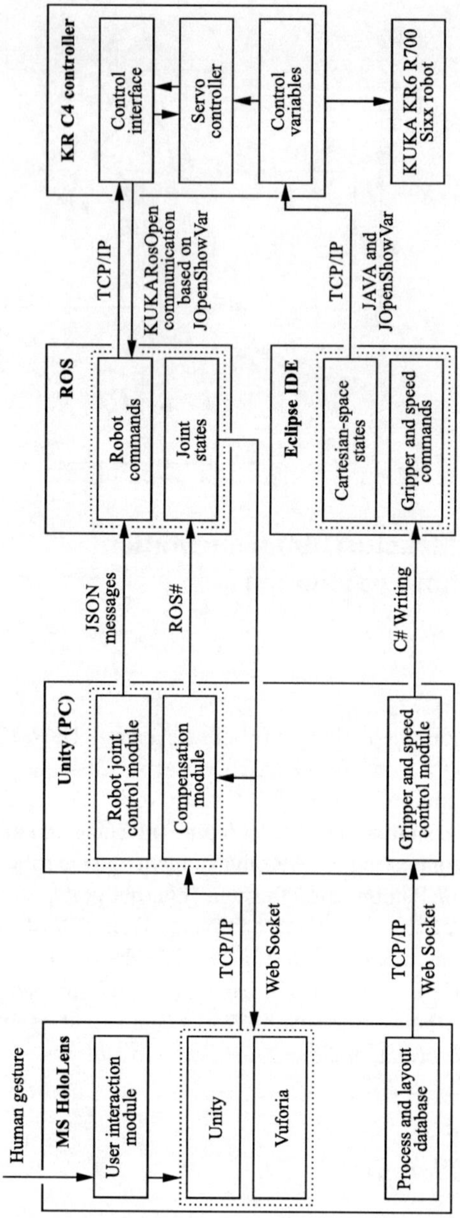

Fig. 16.5 Deployment structure

shown in Fig. 16.5, which is compliant with the system structure design introduced in the previous section.

In this research, Microsoft *HoloLens* is utilised as the AR device to host the interaction module. The AR environment and user interface is developed based on the *Unity* real-time development platform. The *Vuforia* engine is chosen for the model positioning and calibration. The user interaction module and trajectory planning module are developed based on the above-mentioned interaction logic.

After the system is initiated, the virtual robot is automatically generated in the AR environment based on the physical surroundings of the operator. The operator can move the virtual robot in two ways, that is, hand gestures and moving commands via clicking the arrows in different directions. The control arrows and sliders are available in the user interface as shown in Fig. 16.6. The virtual robot can be moved or rotated accordingly. In this way, the virtual robot can be slightly adjusted and calibrated to compensate for the position error caused by the vision devices.

During planning, the operator needs to identify the final position of the robot first. A virtual cube is utilised to define the 6 Degrees of Freedom (DOF) of the robot in the Cartesian space. In the editing mode, the operator can use a hand gesture to move and rotate the virtual cube, and the robot's pose is updated accordingly (Fig. 16.7).

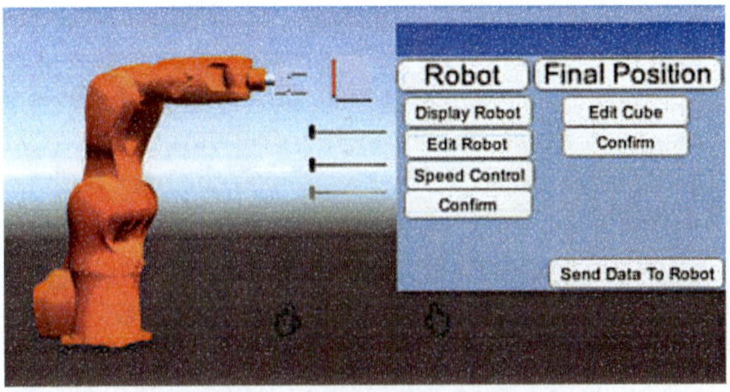

Fig. 16.6 AR environment calibration module

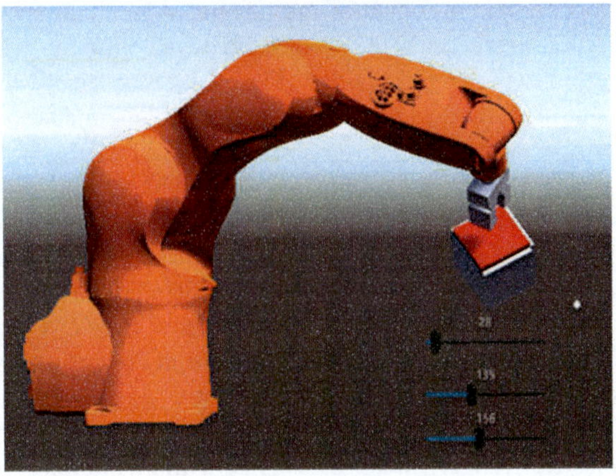

Fig. 16.7 Defining Final position

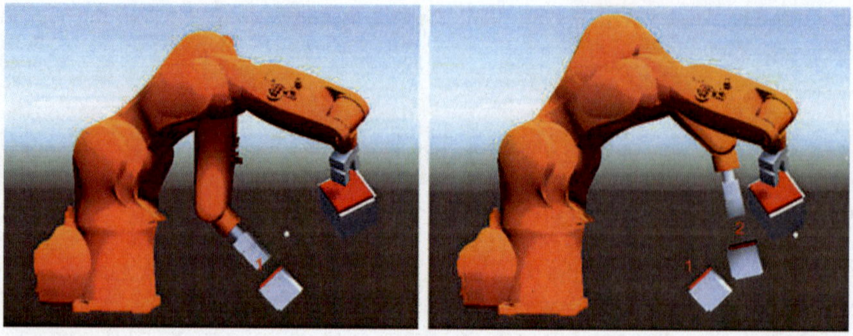

Fig. 16.8 Single (left) and multiple (right) via points

The gripper keeps perpendicular to a specific face on the cube, which is also specified by the operator. When needed, sliders and arrows are also available, which provides a user-friendly environment for robot planning. Then, the operator confirms the final position of the robot, and the planning module moves to the next phase.

After the robot's final position is confirmed, the trajectory planning is realised by the via point module. Via points are added between the initial position of the robot and the final position defined in the previous stage. As shown in Fig. 16.8, the via points are defined via 6-DOF virtual cubes which are similar to the ones of the final pose. Each cube represents one via point in the trajectory and the numbers indicate the sequence that the robot will travel through them. In this way, the operator has the opportunity to define either a simple trajectory with no via points, or a complex path by defining as many via points as needed. Before process execution, the operator still has the chance to modify the defined via points by clicking the *Edit* button. After the via points are confirmed, the robot trajectory is generated by the control module accordingly.

In addition, the operator is able to fine-tune the gripper to approach the target object precisely. The gripper can be rotated by moving the slider in the gripper module. Some common functions are integrated as a command package. For example, the *Pick* button enables the robot to go down, close the gripper to pick up the object and return to the original potion (Fig. 16.9). Conversely, the *Place* button offers an easy control method to place the object to the estimated position including opening the gripper.

After the trajectory plan is confirmed by the user, it is then transferred to the virtual environment hosted in the MS Windows PC via web server-socket protocol. The Unity terminal on the PC interprets the results and generates the robot commands that are comprehensible for ROS. The Unity interface then establishes the communication with ROS via ROS# interface and writes in JSON messages. In parallel, the robot commands in the Cartesian space are documented in the self-contained command data file via C# writing directly.

Next, ROS interprets the commands to be streamed to the KUKA KR C4 controller directly through *KukaRosCommunication* interface based on the *JOpenShowVar*

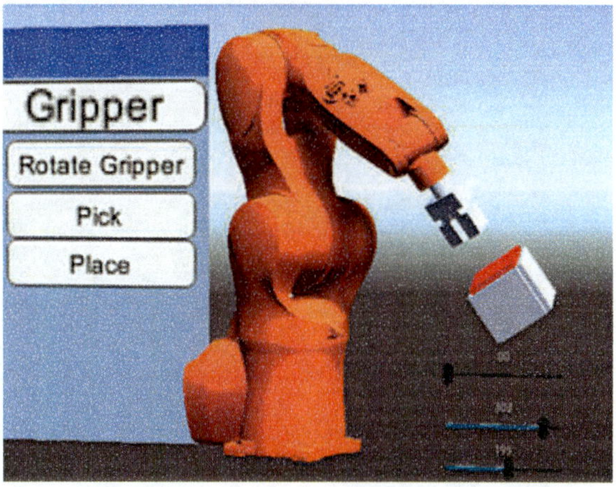

Fig. 16.9 Fine-tuning of the gripper

method. The robot commands are sent to the controller via TCP/IP protocol and manipulate the KUKA KR6 R700 Sixx industrial robot. Eventually, the robot starts to execute the operation accordingly, and Virtual Robot 2 duplicates the physical robot with more details streamed back to the user interaction module on HoloLens, thus forming a closed-loop AR system.

16.4.3 Performance Evaluation

As shown in Fig. 16.10, the user can view both the virtual and real robots through the HoloLens view field, and interact with the functionality panel in the corner. After the connection from the AR device to the robot is established, the user can manipulate the virtual robot via gesture commands to determine the Cartesian TCP pose, trajectory and task. Complex robot trajectories can also be planned via multiple via points defined by the user.

After developing the proposed system, the overall performance was evaluated via quantifiable assessments on the response speed. More than 60 tests have been taken and the response time results are presented in Table 16.1. The launch process connecting the AR environment takes 3,118 ms on average. The initiation process takes more time as there are multiple modules needed to be triggered and connected. After the communication is successfully established, the actual communication interval between the virtual environment and ROS is 85.8 ms on average. The average control latency from ROS to the KUKA robot is 69.6 ms through the TCP/IP communication. Eventually, throughout the whole interactive control loop, the total time interval is 155.4 ms on average. In practice, AR-based planning and control scenarios do not require real-time control and fast system response. Thus, it is a sufficient response speed especially considering the intermedium runtime requirement.

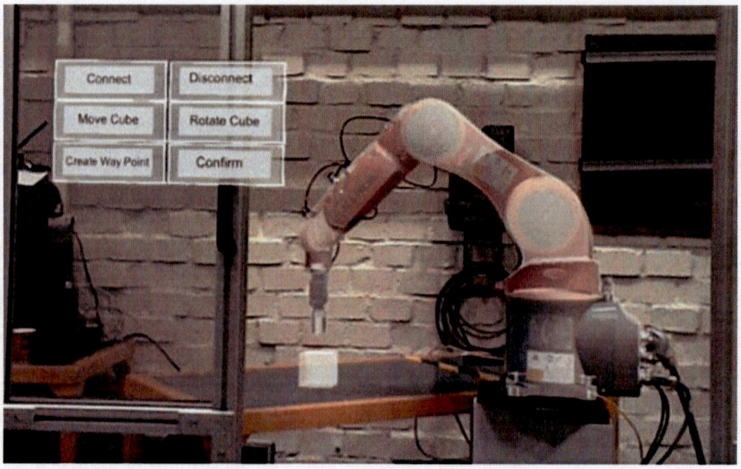

Fig. 16.10 AR-based industrial robot system (the virtual robot is visible in white, and the real KUKA robot in orange)

Table 16.1 System response time

Response Time [ms]	Connection Initiation	Unity to ROS	ROS to Robot	Overall System Latency
Maximum	3824	156	87	243
Minimum	2278	62	50	112
Average	**3118.7**	**85.8**	**69.6**	**155.4**

Regarding the compensation mechanism, multiple random deviations are deployed on each joint from A2 (joint closer to the robot's base, according to KUKA's definition) to A5 (joint closer to the end-effector) and the result is presented in Table 16.2. The compensation mechanism automatically detects the difference between the desired joint angle and the actual one. If the difference is larger than the threshold, the compensation module generates the compensational move command and drives the robot accordingly. From joints A2 to A4, the average deviation input is approximately 2 to 3 degrees, and the compensated position error decreases to

Table 16.2 Compensation results

Average deviation [degrees]	Before compensation	After compensation	Compensate rate (%)
A2	2.341	0.012	99.49
A3	2.679	0.009	99.66
A4	2.009	0.012	99.40
A5	4.006	0.054	98.65

0.012 degrees, which is close to the maximum accuracy of the KUKA KR6 R700 Sixx robot. Overall more than 99.4% of the deviations are compensated. As joint A5 is a smaller joint close to the end-effector, more deviations (average 4 degrees) are applied to the joint to evaluate a more challenging scenario. Eventually, the system is capable of compensating the movement and decreases the deviation to 0.054 degrees on average, in which 98.65% of the position error is eliminated. Hence, the feasibility of the proposed closed-loop compensation mechanism is validated with high automation level and accuracy.

16.5 Discussion and Conclusions

The modern manufacturing industry calls for new systems and technologies to achieve higher efficiency, user-friendliness and safety. AR technology provides an opportunity for a more interactive and friendly manufacturing environment, while HRC solutions combine the flexibility of the human operator and the accuracy of the industrial robot. However, a conventional AR-based approach cannot guarantee the accuracy of the actual robot execution. Hence, in this research, the AR and HRC approaches are integrated. The novel scientific contributions of this work include the comprehensive system structure for AR-based HRC, and the closed-loop compensation mechanism towards higher accuracy, answering the research question in system integration and compensation mechanism, respectively. The proposed approach is validated and evaluated via quantifiable performance assessments focusing on both the system response time and accuracy performances.

In the future, the proposed AR-based HRC system can be further evaluated with more complex manufacturing environments and operations. The safety and ergonomic measures can be specifically taken to assess the operators' experience and acceptance of the proposed approach. The results can be much valuable for future AR and HRC research in a manufacturing environment, especially where the operator wears the AR devices for a longer period of time for daily manufacturing operations. Moreover, the compensation module can be further improved by more advanced algorithms, for example, learning-by-failure mechanisms.

Acknowledgements The authors would like to thank Mingtian Lei and Yongqing Zhao for their contributions to programming and implementation.

References

1. G. Reinhart, U. Munzert, and W. Vogl, A programming system for robot-based remote-laser-welding with conventional optics, *CIRP Ann.—Manuf. Technol.*, **57**(1), pp. 37–40, 2008
2. X. V. Wang, Zs. Kemény, J. Váncza, and L. Wang, Human–robot collaborative assembly in cyber-physical production: Classification framework and implementation, *CIRP Ann.—Manuf. Technol.*, **66**(4), pp. 5–8, 2017 https://doi.org/10.1016/j.cirp.2017.04.101
3. L. Wang, R. X. Gao, J. Váncza, J. Krüger, X. V. Wang, S. Makris and G. Chryssolouris, Symbiotic human–robot collaborative assembly, *CIRP Ann.—Manuf. Technol.*, **68**(2), pp. 701–726, 2019 https://doi.org/10.1016/j.cirp.2019.05.002
4. K. Wegener, W.H. Chen, F. Dietrich, K. Dröder, S. Kara, Robot assisted disassembly for the recycling of electric vehicle batteries. Procedia CIRP **29**, 716–721 (2015)
5. Sk. Ong, Y. Pang, and A. Y. C. Nee, Augmented reality aided assembly design and planning, *CIRP Ann.—Manuf. Technol.*, **56**(1), pp. 49–52, 2007
6. S. K. Ong and Y. Shen, A mixed reality environment for collaborative product design and development, *CIRP Ann.—Manuf. Technol.*, **58**(1), pp. 139–142, 2009
7. S. K. Ong and Z. B. Wang, Augmented assembly technologies based on 3D bare-hand interaction, *CIRP Ann.—Manuf. Technol.*, **60**(1), pp. 1–4, 2011
8. A.Y.C. Nee, S.-K. Ong, Virtual and augmented reality applications in manufacturing. IFAC Proc. **46**(9), 15–26 (2013)
9. S.K. Ong, A.W.W. Yew, N.K. Thanigaivel, A.Y.C. Nee, Augmented reality-assisted robot programming system for industrial applications. Robot. Comput. Integr. Manuf. **61**, (2020)
10. J. E. Solanes, A. Muñoz, L. Gracia, A. Martí, V. Girbés-Juan, and J. Tornero, Teleoperation of industrial robot manipulators based on augmented reality, *Int. J. Adv. Manuf. Technol.*, pp. 1–21, 2020
11. B. Akan, A. Ameri, B. Cürüklü, L. Asplund, Intuitive industrial robot programming through incremental multimodal language and augmented reality, in. IEEE International Conference on Robotics and Automation **2011**, 3934–3939 (2011)
12. S. Jiang and A. Y. C. Nee, A novel facility layout planning and optimization methodology, *CIRP Ann. Manuf. Technol.*, **62**(1), pp. 483–486, 2013
13. S. Makris, G. Pintzos, L. Rentzos, and G. Chryssolouris, Assembly support using AR technology based on automatic sequence generation, *CIRP Ann.Manuf. Technol.*, **62**(1), pp. 9–12, 2013
14. H. Liu, L. Wang, An AR-based Worker Support System for Human–Robot Collaboration. Procedia Manufacturing **11**, 1–9 (2017)
15. S. Makris, P. Karagiannis, S. Koukas, and A.-S. Matthaiakis, Augmented reality system for operator support in human–robot collaborative assembly, *CIRP Ann.—Manuf. Technol.*, **65**(1), pp. 61–64, 2016
16. N. Kousi, C. Stoubos, C. Gkournelos, G. Michalos, S. Makris, Enabling Human Robot Interaction in flexible robotic assembly lines: an Augmented Reality based software suite. Procedia CIRP **81**, 1429–1434 (2019)
17. Cs. Kardos, Zs. Kemény, A. Kovács, B. Pataki, and J. Váncza, Context-dependent multimodal communication in human–robot collaboration, *PROCEDIA CIRP*, **72** pp. 15–20, 2018
18. S. Hashimoto, A. Ishida, M. Inami, and T. Igarashi, Touchme: An augmented reality based remote robot manipulation, in *The 21st International Conference on Artificial Reality and Telexistence, Proceedings of ICAT2011*, 2011, **2**, pp. 61–66
19. J. Lambrecht and J. Krüger, Spatial programming for industrial robots based on gestures and augmented reality, in *Intelligent Robots and Systems (IROS), 2012 IEEE/RSJ International Conference on*, 2012, pp. 466–472, 2012
20. C. Hofmann, T. Staehr, S. Cohen, N. Stricker, B. Haefner, G. Lanza, Augmented Go & See: An approach for improved bottleneck identification in production lines. Procedia Manuf. **31**, 148–154 (2019)

21. I. Malý, D. Sedláček, and P. Leitão, Augmented reality experiments with industrial robot in industry 4.0 environment, in *2016 IEEE 14th International Conference on Industrial Informatics (INDIN)*, 2016, pp. 176–181, 2016
22. J. A. Erkoyuncu, I. F. del Amo, M. Dalle Mura, R. Roy, and G. Dini, Improving efficiency of industrial maintenance with context aware adaptive authoring in augmented reality, *CIRP Ann.—Manuf. Technol.*, **66**(1), pp. 465–468, 2017
23. Ibid. Chapter 10, G. Horváth, G. Erdős, Zs. Kemény, Human–robot collaborative workcell calibration and control
24. KUKA, KR 6 R700 sixx Tenical Data, 2019. [Online]. Available: https://www.kuka.com/-/media/kuka-downloads/imported/6b77eecacfe542d3b736af377562ecaa/0000210361_en.pdf

Chapter 17
Real-World Industrial Demonstrators on Human–Robot Collaborative Assembly

Magnus Holm, Richard Senington, Wei Wang, and Jessica Lindblom

Abstract The development of human–robot collaboration (HRC) is expected to have increasing importance in Industry 4.0 for a growing number of companies. The purpose of this chapter is to address the role and relevance of jointly designed, developed and implemented industrial demonstrators of HRC systems in projects, resulting in an increased knowledge—both for academia and industrial partners—of how to successfully present the obtained research results in an industrial environment. In particular, the chapter focuses on the role of demonstrators and presents three perspectives related to the use of demonstrators in bridging the gap between current knowledge and the work practice on the shop floor. One perspective is the joint process of developing three industrial demonstrators of HRC within the SYMBIO-TIC project, in order to provide the envisioned benefits for the addressed industrial requirements from the companies. Another perspective is how to evaluate the intended operators' perceptions and experiences of these HRC systems from a human's perspective as well as presenting the results obtained from such an evaluation. The last perspective is the voices raised from the industrial project partners' views about jointly building industrial demonstrators as well as the benefits of participating in the research project. The chapter ends with conclusions, an identified research challenge and future work. It also addresses the societal impact of using collaborative robots in industry, and their contributions to society.

17.1 Introduction

It is acknowledged that the essence of research is to explore and generate new kinds of knowledge. Knowledge represents the driving force of academia, which is aligned with knowledge and technology transfer to students, companies and society

M. Holm (✉) · R. Senington · W. Wang
School of Engineering Science, University of Skövde, Skövde, Sweden
e-mail: magnus.holm@his.se

J. Lindblom
School of Informatics, University of Skövde, Skövde, Sweden

L. Wang et al. (eds.), *Advanced Human–Robot Collaboration in Manufacturing*,
https://doi.org/10.1007/978-3-030-69178-3_17

in general. The obtained research results form a foundation for future innovation being keys for manufacturing companies in their aim to reach or to keep a leading market position.

It is essential that the outcome and knowledge gained from research is not withheld by the research community but is communicated to and refined together with industrial partners to increase the possibility of implementation in industry and society. For example, a demonstrator, implementing research approaches and results in close industrial settings, can be used for visualisation and communication of research, demonstrating processes and technologies, technology and knowledge transfer as well as "industrial readiness" tests. A demonstrator, either physical or virtual, can also have a life span beyond the termination of the specific research project enabling knowledge transfer and communication of research, etc., beyond the project.

A research project that involves both academia and companies has a promising starting point. It offers the opportunity to combine the best from different organisations and thus improve the effectiveness of the results obtained. It is important that the obtained research results are not only of interest for academia but also relevant to companies and, by extension, contribute to society at large. One way to achieve this for researchers and companies is to jointly design, develop and build demonstrators that implement and present the obtained research results in an industrial environment. This way of working provides good opportunities for, among other things, evaluating the gap between the current knowledge levels until the technology is ready to be implemented on the shop floor. It also provides valuable information regarding how relevant the research results are; asking whether, to what extent, or how they should be applied in real-world environments outside of experiments, contrived settings, or laboratories. The added value of being able to show and communicate research results through demonstrators to a wider audience, beyond academia and companies' representatives, should not be underestimated.

Establishing a consortium for a research project, engaging academia, research institutions, and companies with very different research experiences and project goals can be a challenging task. Initial planning and interaction between members of the consortia are essential [1, 2]. Shared interests, strong interdependencies, trust and engagement within the consortia also serve to facilitate the project [3–8]. To aim for and to achieve effective collaboration on both individual and organisational levels is crucial for a project's success, fruitful knowledge transfer, dissemination and future implementation of results. Another challenge possibly faced during the setup of a project application is how to engage both large and small companies in a research agenda that is able to meet several expectations and aims, while at the same time performing leading-edge research. In addition, a key facet of the project's agenda is to handle the transfer of knowledge and technology, often bi-directionally between companies and academia, for both individuals and companies covering a spectrum from no prior experience to being well-established and experienced in research cooperation. When establishing a consortium, the scope and objectives of the project are vividly discussed, and many of the issues raised from industrial partners relate to questions such as "What can we gain from the project?", "How can we participate and engage during the project?" and "How can the research results be implemented?".

A viable approach to addressing these challenges is to include demonstrators in close industrial settings especially when performing applied research. Moultrie [9] describes the potential of demonstrators as: "… *(they) provide a key role as 'translator objects' between scientists, their academic community and more widely. This is an important but often overlooked role of demonstrators, to provide interpretation of scientific research and represent it in new ways that have potential to inspire investors, industrialist, academics and the general public*". His most significant finding indicates that if scientists better understood the potential and range of possible demonstrators, a wider array of demonstrators would be actively planned within projects already at the stage of proposals [9]. Gopinath et al. [10] presented a method for designing demonstrators and argued that they represent the cumulative results of the research performed. Yet another, though the very ambitious goal of demonstrator scope, is presented by Oks et al. [11]. They present an architecture for demonstrator design intended to replicate the complete characteristics of an industrial cyber-physical system. When designing the demonstrators in the SYMBIO-TIC project [12], the "replicate level" was not set as fully as advocated by Oks and his colleagues [11]. SYMBIO-TIC is an EU-funded project to facilitate human–robot collaboration (HRC) in assembly [13]. The scope of the demonstrators aimed at validation in near-real industrial settings at the industrial partners' facilities, focusing on different production scenarios at different levels of complexity.

The rest of this chapter is structured as follows. First, the three SYMBIO-TIC demonstrators are briefly presented. Next is a description of the evaluation approach and the obtained results of the collaboration between operators and robots from a human-centred perspective. Then, some views/voices are expressed from industrial partners that participated in the SYMBIO-TIC project. Finally, the chapter ends with conclusions and future work.

17.2 Industrial Demonstrators

There are three demonstrators that were deployed in the SYMBIO-TIC project; the food industry (Demonstrator #1), aerospace industry (Demonstrator #2) and automotive industry (Demonstrator #3), respectively. Each demonstrator verified and evaluated the increasing capabilities of the SYMBIO-TIC solution.

17.2.1 Demonstrator #1: Mixed Packaging of Cheese

Demonstrator #1 was located at Robomotion (ROB) in Germany which is a company specialising in automation and robotic solutions for the food-processing industry including food packaging. The current mixed packaging is a 100% manual process, which is time-consuming and error-prone. Especially during seasonal events, the production volume increases abruptly, which forces the companies to introduce large volumes of temporary employees. However, it is very difficult to recruit occasional labourers. Moreover, the tedious mixed packaging work is undoubtedly unattractive to the young. The industrial issues encountered are listed as follows:

- *Time-consuming and error-prone manual packaging processes increase both time and costs.* Currently, the unpacking of mono boxes, preparation of target boxes, mixed packing of target boxes and palletisation of target boxes are completed by human workers.
- *Diverse customer demands require high flexibility.* Customers change packaging patterns very often, which causes that human workers have to adapt their work programs to the new customer demands frequently.
- *Difficulty of robotisation.* Even though high-cost robots are employed to accelerate the process, the diverse customer demands require flexible and quick programming methods. By comparison, human workers are flexible for various packing tasks, especially for fragile packing which requires gentle handling of products.

In Demonstrator #1, cheese packaging as shown in Fig. 17.1 is adopted to demonstrate the capability of the SYMBIO-TIC solution. The robot will assist the operator to complete mixed packaging of multiple flavours of cheese from several boxes that each has a single flavour of the cheese.

The layout of Demonstrator #1 and its physical installation are shown in Figs. 17.2 and 17.3. There are identical packs of cheese in Box 1 to Box 3, respectively. The robot picks up packs of cheese from Box 1 to Box 3, and puts them in a Buffer to form a stack. The operator moves the cheese stack from Buffer to Mixed Box 1. When Mixed Box 1 is full, it is replaced by Mixed Box 2 manually. Consequently, the operator works with the robot collaboratively to achieve the mixed packaging of

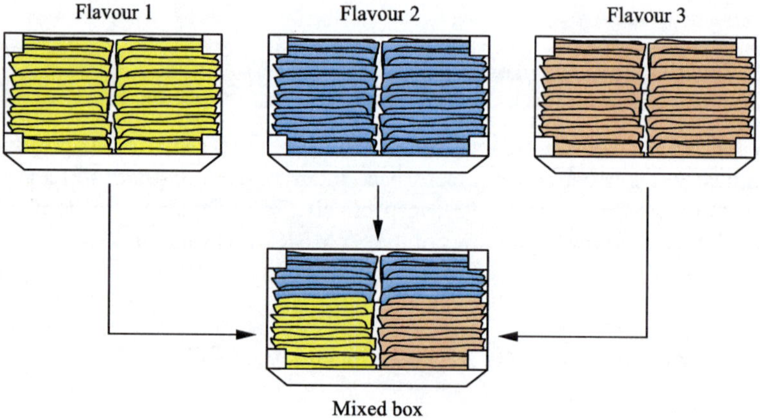

Fig. 17.1 Mixed packaging of multiple flavours of cheese

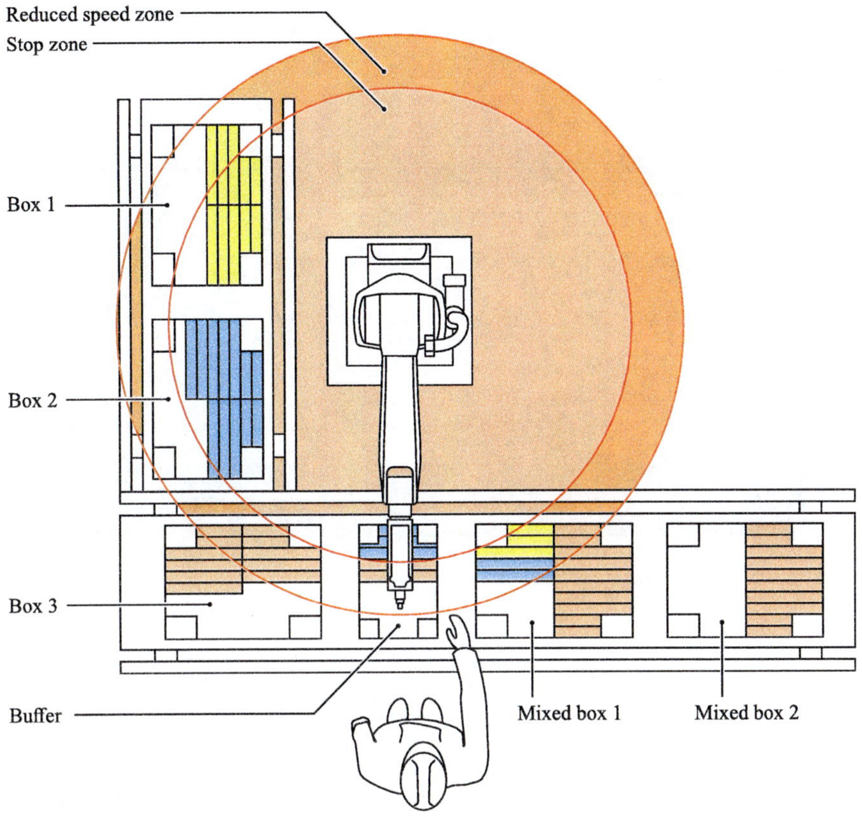

Fig. 17.2 Layout of Demonstrator #1

multiple flavours of cheese. The safety control area of the robot is divided into "Slow-down zone" and "Stop zone". If the operator enters into the "Slow-down zone", the robot will slow down its movement to 20% of the programmed speed. The robot will completely stop when the operator enters into the "Stop zone". The safety zone setup follows common safety regulations, standards and guidelines for collaborative robots. The robot reacts according to the behaviour of the operator.

The Innovations of the SYMBIO-TIC solution are highlighted by a comparison between the current food packaging solution and SYMBIO-TIC solution as showed in Table 17.1.

The Demonstrator #1 results are, among other things, significant savings on time and non-recurrent costs of new and current programs, and the attraction of new clients with cheaper but equally reliable processes. The improvement of the current process turns into a cost reduction in the packaging process. The non-recurrent cost of the existent process is about €250,000 per year for an output of 400 products per minute. Depending on the layout for the final system and the expected costs per system, the company expects a payback time of 2–3 years. The operators obtain a

Fig. 17.3 Physical installation of Demonstrator #1

Table 17.1 Current packaging solution versus SYMBIO-TIC solution

Current packaging solution	SYMBIO-TIC solution
Manual process Packaging by human completely	**Semi-automation process** Packaging by human–robot collaboration
Fixed robot programming In order to meet the requirements of diverse food packaging patterns, the robots have to be re-programmed to adapt to the variations	**Flexible and reusable robot programming** The operations performed by robots are encapsulated in well-defined function blocks that include all information for robot control; the robot programs can be easily revised interactively to reuse for the variations
High demands of human labourers The tedious unpacking and packing work form an unattractive work environment	**Combination of human flexibility and robot automation** The ergonomic situation for the operators is improved, the quality of the packaging is better, and the production capacity can easily be tuned due to HRC

better working environment. The less relevant and failure-prone tasks are left to the robots. The customers of the new system have more flexibility and independence of labour. Especially in the case of high-volume production, the system can assist the operators properly. The system decreases the number of failures in mixing the products, which occur today very often.

17.2.2 Demonstrator #2: Aircraft Wing Rib Assembly

Demonstrator #2 was located at Aciturri (ACI) in Spain that manufactures wing ribs made of carbon fibre for commercial aircraft. The selected assembly process is the assembly of wing ribs as shown in Fig. 17.4. An assembly of a wing rib is a substructure that is mainly composed of a Carbon Fibre Reinforced Polymer (CFRP) rib. Several Titanium "T" Clips called TIE are required to be riveted to the rib perimeter by a special tool called CLECOS. The industrial issues encountered are listed as follows:

- *Time-consuming manual wing rib assembly process increase costs.* Currently, human workers complete most of the wing rib assembly operations, such as setting up jigs, positioning and drilling elements accurately on the jig, and checking key quality characteristics. Meanwhile, these operations require highly-qualified workers.
- *Various wing ribs need special and expensive tooling.* Each product or sub-product needs a specific and expensive jig, and often, jigs are neither reusable between product evolutions nor for different products. For example, the complete pack of ribs of the vertical tailplane of an aircraft needs 14 jigs. Therefore, a flexible and reusable solution is required.
- *Difficulty of robotisation.* Each wing box consists of a series of wing ribs with different specifications, each wing rib requires its own robot programming. Accordingly, flexible and quick programming methods are needed. Meanwhile, the production of aircraft is typically multi-variety with small batch sizes, causing the robots to be underutilised. Moreover, the assembly accuracy of aircraft components is high, which is not met by the positioning accuracy of industrial robots. High accuracy positioning devices and human workers are needed to assist the robots.

The layout of Demonstrator #2 and its physical installation is shown in Fig. 17.5. When an operator is registered by the worker identification system, the SYMBIOTIC system will initialise an assembly task. The operator will pick up a rib and put it on the workbench. Once the rib is placed, the operator will use the measurement tool to scan it to determine its position. This process is called *referencing* the wing

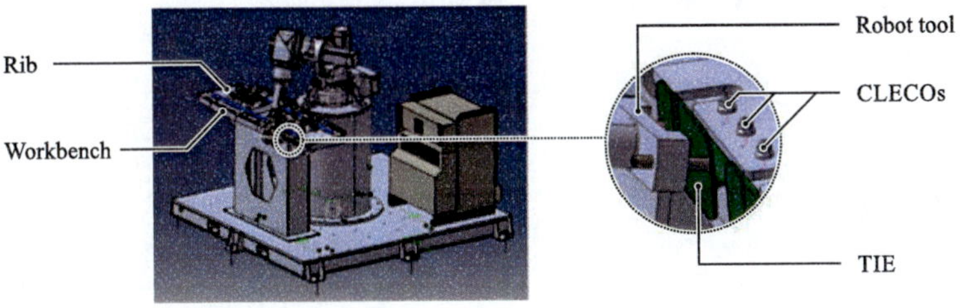

Fig. 17.4 Assembly of aircraft wing rib

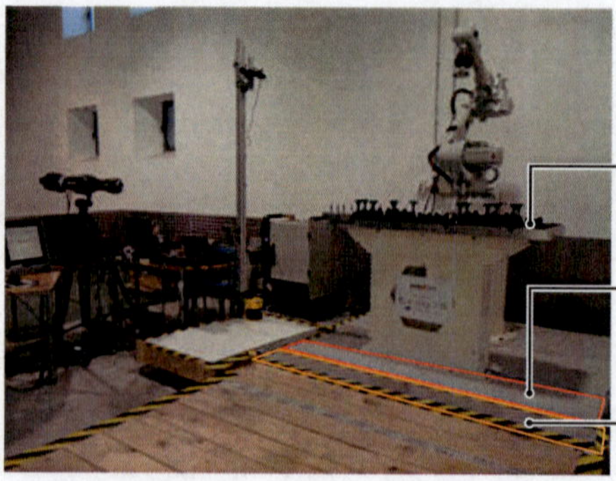

Fig. 17.5 Physical installation of Demonstrator #2 with safety zones marked

rib. After that, the operator will assemble four TIEs on the wing rib in the following order: TIE 401, TIE 601, TIE 501 and TIE 701. The assembly of each TIE is divided into two steps. In the first step, the operator assembles the TIE and fastens it with CLECOs loosely. In the second step, the robot tool approaches the TIE and moves it to an accurate position with the help of a tracking system, and the operator will push the TIE against the robot tool until contacting with the two points, and fasten CLECOs tightly. Finally, the operator will scan the key characteristics for the assembly, and a measurement report will be generated automatically. As in Demonstrator #1 the safety control area of the robot is divided into a "Slow-down zone" and a "Stop zone". If the operator enters the "Slow-down zone", the robot will slow down its movement to 50% of the programmed speed. The robot will completely stop when the operator enters the "Stop zone". The safety zone setup follows common safety regulations, standards and guidelines for collaborative robots. The robot reacts according to the behaviour of the operator.

The Innovations of SYMBIO-TIC solution are highlighted by a comparison between the current wing rib assembly solution and SYMBIO-TIC solution as showed in Table 17.2.

The end-user ACI has confirmed the benefits that are summarised as follows:

Table 17.2 Current wing rib assembly solution versus SYMBIO-TIC solution

Current wing rib assembly solution	SYMBIO-TIC solution
Manual process Assembly by human completely	**Semi-automation process** Assembly by human–robot collaboration
Specific tooling for each wing rib Each wing rib needs a specific and expensive jig that is un-reusable	**Universal tooling for all wing ribs** The introduction of robot and machine vision enables a unified solution for all wing ribs
Demands of high skilled operators The high assembly accuracy of wing ribs requires highly skilled operators	**Decrease the requirements of operators** The introduction of robot and machine vision reduces the required skill level of the operators

- Assembly cost reduction through the elimination of massive tooling.
- Assembly accuracy increased through the introduction of machine vision.
- Increased flexibility through human–robot collaboration.

17.2.3 Demonstrator #3: Automotive Engine Mass Balancing System Assembly

Demonstrator #3 is located at Volvo Car Corporation (VCC) in Sweden, a leading automotive manufacturer in the world. A key unit of the car engine, the mass balancing system (MBS) as shown in Fig. 17.6 is used to demonstrate the capability of the SYMBIO-TIC solution. An MBS mainly consists of an MBS bottom, an MBS top and 2 rods. Both the MBS bottom and top have machined surfaces that match each other. Before they can be assembled together, the machined surfaces are required to be checked by the operator. Rod 1 and Rod 2, which have gear engagement, are located inside of the MBS. Screws are applied to assemble the MBS top and bottom together. The weight of each component and the whole assembly is relatively high. Under conditions of prolonged assembly work, it becomes necessary to use a robot for component handling.

The industrial issues identified are listed as follows:

- *Changing production volume leading to a dynamic manufacturing environment.* Due to the increasing customisation requirements, the production volumes of product variations differ a lot. Accordingly, the demands on the manufacturing resources are dynamic. The use of a fixed allocation of manufacturing resources is too rigid to adapt to the dynamic demands.
- *The workload of robot programming is heavy.* At present, the robot is programmed in a fixed manner. It cannot adapt its behaviour to varying working conditions.
- *The work environment is not attractive.* In the current assembly scenario, the human workers will undertake most of the heavy and repetitive operations, which will impact the human worker's concentration after prolonged work sessions.

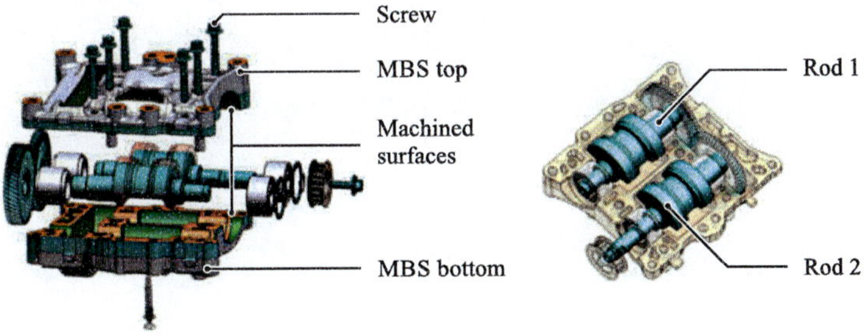

Fig. 17.6 Assembly of a mass balance system

The layout of Demonstrator #3 and its physical installation are shown in Figs. 17.7 and 17.8. An operator (Worker 1), who will be detected by the worker identification system, stands in front of working station 1. The robot will assist the operator in completing the assembly of an MBS collaboratively. The safety of the operator is ensured by a collision-avoidance system. As described before, the robot will move away to avoid hitting the operator instead of stopping completely.

The Innovations of the SYMBIO-TIC solution are highlighted by a comparison between the current MBS assembly solution and the SYMBIO-TIC solution as shown in Table 17.3.

The end-user VCC has confirmed the benefits that are summarised as follows:

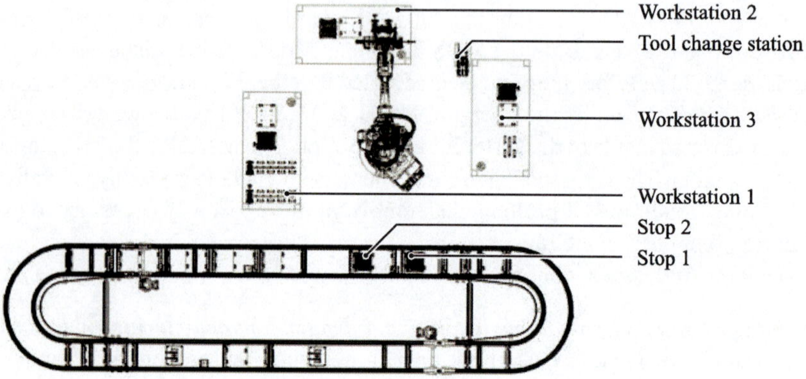

Fig. 17.7 Layout of Demonstrator #3

Fig. 17.8 Physical installation of Demonstrator #3

Table 17.3 Current MBS assembly solution versus the SYMBIO-TIC solution

Current MBS assembly solution	SYMBIO-TIC solution
Manual process Assembly by human completely	**Semi-automation process** Assembly by human–robot collaboration
Fixed planning and scheduling The assembly process is planned and scheduled based on fixed manufacturing resources. If the available manufacturing resources are changed, the whole assembly process needs to be rescheduled	**Dynamic planning and scheduling** Benefitting from the worker identification system, the assembly process is planned and scheduled based on the available manufacturing resources monitoring results. Moreover, the assembly operations are encapsulated in well-defined function blocks which include all information for robot control; the planning and scheduling results can be easily revised interactively without any robot programming effort
Rigid security scenario By setting different kinds of safety zones, it is impossible for the human share the workspace and work with a robot simultaneously	**Flexible security scenario** By setting a safety distance between human and robot, the robot can actively avoid colliding with the human, before continuing with a previous task. Moreover, the robot has a certain working condition perception capability. As a result, the robot allows the human to work with it very closely for collaboration
Traditional robot control The robot is programmed in a fixed manner. It cannot adapt its behaviour to varying working conditions	**Adaptive robot control** Robot control is capable of receiving information from multiple resources, and coordinating and prioritising the various control inputs in order to construct and forward meaningful commands to the robot controller

- *Competitive adaptability for changing production volumes.* Due to mass customisation, even automotive production has evolved into small batch production of multiple variations. The SYMBIO-TIC solution provides a method that can increase the adaptability of the current production system without complete re-configuration.
- *Trustworthy human–robot collaboration environment.* In a traditional manufacturing environment, the human and the robot are kept in strictly separated areas. The SYMBIO-TIC solutions show that even industrial robots can have a certain working condition perception capability that lets the human feel comfortable and trust working with the robot. Meanwhile, the robot can be more fully utilised.
- *Sustainable working environment.* In human–robot collaboration scenarios, the robot will undertake most of the heavy and repetitive operations, which relieves human from high ergonomic loads. Therefore, the SYMBIO-TIC solution creates a more attractive working environment for the young.

17.3 Evaluation Approach: A Human-Centred Perspective

When human operators should collaborate together with robots in the same environment, this collaboration results in additional challenges for the way of evaluating various aspects of the collaboration between them [14]. It has been pointed out that safety is a necessary but not sufficient condition for avoiding accidents, increasing acceptance and trust between humans and robots. The human-centred perspective, having its foundation in the ISO 9241-10:2019 Standard *Ergonomics of human–system interaction* [15] has become invaluable to evaluate the quality of human-system interaction in general. Prior research shows that positive user experiences (UX) of the operators could be correlated to product quality, increased productivity as well as a favourable and safe working environment when collaborating with robots [15–20]. From a human-centred perspective [15, 21, 22], the commonly used term "user-friendliness", is nowadays outdated, and has been replaced by "usability" and "UX". In ISO 9241-10:2019, "usability" is defined as "*the effectiveness, efficiency and satisfaction with which specified users achieve specified goals in particular environments*". Here "effectiveness" is defined as: "*the accuracy and completeness with which specified users can achieve specified goals in particular environments*". "Efficiency" is defined as: "*the resources expended in relation to the accuracy and completeness of goals achieved,*" and "satisfaction" as "*the comfort and acceptability of the work system to its users and other people affected by its use*" [15]. In addition, UX is defined as the "*user's perceptions and responses*[1] *that result from the use and/or anticipated use of a system, product or service*[2]" [15].

17.3.1 Evaluation Framework of Human–Robot Collaboration

In the SYMBIO-TIC project, a comparative analysis that focused on human-centred aspects of human–robot collaboration (HRC) in relevant assembly scenarios was conducted at the three demonstrators. One of the major contributions of the UX field is the established approach of UX testing, which provides valuable support to evaluators and designers in improving their products, software and service. Before presenting the evaluation approach for the three SYMBIO-TIC demonstrators, it should be pointed out that many researchers and practitioners often confuse the human-centred approach of UX testing with user research in form of contrived behavioural experiments. Dumas and Redish [23] argued that both methods are empirical by nature;

[1] "Note to entry 1: Users' perceptions and responses include the users' emotions, beliefs, preferences, perceptions, comfort, behaviours, and accomplishments that occur before, during and after use" [15].

[2] "Note to entry 2: User experience is a consequence of brand image, presentation, functionality, system performance, interactive behaviour, and assistive capabilities of a system, product or service. It also results from the user's internal and physical state resulting from prior experiences, attitudes, skills, abilities and personality; and from the context of use" [15].

focusing on actual behavioural observations conducted in laboratory/demonstrator facilities, sampling of participants, collecting and analysing quantitative and qualitative data, but they significantly differ in other respects. The major difference between them is in their goals, where UX testing seeks to identify problems in the interaction, not to demonstrate or investigate a specific phenomenon and/or aspect [23]. The selection of participants differs, where a valid UX testing purposely selects available participants from the population that should use the system, that is, convenience sampling, and is not striving for random sampling of subjects preferably used in user research. There is a vast UX literature showing the possibility to uncover most major usability problems with rather few participants (approx. 5–7 persons) [21–23]. In UX testing, the collected data are triangulated, where the observations of and the comments from the participants provide valuable insights of what causes the problems in the performance of the actual tasks done in the evaluation scenario. The structured analysis of quantitative and qualitative data combined with UX expertise is necessary to diagnose valid and credible causes of the identified usability and UX problems in order to make improvements and redesign tentative solutions for the problems identified.

A literature review revealed that there is a lack of available evaluation frameworks in HRC from a human-centred perspective Therefore, the envisioned evaluation framework of human–robot collaboration developed by Lindblom and Wang [24] was used as the methodological foundation for the comparative analysis. Their evaluation framework [24] focuses on three UX indicators: "Operator Experience" (OX), "Trust", and "Safety". Each UX indicator consists of several factors, which are being assessed both quantitatively and qualitatively, to investigate and analyse various aspects and performance levels of the human–robot collaboration, as shown in Fig. 17.9 (for more methodological details of how to conduct UX evaluation and UX testing, see references [21, 22, 25]). It is important to emphasise that it is not the operators that are evaluated and tested; it is the interaction quality and the operators' experience of interacting with the robot that is the major focus of study.

From a human-centred perspective, it is stressed that as part of designing, developing and evaluating any interactive technology, in this case, human–robot collaborative systems, it is an identified ambition and need to satisfy not only the human user's (operator's) goals when interacting with the robot to ensure good usability and a positive UX. In addition, there is a need to address the goals of the business producing or investing in the collaborative robot system, goals such as profitability and marketability. Therefore, it is necessary for a human–robot collaboration system to be aligned with an appropriate business case, or a demonstrator of such business case [22]. In the SYMBIO-TIC project, the three demonstrators presented in Sect. 17.2 function as tentative business cases. To identify relevant usage contexts for the evaluation, each demonstrator developed and implemented representative benchmark tasks and scenarios. The scenarios were designed in collaboration with the industrial partners, focusing on typical interaction and collaboration tasks at each demonstrator. The objective was to reflect several levels of complexity of collaboration and task-allocation between the human operator and the robot. The scenarios for each demonstrator were presented in Sects. 17.2.1–17.2.3.

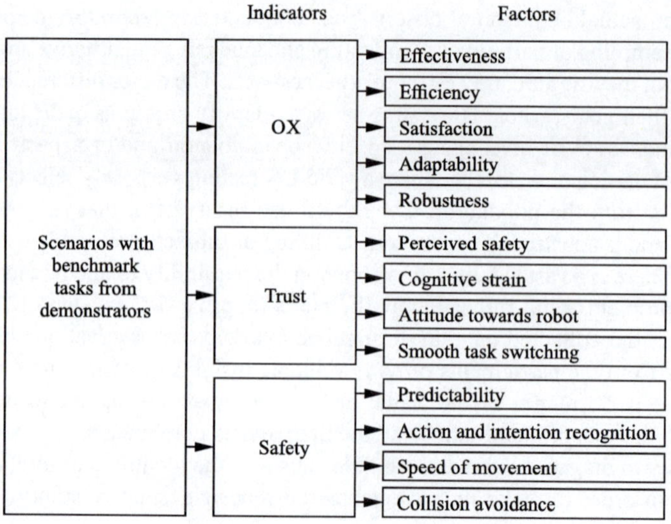

Fig. 17.9 Indicators of an envisioned evaluation framework for HRC [24]

17.3.2 Data Collection and Overall Procedure

During the evaluation, quantitative and qualitative data were collected in the following ways: all scenario sessions were observed as well as video-recorded when participants were interacting with the collaborative robots. A questionnaire was developed based on a modified version of the System Usability Scale (SUS) [26]. The modified SUS questionnaire consisted of 20 questions that covered OX, trust and safety aspects, and it was used for task completion to be filled in after each scenario session. The above questionnaire was complemented with a semi-structured post-test interview with open questions that the participants should answer after they had participated in the scenarios. These interviews were performed by the last author of the chapter and audio-recorded for further analysis. In total, this setup of different data collection techniques (observation, video recording, questionnaires, interview, and audio recording) enabled to collect both quantitative and qualitative data, covering the three UX indicators of OX, trust and safety, in order to triangulate the data to identify several aspects of interaction quality.

The procedure usually went as follows. First, the participants were introduced to the study and had to read through and then sign a consent form. The information provided was that the evaluation should study human–robot collaboration intended for human operators working in a factory. The goal of the study was to make the robot more appealing, intuitive and safe; and that their participation should help the researchers in the SYMBIO-TIC project to achieve this major goal. It was stressed that this was a test of the robot and the quality of the interaction with it, and the participants' skills and abilities were not tested. Second, before the first session started, each participant was informed about the safety rules by a representative from the demonstrator facility. These safety rules had to be followed. Third, the participants performed the first session consisting of doing the tasks in scenario 1. Fourth, after completing scenario 1, the participant went to a desktop and was asked to fill in the modified SUS questionnaire. Fifth, when the participant had completed the questionnaire, the second session usually started, performing scenario 2. Sixth, when completing scenario 2, the participant again went to the desktop and was asked to fill in the same questionnaire as described above, but this time the participant should only focus on scenario 2 when answering the questions. Finally, when the participant had filled in the questionnaire for scenario 2, the semi-structured post-test interview (audio-recorded) was conducted. The interview consisted of open questions regarding both sessions and focused on the participants' subjective experiences of interacting with the collaborative robot. In total, the HRC evaluation for each participant took approximately 30–40 min to complete.

17.3.3 Demonstrator #1: Participants and Obtained Results

In order to evaluate HRC in Demonstrator #1 (as shown in Figs. 17.2–17.3), two scenarios were designed, which are described in Sect. 17.2.2. In total, seven human participants were recruited in-house, and of these two were females and five were males (see Fig. 17.10). The ages varied between 20 and 33 years old. They had no prior experience of collaborating with the robot and they did not work as operators, so they were considered as secondary user group [22, 25].

The results obtained for the OX indicator revealed that all participants (100%) completed all sessions, although some minor errors occurred, showing a very high degree of "effectiveness." However, the "adaptability" was rather low, because the

Fig. 17.10 Illustration of human–robot collaboration at Demonstrator #1

robot did not continue from an appropriate point of action after accidently dropping the cheese package in the second scenario, and the participants reported that they found there were limited ways of communicating with the robot, which is an aspect that needs to be improved.

Results obtained for the trust indicator revealed, for example, that "perceived safety" was accomplished to a high extent, although the quick and jerky movement of the robot close to the participants' head when picking a package in Box 3, often resulted in a quick back-off movement by the participants. Furthermore, "smooth task switching" was accomplished to a moderate extent, because it was not completely obvious in the first session when the robot was finished and when the task was allocated to the human operator. The accomplishment of this factor increased during the second session. Otherwise, no severe trust issues were observed or reported.

The obtained results for the safety indicator revealed, for example, that both "predictability" and "action and intention recognition" factors were accomplished to a moderate extent in the first session, but increased to a higher extent during the second session. The participants had some initial problems understanding the robot's "intentions" via its movements. Based on the observations and the interviews, some participants experienced the robot as behaving unpredictably after dropping the cheese package, because it started its movements from the beginning rather than continuing from where the collaborative task occurred. In other words, the robot behaved "irrationally" because it did not "understand" that the human participant had picked up the cheese package and put it in the buffer. Furthermore, the result obtained for "collision avoidance" was low, because the participants perceived no risk for colliding with the robot. The most discussed factor was the "speed of movement," where the results obtained revealed mixed outputs. On the one hand, the participants appreciated that the robot moved slowly when acting closely to them, but perceived the robot's motions as being too slow during transportation. However, their suggestion to increase the robot's speed when not moving closer to the human operator may be hindered by current safety regulations.

17.3.4 Demonstrator #2: Participants and Obtained Results

In order to evaluate HRC in Demonstrator #2 (as shown in Figs. 17.4–17.5), one scenario in two variants was designed, which is described in Sect. 17.2.3. In total six participants were recruited in-house, and of these two were females and four were males (see Fig. 17.11). The age varied between 23 and 35 years old. They had no prior experience of collaborating with the robot, and they did not work as operators, so they were considered a secondary user group [22, 25].

The obtained results for the OX indicator revealed, for example, that all participants (100%) completed all sessions, although some minor errors occurred, showing a very high degree of "effectiveness". However, the "efficiency" (time on task) was reached to a low extent, and one underlying cause for this negative finding is the layout of the workplace setting with a safety zone in one direction, the workstation

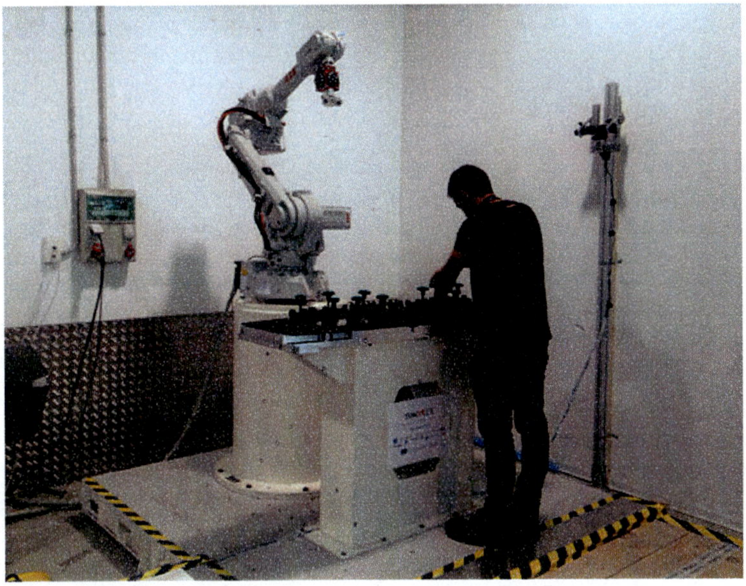

Fig. 17.11 Illustration of human–robot collaboration at Demonstrator #2

bench in another direction, and the robot in a different direction again, resulting in unnecessary distances and interruptions in the performance. The "adaptability" factor was met to a low extent because the instructions on the tablet caused several misunderstandings (included no images, only written text, at the time of the evaluation). The "robustness" factor was satisfied to a moderate extent because the instructions on the tablet resulted in several misunderstandings, and there was, sometimes, no proper place to put down the tablet when doing the assembly, so very often the tablet was placed on top of the rib, which is not a good place to situate it.

The results obtained for the trust indicator revealed, for example, that the "perceived safety" indicator was reached to a very high extent, the participants worked in very close proximity to the robot during assembly, sometimes leaning towards the robot. The "smooth task switching" indicator was reached to a moderate extent because it was not completely obvious—due to the sometimes poor instructions—when the robot was finished with its part and the task switched to the participant.

The results obtained for the safety indicator revealed, for example, that the "action and intention recognition" factor was reached to a high extent, although the participants had some initial problems grasping the robot's "intentions" via its movements only, and the instructions did not always provide necessary assistance in this regard. The "speed of movement" factor was reached to a low extent, because of the experienced slow movements of the robot. A tentative explanation can be the rather wide scope of the safety zone, which resulted in the participants standing too far away from the robot when it moved, implying that the robot's speed can be faster. It was suggested that the robot's speed should be increased when it was not moving closer to the participant. The participants perceived very low risk of colliding with the robot when it was moving because they were situated in the safety zone.

17.3.5 Demonstrator #3: Participants and Obtained Results

In order to evaluate HRC in Demonstrator #3 (as shown in Figs. 17.7–17.8), one scenario was designed which is described in Sect. 17.2.3. One group of 15 participants were university students in higher education from the schools of engineering science or informatics at the University of Skövde, and of these two were females and 13 were males. The ages varied between 21–32 years old. They were considered the secondary user group (beginners), having limited or no prior first-hand experience of assembly or collaboration with robots The other group of 12 participants were professional operators working at Volvo Car Company, and of these were six females, and six were males. The ages varied between 23 and 46 years old. They were considered the primary user group (experts). They had prior extensive experience in assembly and varied experience of collaborating with robots. In total, 27 participants were recruited (see Fig. 17.12) [22, 25]. In order to reduce the first encounter effect, each participant did the same scenario twice.

The results obtained for the OX indicator revealed, for example, that all participants (100%) completed all sessions, although some minor errors occurred, showing a very high degree of "effectiveness". The number of errors was almost totally reduced between the first and the second sessions. The "efficiency" factor was reached to a high extent, decreasing a lot between the first and second sessions although the

Fig. 17.12 Illustration of human–robot collaboration at Demonstrator #3

transportation time for the robot to and from the pallet was not changed. The "satisfaction" factor was reached to a high extent. Participants were very positive to the scenario as such, and the operators from VCC were the most positive from a professional perspective, stressing physical ergonomics aspects and the reduction of heavy lifting. The "adaptability" factor was reached to a moderate extent because the instructions on the tablet caused some minor misunderstandings. For example, the instructions did not present where to place the six screws (the instructions did not include any images, only written text, at the time for the evaluation). Only the professional workers placed the screws in the correct places without any hesitation, showing their greater previous experience.

The results obtained for the trust indicator revealed, for example, that the "perceived safety" indicator was reached to a very high extent The "attitudes towards robot" factor was reached to a very high extent, participants expressed positive attitudes towards using robots in the assembly of heavy parts/components and in repetitive tasks. The "smooth task switching" factor was reached to a very high extent, no misunderstanding occurred regarding when the robot was finished and switching over the assembly task to the participant and the other way around.

The results obtained for the safety indicator revealed, for example, that the "predictability" factor was reached to a high extent, although the instructions on the tablet sometimes misled the participants. The predictability increased significantly between the first and second sessions. The "action and intention recognition" factor was reached to a high extent, although the participants had some initial problems grasping the robot's "intentions" via its movements only because the movement was not always smooth. However, the participant's perception of the robot's actions improved between the first and second sessions. The "speed of movement" factor was reached to a low extent. This may be caused by the rather long transportation time between the pallet and the workstation, even with the participant standing far away from the robot while it moved, implying that the robot's speed can be faster in that regard. The participants appreciated that the robot moved more slowly when being close to the workstation and the participants. The participants and the robot collaborated closely in the shared workspace. The participants neither showed nor expressed any fear of colliding with the robot during assembly. There is a possibility to increase the robot's speed when not moving close to a human operator.

17.3.6 Summary of Evaluation Results and Conclusions

It should be emphasised that no incidents or accidents occurred during the evaluation that concerned safety issues or the actual performance of the scenarios at the three demonstrators. Each and every participant ended and completed the tasks to be carried out in the scenarios and there were no dropouts. All participants, except one, answered all the questions in the task completion questionnaire. Every participant took part in the post-test interview, providing answers to all questions.

A general analysis of the data collected from the task completion questionnaires has been conducted (see Fig. 17.13).

It should be pointed out that questions 1–12 and 19 relate to the OX-indicator. High scores on questions 1, 3, 5, 7, 9, 11, 19 and low scores on questions 2, 4, 6, 8, 10 show positive operator experience, including good usability. High scores on questions 13–18, 20 and low scores on question 12 show a high amount of perceived safety and trustworthiness. The findings obtained for these questions reveal a rather good OX with a mean around 4, despite question 11 ("I felt that there was a number of possible ways of communicating with the robot") where the mean is around 2. Many participants expressed that they did not know how to directly communicate with the robot, besides the tablet, experiencing the robot as acting on its own.

High scores on questions 13–18 and 20 and low scores on question 12 show high perceived safety and high trustworthiness. The findings obtained for these questions reveal that the system was experienced as having a high level of safety and was trustworthy with a mean around 4, despite question 16 ("I felt that the robot could understand my intentions") where the mean is around 3. Low scores on question 12, below 2, also indicate a high level of experienced safety and trust. The highest means were obtained for question 7 ("I would imagine that most people would learn to use this robot very quickly") approximately 4.5. This result should be considered in contrast to the scores on question 4 ("I think that I would need the support of a technical person to be able to use this robot") where females' score is just below 3, although they scored higher than men on question 7. Otherwise, no substantial gender differences were revealed on a general level.

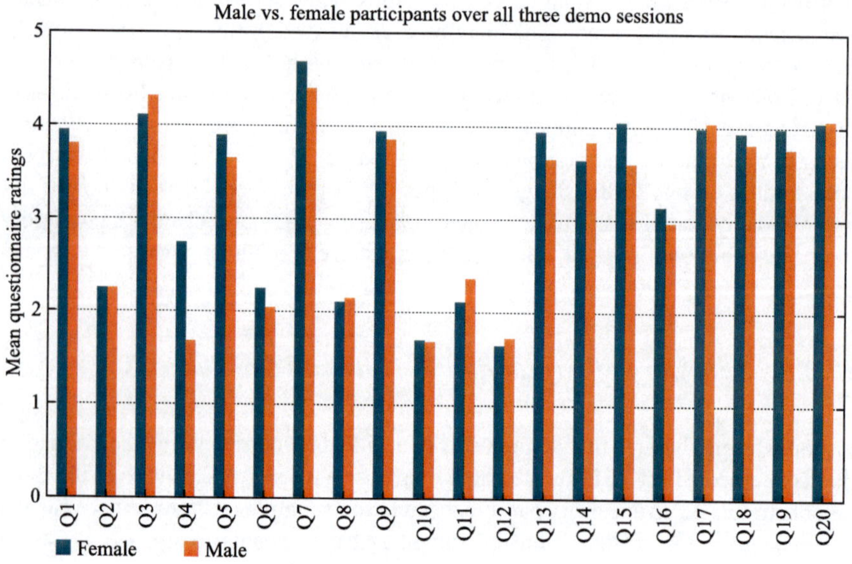

Fig. 17.13 Mean ratings from all 41 participants' sessions (75 sessions in total) at the three demonstrators, divided into 29 males (orange) and 12 females (blue)

The results obtained so far indicate that all participants were generally very positive about using and collaborating with the robot at the three demonstrators. They also expressed a very positive attitude towards using robots in the factories of the future. They mostly emphasised the role of the collaborative robot as a support for doing simple repetitive tasks and lifting heavy objects during assembly processes. It should be mentioned that very few drawbacks were expressed, but some participants were a bit worried that if collaborative robots would be more frequently present on the shop floor in the near future, they may not have as many human workers as teammates which could be negative for the social working climate and their wellbeing. In other words, human teammates are considered to be an important aspect of the working environment. The most relevant and detailed comments on improving the collaboration with the robot in assembly tasks were collected from real operators (in particular Demonstrator #3), stressing the need to focus on the primary user group when doing evaluations. The biggest identified problem, in Demonstrators #2 and #3, was caused by somewhat unclear instructions on the tablets used. The identified problems caused less issues the second time when they interacted with the robot, indicating a kind of familiarisation or learnability of the HRC systems.

The main criticism raised from almost all participants in the three demonstrators concerned the robots' speed—the robots were experienced to move a bit too slowly—although the participants were positive regarding the robot's slower motion when it was moving near to the participants (for safety reasons). However, they still felt that the robots could move faster during transportation, for example, when picking and leaving the MSB on the pallet at Demonstrator #3. In Demonstrator #1, the robot did a quick and jerky movement rather close to the participants and the buffer, which worried some of them. They were surprised and winced backward for a moment. This particular robot movement should be programmed to be much smoother. In addition, it was observed, as well as commented on by the participants, that the safety zones in Demonstrator #1 and Demonstrator #2 sometimes caused overly long interruptions in the workflow. For example, the participant had to move several metres to be situated in the correct zone for the robot to perform its next step, and then the participant had to move in a non-straightforward direction to pick up assembly material, before starting his/her part of the collaboration with the robot.

A difference in performance-based aspects was revealed, being part of the OX indicator, between the first and second sessions in Demonstrator #3. The participants made less errors and performed the joint task much more quickly the second time, which is reflected in the scores obtained from another analysis made from the data of the task completion questionnaires. The main reasons for the increased performance-based aspects may be that they could better predict what the robot would do, and thus experienced increased control and confidence. Hence, there is development potential for the design of the workstation and the distribution of tasks between the operator and the robot in all demonstrators, which is an identified research challenge.

However, how representatives from the industrial partners perceived and experienced participating in the SYMBIO-TIC project has largely been omitted so far in this chapter. This is the topic to which the next section will now turn.

17.4 Views/Voices from the Industrial Project Partners

The analysis of collected data obtained via individual interviews with representatives from the industrial partners of the SYMBIO-TIC project by the first author reveals the importance of thorough discussions about project aims and goals as previously mentioned in the Introduction. Their level of engagement in the project was directly interrelated to how well the project addressed the challenges and questions generated from the production processes and the shop floor, in addition to the needs identified by their customers. The possibility to affect the alignment between research and current work practice motivated their participation in the SYMBIO-TIC project, besides access to the forefront of specific technology and associated knowledge of the researchers concerned.

Based on the interviews, it was revealed that the demonstrators were associated with a prototype or a concept. It shows, in a good way, the outside world what the concept of demonstrators is about, today's aim and possibilities, and what the future could hold. Since the demonstrator is not incorporated into real production systems, the use of demonstrators makes it possible to "challenge" current work practices and technical solutions in a way different than if it was implemented in live production. With the use of demonstrators, it is possible in a beneficial way to show both the ambition and the vision of the SYMBIO-TIC project. Since the concepts illustrated through the demonstrators could be far from what is currently being produced, the demonstrators facilitate broadening both HRC frameworks and ways of thinking. It was also expressed that a demonstrator provides an advantageous physical installation to gather around, both internally and externally, and to show future direction.

In general, representatives from the companies wanted the demonstrators in the project to address those kinds of tasks that could be shared between operator and robot and what this distribution and allocation of shared tasks would actually mean in practice. Is it the case that the operator and the robot only share the same working area but do not cooperate or collaborate—or is joint interaction between the robot and the operator possible? Accordingly, if cooperative, collaborative, or joint interactions occur, how can these interactions be used to both improve the operator's work environment and at the same time improve productivity? The partners also expressed the need to explore the technology used in the demonstrators to find additional and novel ways and methods to implement these in real production settings.

In the interviews, the representatives from the companies stated that they felt that the respective goals set for the demonstrators have been addressed well by them. It was expressed that the SYMBIO-TIC project has been an opportunity to reach even further regarding certain project goals (for the demonstrators), if the demonstrators would have been as near to production-like as decided. The demonstrators were regarded to be well-balanced between being a concept and being "production-like", and the results obtained from the demonstrators were considered to be generalisable and representative of several companies and industries since there are many similar application areas for products of similar size and weight. Hence, the demonstrators showed the possibilities of novel technological solutions. Although the technological solutions obtained, that is, the results of the project, are not yet ready for implementation in the industry right away, the outcome has generated several discussions on HRC and addressed "how," "where," and "when", it can be implemented. It was also revealed that the demonstrators and the technical solutions have also facilitated discussions with customers, highlighting what they can gain from this HRC technology and how it can be implemented and used. As one of the representatives stated: "The customers have had the chance to see if this was interesting for them or not". Hence, the demonstrators have provided the information and knowledge needed to continue the work beyond the SYMBIO-TIC project. The project as such, including the demonstrators, has confirmed that a collaborative environment is an enabler for flexible automation and that it does not replace the operator, instead, it enhances the operator's work. A historical disadvantage of prior robotic installations has been the high cost and the high level of competence required to use the equipment and the robotic systems. These novel collaborative systems can possibly be cheaper to install and it does not require as much expertise to use. Consequently, these systems enable a wider group of people to use robots than before. The possibility of configurability increases as these novel HRC solutions are mobile in a completely different way than the typical industrial robot installations. To summarise, it was pointed out several times that being able to demonstrate these novel HRC concepts brought forth ideas about what "I," that is, the person who sees the demonstrator, could do in his/her own field.

17.5 Conclusions

The purpose of this chapter has been to address the role and relevance of jointly designed, developed and implemented demonstrators of HRC systems in the SYMBIO-TIC project, resulting in an increase in knowledge—for both academia and industrial partners—of how to successfully present the obtained research results in an industrial environment. As stated in the introduction, this way of working provides good opportunities for, among other things, evaluating the gap between the current knowledge levels and work practice on the shop floor.

An identified challenge, both in academia as well as from the company representatives in the SYMBIO-TIC project, is the need to narrow the knowledge gap of the levels of collaboration (LoC), that is, task allocation and distribution of tasks between the human operator and the robot in HRC systems. Some initial steps towards a taxonomy of different levels of collaboration between humans and robots have been conducted by Kolbeinsson et al. [27]. The authors in [27] aimed to develop a taxonomy that facilitates the distribution of workflow and task allocation when the human and the robot each perform the parts of the task that they are good at, thus complementing each other. To address this, it is necessary to keep in mind the intended context, and specify three aspects; what kind of humans and robots are involved, and what kind of collaborations take place. The LoC taxonomy could then be used as a tool in future workstation design and production planning for Industry 4.0.

A more detailed and comparative analysis from the human-centred evaluation in the three demonstrators has been performed but is beyond the scope of this chapter. Furthermore, the collected data from the evaluation will be used in upcoming publications, to further develop and validate the envisioned evaluation framework of safety, trust, and operator experience in different kinds of human–robot collaboration in manufacturing assembly.

The representatives from the industrial partners stated during the interviews that have participated in the project have provided fantastic opportunities to create and expand networks within the area of research, which has resulted in, among other things, several requests to participate in exciting new projects. By participating in such projects, they are at the forefront of technological change and have access to the latest insights, which provides them with a great opportunity to influence the focus of the research being conducted. Generally, the return on investment of participating in research projects is viewed as high, which one of the representatives expressed as follows:

> It has been a fantastic opportunity to be part of the project, to give input. Taking the opportunity to join. It has been fun. Good dialogue that challenges and provides, both from an industrial and an academic point of view. Sometimes the industry pursues development that loses the academic perspective and sometimes the academy becomes too theoretical. The different perspectives need to be mixed to get exciting.

To conclude, the ambition of the project partners is that the final outcomes from the work addressed earlier, that is, the LoC taxonomy, the evaluation framework, and the identified research challenge should, in the long run, function as roadmaps for successful implementation of HRC systems in the industry. If the issues and the challenges raised in this chapter are properly addressed—and there is good reason to believe that they can be—there is a great reason to be optimistic that what is known from using demonstrators can be successfully applied to HRC systems on the shopfloor. As a result, the companies will benefit, and the users of these collaborative robots will benefit even more. Thus, the demonstrators show the societal impact of collaborative robots in industry, which also contributes to society at large.

Acknowledgements The authors would like to thank the industrial partners who contributed to the design, development and implementation of the demonstrators in the SYMBIO-TIC project. The authors also express sincere gratitude towards all the persons that participated in the evaluation of human–robot collaboration at the three demonstrators and provided constructive comments and insights that are valuable for future work. Work presented in this chapter has been financed by SYMBIO-TIC (Project #637107), in the EU Horizon 2020 Programme.

References

1. J.L. Cummings, B.-S. Teng, Transferring R&D knowledge: the key factors affecting knowledge transfer success. J. Eng. Tech. Manage. **20**(1–2), 39–68 (2003)

2. V. Morandi, The management of industry–university joint research projects: how do partners coordinate and control R&D activities? The Journal of Technology Transfer **38**(2), 69–92 (2013)
3. T. Thune, M. Gulbrandsen, Dynamics of collaboration in university–industry partnerships: do initial conditions explain development patterns? The Journal of Technology Transfer **39**(6), 977–993 (2014)
4. C. Battistella, A.F. De Toni, R. Pillon, Inter-organisational technology/knowledge transfer: a framework from critical literature review. The Journal of Technology Transfer **41**(5), 1195–1234 (2015)
5. S.M. Breznitz, M.P. Feldman, The engaged university. The Journal of Technology Transfer **37**(2), 139–157 (2012)
6. Y. Hirose, *"Knowledge transfer from researchers to society: How to offset the cultural gap?" in PICMET '12: Technology Management for Emerging Technologies* (IEEE, Vancouver, Canada, 2012)
7. C. Plewa, N. Korff, C. Johnson, G. Macpherson, T. Baaken, G.C. Rampersad, The evolution of university–industry linkages—A framework. J. Eng. Tech. Manage. **30**(1), 21–44 (2013)
8. M. Holm, M. Frantzén, T. Aslam, P. Moore, L. Wang, A methodology facilitating knowledge transfer to both research experienced companies and to novice SMEs. International Journal of Enterprise Network Management **8**(2), 123–140 (2017)
9. J. Moultrie, Understanding and classifying the role of design demonstrators in scientific exploration. Technovation **43–44**, 1–16 (2015)
10. V. Gopinath, K. Johansen, M. Derelöv, Demonstrators to support research in Industrial safety—A Methodology. Procedia Manufacturing **17**, 246–253 (2018)
11. S.J. Oks, M. Jalowski, A. Fritzsche, K.M. Möslein, Cyber-physical modeling and simulation: A reference architecture for designing demonstrators for industrial cyber-physical systems. Procedia CIRP **84**, 257–264 (2019)
12. EU SYMBIO-TIC Project: https://cordis.europa.eu/project/id/637107
13. L. Wang, R. X. Gao, J. Váncza, J. Krüger, X. V. Wang, S. Makris and G. Chryssolouris, "Symbiotic Human–Robot Collaborative Assembly," *CIRP Annals—Manufacturing Technology*, **68**(2): pp. 701–726, 2019
14. J. Lindblom and R. Andreasson, "Current Challenges for UX Evaluation of Human–Robot Interaction," in *Advances in Ergonomics of Manufacturing: Managing the Enterprise of the Future*, 2016
15. ISO, *Ergonomics of human-system interaction—Part 210: Human-centred design for interactive systems*, in *ISO 9241-210:2019*. ISO: Switzerland, 2019
16. I. Gaudiello, E. Zibetti, S. Lefort, M. Chetouani, S. Ivaldi, Trust as indicator of robot functional and social acceptance. An experimental study on user conformation to iCub answers. Comput. Hum. Behav. **61**, 633–655 (2016)
17. A. Bicchi, M.A. Peshkin, J.E. Colgate, *"Safety for Physical Human–Robot Interaction", in Springer Handbook of Robotics, Springer* (Berlin, Heidelberg, Berlin Heidelberg, 2008)
18. P.A. Lasota, T. Fong, J.A. Shah, A Survey of Methods for Safe Human–Robot Interaction. Foundations and Trends in Robotics **5**(3), 261–349 (2017)
19. I. Maurtua, A. Ibarguren, J. Kildal, L. Susperregi, B. Sierra, Human–robot collaboration in industrial applications. Int. J. Adv. Rob. Syst. **14**(4), 1–10 (2017)
20. M.M.A. de Graaf and S. Ben Allouch, "Exploring influencing variables for the acceptance of social robots," *Robotics and Autonomous Systems*, **61**(12): pp. 1476–1486, 2013
21. D. Benyon, *Designing User Experience: A guide to HCI, UX and interaction design* (Pearson, Harlow, England, 2018)
22. R. Hartson, P. Pyla, *The UX Book: Agile UX Design for a Quality User Experience*, 2nd edn. (Morgan Kaufmann, Cambridge, United States, 2018)
23. J.S. Dumas, J.C. Redish, *A Practical Guide to Usability Testing* (Intellect Books Exeter, UK, 1999)
24. J. Lindblom, W. Wang, *"Towards an Evaluation Framework of Safety, Trust, and Operator Experience in Different Demonstrators of Human–Robot Collaboration", in Advances in Manufacturing Technology XXXII* (IOS Press, Amesterdam, Netherlands, 2018)

25. J. Lindblom, B. Alenljung, and E. Billing, "Evaluating the User Experience of Human–Robot Interaction," in *Evaluation Methods Standardization for Human–Robot Interaction*, Springer, 2020
26. J. Brooke, *"SUS—A quick and dirty usability scale", in Usability Evaluation In Industry* (CRC Press, London, UK, 1996)
27. A. Kolbeinsson, E. Lagerstedt, J. Lindblom, Foundation for a classification of collaboration levels for human–robot cooperation in manufacturing. Production & Manufacturing Research **7**(1), 448–471 (2019)

Chapter 18
Future Research Directions on Human–Robot Collaboration

Lihui Wang, József Váncza, Zsolt Kemény, and Xi Vincent Wang

Abstract Human–robot collaboration (HRC) in manufacturing aims to realise an environment where humans can work side by side with robots in close proximity. In such a collaborative setup, the humans and the robots share the same workspace, the same resources, and in some cases the same tasks. Through HRC, higher overall productivity and better product quality can be achieved. While a set of innovative solutions to manufacturing problems are presented in different chapters of this book based on the symbiosis between humans and robots, this closing chapter aims to highlight remaining challenges and future research directions, and outline continued research efforts and industrial practices of the years to come.

18.1 Introduction

Smart manufacturing, in the era of Industry 4.0, aims for flexible and adaptive manufacturing operations—locally or globally—by using integrated information technology (IT) and artificial intelligence (AI) that can combine advanced computing power with manufacturing equipment [1]. Smart manufacturing depends on the

L. Wang (✉) · X. V. Wang
Department of Production Engineering, KTH Royal Institute of Technology, Stockholm, Sweden
e-mail: lihuiw@kth.se

X. V. Wang
e-mail: wangxi@kth.se

J. Váncza · Zs. Kemény
Research Laboratory on Engineering and Management Intelligence, SZTAKI Institute for Computer Science and Control, Eötvös Loránd Research Network, Budapest, Hungary
e-mail: jozsef.vancza@sztaki.hu

Zs. Kemény
e-mail: zsolt.kemeny@sztaki.hu

J. Váncza
Department of Manufacturing Science and Technology, Budapest University of Technology and Economics, Budapest, Hungary

L. Wang et al. (eds.), *Advanced Human–Robot Collaboration in Manufacturing*,
https://doi.org/10.1007/978-3-030-69178-3_18

timely acquisition, distribution, and utilisation of real-time data from both machines and processes on manufacturing shop floors, and even across product lifecycles. Effective information sharing can improve production quality, reliability, resource efficiency and recyclability of end-of-life products. Smart manufacturing built on digitalisation also aims for better sustainability and contributes to factories of the future. It transcends the traditional notion of intelligent manufacturing in that smart systems and technologies offer easy-to-use, intelligible interfaces towards their human partners or users. Hence, smart manufacturing depends extensively on digitalisation, AI and human-machine interfaces. To better realise and sustain smart manufacturing in the future, it is necessary to thoroughly understand the roles of Industry 4.0 technologies such as human–robot collaboration (HRC), digital twin, cyber-physical systems, Internet of Things, cloud computing and big data analytics. These research fields have been placed at the forefront of the international research agenda.

In the area of industrial robotics, the trend goes to HRC to best utilise the strengths of both humans (flexibility, adaptability, cognition) and robots (durability, repeatability, accuracy) [2]. The underlying technology is the emerging domain of *collaborative intelligence* (CI). According to a survey of Harvard Business Review 'Humans and AI are joining forces' [3], CI is based on *Humans-Assisting-Robots* (train, explain, sustain) and *Robots-Assisting-Humans* (amplify, interact, embody) primitives. In the former case, humans need to apply competences not just related to teaching robots what to do in certain situations (train, e.g., setting thresholds), but also to explain them how such decisions are taken (explain, e.g., risk analysis, fault trees), and to defend such decisions against objections (sustain, e.g., by simulations and forecasts). In the latter case, amplify–interact–embody must be modelled and implemented in such a configurable and self-adaptable way, so that humans are empowered to interact with work objects using a robot as the third arm. In turn, CI should be able to constantly consider train–explain–sustain as well, in a never-ending loop towards human–robot collaboration. Note that the multi-agent view of HRC espoused in the first chapter of this book [10] is particularly fitting to CI. It can, and as we are convinced, it will push the research frontier further.

Within the context of HRC in smart manufacturing in factories of the future, effective HRC will require the symbiosis of human safety, sensor fusion, dynamic planning, adaptive control, and in-situ decision support for operators. In addition to multimodal human–robot communication by gesture, speech, haptics and brainwave, advanced IT and AI technologies as well as CI algorithms will add value to future HRC systems. Any practical HRC applications will also rely on the successful establishment of HRC-related standards and inclusive safety assurance for workers and business owners. In what follows, this closing chapter aims to highlight remaining challenges and future research directions based on a recent HRC literature survey [2], and shed some light on continued research efforts and industrial practices in the years to come.

18.2 Remaining Challenges

In the HRC literature, it is recognised that high robot speed, small distances between a human and a robot, and the lack of a signal communicating the start of the robot motion drastically reduce the human acceptance of the robot. Similarly, a properly designed HRC cell can lead to about 6% time reduction toward task completion, 20% more concurrent motion and 3% less human idle time [2]. An important aspect affecting human propensity to work with a robot companion is the ability of the robot to dynamically change and adapt its operations in line with the human behaviour and actions. The robot should be able to interpret the HRC setting and, in particular, the level of human adaptability, and act properly when powered by CI.

Indeed, when humans can easily adapt, robots can exploit human adaptability to guide humans towards a certain strategy. When a human is stubborn and not willing to comply, the robot companion should respect human choices and optimise its selected actions considering the human preferences as a set of constraints. This leads to the development of the concept of *mutual adaptation*, where the human trust is considered during the definition of the robot strategy.

For a *human–robot team* to accomplish its goal, humans must *trust* that a robotic teammate will protect the interests and safety of every other individual in the team. The level of trust in the team is particularly critical in high-risk situations, such as in HRC assembly. Trust is important in these contexts because it directly affects the willingness of humans to accept any robot-produced information, follow robots' suggestions, and thus benefit from the advantages inherent in an HRC system. In other words, trust affects much the decisions that humans make in uncertain or risky environments [4]. For example, the less an individual trusts a robot, the sooner he or she will intervene as it progresses towards task completion.

While *social and emotional* dimensions are apparently less subject to formal specifications, they, too, have been a focus of research in recent years, but remain a challenge. Regardless of cultural differences in consistent teamwork and management styles, the emotional alignment of the individual with the team has long since been recognised as vital for both performance and sustained mental health [5]. Various modelling approaches allow the mapping of the individual's attitude and group sentiment, both with surveys building a general corpus of conventions and customs within a given working culture, as well as with physiological measurement and machine learning techniques that respond to variations of the individual or deviations from past patterns and cultural average [5]. Both trust and psychological challenges are vital to HRC applications, and deserve more attention in the future.

18.3 Future Research Directions

18.3.1 Necessity of New Industrial Standards of HRC

The rapid emergence and anticipated potential of human–robot collaboration creates such a new scenario of industrial production which requires not only novel scientific insights and technological solutions but also a new regulatory framework. Current standards, such as ISO 10218-1/-2 [6, 7] amended with ISO/TS 15066 [8] for collaborative robotic environments, already take an integrated approach with the compound of control, robot, end-effector, workpiece and task being simultaneously subject to assessment. ISO/TS 15066 (technical specification) distinguishes four modes of collaborative operation in a shared safeguarded area (safety-rated monitored stop, hand guiding, speed and separation monitoring, power and force limiting), taking the first step in removing rigid cell barriers of robotised workstations. As it was suggested in Chap. 2 [9], gesture recognition in HRC also calls for standardisation. Nonetheless, both ISO 10218 and ISO/TS 15066 deal with physical (mechanical) aspects only—mental, emotional and social issues remain largely unspecified.

Chapter 1 [10] proposed the extension of the regulatory framework into the above directions. Considering mental workload, any HRC environment should optimise mental stress to avoid monotony, any forms of underload and overload, fatigue and burnout. It is to be expected that the introduction of collaborative intelligence, especially in complex work environments where human–robot teams operate, will cross over the category limits, requiring a new integrated perspective in assessing, as well as designing such HRC systems. While ISO 10075 does not specifically address such views, the perspectives of the standard are a consistent starting point for further synthesis. This standard largely remains on the level of the individual, higher-level perspectives are rather taken by ISO 26000 on the 'Guidance on Social Responsibility' [11]. Nonetheless, these generic ethical guidelines can only provide a point of departure for formulating more rigorous, technically valid rules for collaborative human–robot teams working in an industrial environment.

Albeit current standards do not yet address important issues of teams equipped with CI, they outline some of the key aspects of a roadmap for socially aware collaborative environments. In the close future, intensive standardisation activities are to be expected for setting the baselines for a safe, ergonomic, mentally and socially satisfying work environment for HRC.

18.3.2 Human Worker Modelling

This book, and Chap. 11 [12] in particular, presented AI-based techniques for recognising human action and predicting human motion trajectory, which are essential in realising effective human–robot collaboration. This work can be extended towards an adaptive strategy in motion trajectory prediction that allows different functional units to be prioritised according to different tasks and relevant human body parts.

However, in CI, the mindset of human workers matters, too. Even though continuous observation of behaviours and models of human disposition and emotion at workplace have been investigated intensively, industry-ripe application to work execution control at assembly workstations is still missing. There is a need to elaborate and populate models of human workforce and develop task execution control and HRC adaptation approaches that can establish individual worker preference profiles, pick up transient changes in the state of the individual worker, and tune both communication and acquired models accordingly.

18.3.3 Applying and Calibrating Digital Twins to HRC

Humans and artificial agents pursue parallel approaches in observation and construction of beliefs about their environment. In a CI setting, at least a partially shared picture of the world (regarding underlying objects, relations and states) has to be maintained by using a digital twin (DT). As it was discussed in Chap. 1 [10], the DT is a virtual representation that is able to closely follow the characteristics and changes of a real work environment. The sharing of a world-view consists in obtaining information about proposed common views via perception, reconstructing a model internally via reasoning, and using it in the assumption that others apparently sharing the same views share a fitting inner model as well. Furthermore, the symbiotic interplay of human and robotic agents requires representing their multimodal and bidirectional channels of communication and control as well.

The digital twin should be calibrated on a regular basis through steps of a systematic workflow to the real environment [13] so that both robot programming and operator instructions can automatically adapt to the actual situation. As typical in factory automation, elements of a collaborative production cell may have a complex but engineered—hence, certainly structured—geometry with some characteristic features. This provides guidance for geometrical calibration, as specific features can be identified via sensing, and partial models can be matched accordingly. One can expect that rapidly developing 2D and 3D imaging, along with graphical processing technologies, remain prevalent in devising essential measurement data for calibration.

18.3.4 Redundant Robot Optimisation

Redundancy—when the robot has more degrees of freedom than needed for performing a task—provides much opportunity for optimisation and adaptation, because, at least in principle, infinitely many joint configurations may result in executing the same task. However, when executing a set of tasks in a complex work environment, which is characteristic to HRC assembly in particular, a number of technological and geometric constraints must be met. The solution requires a bidirectional transition between the task and configuration spaces for generating relevant and collision-free configurations only. The computational complexity of such an approach is still too demanding for supporting real-time robot control. More adaptive and efficient methods and algorithms are in need. Improvement of time performance is also a prerequisite of adaptive collision avoidance (see also Chap. 4 [14]).

18.3.5 Shared Tasks and Team Formation

In today's industrial practice of assembly, collaborative workstations exist already. However, they do not exhibit all characteristics of a working environment hosting multiple collaborating humans and robotic agents. As discussed earlier in this book, advanced methods of perception and sensor fusion, situation recognition, variable task assignment considering also skill levels, dynamic and adaptive planning, load balancing, as well as of human-machine mixed-initiative motion generation are known; however, it is still open how to harmonise autonomous decisions, change levels of autonomy and responsibility, and leave more abstract decisions to humans where conscious judgement is needed. In a multi-agent setting, the agents need to be aware of each other's states (and intentions), as well as the state of objects to successfully work together or safely coexist within the same physical space. For robots, a proper degree of awareness depends on the adequate implementation of communication or information acquisition channels and their introduction into their control. Being more autonomous but less predictable and reliable actors, humans need more support for being aware of the current situation. All these are prerequisites of true shared task execution and adaptive team formation in collaborative environments.

18.3.6 Handling of Exceptions, Emergency and Recovery

Planning is always based on assumptions and refers to future actions. If assumptions are violated, execution may *deviate* from the planned course of actions. Even though such situations are bound to occur in HRC, plan recovery is quite an untouched research area. Depending on the level of deviation, a number of different *recovery approaches* can be taken. For instance, exceptions occur if the flow of processes involved is close to or has passed the nominal limits; however, the processes can still be brought back to normal without interruption, and the situation does not endanger the integrity or safety of human. Emergencies occur when danger of irrevocable damage is imminent. The latter case requires fast and guaranteed mitigation. Therefore, certain modes of communication do not apply to emergency signals. In addition, emergency mechanisms may have direct effect (e.g., emergency power-down by a single circuit breaker) that usually does not occur with communication in the usual sense.

18.3.7 Learning by Doing

Current industrial practice does not explicitly foresee the integration of normal production operations with the handling of exceptions and emergencies and with recovery operations. Even though some methodologies address early fault prevention and detection, this is not necessarily equal to continuous feedback regarding operation results. Not only does this often result in the costly temporal and spatial separation of training from work execution, it also hampers orientation of the worker in ad-hoc work situations, e.g., in construction of individual products, or in the maintenance of poorly documented legacy equipment. Since training and quality check are typically separated from task execution, gradual skill development, live assistance and real-time feedback are obstructed. Industry-proof integrated process, robot and human models adequate in the context of HRC assembly must be elaborated where performance evaluation results (including those of quality checks) are fed back to support the improvement of the overall team. In any learning-by-doing scheme, the safety requirements must be continuously observed.

18.3.8 Adaptive Work Instructions

Human–robot interfaces, especially those conveying work instructions, often provide adaptability to the given worker in discrete steps, namely, by skill level categories. Adaptation to the worker's current (and changing) fitness for the current assembly task is, however, not part of industrial practice. The devices keeping track of the worker's awareness primarily serve safety purposes only. AR-based in-situ decision support to workers in dynamic HRC assembly environments deserves more attention to be both intuitive and free of mental overload. Work instructions need to be adaptive to not only the changing competence level of individual workers but also to declining focus and concentration during the day or within the week.

18.3.9 Programming-Free Robot Control

One potential application of data fusion in HRC assembly that should be actively explored is the multimodal fusion of human commands, for example, integration of both hand gestures and voice, to further improve the human command recognition rate and contribute to programming-free robot control.

Despite the increasing use of gestures and voice commands in HRC for robot control, they are less natural and practical in busy and noisy work environments. Often, such commands are laden by ambiguity, too. Hence, instead of defined gestures and voice commands, recognition and prediction of human motions through deep learning provides better context awareness and less interruption to normal performance induced by signalling gestures. Haptic interaction with collaborating robots, especially for legacy industrial robots with few embedded sensors, is another way to achieve programming-free robot control. As an emerging research direction, brain robotics or brainwave-driven robot control [15] can enhance symbiotic HRC assembly. 'Think ahead' combined with human motion prediction contributes to timely cognition and assistance from the robots, and adds value to the intelligence of an HRC system. Programming-free robot control can be facilitated by the use of event-driven function blocks with smart decision algorithms for detailed adaptive robot control, whereas the high-level commands (e.g., gesture, voice, gaze, haptic, brainwave) serve as the triggering events to the function blocks for task execution. This unique combination paves the way towards multimodal symbiotic HRC in manufacturing.

18.3.10 HRC as Social Activity with Trust

Although rarely applied in industrial production, research has already elaborated approaches for the modelling of emotional and social processes, as well as acquisition of individual or cultural profiles and their tuning to newly perceived changes. Existing results, therefore, provide a solid starting point for the elaboration of similar solutions for industrial applications in the HRC context. What is missing is the formation of mutual models of *trust*. When adding humans to shared robotic environments, this aspect is unavoidably important and deserves attention. This problem generalises to so-called *trustworthy AI*, which is based on the idea that trust provides the solid foundation for all communities. Therefore, individuals, teams, organisations, and even societies will only be able to harness the full potential of AI if trust can be established in their development, operation and use [16].

18.3.11 Social Responsibility of a New Type of Automation

An HRC team forms a meta-society where each team member bears certain social responsibilities for others such as safety and work ethics. The social stability of the team contributes to the level of automation for seamless HRC assembly. In such a new automation environment, mental stress and even psychological discomfort leading to any potential accident can be monitored and diagnosed via the brainwaves of human workers, which can be collected by sensors embedded in a safety helmet. Although an ethical issue, it contributes to safety assurance in workplaces.

18.3.12 Limitations and Further Challenges

The wide interest of research and industry in HRC-related topics is proportional to the increased productivity and flexibility of production lines, as it combines human and robot capabilities [17]. However, the deployment of HRC in wide industrial applications is facing big challenges. Safety is the first concern. Although advanced methods/mechanisms have been developed in the past years, most of these methods are established in the lab environment. The hardware utilised are prototype-level devices that cannot be transferred to industry directly. More mature devices at higher safety integrity level (SIL) are needed to further improve and exploit the research results. Moreover, the feasibility of the HRC solutions has been evaluated well, but the safety performance needs to be assessed systematically. For example, stability, robustness, response time, redundant safety, backup solutions and emergency handling need to be evaluated in a structured and standardised way.

As a result, research on collision avoidance needs to have impact on future standardisation. Current standards focus on distance and force/torque/energy limits to protect humans from severe injuries. However, there is no complete standard for new technologies, e.g., vision-, ultrasonic-, depth-based sensing systems. Standards need to include those new technologies as a part of the robotic solutions.

Additionally, the social impacts of collision avoidance need much attention as the operator needs to interact with the robot in an HRC setting without any wearable protections in daily work. Psychological reactions may change if the operator works with the robot for a long period of time. Sometimes, such changes may not be noticed by the operator himself/herself. Thus, more effort is needed in addressing these challenges, as mentioned in Sects. 18.3.10 and 18.3.11.

18.4 Conclusions

Research on HRC has been active for many years. Despite the advances of HRC in recent years, limitations exist. There is a need to identify the remaining challenges in HRC and highlight future research directions. Within the context of symbiotic HRC in manufacturing, challenges and twelve future research directions have been identified, hoping to outline advances anticipated or needed in the years to come. With the support of the latest technologies of sensing, communication, augmented reality, AI/CI and robot control, HRC research will find its way to practical application on shop floors in smart factories of the future.

References

1. L. Wang, From intelligence science to intelligent manufacturing. Engineering **5**(4), 615–618 (2019)
2. L. Wang, R.X. Gao, J. Váncza, J. Krüger, X.V. Wang, S. Makris, G. Chryssolouris, Symbiotic human–robot collaborative assembly. CIRP Ann. Manuf. Technol. **68**(2), 701–726 (2019)
3. H.J. Wilson, P.R. Daugherty, Collaborative intelligence: humans and AI are joining forces. Harv. Bus. Rev. 7–8 (2018). https://hbr.org/2018/07/collaborative-intelligence-humans-and-ai-are-joining-forces
4. P.A. Hancock, D.R. Billings, K.E. Schaefer, J.Y.C. Chen, E.J. de Visser, R. Parasuraman, A meta-analysis of factors affecting trust in human–robot interaction. Hum Factors J Hum Factors Ergon Soc **53**(5), 517–527 (2011)
5. T.M. Moerland, J. Broekens, C.M. Jonker, Emotion in reinforcement learning agents and robots: a survey. Mach Learn **5**, 1–38 (2017)
6. ISO, 2011, ISO 10218-1: robots and robotic devices—safety requirements for industrial robots—Part 1: Robots
7. ISO, 2011, ISO 10218-2: robots and robotic devices—safety requirements for industrial robots—Part 2: Robot systems and integration
8. ISO, 2016, ISO/TS 15066 robotics and robotic devices—collaborative robots
9. H. Liu, L. Wang, Latest developments of gesture recognition for human–robot collaboration. Ibid. Chapter 2
10. Zs. Kemény, J. Váncza, L. Wang, X.V. Wang, Human–robot collaboration in manufacturing: a multi-agent view. Ibid. Chapter 1
11. ISO, 2010, ISO 26000: guidance on social responsibility
12. R.X. Gao, L. Wang, P. Wang, J. Zhang, H. Liu, Human motion tracking, recognition, and prediction for robot control. Ibid. Chapter 11
13. G. Horváth, G. Erdős, Point cloud based robot cell calibration. CIRP Ann **66**(1), 145–148 (2017)
14. B. Schmidt, Real-time collision detection and collision avoidance. Ibid. Chapter 4
15. A. Mohammed, L. Wang, Brainwaves driven human–robot collaborative assembly, CIRP Ann. Manuf. Technol.**67**(1), 13–16 (2018)
16. S. Thiebes, S. Lins, A. Sunyaev, Trustworthy artificial intelligence. Electron. Mark. (2020). https://doi.org/10.1007/s12525-020-00441-4
17. P. Tsarouchi, S. Makris, G. Chryssolouris, Human–robot interaction review and challenges on task planning and programming. Int J Comput Integr Manuf **29**(8), 916–931 (2016)

Index

L. Wang et al. (eds.), *Advanced Human–Robot Collaboration in Manufacturing*,
https://doi.org/10.1007/978-3-030-69178-3